Scientific Approaches to Consciousness

Carnegie Mellon Symposia
on Cognition

David Klahr, Series Editor

Scientific Approaches to Consciousness

Edited by

Jonathan D. Cohen

Carnegie Mellon University
University of Pittsburgh School of Medicine

Jonathan W. Schooler

University of Pittsburgh

Psychology Press
Taylor & Francis Group

New York   London

First Published by Lawrence Erlbaum Associates, Inc., Publishers
10 Industrial Avenue
Mahwah, New Jersey 07430

Reprinted 2010 by Psychology Press

Lawrence Erlbaum Associates, Inc., Publishers
10 Industrial Avenue
Mahwah, New Jersey 07430

Cover design by Cheryl Minden

Library of Congress Cataloging-in-Publication Data

Scientific approaches to consciousness / edited by Jonathan D. Cohen, Jonathan W. Schooler.
 p. cm.
 Includes bibliographical references and index.
 ISBN 0–8058–1471–X (c : alk. paper). — ISBN 0–8058–1472–8 (p : alk. paper)
 1. Consciousness. 2. Cognition. 3. Human information processing. I. Cohen, Jonathan D. II. Schooler, Jonathan W.
 BF311.S3876   1996   96–15490
 153—dc20   CIP

10 9 8 7 6 5 4 3

*This volume is dedicated to
the memory of our grandfathers,
Samuel Rothstein and Samuel Schooler,
who have been enduring sources of inspiration
in our personal and professional lives.*

Contents

Preface

The chapters in this volume are based on the formal talks and prepared discussions that were delivered at the 25th Carnegie Mellon Symposium on Cognition. This series now spans 30 years, covering topics as diverse as visual information processing, problem solving, animal cognition, cognitive development, and cognition and affect, and has involved many of the central figures in modern psychological research. Interestingly, however, *consciousness*—an area that might be considered central to an understanding of mental processing—has received little attention in this series. Indeed, as we note in the Introduction, the topic of consciousness has generally been avoided as a target of scientific scrutiny. The purpose of this symposium was to explore how far we have come in studying phenomena related to consciousness and to consider whether it is now possible to study this topic directly in a scientific fashion. The symposium included a broad sampling of such research, from leaders in the fields of cognitive psychology, neuropsychology, and neuroscience; these are reflected in this volume. Given the breadth of the topic, however, we were unable to represent all areas of scientific research that bear on issues of consciousness. Important topics are missing, both at the psychological level (e.g., working memory) and the biological level (e.g., the role of neural oscillations). Nevertheless, we feel that the contributions in this volume serve to illustrate the level of interest and sophistication with which the topic of consciousness is now being met within these fields.

We are deeply indebted to a number of individuals and organizations, who helped to make both the symposium and this volume possible. First, we would like to thank the authors for the superb symposium talks that they gave and the excellent chapters they have contributed to this volume. We would also like to express our gratitude to the Fetzer Institute for their generous support of the symposium. The topic of consciousness has only recently begun to recapture the interest of the scientific community, and funding support for research activities in this area is still limited. We are therefore deeply appreciative of the vision and foresight shown by the Fetzer Institute in their support of this event. We are also grateful to the Department of Psychology at Carnegie Mellon University and the many staff within the department who worked to make the event a success. In particular, we thank Theresa Kurutz, who has continued to contribute her effort toward the compilation of this volume, and Betty Boal, who was honored at the event itself for her many years of service to the Department and in particular to the Carnegie Mellon Symposium series. David Klahr deserves our thanks for his advice on how to organize the symposium and edit the book, as do the students who helped to actually organize and run the event, including Kate Brennan, Todd Braver, and Terri Huston. We are grateful to Stephen Fiore for his invaluable effort in compiling the subject index. Finally, we would like to acknowledge the many participants who made the proceedings themselves a stimulating and informative experience for all. This includes the five individuals who were the recipients of a set of competitive fellowships to attend the symposium made possible by the Fetzer Foundation—Maria Brandimonte, Guven Guzeldere, Katherine McGovern, Felicia Pratto, and Suparna Rajaram—as well as our esteemed colleagues, both local and from afar, who attended the symposium and contributed to its success.

We hope that this volume which, directly or indirectly, reflects the efforts of all of these people, will help fuel the growing interest in the topic of consciousness within the scientific community, and that the coming years will see the fulfillment of the promise of this volume: an understanding of consciousness grounded in science.

—*Jonathan Cohen*
—*Jonathan Schooler*

Introduction

CHAPTER 1

Science and Sentience:
Some Questions Regarding the
Scientific Investigation of Consciousness

Jonathan D. Cohen
Carnegie Mellon University
University of Pittsburgh School of Medicine

Jonathan W. Schooler
University of Pittsburgh

The topic of consciousness has occupied philosophers and metaphysicians for centuries. What distinguishes the scientific approach to this topic, however, is a methodology—a set of tools that permit the empirical and theoretical study of a phenomenon in an objective, reproducible way. The relation between this methodology and the phenomenon of consciousness has a restless and controversial history. The topic of consciousness was central to psychology at its inception as a scientific enterprise. However, it rapidly fell into disfavor, in part because the method that seemed most appropriate for studying it (introspection) was abandoned as a scientific method. During the 1960s there was a brief reawakening of the topic. This was probably due in part to sociological factors, but almost certainly it had something to do with the advent of a set of empirical and theoretical methods, now known as cognitive science, that provided a scientific way of asking and answering questions about phenomena such as memory, language, and attention—phenomena clearly related to consciousness.

The reason why the topic of consciousness fell back out of favor is not entirely clear. Perhaps, again, it had to do with sociological factors. Possibly its association with the excesses and pop psychology that flourished as part of the 1960s counterculture revolution cost the topic its scientific credibility. Or, perhaps it was that early cognitive scientists knew, consciously or unconsciously, that additional groundwork was necessary before this topic could be productively addressed. Instead of taking consciousness head on, scientists chose to see how far the new tools of cognitive science could take them, each

3

in their own direction. Thus, it seemed possible to make solid scientific headway on phenomena such as strategic versus automatic processing, allocation
of attention, working memory, subliminal processing, implicit memory, and
differences between information processes in the awake and dreaming states,
all without directly addressing the vexing questions that surround consciousness qua consciousness. These lines of research share a common assumption
of a particular quality, state, or process—awakeness, awareness, alertness—
that seems closely related to consciousness, but do not address the question
directly. Newell (1990) made this point with characteristic clarity in his book
about the Soar architecture:

> SOAR provides a theory of awareness. Soar is aware of something if its deliberate
> behavior can be made to depend on it. In this sense, awareness is operationally
> defined and fundamental, and it is a much-used notion throughout cognitive psy-
> chology. But consciousness can be taken to imply more than awareness . . .
> namely, the phenomenally subjective. It can mean the process, mechanism,
> state (or whatever) that establishes when and what an honest human would
> claim to be conscious of, both concurrently or retrospectively. Soar does not
> touch the phenomena of consciousness, thus designated. Neither does much
> else in cognitive psychology. That it seems out of reach at the moment might
> justify pushing the issue into the future, until additional regularities and phe-
> nomena accumulate, but the challenge remains. (p. 433)

Accordingly, researchers may have been reluctant to directly take on the
issue of consciousness because it is such a highly subjective phenomenon. Another aspect of the challenge was captured by James (1890/1952), when he
articulated the difficulty of examining consciousness directly:

> As a snow-flake crystal caught in the warm hand is no longer a crystal but a
> drop, so, instead of catching the feeling of relation moving to its term, we find
> we have caught some substantive thing, usually the last word we were pro-
> nouncing, statically taken, and with its function, tendency, and particular mean-
> ing in the sentence quite evaporated. The attempt at introspective analysis in
> these cases is in fact like seizing a spinning top to catch its motion, or trying to
> turn up the gas quickly enough to see how the darkness looks. (p. 158)

Thus, consciousness is not only subjective, but also ephemeral. The question
then becomes: How can we come to examine and understand the snowflake
without melting it in the process?

We have no illusions that this volume will resolve all of the difficulties associated with the scientific investigation of consciousness, although we hope
it will make progress on a few. Most importantly, however, the goal is to explore the extent to which consciousness can be the target of direct scientific
inquiry, that is, to get on the table some of the relevant scientific work and to
consider the degree to which this research can help inform our understand-

ing of consciousness. Put simply: Is it now possible to study consciousness directly in a scientific way?

With this goal in mind, we open this volume by raising a few questions and issues that any scientific account of consciousness must ultimately address. Some are addressed head-on in this volume, whereas others remain as challenges for the future.

HOW CAN WE IDENTIFY CONSCIOUSNESS?

How do we identify consciousness, either within ourselves or in others? What are our criteria? One way to begin is to consider the extremes: that is, by identifying processes or phenomena we consider to be nonconscious, and those we take to be clearly indicative of consciousness. There are at least two dimensions or contrasts that we can focus on in this respect. Consider, for example, the differentiation of reportable and unreportable experiences. At the one extreme, we have unreportable cognitive processes such as implicit memory and learning and subliminal perception. At the other extreme, we have reportable and some highly self-reflective processes such as explicit memory and metacognition. Automaticity is another dimension that seems relevant—that is, the distinction between automatic versus attentionally dependent processes with completely automatized processes, such as typing or driving a car at one extreme, and tasks heavily demanding attention (e.g., playing chess) at the other.

How does consciousness relate to the dimensions of reportability and attention? It seems, at least intuitively, that an important criterion we use in identifying consciousness in others is the ability to report the experience of consciousness. And this, in turn, seems to hinge on the ability for self-reflection. Specifically, to what extent are self-reflection (cf. Hobson, chapter 19; Kihlstrom, chapter 24; Johnson & Reeder, chapter 13) and the reportability of experience (cf. Reber, chapter 8; Dulany, chapter 10; Merikle & Joordens, chapter 6) useful or criterial for identifying consciousness? Related to the question of reportability is the question about the relation between language and consciousness. If reportability is a critical component of consciousness or of our ability to identify it, then what is the relation between language of any type and consciousness (see Reber, chapter 8, and Dulany, chapter 10, for contrasting discussions of this issue)? If there are potential limitations to the use of reportability as a criterion for consciousness, then what other types of criteria might we use to determine or infer when conscious processes are operating (see Jacoby, Yonelinas, & Jennings, chapter 2; Merikle & Joordens, chapter 6; and Greenwald & Draine, chapter 5, for possible alternatives to a simple reportability criterion for consciousness)? Attention also seems to be related to consciousness in some important way, but exactly how? Is attention "the gateway to consciousness," as has often been assumed, or is its relation to consciousness more complex (see Shiffrin, chapter 3, for one response to this question)? Finally, how do attention, reportability, and self-reflection relate to one another?

These questions all focus on the identification of consciousness. It seems unlikely that we will make much scientific progress in understanding a phenomenon until we can reliably identify it. Assuming this is possible, several additional questions arise.

WHAT IS THE RELATIONSHIP BETWEEN CONSCIOUS AND UNCONSCIOUS PROCESSES?

As suggested earlier, one way of approaching the identification of consciousness is to focus on its distinction from unconsciousness. That is, how do conscious and unconscious processes differ? Do, as a number of contributors suggest (e.g., Jacoby et al., chapter 2; Rajaram & Roediger, chapter 11; Merikle & Joordens, chapter 6; Wegner, chapter 14; Dulany, chapter 10), conscious and unconscious processes involve qualitatively different operations? If conscious and unconscious processes do involve distinct operations, then how do these systems interact? Are they truly independent processes (cf. Jacoby et al.), or do they impact one another (cf. Wegner)?

The suggestion that conscious and unconscious processes may be qualitatively different raises important questions about the nature of unconscious processing. How cognitively sophisticated are unconscious processes? Are unconscious processes capable of deriving and representing complex rules (cf. Lewicki, Czyzewska, & Hill, chapter 9; Reber, chapter 8), or must complex symbolic thought be limited exclusively to the domain of consciousness (cf. Dulany, chapter 10)? Determining the limitations of unconscious processes may help to reveal the attributes that make consciousness unique (cf. Baars, Fehling, LaPolla, & McGovern, chapter 22).

WHAT ARE THE MECHANISMS UNDERLYING CONSCIOUSNESS?

Ultimately, any satisfactory scientific account of consciousness will require an articulation of the mechanisms that underlie consciousness. Traditional cognitive psychological and computational approaches have suggested that higher cognitive processes—what many refer to as "controlled" processes—rely on a central executive, and this would seem to be a natural candidate for the seat of consciousness. Dennett (1991) referred to this as the Cartesian theater, that is, the arena in which "it all comes together and consciousness happens" (p. 39).

This approach has intuitive appeal and helps explain some of the central observations about conscious experience, including its limited capacity and its apparently sequential or serial nature. Nevertheless, neurobiological and neuropsychological data, as well as recent research within cognitive science and psychology, have suggested an alternative view—a more distributed view of neural and cognitive processing. Further, there are now a number of theories of consciousness to go with this approach. For example, Minsky's (1985) Society of Minds and Dennett's Multiple Drafts theory, as well as two presented in this volume (Baars, Fehling, LaPolla, & McGovern's Global Workspace hypothesis; and Kinsbourne's distributed systems view), have the appeal of fitting better with what we know about neural organization. However, what they seem to lack, at least up to this juncture, is the ability to account for the phenomenological unity of consciousness. How do cooperation, seriality, sequentiality, and the focality of consciousness actually come about within a distributed system? This has yet to be fully specified or demonstrated. It also raises a closely related question: Should consciousness be treated as a unitary construct? Specifically, does it represent the operation of a delineable set of identifiable processes (as Hobson, chapter 19, and Mandler, chapter 26, seem to suggest), a particular type of interaction among some or all processes (as suggested by Farah, O'Reilly, & Vecera, chapter 17; Kinsbourne, chapter 16), or is it an emergent and irreducible property of the processing system as a whole? In particular, does consciousness, as a construct, promise to add anything to our understanding of neurobiological and/ or psychological processes that a complete account of each component processing system, and the interaction among them, would not otherwise provide? This question leads naturally to questions about the conditions under which consciousness may arise.

WHERE AND WHEN DOES CONSCIOUSNESS ARISE?

Suppose we could fully specify the mechanisms underlying consciousness— either computationally, psychologically, or even neurally. It is then tempting to ask whether or not these would be sufficient to support consciousness. For example, could we ever consider a computer to be conscious? Two related questions should also be considered. First, are any nonhuman animals conscious (if so, at what phylogenetic level)? Second, are infants conscious (if not, at what age does consciousness emerge)?

We think questions about what can be conscious are critical if only because they are so inescapably asked, and once asked, can become vexing. It is important, therefore, to define the scope of this volume with respect to such questions. We think these can lead in one of two directions. One raises the issue of dualism, that is, whether a physical mechanism is sufficient to produce consciousness, or whether some additional ingredient (e.g., some ectoplasm or spirit) is necessary for consciousness to occur. This volume does not address the issue of dualism, not because we believe it is an illegitimate, uninteresting, or even unimportant issue, but because we believe it is not an issue that science can profitably address. In our view, scientific questions are by their very nature directed at the physical, observable world, and therefore their scope should be limited to this domain. We realize this is subject to philosophical debate, and the very terms *physical* and *observable* are themselves subject to interpretation. However, as a matter of procedural course, we believe the scientific community can agree about what is and is not observable and that, by definition, the elusive twin in the dualist view—the ghost in the machine—is neither physical nor observable. The challenge for us is to see how far we can get in our understanding of consciousness without invoking dualism. Referring back to James' observation, our goal as scientists is the description and explanation of the snowflake without melting it.

Keeping this in mind, we can more adequately consider the following question: If we could fully specify the mechanisms and processes underlying consciousness in humans, then would reproducing or identifying these in a nonhuman be sufficient to produce (or attribute) consciousness in (to) that system? We would like to suggest that this question might be answerable only by a Cartesian version of the Turing test: If an entity that we assume a priori has consciousness—that is, a human being—interacts with the system under question, and cannot distinguish between its possession of consciousness and that of another human being, then the mechanisms we have specified are indeed sufficient to produce consciousness, at least as far as it is identifiable and studiable in a scientific fashion.

Once we have a system that passes the Cartesian Turing test, would we now be in the position to derive a complete set of criteria for consciousness? That is, could we now just figure out which components of the system are critical for allowing it to pass the test? Here, we would like to offer the following monition. Even if we have successfully identified consciousness in an entity, it does not follow that there is a distinct, specifiable set of criteria that precisely delimit the phenomenon. By analogy, we can say unequivocally that at noon it is day and at midnight it is night. But at what point does the night become the day? Our point is that it may be less useful to identify a rigid set of criteria for consciousness than it is to identify the variable or variables that form the continuum or dimensions along which consciousness lies. In other words, we would like to suggest that one key to success in understanding consciousness will lie in our ability to identify the relevant dimensions of neurobiological, psychological, and computational processing that define a continuum between conscious and unconscious processes (cf. Farah et al., chapter 17; Hobson, chapter 19; Johnson & Reeder, chapter 13; Kinsbourne, chapter 16; Reber, chapter 8).

In conceptualizing consciousness as involving the interplay of continuous dimensions, we must be cautious to avoid some potential traps that can ensnare discussions of psychologically contintuous dimensions. In particular, positing continuous dimensions to consciousness does not necessarily rule out the possibility that the extremes along certain dimensions may rely on qualitatively different processes. For example, our perception of the transition from day to night may seem quite continuous, even though neurobiologically it entails a gradual shift in the relative contributions of two different, and qualitiatively distinct processing mechanisms: rods and cones. Thus, qualitatively different mechanisms may work together to instantiate processing along a continuous dimension. In short, it is quite possible that consciousness lies along a continuum while conscious and unconscious processes draw on distinct processes.

A MULTIDISCIPLINARY APPROACH

Ultimately, answers to the questions raised here require the contributions of scientists addressing the issue of consciousness at a variety of different levels of analysis. We need to consider research addressing the low levels of consciousness associated with domains such as automaticity, subliminal perception, and implicit learning, and research addressing the high levels of consciousness associated with domains such as attention and metacognition. We also need to consider a variety of different research approaches, including the experimental, clinical, neurobiological, neuropsychological, philosophical, theoretical, and computational. By considering the contributions of scientists examining aspects of consciousness at different levels and from the vantage of multiple disciplines we believe it will indeed be possible to address questions about consciousness in a direct and scientific fashion. We hope this volume will contribute productively to this effort.

ACKNOWLEDGMENTS

The writing of this chapter was supported by grants to both authors from the National Institute of Mental Health. Steve Fiore provided helpful comments on an earlier draft.

REFERENCES

Dennett, D. (1991). *Consciousness explained.* Boston: Little, Brown.

James, W. (1952). *Great books of the western world: The principles of psychology.* Chicago: Encyclopedia Britannica. (Original work published 1890)

Minsky, M. (1985). *The society of mind.* New York: Simon & Schuster.

Newell, A. (1990). *Unified theories of cognition.* Cambridge, MA: Harvard University Press.

Attention and Automaticity

CHAPTER 2

The Relation Between Conscious and Unconscious (Automatic) Influences: A Declaration of Independence

Larry L. Jacoby
Andrew P. Yonelinas
Janine M. Jennings
McMaster University

Research on unconscious processes has long been plagued by theoretical and methodological problems. Consequently, the unconscious was banished, along with consciousness, by radical behaviorists, and has only recently regained respect as a research topic. The resurgence of interest in conscious and unconscious processes is largely due to findings of dissociations between performance on direct and indirect tests of memory and perception. Effects of the past in the absence of remembering, and perceptual analysis in the absence of conscious seeing arise from studies of patients with neurological deficits. Warrington and Weiskrantz (1974) found that amnesics showed little evidence of memory for an earlier-read word list when asked to recall or recognize those words (a direct memory test). However, the amnesics used those words to complete word fragments (an indirect test) more often than if the words had not been seen earlier (see Moscovitch, Vriezen, & Gottstein, 1993, for a review of related research). Similar memory dissociations are evident in people with normal functioning memory (for a review, see Roediger & McDermott, 1993). The form of dissociations found for memory is comparable to dissociations taken as evidence for unconscious perception. For example, Marcel (1983) flashed words for a duration so brief that subjects could not "see" them, but could show effects of those words on a lexical decision task used as an indirect test of perception.

Empirical advances derived from the indirect versus direct test distinction have proceeded without confronting many of the methodological and conceptual issues that plagued earlier investigations of unconscious processes.

However, those issues have now resurfaced. The major difficulty for drawing a distinction between conscious versus unconscious processes is the problem of defining each type of process. Essential here is the relation of processes to tasks (Dunn & Kirsner, 1989). Typically, unconscious processes are equated with performance on indirect or implicit tests and conscious processes with performance on direct or explicit tests. However, this form of definition is problematic because conscious processes may contaminate performance on indirect tests (e.g., Holender, 1986; Reingold & Merikle, 1990) and, less obviously, unconscious processes might contaminate performance on direct tests (Jacoby, Toth, & Yonelinas, 1993). In addition, mapping processes onto test performance overlooks an essential aspect of defining unconscious and conscious processes; which is specifying the relation between them.

This chapter provides an overview of research done within our process-dissociation framework (e.g., Jacoby, 1991; Jacoby et al., 1993) for separating conscious and unconscious influences. Rather than equating processes with tasks, as is done by the direct versus indirect test distinction, our strategy has been to gain estimates of the contributions of each type of process to performance on a single task and show the dissociative effects of variables on those estimates. In order to distinguish between conscious and unconscious processes, we need to make an assumption about their relation. Our work has been based on the assumption that conscious and unconscious influences are independent of one another.

The goals of this overview are as follows: First, we highlight the purpose of the process-dissociation procedure and then describe an experiment using that procedure to examine age-related differences in cognitive control. Second, we consider potential assumptions for the relation between unconscious and conscious processes, specifically, independence, redundancy, and exclusivity. Third, we summarize evidence to support the choice of the independence assumption and, finally, provide strong evidence against the alternatives. Throughout, we treat the contrast between unconscious versus consciously controlled processes as identical to the contrast between automatic versus controlled processes. Later, we justify doing so and discuss how our approach offers a solution to the problem of defining conscious and unconscious processes.

PROCESS-DISSOCIATION PROCEDURE

As described earlier, dissatisfaction with indirect tests of memory and perception has centered on the possibility of their contamination by contributions of aware, intentional processes. However, there is a more serious problem: Automatic processes operating in isolation may be qualitatively different from those operating in the context of consciously controlled processes. Consider the commonplace claim that you should get an individual drunk to learn what that person really believes. Drunkenness is treated as a pure measure of automaticity or true belief. The "contamination" problem is to question how drunk a person has to be before his responding is no longer contaminated by consciously controlled processing. Even if one could achieve an uncontaminated test, there is the more serious "qualitative difference" problem of whether or not the test reveals a person's "true" beliefs or only what that person believes when drunk. Most likely, some of one's beliefs when drunk are qualitatively different from one's beliefs when sober. Automatic influences in the context of consciously controlled processes, like true beliefs when sober, are of great interest. Because the indirect versus direct test distinction identifies processes with tasks, it provides no means of measuring automaticity in the presence of consciously controlled processing.

An objective means of measuring conscious control is necessary to separate controlled and automatic influences. The process-dissociation procedure measures conscious control by combining results from a condition for which automatic and consciously controlled processes act in opposition, with results from a condition for which the two types of process act in concert. The measure is the very commonsensical one of the difference between performance when one is trying to as compared with trying not to engage in some act. The difference between performance in those two cases reveals the degree of cognitive control. The chapter illustrates the process-dissociation procedure in conjunction with the results of an experiment done to show an age-related effect on recollection—a consciously controlled use of memory.

Jacoby et al. (1993) used an Inclusion/Exclusion procedure with a stem-completion task to separate recollection from automatic influences of memory. Jacoby (1992) used a similar procedure to examine age-related differences in memory performance. In that experiment, words were presented for study and then tested by presentation of their first letters as a cue for recall (e.g., motel; mot--). Study and test items were intermixed, and both the number of items intervening between the study presentation of a word and its test (spacing) and the nature of the test were varied. For an inclusion test, the word stem was accompanied by the message "old" and subjects were instructed to use the stem as a cue for recall of an old word or, if they could not do so, to complete the stem with the first word that came to mind. An inclusion test corresponds to a standard test of cued recall with instructions to guess when recollection fails. For an exclusion test, a word stem was accompanied by the message "new" and subjects were instructed to use the stem as a cue for recall of an old word but to *not* use a recalled word as a completion for the stem. That is, subjects were told to exclude old words and complete stems only with new words. The two types of tests were randomly intermixed.

When an inclusion or exclusion test immediately followed presentation of its completion word (0 spacing), performance of the elderly and of the young was near perfect. This finding is important because it shows that the elderly were able to understand and follow instructions. In contrast, when a large number of items intervened between the presentation of a word and its inclusion or exclusion test (48 spacing), the elderly performed much more poorly than did the young (Table 2.1).

For the exclusion test, elderly subjects were more likely to mistakenly complete a stem with an old word than were younger subjects. In the exclusion test, effects of automatic influences of memory for earlier reading a word should be opposed by recollection. The poorer performance of the elderly on the exclusion test can be explained as resulting from a deficit in recollection as can their poorer performance on the inclusion test. Placing recollection and automatic influences in opposition, as was done by the exclusion test, can provide evidence of the existence of the two types of processes (Jacoby, et al., 1993). However, it is necessary to combine results from the exclusion and inclusion tests to estimate the separate contributions of consciously controlled and automatic processes.

For an inclusion test, subjects could complete a stem with an old word either because they recollected the old word, with a probability R, or because even though recollection failed $(1 - R)$, the old word came automatically to mind (A) as a completion: $Inc = R + A(1 - R)$. For an exclusion test, in contrast, a stem would be completed with an old word only if recollection failed and the word came automatically to mind: $Exc = A(1 - R)$. Thus, the difference between the inclusion (trying to use old words) and exclusion (trying not to use old words) tests provides a measure of the probability of recollection. Given that estimate, the probability of an old word automatically coming to mind as a completion can be computed. One way of doing this is to divide the probability of responding with an old word for an exclusion test by $(1 - R)$: Exclusion$/(1 - R) = A(1 - R)/(1 - R) = A$.

The estimates (Table 2.1) provide evidence that the elderly suffered a deficit in recollection as compared to younger participants, but that auto-

TABLE 2.1
Probability of Correct Fragment Completion and Estimates of
Conscious and Unconscious Influences of Memory as a Function of Age

	Young	Old
Test		
Inclusion	.70	.55
Exclusion	.26	.39
Estimates		
Conscious	.44	.16
Unconscious	.46	.46

Note: The baseline completion for items not presented was .36.

matic influences of memory were unchanged. The estimates of automatic influences were well above the baseline probability of completing a stem with a target word when that word had not been earlier presented (.36). The difference between estimated automatic influences and baseline performance serves as a measure of automatic influences of *memory*—the effect of studying a word on the probability of its later coming to mind automatically as a completion for a stem.

If one makes the assumption that the base rate and unconscious influences of memory are additive, we can subtract base rate to compute automatic influences of memory. However, this approach is only necessary when base rates across a manipulation (e.g., age) are not equal, and is the same as the standard procedure of measuring "priming" as the difference between performance on old and new items (e.g., Roediger & McDermott, 1993). Alternatively, we could use signal-detection theory to take differences in base rate into account, and U would be replaced by $P(U > Cr)$ with Cr being the criterial level of strength required for a response reflecting unconscious memory influences to be produced. We discuss this use of signal-detection theory later.

The aforementioned results show age-related differences in memory to be very similar to effects produced by dividing attention during the study presentation of a word. Jacoby et al. (Exp. 1b, 1993) used the same materials as Jacoby (1992) but tested only young participants. Results showed that divided, as compared to full, attention during study reduced the probability of recollection (.00 vs. .25) but left automatic influences unchanged (.46 vs. .47). For the divided- versus full-attention experiment, study and test were in separate phases rather than intermixed, and, on average, the spacing between study of a word and its test was a bit longer than 48 intervening items. Nonetheless, the correspondence between age-related differences in memory and effects of full versus divided attention supports Craik and Byrd's (1982) claim that dividing attention during study can mimic the memory effects of aging.

The process-dissociation procedure has the advantage of allowing one to examine automatic influences operating in the context of consciously controlled processes (and vice versa) while eliminating problems of contamination. For example, had one relied on a test of cued recall to measure recollection, as is typically done, the probability of recollection would be overestimated because of failure to take automatic influences of memory into account (Jacoby et al., 1993). By separating the contributions of automatic and controlled processes, it was possible to show that aging and divided attention produced process dissociations, reducing recollection but leaving automatic influences invariant. Results from the manipulation of full versus divided attention provide a case that corresponds to commonplace claims about drunkenness. Divided attention, like the supposed effect of drunkenness, totally eliminated contributions of controlled processes ($R = 0$) to produce a pure test of automatic influences. Further, automatic influences measured by that process-pure test almost perfectly predicted the contribution of automatic influences when controlled processes were in play. That is, automatic influences operating in the presence of recollection (after full attention) were the same as those operating in its absence (after divided attention).

Without the process-dissociation procedure, one could not know that, in this case, dividing attention did achieve a pure test of automaticity nor could one know that automaticity was the same across different levels of cognitive control. Neither conclusion could be reached by contrasting indirect and direct test performance. However, the equations used to estimate the separate contributions of automatic and controlled processes rest on the assumption that the two types of processes act independently. If one refuses to hold the independence assumption, the apparent process dissociations must be dismissed as chance. The next section considers alternatives to the independence assumption.

THE RELATION BETWEEN CONSCIOUS AND UNCONSCIOUS INFLUENCES

As already mentioned, one cannot distinguish between conscious and unconscious processing without making an assumption, at least implicitly, about the relation between the two types of processes. There are three fundamental relations: *exclusivity, redundancy,* and *independence* (see Jones, 1987). Unfortunately, each holds a good deal of intuitive appeal.

When describing subjective experience, an exclusivity view seems an obvious choice for the relation between conscious and unconscious processing. By that view (Fig. 2.1a), one is either conscious or unconscious of a type of influence. That is, the two processes are mutually exclusive. An historical example of the exclusivity assumption is the Freudian notion that memories reside either in the conscious or the unconscious. More recently, following Tulving (1983), Gardiner and associates (Gardiner, 1988; Gardiner & Java, 1991; Gardiner & Parkin, 1990) applied the exclusivity assumption in their Remember/Know procedure to differentiate among the subjective experiences accompanying memory.

It seems equally appealing to claim that consciousness is but the "tip of the iceberg" (e.g., Baars, 1988) so that events that gain consciousness comprise a small subset of those that are unconsciously processed. By this redundancy assumption (Fig. 2.1b), conscious processing can emerge out of unconscious processing as described in threshold models of perception. That is, items are processed unconsciously until they reach a particular threshold, at which point they become conscious. As applied to the development of automaticity, the notion is that after extended practice, the execution of skills becomes au-

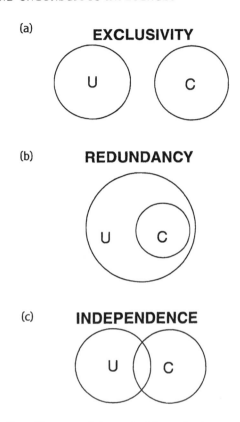

FIG. 2.1.   Venn Diagrams of the exclusivity, redundancy, and independence relations between conscious (C) and unconscious (U) influences.

tomatic or unconscious by receding below the threshold of consciousness.
The redundancy assumption has also been applied to the study of memory,
and provides the underpinning for Generate/Recognize models (e.g., Jacoby
& Hollingshead, 1990; Joordens & Merikle, 1993), which hold that items that
can be recognized are a subset of those that are generated. As described later,
recognition is identified with a conscious influence of memory, whereas generation is treated as reflecting unconscious influences.

    An alternative approach is to assume that conscious and unconscious in-
fluences act fully independently of one another (Fig. 2.1c). Conscious processing can occur with or without unconscious processing and vice versa. This
assumption differs from redundancy, which requires that the conscious cannot occur without the unconscious, and exclusivity, which specifies that the
two can never co-occur. Logan's (1988) view of automaticity is based on an
independence assumption and provides a contrast to the redundancy view of
automatization described earlier. Automaticity reflects memory for instances

in which a skill was earlier applied. Such unconscious influences of memory provide a basis for responding that is independent of the consciously controlled application of an algorithm. In other cases, independence might arise from the contribution of two separate structures to performance. The phenomenon of blindsight, for example, has been understood by postulating independent visual systems (Weiskrantz, 1986). In favor of the independence assumption, Jones (1987) noted the general consensus that "on a criterion of parsimony, processes should be assumed to be unaffected by each other's presence until a demonstration to the contrary occurs" (p. 230).

ESTIMATING CONSCIOUS AND UNCONSCIOUS INFLUENCES

The importance of the relation between conscious and unconscious processes becomes obvious when one attempts to estimate their separate effects. This can be seen by considering a more general form of the equations used by Jacoby (1992) and Jacoby et al. (1993) to separate consciously controlled and automatic (unconscious) influences of memory. For an inclusion test, a stem could be completed with an old word either because of conscious recollection (C) or because of unconscious influences (U) of the same sort that would be revealed by an indirect test. For an exclusion test, a stem could be completed with an old word on an exclusion test only if the word is produced because of unconscious influences and is not recollected as being old.

The choice among assumptions about the relation between conscious and unconscious influences can be clarified by writing equations for performance on inclusion and exclusion tests in a general form that follows Jones (1987):

$$\text{Inclusion} = C + U - (C \cap U) \qquad (1)$$
$$\text{Exclusion} = U - (C \cap U) \qquad (2)$$

These equations describe the probability of completing a stem with an old word as reflecting the separate contributions of conscious (C) and unconscious (U) influences. It is the definition of the intersect between the two types of influence, ($C \cap U$), that differentiates assumptions about their relation. If independence is assumed, then ($C \cap U$) is defined as CU. But, if redundancy is assumed, then ($C \cap U$) is defined as C. That is, by an independence model, the effect of conscious recollection is symmetrical in that conscious recollection increases the probability of completing a stem with an old word for the inclusion test but decreases that probability for the exclusion test. By the redundancy assumption, on the other hand, conscious recollection only plays a role for the exclusion test where it can be used to avoid responding with old words. Finally, by an assumption of exclusivity, ($C \cap U$) is defined as 0. Thus, conscious recollection adds to performance in the inclusion test but does not contribute to performance in the exclusion test.

Subtracting Exclusion from Inclusion provides an estimate of C. As can be seen by substituting the different definitions of $(C \cap U)$ into Equations 1 and 2, this is true regardless of which assumption about the relation between C and U is adopted. However, the manner in which U is estimated differs depending on the assumed relation between C and U. By a redundancy model, performance on an inclusion test provides a process-pure estimate of U, whereas by an exclusivity model such an estimate is provided by performance on an exclusion test. An independence model, in contrast, does not treat either of the two tests as yielding a pure estimate of unconscious influences. Rather, once an estimate for C has been gained, then U can be determined algebraically as performance on the Exclusion test divided by $(1 - C)$. The following sections provide arguments for choosing the independence assumption.

DEFINING CONSCIOUS AND UNCONSCIOUS PROCESSES BY AN INDEPENDENCE MODEL

Specifying the relation between conscious and unconscious processes achieves much with regard to their definition. Our strategy has been to try to arrange conditions such that the two types of processes independently contribute to performance. Consciously controlled processes are then defined as those producing a difference in performance that reflect current intentions. For example, the difference between the probability of producing an old word for an inclusion as compared to an exclusion test is taken as measuring intentional, consciously controlled use of memory. Automatic or unconscious processes, in contrast, are defined as producing the same effect regardless of whether the effect is in concert or in opposition to current intentions. Thus, by the process-dissociation procedure, conscious and unconscious processes are defined in terms of their differential relation to intention. This relational definition contrasts with attempts to define the two types of processes by identifying them with different tasks or experimental conditions.

Identifying conscious and unconscious influences with direct and indirect tests, respectively, also defines the two types of processes with reference to intention. However, unlike the process-dissociation procedure, the validity of that approach relies on achieving process-pure tests. Other techniques have also been used as an attempt to fully eliminate consciously controlled processes to produce pure measures of unconscious influences or automaticity. *Automaticity* has been defined as a typically rapid basis for responding that does not require attentional capacity and is privileged, for example, in being relatively uninfluenced by the effects of aging or drugs such as alcohol (e.g., Hasher & Zacks, 1979; Posner & Snyder, 1975; Schneider & Shiffrin, 1977). Each of the criteria in this definition has been used as an attempt to produce a pure measure of automaticity. It is likely that indirect tests (dividing atten-

(dividing attention, requiring fast responding, etc.) limit the contribution of consciously controlled processes. However, as discussed for the effects of alcohol, none can be relied on to always fully eliminate conscious processing.

Our definition of conscious and unconscious processes rests on the independence of their contribution. What evidence can we gain to show our attempt to achieve conditions that produce independence are successful? One source of support would be to show that variables traditionally identified with cognitive control produce process dissociations by affecting consciously controlled processing while leaving unconscious influences invariant. In this light, finding invariance in the unconscious use of memory across a group of manipulations (e.g., dividing attention and aging) that influences conscious memory provides evidence for independent processes (Jacoby et al., 1993).

Process Dissociations Supporting Independence

Results from some of our experiments showing process dissociations produced by variables identified with cognitive control are summarized in Table 2.2, where changes in conscious (Δ Consc) and unconscious (Δ Uncon) influences are presented as a function of attention, retrieval speed, and age.

We have found invariance in unconscious influences of memory accompanied by changes in conscious memory from comparing full versus divided attention for word stem completion, recognition memory, and fame judgment tasks. Conscious and unconscious perception also exhibit this pattern of results. Beside attentional manipulations, other variables used as indices of automaticity, such as retrieval speed and age have been examined. Decreasing retrieval time in recognition by forcing subjects to respond before a signal reduces conscious memory processing, but leaves the unconscious use of memory intact. Similarly, aged subjects show large deficits in conscious memory but no impairments in unconscious memory. The results in Table 2.2 are strikingly consistent. The average change in unconscious influences across all of the experimental conditions where we expect invariance is only .002, and the change in conscious memory is .24.

Process dissociations produced by manipulations identified with automaticity are so consistent that they must be explained by any adequate model. The invariance in U across large effects on C (Table 2.2) is expected if C and U are independent. However, from the perspective of either an exclusivity or a redundancy model, this invariance must be seen as a curious accident reflecting a delicate balance between offsetting effects. For example, suppose one estimated conscious and unconscious influences using the independence model whereas, in reality, the redundancy model holds true. This would result in an underestimation of unconscious influences. To hold a redundancy model and explain the invariance in U estimated by the independence model, it must be argued that the "apparent" cases of invariance shown in Table 2.2 each reflect a delicate balance between the underestimation of U produced

TABLE 2.2
Changes in Conscious and Unconscious Influences as a
Function of Attention, Retrieval Speed, and Age

Experiment	Condition	Δ Consc	Δ Uncon
	Attention—Full vs. Divided		
Stem Completion	Exp. 1A	.20	.00
	Exp. 1B	.25	.01
Cued Word Completion	Related Cue	.24	−.03
	Urelated Cue	−.01	.01
Cued Fragment Completion	Related Cue	.17	−.01
	Unrelated Cue	.05	−.01
Fame Judgment		.26	−.05
Recognition (Yes/No)	Related Cue	.24	.00
	Unrelated Cue	.20	.00
Perceptual Task	Exp. 2	.42	.01
	Exp. 3	.64	−.02
	Retrieval Speed—Slow vs. Fast		
Recognition (Yes/No)	Short Lists	.09	.03
	Long Lists	.08	.04
Recognition (Yes/No)	Sem Encoding	.22	−.04
	Nonsemantic	.13	.02
	Age Effects— Young vs. Elderly Adults		
Fame Judgment		.29	−.06
Recognition (Forced Choice)	Read	.33	.04
	Anagram	.21	−.01
Continuous Stem Completion	Lag 12	.45	.03
	Lag 48	.28	.00
Average Difference:		.24	.002

Note: The representation of data in this table is very conservative. For example, for the fame judgment experiments, changes in unconscious influences (−.05 and −.06) were due to differences in base rates between conditions. After correcting for base rates, the respective changes in unconscious influences are −.02 and .00. We mention this because problems with base rate along with floor and ceiling effects are the most common problems one is likely to encounter when using the inclusion/exclusion procedure.

by violating the independence assumption, and a near perfectly offsetting effect of the manipulated variable. Given that the invariance in U has been found across a wide range of estimates of C, the required balance becomes even more delicate. We feel that the burden of proof is on those who propose such delicate balances (e.g., Joordens & Merikle, 1993); that is, it must be explained why balances that mimic independence occur so frequently.

As further evidence for independence, Jacoby et al. (1993) showed that a read versus generate manipulation produced opposite effects on C and U. Generating, as compared to reading a word, enhanced conscious recollection but produced less unconscious influences on stem-completion performance. The pattern of results converges with findings of task dissociations. Generating a word at study produces better conscious memory, measured by a direct test, than does reading a word (e.g., Jacoby, 1978; Slamecka & Graf, 1978), whereas the opposite is true for an indirect test of visual word recognition (Jacoby, 1983; Winnick & Daniel, 1970). These results show that invariances found with U do not reflect general insensitivity of that measure, because there are manipulations that have an effect.

Comparisons with Indirect Test Performance Provide Converging Evidence

We gain additional support for our independence model by further building on findings of task dissociations. Although sometimes contaminated by effects of conscious processes, it seems likely that performance on indirect tests primarily reflect unconscious influences. The majority of findings from indirect tests provide converging evidence for conclusions based on the use of the independence model. For example, manipulations of attention (e.g., Koriat & Feuerstein, 1976), levels of processing (e.g., Jacoby & Dallas, 1981), and aging (e.g., Light & Albertson, 1989) have large effects on direct test performance but little or no effect on indirect test performance. Similarly, we find those manipulations influence C but not U.

Although estimates of U are generally in agreement with results from indirect tests, the two can diverge, presumably because C can contaminate performance on indirect tests. As an example, consider the effects of levels of processing. One way of accounting for the small effects that are obtained (e.g., Challis & Brodbeck, 1992) is to claim that performance on the indirect test is contaminated by conscious influences of memory of the same sort as measured by a direct test (Roediger, Weldon, Stadler, & Riegler, 1992). Further, if it were not for such contamination, no effects of levels on indirect test performance would be found. Recent work using the process-dissociation procedure supports this claim. Toth, Reingold, and Jacoby (1994) used a stem-completion task and showed that estimates of conscious recollection are higher after deep than shallow processing (.23 vs. .03), but unconscious influences are near identical (.45 vs. .44). This finding of invariance is similar to those found with manipulations of attention (see Table 2.2).

The issue of contamination is crucial for choosing among models. By the redundancy assumption, performance on an indirect task or even an inclusion condition is treated as a pure measure of U. However, if that task is contaminated by C, then any conclusions will be untenable. The next section describes a series of experiments to show that cross-modality transfer in performance on an indirect test can arise from the contaminating effects of conscious use of memory.

COMPARING INDEPENDENCE
WITH REDUNDANCY AND EXCLUSIVITY

Cross-Modality Transfer as a Testing Ground for the
Relation Between Conscious and Unconscious Influences

Experiments using indirect tests such as perceptual identification and fragment completion have generally shown effects of changing modality between study and test. For visual word identification, reading a word substantially increases its later identification, whereas hearing a word can confer little or no advantage in later identification performance (Jacoby & Dallas, 1981; Morton, 1979). For visual stem- and fragment-completion performance, reading a word also does more to increase the probability of its later being given as a completion than does hearing a word. However, words that were earlier heard are often found to enjoy a large advantage over new words (Blaxton, 1989; Graf, Shimamura, & Squire, 1985; Roediger & Blaxton, 1987). This cross-modality transfer has been taken as evidence for the existence of an abstract, modality-free representation of words that can be temporarily primed (Kirsner & Dunn, 1985).

Conclusions drawn about the basis for cross-modality transfer rest on assumptions about the relation between conscious and unconscious influences of memory, and the mapping of forms of memory onto types of test. For the conclusion that cross-modality transfer on an indirect test reflects the priming of an abstract representation, it must be assumed that an indirect test serves as a pure measure of automatic or unconscious influences. Alternatively, cross-modality transfer could arise from the contamination of performance on the indirect test by intentional uses of memory (Jacoby et al., 1993). Concerns of this sort are illustrated by results from a series of experiments done to examine cross-modality transfer, and to use findings of transfer as a testing ground for assumptions about the relation between conscious and unconscious influences of memory. The independence assumption is first contrasted with the redundancy assumption, then with the exclusivity assumption.

A series of experiments was done to examine cross-modality transfer. In the first phase of each of those experiments, a long series of words was presented with half being read and the other half being heard. Subjects read words aloud and repeated heard words to ensure perception of the presented words. Memory was tested by presenting word fragments that subjects were to complete. Word fragments rather than word stems were used because cross-modality transfer found using word stems may result from subjects pronouncing the stem and, thereby, gaining access to memory for earlier hearing the word (Donaldson & Geneau, 1991). That is, when word stems are used, transfer from heard words may not be truly cross-modal but, rather, have an aural basis because of stem pronunciation. To eliminate such transfer, unpronounceable word fragments were used. Each fragment allowed only a single solution. Test conditions were varied to reflect different assumptions about the relation between conscious and unconscious influences of memory.

Redundancy versus Independence: The Indirect Test as a Pure Measure of Unconscious Influences

The redundancy model treats consciousness as the result of processing that follows unconscious influences. For example, conscious awareness that a word was presented earlier might only be gained after the word came to mind as a completion for a fragment. Models of this sort (Jacoby & Hollingshead, 1990; Jacoby et al., 1993; Joordens & Merikle, 1993) can be classified as generate-recognize models of memory, and estimates of unconscious influences differ depending on the specific model used. For example, Joordens and Merikle used performance on an *inclusion* test as an estimate of unconscious influences, whereas Jacoby and Hollingshead used an *indirect* test. Given the argument against equating unconscious influences with indirect test performance is the potential contamination from conscious processes (e.g., Holender, 1986; Reingold & Merikle, 1990), this problem is increased with an inclusion test where subjects are instructed to intentionally use memory.

Cross-Modality Transfer as Measured by an Indirect Test.   For an indirect test (Exp. 1), fragments that could be completed with an old word were intermixed with fragments that could only be completed with a new word. Subjects were instructed to complete as many of the fragments as they could. No mention was made of the relation between study and test. Results from that indirect test (Table 2.3) provide evidence of modality-specific transfer by showing that words that were earlier read were more likely to be given as a com-

TABLE 2.3
Probability of Correct Fragment Completion in Experiment 1

Study		
Read	Heard	New
.61	.48	.24

pletion than were words that were earlier heard. However, there was also a substantial amount of cross-modality transfer as shown by the advantage of words that were heard over new words. This pattern of results is the same as reported by others (e.g., Graf et al., 1985).

Cross-Modality Transfer as Measured by Process-Dissociation Procedure. The next question asked was whether we still see cross-modality transfer in unconscious processes when the process-dissociation procedure is used to eliminate contamination. In Experiment 2, the study procedure and the fragments presented at test were the same as for the previous fragment-completion experiment. Instructions for the inclusion and exclusion tests, as well as other details of the test procedure, were the same as described by Jacoby et al. (1993).

Results for the inclusion and exclusion tests along with estimates of conscious and unconscious influences based on the independence assumption are displayed in Table 2.4. Those estimates show that study modality had an effect on both conscious and unconscious influences. Words that were read were more likely to be recollected as a completion for a fragment, a conscious use of memory, than were words that were heard, and were also more likely to produce an unconscious use of memory. This unconscious influence produced by words that were read may have had the same basis as priming observed on the indirect test. However, unlike performance on the indirect test, the process-dissociation data provide no evidence of cross-modality transfer. The estimated unconscious influences for words that were heard did not differ significantly from the baseline provided by performance on fragments that could only be completed with a new word.

Comparing results across the two experiments reveals a striking similarity

TABLE 2.4
Probability of Correct Fragment Completion and Estimates
of Conscious and Unconscious Influences of Memory
as a Function of Study Processing in Experiment 2

	Study	
	Read	Heard
Test		
Inclusion	.63	.46
Exclusion	.27	.24
Estimates		
Conscious	.36	.23
Unconscious	.39	.29

Note: The baseline completion rate for items not presented at study was .26.

between results for the indirect test (Table 2.3) and results for the inclusion test (Table 2.4). The near-identity of those results would be explained differently depending on whether one assumed redundancy or independence between unconscious and conscious memory. To uphold a redundancy model one could argue that instructing subjects to use fragments as cues for recall, as was done for the inclusion test, did nothing but add a recognition-memory check to the generation processes relied on for the indirect test (Jacoby & Hollingshead, 1990). That recognition-memory check would be irrelevant for the inclusion test, because subjects were instructed to complete fragments with any word that came to mind if unable to recall an old word. If the redundancy model holds, the lack of cross-modality transfer revealed by applying the independence model can be dismissed as resulting from the underestimation of U produced by violation of the independence assumption.

Alternatively, by the independence model, the near-equivalence of performance on the indirect and inclusion tests can be produced from the indirect test being functionally equivalent to the inclusion test in its reliance on intentional use of memory. That is, although subjects were not informed about the relation between study and the indirect test, they may have "caught-on" and intentionally used memory to complete word fragments just as did subjects given an inclusion test. If so, cross-modality transfer effects found on the indirect test likely result from contamination by intentional use of memory.

Converging Evidence for Independence.   Jacoby et al. (1993) showed that divided, as compared to full, attention during study reduced subjects' ability to recollect those items but left unconscious influences of memory unchanged. Arguing from that result, manipulating full versus divided attention during study provides a potential means for choosing between the different assumptions. In particular, if cross-modality transfer observed on an indirect test is due to contamination by intentional use of memory, and if dividing attention during study reduces the ability to later use intentional memory, then cross-modality transfer should be eliminated on an indirect test by dividing attention during study. The rationale is the same as described earlier for the effects of drunkenness. If the indirect test could be made process pure, performance on that test should match the estimate of automatic influences gained by the process-dissociation procedure.

In an experiment done to examine that possibility (Exp. 3), subjects heard a long list of words and repeated aloud each word in the list immediately after its presentation. While hearing and repeating words, subjects in a divided-attention condition engaged in the additional task of searching through numbers that were rapidly presented visually for runs of three odd numbers in a row (e.g., 9, 7, 3). They were instructed that this visual search task was their primary task, and repeating the words should be done rather automatically to not interfere with the search task. Subjects in a full-attention condition did

TABLE 2.5
Probability of Correct Fragment Completion
Under Conditions of Full and Divided Attention in Experiment 3

	Study	
Attention	Heard	New
Full	.50	.30
Divided	.36	.30

not engage in the search task while hearing and repeating words. The indirect test of fragment completion was the same as described previously.

Results on that test (Table 2.5) showed that cross-modality transfer was radically reduced after divided, rather than full, attention was given to hearing and repeating words. The reduction in transfer is consistent with the suggestion that cross-modality transfer relies on intentional use of memory, which was reduced after attention was divided during study. Further evidence that recollection played a role was gained by questioning subjects after the fragment-completion test. All subjects in the full-attention condition claimed to have noticed the relation between study and test and to have intentionally used memory for the earlier heard words to complete fragments. In the divided-attention condition, subjects were assigned to an "aware" or "unaware" group on the basis of their self-reports. Subjects in the unaware group either claimed to have been unaware of the relation between study and test or to have become aware of that relation very late in the test, whereas those in the aware group claimed to have become aware early on and to have intentionally used memory to respond. Separating fragment-completion performance for these two groups revealed that only aware subjects showed cross-modality transfer, completing more fragments with earlier heard words than with new ones (.40 vs. .29). For unaware subjects, the probability of completing fragments with earlier-heard words was nearly identical to that of completing fragments with new words (.32 vs. .31).

To summarize Experiment 3, dividing attention during study provides evidence that performance on the indirect test of fragment completion is sometimes badly contaminated by an intentional use of memory. After dividing attention, cross-modality transfer is radically reduced and is near zero when self-reported awareness is taken into account. These results are identical to those estimated using the independence assumption. That is, performance on an indirect test that is unlikely to be contaminated by intentional uses of memory almost perfectly matches the estimate of automatic influences gained by relying on the independence assumption. Similarly, Toth et al. (1994) found that indirect test performance after shallow, but not after deep, processing almost perfectly matched U. These findings of correspondence between "uncontaminated" indirect test performance and U converge with findings of process dissociations produced by manipulating attention and level of processing. Just as earlier described for the pure test of automatic influences produced by dividing attention, results from uncontaminated indirect tests almost perfectly predict automatic influences operating in the presence of consciously controlled processes. However, we could not know this without the process-dissociation procedure.

The effects of dividing attention on later cross-modality transfer, as measured by an indirect test, converge with results reported by Weldon and Jackson-Barrett (1993). They were able to eliminate picture-to-word transfer on an indirect test by having subjects divide attention during encoding (Exp. 2) or by requiring rapid responding (Exp. 3). Weldon and Jackson-Barrett interpreted their divided-attention results as showing that picture-to-word priming relies on covert naming of a picture during study. However, the results might be better interpreted as showing that picture-to-word transfer reflects contamination of performance on the indirect test by intentional use of memory. Such contamination is decreased when recollection is made less likely by dividing attention during encoding, or requiring fast responding at test (see Toth & Reingold, in press, for further arguments to support this interpretation). Of course, we cannot be certain that all cross-form transfer on indirect tests reflects contamination. However, there is reason to suspect that when contamination is removed, there will be less need to postulate abstract, modality-free representations (cf. Kirsner & Dunn, 1985).

We are unable to specify conditions that ensure an indirect test will be uncontaminated. The problems are the same as would be faced by an attempt to specify a criterial level of drunkenness to achieve a pure test of automaticity. There are likely individual differences in response to drunkenness, divided attention, instructions for an indirect test, and so forth. We have been able to show that changing conditions so that they are less hospitable to conscious control increases the similarity between indirect test performance and our estimates of U. Findings from this series of experiments are problematic for any redundancy model that views indirect or inclusion tests as a pure measure of unconscious processing. Moreover, a redundancy model is unable to account for the findings of process dissociations produced by manipulations traditionally identified with cognitive control, or for the match between estimated U and performance on an uncontaminated indirect test.

Independence versus Exclusivity:
Remember/Know Judgments

Tulving (1983), along with Gardiner and colleagues (Gardiner, 1988; Gardiner & Java, 1991; Gardiner & Parkin, 1990) emphasized differences in subjective experience by requiring subjects to judge whether they "remember" an item was presented in an earlier list or only "know" that the item was in the list. Subjects are instructed to respond Remember if they can consciously recollect aspects of the study episode, such as the item's appearance or an association it brought to mind. They are instructed to respond Know if they find the item familiar but they cannot recollect any details of its prior occurrence. A large number of experiments have been conducted to show differential effects of manipulated variables on Remember and Know judgments (see Gardiner & Java, 1993, for a review).

Because subjects are only allowed to make one response to each item, it is unarguable that the R and K responses are mutually exclusive, just as are Yes and No responses in a standard recognition memory test. More importantly, Gardiner and colleagues assumed that the underlying processes are mutually exclusive, and treat R and K responses as providing pure measures of recollection and familiarity. For example, if a variable is found to have an effect on the proportion of K responses, they concluded that the variable has an effect on familiarity.

If, on the other hand, the two processes are independent, then how do we interpret Remember and Know responses? Remember responses should provide a relatively pure measure of conscious recollection. This is true as long as subjects respond R when, and only when, they recollect. If they mistakenly respond Remember to items that are only familiar, our estimates of recollection will be inflated, as will false Remember responses to new items. Fortunately, in most Remember/Know experiments the probability of falsely remembering a new item is very low, suggesting that Remember provides an adequate measure of recollection.

In contrast, Know responses will not provide a pure measure of familiarity. Rather, they only reflect familiarity in the absence of recollection ($F(1 - R)$). In fact, subjects are instructed to respond Know only if the item is recognized as familiar but is not recollected. If the two processes are independent, there will be some proportion of items that are both familiar and recollected (a possibility not allowed by the exclusivity assumption). For these items, subjects will respond Remember even though the items are also familiar. Consequently, the proportion of Know responses underestimates the probability that an item is familiar. To determine the probability that an item is familiar (F), one most divide the proportion of Know responses by the opportunity the subject has to make a Know response ($1 - R$). That is, $F = K / (1 - R)$. We refer to this method for measuring recollection and familiarity as the Independence Remember/Know Procedure (IRK) to distinguish it from approaches relying on the exclusivity assumption.

To reiterate, according to both independence and exclusivity, Remember responses provide a measure of conscious recollection. However, the two assumptions differ in their estimation of familiarity. By the exclusivity assumption, the proportion of K responses provides a pure measure of familiarity, whereas by the independence assumption, familiarity is calculated by dividing the proportion of K responses by $1 - R$.

In the following experiments, we contrast the exclusivity and independence assumptions using the Remember/Know procedure. Experiment 4 examines cross-modality transfer in a fragment completion task. In Experiment 5, we examine the effects of size congruency on recognition performance, comparing our results to those obtained by Rajaram and Roediger (Chapter 11, this volume). In a final experiment, we conduct an ROC analysis to further compare the independence and exclusivity assumptions.

Fragment Completion: Cross-Modality Transfer.   In Experiment 4, subjects were presented with the same study and test list as described in the previous section, and were instructed to use the word fragments as cues for the recall of words they read or heard at study. After completing each fragment, subjects were to indicate whether they remembered the completion word as one presented earlier, did not remember but knew the completion word was presented earlier, or thought that the completion word had not been presented earlier (New). Definitions of Remember and Know given to subjects were adapted from the instructions used by Gardiner.

First consider the results using the exclusivity assumption. Examination of Table 2.6 shows that subjects were more likely to remember words they had read as compared to words they had heard. Thus changes in modality between study and test decreased conscious recollection. However, by the exclusivity assumption there was no evidence of unconscious memory for either heard or seen words. That is, for items receiving a Know response there was no advantage for heard or seen words over new words. Given one would expect some unconscious influence of memory on a fragment completion task, based on the indirect test literature already described, these results are puzzling.

One question that may arise here is whether Know judgments are truly unconscious, given subjects indicate some memory for an item. We treat Know responses as a measure of unconscious memory because subjects do not recollect any details of having seen an item. The notion is that, in some cases,

TABLE 2.6
Probability of Responding Remember, Know, or New
as a Function of Study Processing in Experiment 4

Response	Study		
	Read	*Heard*	*New*
Remember	.36	.24	.04
Know	.14	.10	.13
New	.11	.10	.15

TABLE 2.7
Estimates for Conscious and Unconscious Influences
Calculated by Applying the Independence Assumption to
the Remember/Know Procedure (IRK) in Experiment 4

	Study	
Estimates	Read	Heard
Conscious	.36	.24
Unconscious	.40	.27

Note: The baseline completion rate for items not presented at study was .32.

items may come to mind very fluently, and subjects may attribute this fluency
to prior presentation, responding Know. However, the degree of fluency associated with unconscious influences may differ such that some items come
to mind but result in a New response. Therefore, to assess familiarity one may
want to examine both Know and New responses combined. However, doing so
does not change the conclusion, because there was no advantage for old words
over new words for items eliciting either a Know or a New response. The notion that Know and New items differ in terms of the strength or fluency is discussed later.

How does the pattern of results (Table 2.6) change if the independence,
rather than the exclusivity, assumption is adopted? Results, recomputed with
the independence assumption, are shown in Table 2.7. To calculate C and U,
completions that subjects remember are equated with conscious recollection,
whereas all occasions on which a fragment was completed with old words that
were not remembered were used to estimate unconscious memory ($U = (K + N)/(1 - R)$).

Examination of Table 2.7 shows that both conscious and unconscious influences were greater for read words than heard items. What is most striking
though, is the similarity of the results in Table 2.7 to those gained by using
the Inclusion/Exclusion test procedure to estimate conscious and unconscious influences (Table 2.4). The probabilities of recollection for read and
heard words are nearly identical to the probabilities of responding Remember when completing fragments with those words. The estimates of unconscious influences gained by the two procedures are also comparable. Clearly,
the choice between the assumptions of exclusivity and independence (Table
2.6 vs. Table 2.7) was more important than the choice between the Inclusion/
Exclusion and the Remember/Know approaches, given the independence assumption for both procedures.

In retrospect, it should not be surprising that Remember/Know and In-
clusion/Exclusion procedures produced near-identical results. The rationale

underlying the Inclusion/Exclusion procedure holds that recollection serves as a basis for control. It was argued that if subjects recollected a word as earlier presented, they could either include or exclude that word as a fragment completion, when instructed to do so (e.g., Jacoby et al., 1993). If recollection serves as a basis for control, then it should be identifiable by self-report.

When the treatment of data is based on the independence assumption, results from the Remember/Know procedure agree with those from the Inclusion/Exclusion procedure in showing no cross-modality transfer in unconscious influences of memory. Indeed, the estimate of unconscious influences for words that were heard is slightly, though not significantly, below the baseline for new words (.27 vs. .32). The major difference between the two sets of results is in baseline performance. If one simply averages the baselines, which is legitimate because the same materials were used in the two experiments, one finds that estimates of unconscious influences for heard words are near identical to baseline for both procedures.

Recognition Memory: Parallel Effects of Size Congruency.   Results described in the last section showed that the Remember/Know and Inclusion/Exclusion procedures produce parallel results when the independence assumption is adopted. In contrast, Gardiner and his associates (Gardiner, 1988; Gardiner & Java, 1991; Gardiner & Parkin, 1990) relied on the exclusivity assumption to show that many manipulations produce dissociations between Remember and Know responses. Among the dissociations reported are some that are difficult to interpret. For example, in an investigation of recognition-memory performance, Rajaram and Roediger (Chapter 11, this volume) found that manipulating size congruency produced a dissociation between Remember responses, which were greatest when study and test stimuli were congruent in size, and Know responses, which produced the opposite pattern of results. That finding appears curious if Know responses are identified with the use of familiarity as a basis for recognition in a two-factor theory of memory (e.g., Mandler, 1980). The claim would have to be that increasing similarity (size congruency) decreased familiarity.

We investigated the effects of size congruency, and contrasted results based on the exclusivity assumption with results gained using the independence assumption (Yonelinas & Jacoby, 1995b). Subjects studied a list of randomly generated shapes and were then given a recognition-memory test that included old shapes mixed with new shapes. Half of the old shapes were the same size they had been during study, whereas the other half were either larger or smaller in size. Similar to the Remember/Know procedure described earlier, subjects made recognition decisions by responding Recollect if they could consciously recollect having seen a test item in the study list, Know if they felt the item was in the study list but they could not recollect it, or New if they felt the item had not been seen before. Despite the fact that the sub-

TABLE 2.8
The Effects of Size Congruency on Recollect and Know Responses,
and Estimates of Familiarity in Experiment 5

| | Size | | |
	Congruent	Incongruent	New
Recollect	.45	.30	.08
Know	.36	.41	.36
Familiarity	.65	.58	

jects were told that the size of the shapes would vary and size was irrelevant, they recognized more of the size congruent objects than the size incongruent objects (see Table 2.8).

Using the Recollect responses as a measure of conscious memory, we find that size congruency increased conscious influences. Estimates of familiarity were then derived using independence as well as the exclusivity assumptions. Adopting the exclusivity assumption, whereby Know responses are treated as a pure measure of familiarity, leads to the strange conclusion that size congruency *decreased* the familiarity of old shapes. Furthermore, the estimated familiarity of same-size shapes was no greater than that of new shapes (.36 and .36). Only when shapes changed size was an advantage in familiarity for old shapes found. These results are odd because it is common to assume that increasing similarity, as produced by size congruency, increases familiarity. Indeed, that was found when the independence assumption was used to estimate familiarity. By the independence assumption, old shapes were more familiar than new shapes, and old items that were the same size at study and test were more familiar than those for which size changed.

The exclusivity assumption, in effect, forces the odd results found for familiarity. Because of the assumption that items can be either recollected or familiar, high recollection levels associated with old size congruent shapes push the level of Know responses down. That is, the measure of familiarity is influenced as much by factors influencing recollection as by factors influencing familiarity per se.

Similar problems produced by the exclusivity assumption can be seen in examinations of the effect of aging. Using the R/K procedure, Parkin and Walter (1992) found that older adults showed poorer recollection than younger adults, but use of Know responses increased. Although it may be comforting to think that deficits in recollection are offset by improvements in familiarity, this pattern of results appears to be an artifact of the R/K procedure, as was the size congruency effect. That is, the increase in Know responses shown by the elderly does not reflect an increase in familiarity. Instead, young adults are showing a decrease in Know responses forced by their higher level of Remember responses. In support of this possibility, the results of Jacoby (1992) described earlier, and Jennings and Jacoby (1993) showed that despite decreases in recollection with age, familiarity remained intact. These results converge with those of several experiments that show indirect test performance, which presumably relies on processes similar to familiarity, is not greatly influenced by aging (e.g., Light & Albertson, 1989).

The same problem described earlier also produces inconsistencies across other Remember/Know experiments. For example, Gardiner (1988) found that deeper levels of processing increased the probability of Remember responses without influencing the probability of Know responses, and concluded that depth of processing selectively influences the process that supports remembering. However, Rajaram (1993) found that deeper processing led to an increase in Remember responses accompanied by a significant decrease in Know responses. These inconsistencies are a product of the exclusivity assumption. If, as held by the independence assumption, Know responses reflect familiarity and recollection, then large effects on recollection will tend to artifactually produce opposite effects on the Know responses. Know responses remained constant in the Gardiner study but changed in the Rajaram study because of differences in the magnitude of levels of processing effect on recollection. In the Rajaram study, manipulating levels of processing led to a difference of .34 in recollection versus the .20 difference shown by Gardiner.[1] Further problems for the exclusivity assumption are considered in the next section.

Signal Detection Theory in a Dual-factor Model. We have argued elsewhere (Jacoby et al., 1993; Yonelinas, 1994) that the unconscious influences of memory, such as familiarity, are well described by signal-detection theory, and a weakness of earlier applications of signal-detection theory was the failure to separate unconscious influences from the effects of recollection. In order to test these notions we collected confidence judgments in a Remember/Know recognition task, and plotted a receiver operator characteristic (ROC). The ROC allowed us to determine whether familiarity was related to false alarm rates in a manner that would be predictable by signal-detection theory. It also allowed us to further evaluate the independence and exclusivity assumptions.

---

[1]Using the IRK procedure with Gardiner's (1988) and Rajaram's (1993) data showed that both recollection and familiarity increased with levels of processing. Such results are in agreement with those found using the process-dissociation procedure with a recognition task (Jacoby & Kelley, 1991; Toth, 1992). Note that the effects of levels of processing in recognition do not parallel those found with cued recall, where levels of processing had no effect on the unconscious uses on memory (see Toth et al., 1994). This is reasonable given that retrieval cues differ across the two tests (see Jacoby et al., 1993). By in large, we find that dissociations are more easily obtained between conscious and unconscious influences in cued recall than in recognition.

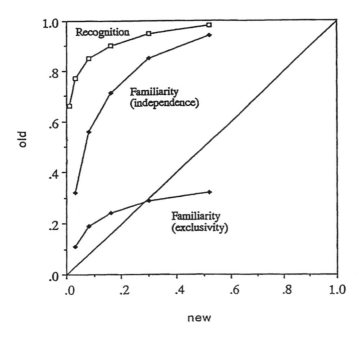

FIG. 2.2.   Receiver operating characteristics for recognition performance and for familiarity derived using the independence and exclusivity assumptions.

We used a repeated study–test procedure in which each study list was immediately followed by a recognition memory test for the words in that list. For
each test item, subjects reported whether they could recollect the item, and
if they could not, they used a 6-point scale to rate the item's familiarity in the
experimental context. That is, they rated their confidence that the item was
in the study list, although they could not recollect any details of its presentation. We plotted correct recognitions against false alarms across all levels of
confidence, treating recollected items as the most confident level (see Fig.
2.2). The obtained recognition function is typical of those found for recognition memory (see Ratcliff, Sheu, & Gronlund, 1992). The function is curvilinear, and exhibits a skew or asymmetry so the curve does not approach the
zero–zero intercept but seems to intersect the y-axis.

Estimates of familiarity were derived using the independence and exclusivity assumptions, and are plotted in Fig. 2.2. The independence assumption
produces a curvilinear familiarity function that increases gradually as a function of false alarm rate. The discriminability (measured by d') afforded by familiarity remained constant as false alarm rate increased. Such a function
would be expected if subjects were relying on assessments of item familiarity
as would be described by signal-detection theory. In fact, the familiarity func-

tion no longer has the asymmetry that was apparent in the overall recognition curve that included both familiarity and recollection. In fact, Yonelinas (1994) argued that the asymmetry in the ROC reflects the contribution of recollection to recognition performance.

It is informative to examine the estimates of familiarity in light of the exclusivity assumption. As can be seen in Fig. 2.2, estimates of familiarity increase slightly as a function of false alarm rate. However, when the false alarm rate increases, the function falls below the diagonal. This would lead one to conclude that new items become more familiar than old items! We have found this unreasonable pattern of results before. In fact, it seems to be a by-product of the exclusivity assumption. Similar to our size congruency experiment, the measure of familiarity based on the exclusivity assumption produces unreasonable results because it reflects the level of recollection as much as familiarity per se. Because of results of this sort, we think the exclusivity assumption can be dismissed as a viable characterization of the relation between conscious and unconscious influences of memory.

Applying signal-detection theory to describe unconscious influences also sheds light on the difference between Know and New judgments. Both judgments reflect the familiarity or strength of an item and differ only in that the Know judgment is made if the strength is above some criterion. To gain an estimate of unconscious influences, it is justifiable to ignore the difference between the two types of judgments as done earlier. Elsewhere (e.g., Jacoby, Kelley, & Dywan, 1989) we argued that familiarity reflects a memory attribution, and the strength or fluency that results in that attribution can be misattributed to other sources, such as famousness or pleasantness.

In contrast to the manipulations that are found to influence C but to leave U in place (e.g., Table 2.2), the ROC analysis showed that familiarity was influenced by response criterion. Similar ROC studies using the process–dissociation procedure have shown that relaxing response criterion increased the proportion of items accepted on the basis of familiarity but did not influence estimates of recollection (Yonelinas, 1994). Other manipulations, such as varying stimulus–response probabilities, are also found to selectively affect unconscious influences (Jacoby & Hay, 1993; Yonelinas & Jacoby, 1995a). Such dissociations provide further support for the independence assumption.

CONCLUDING COMMENTS

Our approach to defining conscious and unconscious processes differs in important ways from those taken by others. Eriksen (1960), for example, operationally defined awareness in terms of performance on tasks measuring perceptual discriminations, and sought evidence of perception under conditions establishing chance discrimination. Eriksen's approach relied on signal-detection theory (Swets, Tanner, & Birdsall, 1961) to separate discriminability from criterion differences. However, discriminative responding may be sensitive to both conscious and unconscious processing so that equating awareness with discriminative responding defines unconscious perception out of existence (Bowers, 1984). Later researchers abandoned the single-process or "strength" view of perception and memory, represented by signal-detection theory, and used task dissociations as evidence for the existence of unconscious processes. Conscious and unconscious processes are identified with performance on direct and indirect tests, respectively. However, that approach also suffers from the criticism that tasks may not be process pure. Effects on performance of an indirect test that are taken as evidence of unconscious perception or memory might truly reflect intentional, aware perception or memory that is undetected by the experimenter (e.g., Holender, 1986; Jacoby, 1991; Reingold & Merikle, 1990).

Because of the process-pure assumption made by other approaches, controversy has surrounded claims of the existence of unconscious processes. In contrast to Eriksen's discriminative responding approach, we define unconscious influence into existence. We begin with a two-factor model that assumes the separate existence of conscious and unconscious processes, and holds that they independently contribute to performance. We then construct tests and conditions in ways aimed at satisfying the assumptions of that model. Doing so allows us to separate the contributions of conscious and unconscious processes within a task and, thereby, avoid the process-pure assumption made when the different types of processes are identified with different tasks. As important, our process-dissociation approach allows us to examine unconscious processes operating in the context of conscious processes, which cannot be done using the direct–indirect test approach. We go beyond attempts to prove the existence of unconscious processes to search for process dissociations that show the differential effect of manipulations on the separate contributions of conscious and unconscious processes.

A prerequisite for separating the contributions of conscious and unconscious processes is that an assumption be made about their relation. It is important to note that we do not claim that conscious and unconscious processes are always independent. Rather, we have spent considerable effort constructing conditions meant to meet the independence assumption along with other assumptions underlying the process–dissociation approach. To get neat data, it is important to avoid floor and ceiling effects. For example, if performance in an exclusion condition is perfect so that old items are never given as a response (Exclusion = 0), our equations will necessarily estimate unconscious influences as being zero. Such floor effects can mask invariances that would otherwise be found (Jacoby et al., 1993). Another important concern involves the instructions given at the time of test. For the experiments described here, instructions for both the inclusion and exclusion tests strongly encouraged subjects to use word fragments as cues for direct retrieval of studied items. However, we have unpublished data to show that changing these direct retrieval instructions to encourage a Generate/Recognize strategy results in violation of the independence assumption. Given that it is easy to violate the assumptions underlying our process-dissociation approach, how can we be certain that we have been successful in our attempts to satisfy those assumptions? Why should one bother to attempt to construct situations for which conscious and unconscious processes independently contribute to performance? We answer those questions in the next two sections.

Support for the Independence Assumption

The independence assumption has the advantage of revealing invariance in unconscious influences across manipulations that produce a large effect on conscious recollection (Table 2.2). Importantly, the conditions that produce such process dissociations are those that have traditionally been identified with automaticity, such as divided attention, aging, and speeded responding. These results often converge with those of indirect tests. For example, the read/anagram manipulation produces opposite effects on conscious and unconscious influences, and a corresponding dissociation between performance on direct and indirect tests (Jacoby et al., 1993). However, results for estimates of the unconscious and indirect tests do sometimes differ, as illustrated by the effects of dividing attention on cross-modality transfer (Tables 2.3 and 2.4).

Comparisons across our experiments examining cross-modality transfer show that radically different conclusions can be drawn depending on the assumed relation between unconscious and conscious memory influences. Based on the assumption of independence, one would conclude that changing modality between study and test reduced both conscious recollection and unconscious influences of memory. In addition, estimates of unconscious processes showed no evidence of cross-modality transfer (Table 2.4). In contrast, based on the redundancy assumption (unconscious influences are identified either with performance on an indirect test or the inclusion test), one would conclude that manipulating study modality affected the magnitude of unconscious influences, but that a substantial amount of cross-modality transfer occurred (Table 2.3). However, when we divided attention at study, to reduce the likelihood of conscious contamination of the indirect test we found little evidence of cross-modality transfer (Table 2.5). As mentioned, we are not claiming that indirect tests are always badly contaminated or that it is impossible to create conditions under which redundancy may hold. However, for our conditions meant to satisfy the independence assumption, the data strongly favor the independence assumption over the redundancy assumption. Adoption of the redundancy assumption requires the sacrifice of our consistent findings of process dissociations produced by manipulations traditionally associated with cognitive control as well as the converging evidence gained from "uncontaminated" indirect tests of memory.

The exclusivity assumption proved to be even less satisfactory than the redundancy assumption. As shown here, several conclusions based on exclusivity are problematic. By an exclusivity assumption, there was no evidence of unconscious influences in word fragment completion (a task known to reveal unconscious memory). In addition, changing the size of objects between study and test led to an unexpected increase in familiarity on a recognition task. Finally, relaxing response criterion in an ROC task led to the conclusion that old items were less familiar than new ones. Adoption of the independence assumption with the Remember/Know procedure (IRK) produced a more reasonable pattern of results. IRK showed evidence for unconscious influences in fragment completion, familiarity increased with size congruency, and familiarity was well described by signal-detection theory. Because of these results, the exclusivity assumption can be dismissed as a viable characterization of the relation between conscious and unconscious processes. We emphasize that it is only the treatment of Know responses that is problematic for exclusivity in the Remember/Know procedure. Applying the independence assumption to Remember/Know data (IRK) produces results that converge with those from the Inclusion/Exclusion procedure.

What is the relation between the IRK and the Inclusion/Exclusion procedures? Both qualify as process-dissociation procedures in that their goal is the within-task separation of conscious and unconscious processing. However, the IRK procedure, by its reliance on subjective reports, identifies consciousness with awareness. The Inclusion/Exclusion procedure, in contrast, defines consciousness with reference to intentional control of responding. In everyday life, people often make conscious awareness a prerequisite for intentional actions (Kelley & Jacoby, 1990; Marcel, 1988) and, so, it should not be surprising to find agreement between measures of awareness and intentional responding. However, one should not expect to always find such agreement. The possibility of dissociations between awareness and control suggests parallels with observations of frontal lobe patients. One of the most interesting findings with those patients is their deficit in controlled responding despite awareness of the information that would allow such control to operate. For example, on the Wisconsin Card Sorting Task, frontal patients can often explicitly state the principles underlying the task, thus showing awareness, yet fail to utilize these principles in their actual performance (Stuss & Benson, 1984). Comparisons between results from IRK and Inclusion/Exclusion procedures may be useful for examining this "dysexecutive syndrome."

Reaping the Benefits of Independence

The preceding sections have illustrated empirical and theoretical support for the independence assumption, but perhaps even stronger support lies in the assumption's ability to reconceptualize old problems and offer new solutions. The independence assumption has provided us with tools to investigate practical problems, such as action slips, subjective report of cognitive failures, and memory rehabilitation. Action slips offer a powerful example of the interplay between automatic and consciously controlled processes in daily life. James (1890) observed that "very absent-minded persons in going to their bedroom to dress for dinner have been known to take off one garment after another and finally to get into bed" (p. 115). Action slips arise when habit (automatic) and intention (conscious control) are in opposition, just as conscious and unconscious influences of memory were in opposition in our exclusion condition. Exploiting this similarity, we have been able to use the process-dissociation procedure to examine habits and action slips in the lab (e.g., Yonelinas & Jacoby, 1995a). We have shown that there is an increase in action slips in the aged, as well as under conditions of divided attention, both of which are produced by a reduction in conscious control (Jacoby & Hay, 1993).

An obvious question that arises when measuring action slips in the lab is whether these measures correspond to performance in real life. Typically, this issue has been addressed by giving subjects a questionnaire about real-life memory experiences, and comparing self-reported memory failures with lab performance. However, such investigations have found only modest correlations between questionnaires and lab tasks, ranging from .2 to .3 (Herrman, 1982). Given action slips and memory deficits tend to result from deficits in consciously controlled processing, the problem with the correlational data could stem from a failure to separate lab performance into automatic and consciously controlled components. For example, preliminary work by Jennings and Hay examined the relation between lab tests of memory and questionnaire measures with older adults. They too found that the correlation between self-reported memory failure and overall recognition was quite modest at .35. However, when they used the process-dissociation procedure, they found that the correlation between self-report and a measure of conscious control was substantially higher, at .58, whereas the correlation with automatic influences was near zero. Jennings and Hay's data suggest that performance in daily life is predictable from the lab when one focuses on conscious control.

Perhaps our most ambitious application of the independence assumption goes beyond measuring memory performance, investigating the process-dissociation procedure as a means for rehabilitating memory in older adults. Generally, memory rehabilitation with older adults has relied on general practice or on teaching elaborate encoding strategies (for a review, see Kotler-Cope & Camp, 1990). However, our goal is to train intentional, consciously controlled memory processes (Jacoby, Jennings, & Hay, in press). The rationale of our training procedure builds on the observation that elderly adults repeatedly ask the same question, or tell the same story more than once to the same audience. Reasoning that the shorter the interval between mistakenly repeating oneself, the poorer one's recollection, we train people to accomplish consciously controlled use of memory over increasingly long intervals. Using an exclusion task, we placed the elderly in a situation where recollection was easy (a short time interval between presentation and test) and gradually increased the difficulty of the memory setting (length of the time interval) to shape recollection. Slowly moving the elderly from a level where they could perform competently allowed them to adapt their recollective process to more demanding test intervals. Preliminary results are promising, showing evidence of marked improvement in conscious recollection after only five training sessions.

Can Independence Be Proven?

We find our data beautiful (e.g., Table 2.2) and the directions for our research quite exciting. However, others (Graf & Komatsu, 1994; Joordens & Merikle, 1993) are willing to dismiss our findings of near-perfect invariance produced by manipulations traditionally associated with cognitive control as well as our converging evidence from "uncontaminated" indirect tests as reflecting lucky accidents. They pointed out that we are making assumptions and asked that we either prove the validity of the independence assumption or invent an approach not requiring that assumptions be made. We believe both demands are unreasonable but will not digress into a discussion of philosophy of science (for responses to critics, see Jacoby, Toth, Yonelinas, & Debner, 1994; Toth, Reingold, & Jacoby, 1995). Perhaps there is so much critical attention to our assumptions because we have highlighted our assumptions rather than overlooking them or not acknowledging them as such (see Reingold & Toth, in press). We have done so because we believe that specifying the relation between conscious and unconscious processes is a necessary part of their definition.

A proposal that conscious and unconscious influences are independent will likely be met with scepticism by those who have followed Hintzman's (e.g., Hintzman & Hartry, 1990) criticisms of claims for stochastic independence between direct and indirect tests. He argued that little of theoretical importance can be learned by testing the stochastic independence of tasks because measures of association are limited by suppressor variables such as item differences. However, there are important differences between assuming independence of processes and claiming to have proven independence of performance on direct and indirect tests. First, we are not reliant on the use of successive tests of the same items as are those who rely on contingency analyses (e.g., Tulving, Schacter, & Stark, 1982) and, so, we avoid problems produced by prior testing of an item influencing performance on its later test. Second, because our claim is independence of processes, finding correlations between performance on tasks is irrelevant. For example, finding item differences that influence the correlation between word-fragment completion and recognition-memory performance (Hintzman & Hartry, 1990) might reflect the importance of item differences for recollection. Even if the probability of recollection when completing a fragment was highly correlated with the probability of using recollection as a basis for recognition, recollection could be independent from unconscious influences of memory for both tasks.

The greatest obstacle for research contrasting conscious and unconscious processes has been defining the two types of processes. Earlier approaches have attempted to solve this problem by identifying processes with distinctions between tasks such as the distinction between indirect and direct tests of memory. Because tasks cannot be relied on to provide a pure measure of a process, it is better to separate the effects of processes within a task. However, to do so, one must adopt an assumption about the relation between conscious and unconscious influences. The choice among assumptions is important because specifying the relation between conscious and unconscious processes accomplishes a major part of the task of defining the two types of processes. For our situations, we declare that conscious and unconscious processes are independent. Even if independence is not total, just as independence resulting from political declarations is not complete, its declaration is still useful for highlighting dissociations and for setting future directions.

REFERENCES

Baars, B. J. (1988). *A cognitive theory of consciousness.* Cambridge, England: Cambridge University Press.

Blaxton, T. A. (1989). Investigating dissociations among memory measures: Support for a transfer appropriate processing framework. *Journal of Experimental Psychology: Learning, Memory, and Cognition, 15,* 657–668.

Bowers, K. S. (1984). On being unconsciously influenced and informed. In K. S. Bowers & D. Meichenbaum (Eds.), *The unconscious reconsidered* (pp. 227–273). New York: Wiley.

Challis, B. H., & Brodbeck, D. R. (1992). Level of processing affects priming in word fragment completion. *Journal of Learning, Memory, and Cognition, 18,* 595–607.

Craik, F. I. M., & Byrd, M. (1982). Aging and cognitive deficits. The role of attentional resources. In F. I. M. Craik & S. Trehub (Eds.), *Aging and cognitive processes* (pp. 191–211). New York: Plenum.

Donaldson, W., & Geneau, R., (1991, November). *Cross-modal priming in word fragment completion.* Paper presented at the meetings of the Psychonomic Society, San Francisco, CA.

Dunn, J. C., & Kirsner, K. (1989). Implicit memory: Task or process? In S. Lewandowsky, J. C. Dunn, & K. Kirsner (Eds.), *Implicit memory: Theoretical issues* (pp. 17–31). Hillsdale, NJ: Lawrence Erlbaum Associates.

Eriksen, C. W. (1960). Discrimination and learning without awareness: A methodological survey and evaluation. *Psychological Review, 67,* 279–300.

Gardiner, J. M. (1988). Functional aspects of recollective experience. *Memory and Cognition, 16,* 309–313.

Gardiner, J. M., & Java, R. I. (1991). Forgetting in recognition memory with and without recollective experience. *Memory and Cognition, 19,* 617–623.

Gardiner, J. M., & Java, R. I. (1993). Recognizing and remembering. In A. Collins, M. A. Conway, S. E. Gathercole, & P. E. Morris (Eds.), *Theories of memory*. Hillsdale, NJ: Lawrence Erlbaum Associates.

Gardiner, J. M., & Parkin, A. J. (1990). Attention and recollective experience in recognition memory. *Memory and Cognition, 18*, 579–583.

Graf, P., & Komatsu, S. (1994). Process dissociation procedure: Handle with caution! *European Journal of Cognitive Psychology, 6*, 113–129.

Graf, P., Shimamura, A. P., & Squire, L. R. (1985). Priming across modalities and priming across category levels: Extending the domain of preserved function in amnesia. *Journal of Experimental Psychology: Learning, Memory, and Cognition, 2*, 386–396.

Hasher, L., & Zacks, R. T. (1979). Automatic and effortful processes in memory. *Journal of Experimental Psychology: General, 108*, 356–388.

Herrman, D. J. (1982). Know thy memory: The use of questionnaires to assess and study memory. *Psychological Bulletin, 92*, 434–452.

Hintzman, D. L., & Hartry, A. L. (1990). Item effects in recognition and fragment completion: Contingency relations vary for different subsets of words. *Journal of Experimental Psychology: Learning, Memory, and Cognition, 16*, 955–969.

Holender, D. (1986). Semantic activation without conscious identification in dichotic listening, parafoveal vision, and visual masking: A survey and appraisal. *Behavioral and Brain Sciences, 9*, 1–23.

Jacoby, L. L. (1978). On interpreting the effects of repetition: Solving a problem versus remembering a solution. *Journal of Verbal Learning and Verbal Behavior, 17*, 649–667.

Jacoby, L. L. (1983). Remembering the data: Analyzing interactive processes in reading. *Journal of Verbal Learning and Verbal Behavior, 22*, 485–508.

Jacoby, L. L. (1991). A process dissociation framework: Separating automatic from intentional uses of memory. *Journal of Memory and Language, 30*, 513–541.

Jacoby, L. L. (1992, November). *Strategic versus automatic influences of memory: Attention, awareness, and control*. Paper presented at the 33rd annual meeting of the Psychonomic Society, St. Louis.

Jacoby, L. L., & Dallas, M. (1981). On the relationship between autobiographical memory and perceptual learning. *Journal of Experimental Psychology: General, 3*, 306–340.

Jacoby, L. L., & Hay, J. F. (1993, November). *Action slips, proactive interference, and probability matching*. Paper presented at the 34th annual meeting of the Psychonomic Society, Washington, DC.

Jacoby, L. L., & Hollingshead, A. (1990). Toward a generate/recognize model of performance on direct and indirect tests of memory. *Journal of Memory and Language, 29*, 433–454.

Jacoby, L. L., Jennings, J. M., & Hay, J. F. (in press). Dissociating automatic and consciously controlled processes: Implications for diagnosis and rehabilitation of memory deficits. In D. J. Herrman, M. K. Johnson, C. L. McEvoy, C. Hertzog, & P. Hertel (Eds.), *Basic and applied memory research: Theory in context*. Hillsdale, NJ: Lawrence Erlbaum Associates.

Jacoby, L. L., & Kelley, C. M. (1991). Unconscious influences of memory: Dissociations and automaticity. In D. Milner & M. Rugg (Eds.), *The neuropsychology of consciousness* (pp. 201–233). London: Academic Press.

Jacoby, L. L., Kelley, C. M., & Dywan, J. (1989). Memory attributions. In H. L. Roediger & F. I. M. Craik (Eds.), *Varieties of memory and consciousness: Essays in Honour of Endel Tulving*, (pp. 391–422). Hillsdale, NJ: Lawrence Erlbaum Associates.

Jacoby, L. L., Toth, J. P., & Yonelinas, A. P. (1993). Separating conscious and unconscious influences of memory: Attention, awareness, and control. *Journal of Experimental Psychology: General, 122*, 139–154.

Jacoby, L. L., Toth, J. P., Yonelinas, A. P., & Debner, J. A. (1994). The relationship between conscious and unconscious influences: Independence or redundancy? *Journal of Experimental Psychology: General, 123*, 216–219.

James, W. (1890). *Principles of psychology*. New York: Henry Holt.

Jennings, J. M., & Jacoby, L. L. (1993). Automatic versus intentional uses of memory: Aging, attention and control. *Psychology and Aging, 8*, 283–293.

Jones, G. V. (1987). Independence and exclusivity among psychological processes: Implications for the structure of recall. *Psychological Review, 94*, 229–235.

Joordens, S., & Merikle, P. M. (1993). Independence or redundancy? Two models of conscious and unconscious influences. *Journal of Experimental Psychology: General, 122*, 462–467.

Kelley, C. M., & Jacoby, L. L. (1990). The construction of subjective experience: Memory attributions. *Mind and Language, 5*, 49–68.

Kirsner, K., & Dunn, J. (1985). The perceptual record: A common factor in repetition priming and attribute retention? In M. I. Posner & O. S. M. Marin (Eds.), *Mechanisms of attention: Attention and performance* (Vol. 11, pp. 547–565). Hillsdale, NJ: Lawrence Erlbaum Associates.

Koriat, A., & Feuerstein, N. (1976). The recovery of incidentally acquired information. *Acta Psychologica, 40*, 463–474.

Kotler-Cope, S., & Camp, C. J. (1990). Memory intervention in aging populations. In E. A. Lovelace (Ed.), *Aging and cognition: Mental processes, self-awareness and interventions* (pp. 231–261). North Holland: Elsevier Science.

Light, L. L., & Albertson, S. A. (1989). Direct and indirect tests of memory for category exemplars in young and older adults. *Psychology and Aging, 4*, 487–492.

Logan, G. D. (1988). Toward an instance theory of automatization. *Psychological Review, 95*, 492–527.

Mandler, G. (1980). Recognizing: The judgment of previous occurrence. *Psychological Review, 87*, 252–271.

Marcel, A. J. (1983). Conscious and unconscious perception: Experiments on visual masking and word recognition. *Cognitive Psychology, 15*, 197–237.

Marcel, A. J. (1988). Phenomenal experience and functionalism. In A. J. Marcel & E. Bisiach (Eds.), *Consciousness in contemporary science* (pp. 121–158). Oxford: Clarendon Press.

Morton, J. (1979). Facilitation in word recognition: Experiments causing change in the logogen model. In P. A. Kolers, M. E. Wrolstal, & H. Bonma (Eds.), *Processing of visible language* (Vol. 1, pp. 259–268). New York: Plenum.

Moscovitch, M., Vriezen, E. R., & Gottstein, J. (1993). Implicit tests of memory in patients with focal lesions or degenerative brain disorders. In H. Spinnler & F. Boller (Eds.), *Handbook of neuropsychology* (Vol. 8, pp. 133–173). Amsterdam: Elsevier.

Parkin, A. J., & Walter, B. (1992). Recollective experience, normal aging, and frontal dysfunction. *Psychology and Aging, 7*, 290–298.

Posner, M. I., & Snyder, C. R. R. (1975). Attention and cognitive control. In R. L. Solso (Ed.). *Information processing in cognition: The Loyola symposium* (pp. 55–85). Hillsdale, NJ: Lawrence Erlbaum Associates.

Rajaram, S. (1993). Remembering and knowing: Two means of access to the personal past. *Memory and Cognition, 21*, 89–102.

Ratcliff, R. M., Sheu, C. F., & Gronlund, S. D. (1992). Testing global memory models using ROC curves. *Psychological Review, 3*, 518–535.

Reingold, E. M., & Merikle, P. M. (1990). On the inter-relatedness of theory and measurement in the study of unconscious processes. *Mind and Language, 5*, 9–28.

Reingold, E. M., & Toth, J. P. (in press). Process dissociations versus task dissociations: A controversy in progress. In G. Underwood (Ed.), *Implicit cognition*. Oxford, England: Oxford University Press.

Roediger, H. L., & Blaxton, T. A. (1987). Effects of varying modality, surface features, and retention interval on priming in word-fragment completion. *Memory and Cognition, 5*, 379–388.

Roediger, H. L., & McDermott, K. B. (1993). Implicit memory in normal human subjects. In

H. Spinnler & F. Boller (Eds.), *Handbook of neuropsychology* (Vol. 8, pp. 63–131). Amsterdam: Elsevier.

Roediger, H. L., Weldon, M. S., Stadler, M. L., & Riegler, G. L. (1992). Direct comparison of two implicit memory tests: Word fragment and word stem completion. *Journal of Experimental Psychology: Learning, Memory, and Cognition, 18*, 1251–1269.

Schneider, W., & Shiffrin, R. M. (1977). Controlled and automatic human information processing: I. Detection, search and attention. *Psychological Review, 84*, 1–66.

Slamecka, N. J., & Graf, P. (1978). The generation effect: Delineation of a phenomenon. *Journal of Experimental Psychology: Human Learning and Memory, 4*, 592–604.

Stuss, D. T., & Benson, D. F. (1984). Neuropsychological studies of the frontal lobes. *Psychological Bulletin, 95*, 3–28.

Swets, J. A., Tanner, W. P., & Birdsall, T. G. (1961). Decision processes in perception. *Psychological Review, 68*, 301–340.

Toth, J. P. (1992). *Familiarity is affected by prior conceptual processing: Differential effects of elaborative study and response–signal delay on separable processes in recognition memory*. Unpublished data and manuscript.

Toth, J. P., & Reingold, E. M. (in press). Beyond perception: Conceptual contributions to unconscious influences of memory. In G. Underwood (Ed.), *Implicit cognition*. Oxford, England: Oxford University Press.

Toth, J. P., Reingold, E. M., & Jacoby, L. L. (1994). Towards a redefinition of implicit memory: Process dissociations following elaborative processing and self-generation. *Journal of Experimental Psychology: Learning, Memory, and Cognition, 20*, 290–303.

Toth, J. P., Reingold, E. M., & Jacoby, L. L. (1995). A response to Graf and Komatsu's (1994) critique of the process–dissociation procedure: When is caution necessary? *European Journal of Cognitive Psychology, 1*, 113–130.

Tulving, E. (1983). *Elements of episodic memory*. New York: Oxford University Press.

Tulving, E., Schacter, D. L., & Stark, H. A. (1982). Priming effects in word-fragment completion are independent of recognition memory. *Journal of Experimental Psychology: Learning, Memory, and Cognition, 8*, 336–342.

Warrington, E. K., & Weiskrantz, L. (1974). The effect of prior learning on subsequent retention in amnesic patients. *Neuropsychologia, 12*, 419–428.

Weiskrantz, L. (1986). *Blindsight: A case study and implications*. Oxford, England: Oxford University Press.

Weldon, M. S., & Jackson-Barrett, J. L. (1993). Why do pictures produce priming on the word-fragment completion test? A study of encoding and retrieval factors. *Memory and Cognition, 21*, 519–528.

Winnick, W. A., & Daniel, S. A. (1970). Two kinds of response priming in tachistoscopic recognition. *Journal of Experimental Psychology, 84*, 74–81.

Yonelinas, A. P. (1994). Receiver-operating characteristics in recognition memory: Evidence for a dual-process model. *Journal of Experimental Psychology: Learning, Memory, and Cognition, 20*, 1341–1354.

Yonelinas, A. P., & Jacoby, L. L. (1995a). Dissociating automatic and controlled processes in a memory-search task: Beyond implicit memory. *Psychological Research, 57*, 156–165.

Yonelinas, A. P., & Jacoby, L. L. (1995b). The relation between remembering and knowing as bases for recognition: Effects of size congruency. *Journal of Memory and Language, 34*, 622–643.

CHAPTER 3

Attention, Automatism, and Consciousness

Richard M. Shiffrin
Indiana University

A good part of my professional life has been spent exploring data and theory bearing on two related dichotomies: short-term memory/long-term memory, and attention/automatism. These dichotomies provide the most obvious choices for the mapping of consciousness to concepts in cognition. But, up to now, I have largely refrained from discussing these possibilities. This chapter focuses on the possible relation between attention and consciousness.

The relations between attention, automatism, and consciousness were a fundamental concern of the field of philosophy/psychology around the turn of the century (e.g., James, 1890, 1904; Pillsbury, 1908), and have remained an abiding topic of interest in these fields to the present day, probably because attention heightens consciousness of events and thoughts. Indeed, there is little question that when a high level of attention is directed at some event, the event will be high in consciousness, and when something is at a high level of consciousness, it will tend to hold attention. Numerous books and articles on this theme appear on a regular basis (not always expressing novel views). However, the test of the relation lies not in the cases that are clear and uncontroversial, but in the possibility that it will clarify the ambiguous cases that abound in our cognitive systems.

This chapter is concerned with the degree to which we can implicate attentive processing in consciousness, and conversely, identify automatic processing with the failure to reach consciousness. One of the difficulties of drawing such analogies lies in the fact that both conceptual frameworks are imprecisely defined, though for somewhat different reasons. In the case of consciousness, we have a large collection of subjectively defined experiences and perceptions constituting a rather fuzzy, and not necessarily internally consistent, representation. In the case of attention, we have an enormous empirical and theoretical literature, but no one theory comes very close to encompassing the range of findings, and there is no accepted method for distinguishing attentive and automatic processing.

To elaborate on this last point, there is no simple and generally agreed on set of necessary or sufficient conditions, and no generally accepted empirical test for distinguishing attentive and automatic processes. Such criteria are sometimes available in certain restricted paradigms, like search tasks. The difficulty of generating universally applicable criteria may well be rooted in the fact that the concepts of attention and automatism are omnipresent in cognition, producing an enormous range of domains of applicability. Nonetheless, in order to make headway, a collection of empirical and theoretical criteria for distinguishing attentive and automatic processing are outlined. These criteria were presented in Shiffrin (1988; see also Shiffrin & Schneider, 1977). The criteria are rooted in a theory distinguishing the two forms of processing. Though controversial in a number of features, this conceptual framework is shared in large part by researchers going back to the 1800s, and accepted (with some caveats concerning details) by a majority of present theorists. With these criteria as a reference point, this chapter discusses relations with the more subjectively defined concept of consciousness. Keep in mind that consciousness is often viewed through the window of attention. Something unconscious before attentional focus may become conscious during or after. This fact is in good part responsible both for the appeal of the attention/consciousness mapping, as well as for the conceptual fuzziness of the consciousness construct.

A framework for human information processing in terms of automatic and attentive components is described elsewhere (Shiffrin, 1988), and thus it is not repeated here. The crucial point is that tasks are never wholly automatic or attentive, and are always accomplished by mixtures of automatic and attentive processes. With this in mind, consider the following criteria for distinguishing attentive from automatic processing, and the relation of each to consciousness. The criteria are listed in Table 3.1.

CAPACITY AND RESOURCE USE

It is generally agreed that attentive processing requires a share of limited processing resources, and a process that does not require resources is automatic.

In practice, resource demands are often measured by interference tasks, because two attentive processes must interfere with each other in any situation where system capacity is stressed and the subject tries to employ both.

TABLE 3.1
Criteria for Distinguishing Attentive and Automatic Processing

Capacity and resource use
Preparation
Rate of learning and unlearning
Depth of processing
Modifiability
Effort
Precedence and speed
Awareness and consciousness
Control and intentionality
Level of performance and parallel processing
Memory effects

In addition, the subject can eliminate the interference at will, by choosing to stop one or the other process. On the other hand, a process is surely automatic if interference is produced despite attempts to drop or ignore the process.

It is not clear how this distinction bears on consciousness. Consciousness is surely limited in content and extent, but the boundaries are not clearly obtainable. Nonetheless, we might ask whether stimuli that produce interference can be defined as conscious, and stimuli that do not as unconscious.

As discussed in other chapters (e.g., Greenwald; Merikle & Joordens), there are numerous demonstrations that stimuli failing to reach awareness, in the sense that they cannot be identified and perhaps cannot even be reported to have been present, nevertheless produce quite measurable effects on performance, both positive and negative. In visual studies, such stimuli are typically produced by using a brief presentation followed by a mask. In such cases, the apparently below-threshold stimulus can produce priming effects on the response to an immediately following stimulus, both positive and negative, depending on the relation between the prime and target. We would not want to define such primes as conscious.

There is another sense in which this correspondence may fail: Stimuli at an unattended location (or on an unattended perceptual dimension) can call attention automatically and indirectly produce interference as a result (e.g., a sudden loud noise). Such calls for attention can be trained, as shown by Shiffrin and Schneider (1977) and MacCleod and Dunbar (1988), and are often found in vision when one part of a display is quite perceptually different from all the other parts of the display, as summarized by Yantis (1993).

Certainly, automatic processes that call attention produce effects on consciousness. For example, in the study reported by Shiffrin and Schneider (1977), subjects were instructed to attend to two characters along a specified diagonal of a four-character display, ignoring the irrelevant diagonal. When characters previously trained to attract attention appeared on the ignored diagonal, subjects had an increased tendency to miss targets on the relevant diagonal. This performance criterion suggests subjects were unaware of the missed targets. However, the implications for consciousness are not this clear: It is possible that subjects would claim they were continuously conscious of the stream of characters on the relevant diagonal, but simply had too little time to carry out the decision operations necessary to produce a correct response. Although we did not systematically ask subjects for verbal reports, such reasoning is bolstered by the fact that in some of our studies using successively presented displays of four characters, many targets were missed at speeds of 800 ms per display, which is sufficient time to read aloud the four characters in each display. This occurred in conditions where the subject had to search each display for multiple targets. Most people would be reluctant to claim lack of conscious perception of the displayed characters in such a situation. This example also illustrates the oft-made point that consciousness of one quality of a stimulus (in this example, the name of a character) may take place when consciousness of other qualities does not occur (in this example, the classification of the character as a target).

Conversely, one can ask whether some process not requiring capacity (measured by a lack of interference) is necessarily unconscious. Here the answer is surely no. For example, Schneider and Fisk (1982) showed that the requirement to detect a well-trained target on one diagonal of a four-character display did not interfere with an attention-demanding simultaneous task: detection of not-so-well-trained targets on the other diagonal of the display. This situation differs from the previous one because this result occurred at times when the well-trained target did not actually appear. The subjects were conscious of the characters on the automatic diagonal at some level, and were effectively classifying them as nontargets, so the lack of interference seems a poor or arguable criterion for consciousness. More generally, and loosely, would anyone want to claim lack of consciousness of the ordinary evident contents of our sensory environment (say the visual field) because some task was not hindered (e.g., if we perform an auditory task as well with eyes open as closed)?

To look at this another way, it is clear that many stimuli of which we are aware do not produce interference in a given task setting (e.g., the shape of the CRT screen used to display the materials for an experiment). Thus, the claim would have to be that in order to be classified as conscious, such stimuli would have to be able to produce interference in some as-yet-unknown task to be discovered later. Aside from the issue of appropriate experimental controls and interpretation if interference in a task variation were to be found, it is obvious that such a criterion for consciousness could never be of any practical help, because one could never be sure that the set of possible task variations with which to demonstrate interference had been exhausted. Thus, any very tight link between an interference criterion and the presence or absence of consciousness would not seem possible.

Nonetheless, there are incontrovertible links between attention and consciousness: Not only do both exhibit capacity limitations, but these limitations are positively correlated. If nothing else, the positive correlations are produced by the fact that attention can be used to move information into and out of the forefront of consciousness. So we may be unaware of a minor leg pain, or our breathing, when attending to an interesting article (perhaps not this one), become quite aware of these when asked to attend to them, and become unaware again when our attention returns to another domain. We can probably not say much more than this: Despite a great deal of research, the measurement of attentional capacity is still in its infancy, and questions concerning as whether there are one or several capacity limits are still being investigated. To draw precise comparisons between attentional capacity and the even more imprecise notion of limited consciousness is not presently possible.

PREPARATION

Suppose a process is triggered by some stimulus event. If the presence of that event triggers the process, but no attentive process or preparation is required to do so, the process might be considered automatic (e.g., a reaction to a sudden noise). However, such a process or its immediate consequences might well be entered into consciousness, so a process judged to be automatic by this criterion could not be classified unconscious, at least without splitting millisecond-length hairs.

RATE OF LEARNING AND UNLEARNING

Automatic processes are generally learned gradually, or, if already learned (or innate) are difficult (or impossible) to unlearn or modify. Attentive processes can be turned on or off at the whim of the subject. Although cases exist where learning of an automatic process can occur very quickly (e.g., a learned aversion to a smell/taste that produces an immediate and strong, noxious, systemic reaction), subsequent unlearning will generally be slow. Earlier (Shiffrin, 1988) I considered the rate of learning criterion without giving sufficient consideration to rate of unlearning. At the present time, the slow rate of unlearning provides the single best criterion for an automatic process.

There are certainly interesting relations between rate of learning and consciousness. It seems that extensive consistent practice often produces two correlated effects: a learning of an automatic behavior, and a tendency for the behavior to drop out of consciousness. For example, when learning to drive, pressing the accelerator may require attention and consciousness, but after learning this activity takes place automatically and unconsciously. The relation between attention and consciousness early in learning is at least partly due to the fact that high degrees of attention do tend to produce consciousness (as described earlier). Training sometimes produces an automatic call for attention, and an associated conscious perception, but sometimes does not produce an automatic call for attention. In this latter case, there does seem to be a tendency for the trained stimuli to return to the state of consciousness that held prior to training. This state will be less conscious than that during early stages of training, and will sometimes be unconscious.

Nonetheless, rates of learning and unlearning are almost useless as a basis for defining conscious activity. First, as a practical concern, such a criteria defines the status of a process in terms of its past history of training, or the possible future course of training, rather than its present behavior. It would be difficult, to say the least, to define something as unconscious because its use in some task is unlearned slowly. More important, such criterion cannot be applied universally: Some automatic processes are learned quickly (e.g., the taste aversion reaction used in the previous example), and may or may not be conscious. Even slowly learned automatic processes sometimes involve calls to attention and hence may produce consciousness. Conversely, slow unlearning is not helpful, because the process unlearned slowly may or may not have been conscious (in part depending on whether a call to attention had been involved).

DEPTH OF PROCESSING

One view of processing of sensory inputs limits automatic processing to informationally and temporally early stages of processing. Such automatic processing is sometimes termed *pre-attentive* because it is followed by attentive selection. Often the automatic processing in such theories is limited to primitive and simple sensory features. This is not the prevailing view; the evidence points to automatic and attentive processing co-existing in parallel, with sometimes deep processing of information without attention. Nonetheless, does either view provide a useful approach to the problems of consciousness?

The pre-attentive approach would presumably lead one to ascribe unconsciousness to early stages of sensory processing, and consciousness to late stages of processing. However, a great deal of research has shown that a stimulus can be presented, processed to late or deep stages, and yet be out of the subject's awareness. A typical study involves the presentation of a word masked so quickly that subjects are near chance reporting whether anything had been presented, and hence is not in consciousness. Yet the meaning of the masked word can affect responding to a related word that is subsequently presented, showing that the word had been processed to a deep level (as in the work by Marcel, 1983a, 1983b; Fowler, Wolford, Slade, & Tassinary, 1981; see also Merikle, 1992). In other cases, the processing can be shown to be deep because abstracted qualities of a stimulus are involved, as when reports of "liking" a presented stimulus are modified by the masked stimulus (e.g., Niedenthal, 1990).

Such findings show that deep processing does not always produce conscious awareness. Can it be argued that shallow enough processing never leads to awareness? This question is either unanswerable, or the answer is clearly no. We are often conscious of very early stages of stimulus processing, such as color or shape, even if the awareness itself arises at subsequent levels of processing. We also know that very primitive sorts of processing can attract attention (see Yantis, 1993, for one summary). Making this whole issue even more difficult to assess is the role played by attention in studies of subliminal perception: Although studies show that stimuli can affect subsequent behavior even when their existence is not reported, in most studies the subject is trying to attend to the region in sensory space containing the subliminal stimulus. It would be useful to have well-controlled studies examining the subliminal effects of stimuli appearing in sensory regions that are not attended.

A correlated question concerns stimuli that do appear to enter consciousness: At what level are they conscious? An extreme example is, say, a fragmented picture of a Dalmatian: Before the "dog" appears to perception, one is conscious of patches of black on a white background; afterward, one has a fairly clear visual percept of a particular dog, as well as the awareness of the lower level patches. (This could be viewed as another case of the early/late dichotomy, the "dog" entering consciousness late, whether due to automatic processing, attentive processing, or both.) Thus, the depth or level of processing does not provide a good basis for predicting what is and is not conscious.

The distinction between consciousness and automaticity, and how these vary with depth of processing, is perhaps best brought out by research carried out to study perceptual unitization (Lightfoot & Shiffrin, 1992). This study trained initially novel characters, characters whose encoding required attention, to the point where they were encoded automatically as single, integral, units. But, how do these training changes affect conscious perception? To make the point clearly, the tasks and results are described in some detail.

Examples of the stimuli are shown in Fig. 3.1. Subjects were trained to search for such stimuli in a visual field of many similar stimuli. We arranged matters so a conjunction of at least two features was required to carry out each comparison (each pair of stimuli shared exactly one feature). The rate of search is initially slow, as shown in Fig. 3.2.

Each character comparison takes about 250 ms, as if three separate comparisons (one for each feature) were occurring at rates of about 80 ms per

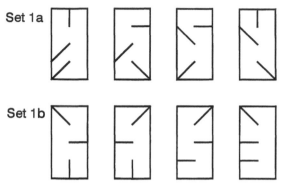

FIG. 3.1. Examples of novel stimuli used in the search task of Lightfoot and Shiffrin (1992). One row of stimuli was used in blocks of trials under varied mapping training: One stimulus was chosen at random in each trial to be the target, and the others were distractors. The other row was used in blocks of trials with consistent mapping training: One stimulus was always the target, and the others always distractors.

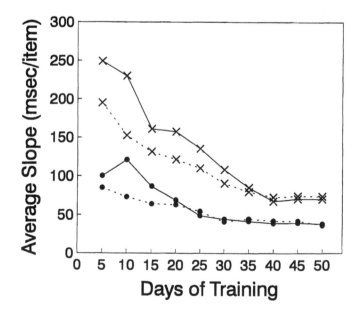

FIG. 3.2. Slopes (given as ms/ display item) of the function relating search time to display size for the stimuli and conditions illustrated in Fig. 3.1, as a function of days of training.

feature. Over 35 sessions of training (800 trials per session), search rates dropped to 60 ms to 70 ms per comparison, as if a single comparison per character was taking place. After training, response time showed the typical linear functions with 2:1 slope ratios (and consistent and varied training made no difference). Thus, at asymptote, the usual limited-capacity, terminating, attentive, search was being utilized. The improvement shown in Fig. 3.2 was not just due to generalized practice. When Ss were switched after training to a new set of conjunction stimuli, sharing no features with the first set, their performance reverted back to that seen at the outset of training. What had been learned was specific to these stimuli.

We carried out a number of investigations to establish that the operative features for these stimuli, before training, were the three internal line segments. For example, we transferred to stimuli that did or did not contain the external frames, and were able to show that conjunctions of the frames and the internal segments were not being used. Quite a number of transfer tasks were then carried out to demonstrate that after training a comparison involved a single holistic step. For example, the task was switched to one still using trained stimuli, but with the target sharing no feature with any distractor. For three sessions or so, performance did not change at all, as if single holistic comparisons were being used, suggesting the training was not specific to particular pairings, and that performance depended only on the use of integral units as stimuli. The fact that performance did not improve suggests that single feature search, which produces better performance in this situation, is not being utilized at the outset of the transfer trials. If subjects had been using feature comparisons prior to transfer, rather than holistic unit comparisons, then there should have been an immediate improvement. The idea, then, is that training produced automatic classification of such stimuli as particular known units.

Suppose we now ask, what is in consciousness when a display of known or unknown characters is presented to the visual field? If trained characters are presented, then the coded, unitary, representations are likely in consciousness, starting from a point relatively early in processing, even when attention is not aimed at a particular stimulus, analogous to what we imagine to be the case when alphabetic characters are presented. If untrained characters are presented, as is the case early in training, then the case may be more ambiguous. Consider first the encoding given to the target presented prior to the trial; it has not yet been trained and thus is almost certainly encoded (slowly) with attentional processing. It seems likely that such attentional processing produces a perception of the target as a unit This view is suggested by the work, say, of Ankrum and Palmer (1991), who showed that unknown figures acted as unitary objects in short-term memory for the purpose of part–whole matching.

Consider next the encoding of the (untrained) display characters early in training. They appear to the subject's consciousness to be unitary characters, or wholes. However, the evidence collected suggests they are compared feature by feature—that what is conscious are separate features placed together in a particular configuration rather than a single unit. This distinction is not obvious to the subject, and only shows up in the observed behavior. In one sense this does no more than provide another demonstration of the failure of the attentive–automatic distinction to map onto the concept of consciousness. Looked at another way, however, the result suggests that researchers in consciousness must carefully specify the "level of awareness" of the stimuli that appear to be conscious.

In summary, depth of processing does not provide a promising vehicle for distinguishing consciousness from unconsciousness (just as depth of processing should not be used as a criterial attribute for distinguishing automatic and attentive processes; e.g., Shiffrin, 1988). Because unaware stimuli can produce fairly deep processing as measured by subsequent responses to other stimuli, deep processing can be unconscious. Conversely, shallow processing can at least sometimes be conscious, although it is generally not possible to be sure whether one is aware of the processing itself, or of the attention to the results of the processing.

MODIFIABILITY OF AUTOMATIC PROCESSING

As discussed earlier, automatic processes are either difficult to learn, or to unlearn, or both. Perhaps all automatic processes can only be unlearned quite slowly. However, the degree of response that is engendered in such cases, at a given level of learning, is often modifiable through the application of attention. Interpretation of such findings is made difficult because we generally do not know what subprocesses are being affected by attention. Thus, modifiability of processing is not of much use in distinguishing automatic from attentive processing. For similar reasons, this criterion could not be of help if applied to the question of consciousness.

As a theoretical point, however, one might ask: What would be the relation of modifiability of component processes to consciousness, assuming that techniques were available to assess at a fine enough grain of analysis the automaticity of component processes? Presumably, we would learn that automatic component processes were sometimes slow to develop and always difficult or slow to modify. Would those processes be unconscious and, conversely, the remaining components conscious? I rather doubt this would be a useful view because consciousness is likely a quality characteristic of macroevents rather than microprocesses. A good example is found in memory search (e.g., Sternberg, 1966), one of the tasks for which information about the status of component processes is available. There is good reason to believe that in memory search tasks where targets and distractors exchange roles from trial to trial, the comparison stage is an attention demanding process (e.g., Schneider &

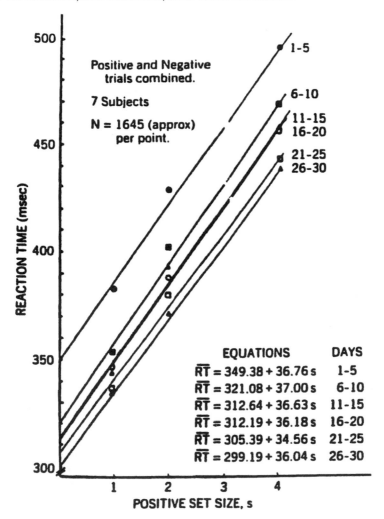

FIG. 3.3.   Mean response time as a function of positive set size, as a function of days of training, using varied mapping memory search. From Kristofferson (1972). ©1972 by Canadian Psychological Association. Reprinted with permission.

Shiffrin, 1977). This conclusion is based on the high slopes of the search function. In fact, the inconsistent mapping prevents automatization of this stage, judging from the fact that the slopes do not decrease with practice. However, other components of the task (perhaps such as the manual response) do improve with practice, because the intercept of the search function drops. These components are undoubtedly becoming increasingly automatic. These slope and intercept results are nicely illustrated by Kristofferson (1972), as indicated in Fig 3.3. On the other hand, consistent mapping leads to a lowering of slopes, probably due to what Schneider and Shiffrin (1977) called the development of automatic categorization. Although the automatization of memory search is certainly accompanied by a considerable reduction in perceived effort and task difficulty, it does not seem that the task itself is any less conscious to the subject. The general problem is that these changes and lack of changes in automaticity of the component processes are not reflected in changes in task consciousness. Memory search happens quite rapidly and consciousness of the component processes does not seem to be available to the subject.

Of course, the import of these results goes beyond the specific issue of modifiability. The fact that components of a task can vary in degree of automaticity without affecting consciousness provides a compelling reason to doubt the analogy between attention/automatism and conscious/unconscious.

EFFORT

Attentive processing is sometimes said to be *effortful,* a concept relating in part to the notions of capacity limitations (see Kahneman, 1973). In the attempts to use this concept as a criterial measure for distinguishing automatic and attentive processing, one runs into some major problems: first, the issue of finding an appropriate method to measure effort. One approach is to define effort by subjective report. This approach does not map onto the consciousness dimension particularly well, as illustrated by the memory search findings discussed in the previous section. More simply, we often seem aware of things that do not require effort, such as the contents of the visual or auditory environment. Another approach defines effort in terms of capacity utilization: Effortful activity should reduce capacity to carry out simultaneous tasks. In other words, effort is defined functionally through observations of interference. A major problem with this approach has been discussed earlier: Automatic processes can call attention, thereby producing interference, and producing what may seem to be effortful results.

Additional problems arise when trying to use effort to classify states of consciousness. Is effort required to produce consciousness? I suspect almost everyone would answer no, in part because of a belief that automatic processing can be used to enter information into consciousness. Other questions relating to this issue could be raised (e.g., Does maintaining information in consciousness require effort?) It is doubtful that the edges and measures of consciousness are clearly enough defined to answer such questions.

PRECEDENCE AND SPEED

More often than I would like to admit, researchers have proposed or assumed implicitly that automatic processes are faster that attentive processes, and precede attentive processes in time and processing order. This is a misconception caused largely by the fact that attentive processes tend to be spread out in time (due to limited capacity), so an automatic process operating in parallel has a speed advantage. However, for any given automatic process operating in a task in which the processing load is low, attentive processing can operate as efficiently, or more efficiently. Schneider and Shiffrin (1977) gave many demonstrations of this point in the context of visual and memory search tasks. For example, with a load of one item, attentive search (produced by varied mapping) results in performance measured in response time and accuracy as good as, or better than, the performance seen in automatic search (produced by consistent mapping). One would expect such demonstrations to be even more evident in cases where the automatic process is not so well learned, but the subject induced to utilize it anyway, perhaps by the presence of a dual task (e.g., see the early stages of training in the study of Schneider & Fisk, 1982).

Regardless of the value of the processing speed criterion for distinguishing automatic from attentive processing, one can ask whether processing speed is related to consciousness (e.g., Are fast thoughts unconscious, and slow ones conscious?). If it becomes possible to define these terms well enough to permit investigation, then this will prove to be a question worth pondering.

AWARENESS AND CONSCIOUSNESS

We need not dwell on this criterion because it is the subject of this chapter. Shiffrin (1988) did not try to distinguish these terms when considering possible criteria for distinguishing automatic from attentive processing. Although other authors in this volume (e.g., Dulany; Johnson & Reeder) discuss the difference between awareness and consciousness, I have not found such a distinction helps to establish an analogy between attention/automatism and either awareness/unawareness or conscious/unconscious.

CONTROL AND INTENTIONALITY

When a process operates on the basis of preset parameter values and/or parameter values set by the environmental input, it may be characterized as automatic; if the parameter values are set by choice of the subject, then the process may be characterized as attentive. This has not proved a useful criterion in practice, partly because the environmental stimuli generally include the results of prior attentive processes, partly because automatic processes can call attention, and partly because some attentive processes operate so quickly that moment-to-moment control is unlikely.

The relation to consciousness is even less clear. Control of behavior is sometimes conscious and sometimes unconscious (e.g., walking vs. mountain climbing). Also, the process of controlling behavior, and controlling attention, is sometimes conscious and sometimes not (e.g., moving visual attention systematically across a field in a visual search vs. attention moving to a location containing a sudden evident movement).

LEVEL OF PERFORMANCE AND PARALLEL PROCESSING

Level of performance was discussed earlier (in the section on Precedence and Speed) and therefore it need not be repeated here. This section points out that consideration of parallel versus serial processing does not help us make an appropriate distinction in either the attentive or consciousness domain.

There can be no doubt that much automatic processing is limited, even if parallel in character: To take an obvious example, two different stimuli may each be trained to call for attention automatically, but a limitation will appear if both stimuli appear simultaneously, one to the right and one to the left of fixation. Sometimes parallel processing produces an advantage for automatic processing, but not always (as discussed in an earlier section). Attentive processing is limited in capacity, but not necessarily serial in character. Thus the distinction between parallel and serial processing does not map well onto the distinction between automatic and attentive processing.

It is not evident how one asks whether consciousness is serial or parallel in character. Thoughts high in consciousness often seem serial, probably because they are associated with language, but at other times consciousness seems parallel, as when we attend to the visual scene before us. So the distinction between parallel and serial processing does not seem to map well onto the distinction between the conscious and unconscious.

MEMORY EFFECTS

One can discourse at length on possible links between consciousness and memory, on the identification of short-term memory with conscious thought, on the possible consciousness of information in very short-term sensory memories, on whether one should assess the presence of memories using implicit or explicit tests, and on ways to distinguish memory from conscious perception. Such questions would be the subject of some other chapter (e.g., see Jacoby, Yonelinas, & Jennings, in press). For the moment, consider only the possibility that attentive processes can be identified because they invariably leave a residue that can be found in explicit, episodic, memory tests, whereas automatic processes may not necessarily leave a memory trace that can be found in explicit memory tests, and/or may always leave a trace that can be found in implicit memory tests.

One key question, then, is whether or not one can be certain that attentive processes leave explicit memory traces, in at least some sort of memory (we know that retrieval failure, amnesias, and forgetting can occur, so that a memory at one point in time may be missing at another). This is not easy to answer, especially for stimuli near or below threshold, or stimuli embedded in massive amounts of other information. So this criterion is not used in practice to distinguish automatic from attentive processes.

The relation to consciousness is not very straightforward. The subliminal perception literature is concerned with demonstrations that unaware stimuli affect memory only on implicit tests. We might therefore ask: Is attention necessary to produce explicit memory? This is not established, because attention is typically given in at least small amounts to all stimuli in the perceptual surround; thus, we really know little about the memorial fate of unattended stimuli, whether tested implicitly or explicitly.

A second key question is whether or not implicit memories can occur without attention, and if so, do they always do so? The subliminal attention literature does not tell us, because the subject is trying to attend to the sensory region containing the subliminal stimulus. A rather different approach is found in the process-dissociation technique of Jacoby and his colleagues (e.g., Jacoby et al., this volume). Although impressive indirect evidence has been collected in many studies that one component of storage is automatic and perhaps unconscious, and results in information storage accessible through implicit tests, all the studies occur in situations where attention is certainly being given at study to the stimuli in question. The same point makes it difficult to say much about consciousness and unconsciousness: At the time of storage, the stimuli are certainly given conscious processing. At the time of implicit test, the subject is sometimes clearly unaware that the test stimulus had been studied, but it is not clear whether this is always the case. It is also clear that one could not use a positive indication of memory using an implicit test to infer whether original storage had been carried out using attention.

In summary, there is no good case for relating the distinction between attentive and automatic processing to that between the conscious and unconscious. Despite the positive correlation between attention and consciousness, the mapping between the two conceptual frameworks is quite poor. This conclusion in no way detracts from the strong positive correlation between these frameworks, and it would be interesting in some other setting to explore or model the nature of the connection. Steps in this direction can be seen, for example, in Jacoby et al.'s chapter.

REFERENCES

Ankrum, C., & Palmer, J. (1991). Memory for objects and parts. *Perception and Psychophysics*, *50*(2), 141–156.

Fowler, C.A., Wolford, G., Slade, R., & Tassinary, L. (1981). Lexical access with and without awareness. *Journal of Experimental Psychology: General, 110*, 341–362.

James, W. (1890). *The principles of psychology*. New York: Holt.

James, W. (1904). Does "consciousness" exist? *Journal of Philosophy, Psychology and Scientific Methods, 1*(18), 477–491.

Kahneman, D. (1973). *Attention and effort*. Englewood Cliffs, NJ: Prentice-Hall.

Kristofferson, M. W. (1972). Effects of practice on character classification performance. *Canadian Journal of Psychology, 26*, 54–60.

Lightfoot, N., & Shiffrin, R. M. (1992). On the unitization of novel, complex visual stimuli. In *Proceedings of the 14th Annual Conference of the Cognitive Science Society* (pp. 277–282). Hillsdale, NJ: Lawrence Erlbaum Associates.

MacLeod, C. M., & Dunbar, K. (1988). Training and Stroop-like interference: Evidence for a continuum of automaticity. *Journal of Experimental Psychology: Learning, Memory, and Cognition, 14*, 126–135.

Marcel, A. (1983a). Conscious and unconscious perception: An approach to the relations between phenomenal experience and perceptual processes. *Cognitive Psychology, 15*, 238–300.

Marcel, A. (1983b). Conscious and unconscious perception: Experiments on visual masking and word recognition. *Cognitive Psychology, 15*, 197–237.

Merikle, P. M. (1992). Perception without awareness: Critical issues. *American Psychologist, 47*, 792–795.

Niedenthal, P.M. (1990). Implicit perception of affective information. *Journal of Experimental Social Psychology, 26*, 505–527.

Pillsbury, W. B. (1908). *Attention*. New York: Macmillan.

Schneider, W., & Fisk, A. D. (1982). Concurrent automatic and controlled visual search: Can processing occur without resource cost? *Journal of Experimental Psychology: Learning, Memory, and Cognition, 8*, 261–278.

Schneider, W., & Shiffrin, R. M. (1977). Controlled and automatic human information processing: I. Detection, search, and attention. *Psychological Review, 84*, 1–66.

Shiffrin, R. M. (1988). Attention. In R. A. Atkinson, R. J. Herrnstein, G. Lindzey, & R. D. Luce (Eds.), *Stevens' handbook of experimental psychology: Vol. 2. Learning and cognition* (pp. 739–811). New York: Wiley.

Shiffrin, R. M., & Schneider, W. (1977). Controlled and automatic human information processing: II. Perceptual learning, automatic attending, and a general theory. *Psychological Review, 84*, 1127–1190.

Sternberg, S. (1966). High-speed scanning in human memory. *Science, 153*, 652–654.

Yantis, S. (1993). Stimulus-driven attentional capture. *Current Directions in Psychological Science, 2*(5), 156–161.

CHAPTER

Consciousness as a
Message Aware Control Mechanism
to Modulate Cognitive Processing

Walter Schneider
Mark Pimm-Smith
University of Pittsburgh

This volume illustrates that there are many theoretical interpretations of consciousness. In this chapter we propose a framework for the interpretation of consciousness in which it serves a critical function of modulating cognitive processing in a hybrid architecture (Schneider & Detweiler, 1987). We also interpret the chapters by Shiffrin (chapter 3, this volume) and Jacoby, Yonelinas, and Jennings (chapter 2, this volume) in the context of this framework.

We look at the issue of consciousness from the perspective of what information-processing advantages it might provide a cognitive organism. We suggest that consciousness interacts heavily with attentional processing. We take the perspective that consciousness may be an evolutionary extension of the attentional system to modulate cortical information flow, provide awareness, and facilitate learning particularly across modalities. We interpret consciousness as a *message aware control system* that can monitor the content of messages transmitted between processing modules.

We begin by addressing what the information-processing roles of consciousness are and how they might be implemented in a modular cognitive architecture, and subsequently relate our framework to the Shiffrin and Jacoby commentaries (this volume).

Cognitive functioning in the real world involves a complex coordination of many information streams. Schneider and Detweiler (1987, 1988) interpreted cognitive processing as occurring in a modular architecture of connectionist modules. *Modules* are organized into groups within *levels,* and separate sets of levels constitute multiple *regions.* Figure 4.1 illustrates the levels and re-

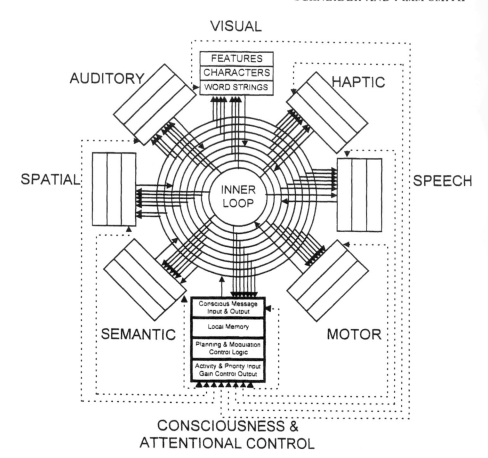

FIG. 4.1. Diagrammatic representation of information and control flow in
modular architecture. The *levels* are cortical stages of processing containing
many modules. They are represented as thin rectangles grouped into *regions*.
For example, the Visual region includes levels for processing features, characters, and word strings. The regions feed information vectors (solid arrows) into
an *inner loop* that connects to other regions on the inner loop. Control information is conveyed via connections sending activity and priority signals to the
attention controller and the attentional controller sending a scalar attention
modulation signal back to the modules (dotted arrows). *Consciousness* in the
model is the set of codes received by the attentional controller of information
on the inner loop. The controller has local memory and logic to allow planning
and modulation of the control hardware.

gions in the hybrid architecture. Each region (e.g., vision, speech, motor control) is assumed to have a small number of levels. For example, the visual region contains multiple levels (e.g., V1, V2, V4, TEO, IT for object processing) and in a word recognition process would encode visual features, simple character level objects, and word strings. There are a large number of modules, perhaps in the range of a hundred thousand.[1] There are many levels—perhaps a thousand—and multiple regions. There are potentially hundreds of thousands of inputs that must be routed to converge on individual points (e.g., a human can be instructed to attend to hundreds of visual or auditory locations and have a stimulus at a given location activate a single finger movement). To limit the amount of cross talk, modules are organized into levels and regions. Lower level processing can occur in parallel without interference (e.g., visual features into visual objects and sounds into phonemes without interference). However, the outputs from different regions must converge on an *inner loop* of processing that transfers information between regions.

When information converges there is a serious cross-talk problem requiring attention to limit interference and to optimize information transmission.[2] Much of the work in attentional theory (e.g., Broadbent, 1958; Laberge, Carter & Brown, 1992; Olshausen, Anderson, & Van Essen, 1993; Schneider & Shiffrin, 1977; Treisman, 1960) deals with the need to selectively filter information to moderate the cross-talk problem.

Given the large number of modules involved, the efficiency of cortical processing is strongly determined by the efficiency of the attentional control system. With thousands of potential inputs, the rate at which the most important inputs are attended influences response efficiency of the system.

The attentional control system may have varied levels of awareness of the information it is trying to control. This chapter equates degrees of consciousness with the "level of awareness" of the information that the attentional control system has available to it to influence the control operations in the net-

---

[1] The model utilizes physiological data to provide "order of magnitude" estimates of the number of modules involved. An estimate of the number of modules would relate to the number of cortical hyper-columns (Szentagothai, 1978) that are about a square millimeter in size. The human cortex has an estimated 250,000 mm² of surface area suggesting 250,000 modules. The number of levels may be estimated by the Felleman and Van Essen (1991) visual system data, estimating the average size of visual areas in the monkey, and relating that size to the equavalent sizes in the human and then dividing the human cortex number by the estimated size of a level (250,000 mm² divided by 170 mm² and assuming an average of 2 replications of an area for 735 levels). The number of regions would include the basic input and output modalities and and internal representational systems (e.g., object vision, location, audition, semantic memory, motor output, context), perhaps in the range of 10–30. For the following discussion, the exact numbers are not critical, but the relative order of magnitude provides a perspective of the number of modules that need to be controlled.

[2] When multiple information streams converge, information transfer increases the efficiency of serial attention by reducing the cross-talk problem which produces low signal to noise and slower message transfer (see Shedden & Schneider, 1990, 1991).

work. The more aware the system is, the quicker the critical data is attended. However, the more information to which the attentional control system must be aware, the more it becomes the bottleneck. One must be cautious of falling into the *homunculus trap*, that is, putting off all the processing to higher levels of processing. This does not simplify the problem, but merely pushes it up to a higher stage. So the attentional control system must be aware of the messages, but the information must be greatly simplified and preprocessed by the modules or the attentional control system itself becomes the bottleneck.

Attentional control can occur at various levels of awareness of the information contained in the modules. Providing more awareness requires more hardware to transfer the information and resolve competing requests. In order to illustrate the benefits of greater message awareness, it is necessary to illustrate how many channels must be scanned until a critical digit is detected. Assume that the visual system has 100 channels and 50 characters are presented, of which 40 are letters and 10 are numbers. The subject is looking for a specific digit (e.g., "7") and must determine if a channel is a "7," and the memory comparison requires .1 second.

A *no-awareness* controller would have to query each channel to determine if it contained a target. The attentional controller would need to select channels for output, blind to whether the channel contained any information. The just *activity aware* controller would receive from each channel an indication that the channel contained some stimulus, perhaps coding the intensity of the stimulus. The controller would then scan only the active channels where characters are present. To receive activity reports, the controller must have inputs indicating the activity of each channel and a selection mechanism that can sequence through the channels based on activity. The dotted lines in Fig. 4.1 illustrate the potential activity reports coming from each region to the processing controller and the attentional modulation from the controller returning to each region. Attention is assumed to gate the output of the modules (see Schneider & Detweiler, 1987, 1988). The activity report and attentional modulation are scalars conveying the degree of activity or the amount of attention but not the specific content of the message. Data showing humans can skip processing empty sensory channels (e.g., Schneider & Shiffrin, 1977, Exps. 1 & 2; Yantis & Johnson, 1990) supports the existence of activity awareness in the control system. The *priority aware* controller would receive from each channel a priority signal indicating how important the content of the channel is for further processing. The channel must have local memory that would associate the signal in the channel with a priority signal that would be transmitted to the controller. The connections of the priority signal are scalars and are also represented in Fig. 4.1 as the dotted lines. The scalar priority code can inform the controller which module contains the signal with the highest priority (e.g., your name in the cocktail party effect) even though it may have less activity than other modules. The data on consistent mapping training producing automatic attention responses (e.g., Schneider & Shiffrin, 1977) shows that the consistent target stimuli are attended before other stimuli, which supports the existence of a priority-based signal. The inability to ignore consistent targets (Shiffrin & Schneider, 1977, Exp. 1) illustrates that the priority tag is not message specific and cannot be easily changed. A *target aware* controller would require hardware to input message codes from all the channels and resolve which had the target character. This requires a massive increase in hardware and risks falling into the homunculus trap (requiring hardware to process all the data in all the channels). The slow serial processing of humans when searching for varied mapping targets suggests that humans do not have a parallel target aware control system (see Shiffrin & Schneider, 1977).

There is evidence for limited feature awareness of the controller. The feature detection studies (Treisman, 1986; Wolfe, 1994) suggest multiple activity maps (e.g., color, shape, size) that can limit the search to the channels with the desired primitive feature. The ability to change from trial to trial what feature is being searched, suggests some rapid establishment of the limited feature search ability. One approach would entail establishment of a *temporary feature map* (e.g., generate and monitor a map of "green" activity on one trial and "blue" on the next). A second approach would be to implement a *top-down projection* of the target feature from a high-level module to the channels so each channel can determine if the stimulus-evoked pattern matches the top-down pattern (e.g., project the target search of a "green" stimulus on one trial and a "blue" stimulus on the next). Codes representing the stimulus would be in each of the modules and a match would produce a high activity or priority code. The number and complexity of the features that could be monitored in this way would depend on the amount of local memory in the modules and the accuracy with which the back projection could activate codes in the modules. In the temporary feature map case, the controller is aware of the feature map that is being monitored. In the top-down projection case, the controller is aware of the code being projected down and the activity reports of the modules.

Table 4.1 illustrates the relative processing time advantages of the different levels of awareness. There is strong evidence for activity- and priority-based selection. The frequency of reports of serial processing in the literature (e.g., Briggs, 1974; Treisman, 1986) suggest that parallel message awareness does not occur in many search situations.

Table 4.1 shows that within a processing region there is high activity- and priority-based attentional control and only very limited message-based awareness. This type of awareness is very important for rapidly identifying the most important channel to be attended (based on activity and priority reports) and transferring that data up the processing hierarchy. In the previous example, such awareness could speed up processing relative to scanning each of the individual channels by multiple orders of magnitude.

TABLE 4.1
Example Levels of Awareness

Level of Awareness	Channels to Scan	Max. Channels Scanned	Max. RT	Required Hardware	Scanning Data
None	All channels	$N_{channels}$	10.0 sec	None	Not supported
Activity	Channels with characters	N_{active}	5.0 sec.	Scalar Activity	Supported
Priority	High priority digit channels	$N_{priority}$	1.0 sec.	Scalar Priority	Supported
Message	Channels with signal	N_{target}	.1 sec	Vector signal	Limited situations

Note: Assuming $N_{channels} = 100$, $N_{active} = 50$, $N_{priority} = 10$, $N_{target} = 1$, and scan time = 100 ms.

A controller with only activity and priority awareness is mostly message blind and cannot easily alter attentional processing to accomplish a goal that does not translate into simple feature search. To illustrate, assume the organism wanted to satisfy a "find food" goal. With only *activity awareness* the controller would be limited to accomplishing a module-based goal such as "reduce the activity of the eating module," which presumably would be more active during a period of hunger. If food stimuli produce more activity (e.g., food stimuli are brighter) or have higher priority (e.g., food stimuli pop out of the perceptual field due the consistently attending to them in the past), food stimuli would be attended to more rapidly than most other stimuli in the visual field. However, if food stimuli are in visual fields with other high activity or priority stimuli, the controller could not distinguish which signal to attend solely based on the priority and activity maps. If the controller monitors the activity of a food module, it could switch attention between the high priority stimuli to determine which stimulus most activated the food module. Having identified a food-related message, based on the increased activity in the food module, the controller could then see if that stimulus would activate the motor regions to produce motor movements related to eating. The differentiation of the goal-seeking behavior of the organism would be quite impoverished, limited to differentiation of the modules in the processing system. Assuming there is an order of magnitude of a thousand processing levels (see footnote 1), the activity and priority system could only accomplish broad goals but not limited goals (e.g., seek "food" but not seek "chocolate covered almonds").

An attentional controller that has only message and priority reports is "aware" of which modules have high activity and priority messages but is completely blind to the specific content of the messages. The knowledge of which modules have high priority signals allows the controller to gate the information at lower levels so that the important messages get processed at upper levels.

A message aware controller provides the attentional control system with the ability to modulate attention to accomplish sharply differentiated goals. If the controller can monitor message traffic and determine which messages relate to a specific goal, modules that contain messages that are germane to the goal can have their messages transmitted more widely to accomplish the goal. For example, if the controller is seeking "chocolate covered almonds," then it could sequentially scan the high activity/priority messages to determine if the message transmitted to a controller activates codes that relate to the goal. If the controller is "aware" that a module contains a "chocolate" related message, it could control the network to get more detailed input (e.g., foviate the stimulus), focus attention on that stimulus location, and evoke the lexical code to see if it matches the goal. If the code matches the goal, then the controller could activate motor programs to grab the foviated stimulus and then eat it.

Building a message aware controller requires greatly limiting the information coming to the controller to avoid the homunculus trap of needing to deal with all of the information. The information load is minimized by first having it only have access to inner-loop message traffic and second by providing symbolic codes that code only the basic characteristics of the full message. Only a small proportion of the modules transfer on the inner loop. By having the controller only monitor inner-loop messages, connections need only be made to inner-loop modules. Monitoring inner-loop traffic has another more important characteristic, that much of the traffic on the inner loop is already serialized (see Schneider & Detweiler, 1987, 1988; Schneider & Oliver, 1991) to minimize interference between regions. This makes the inner loop an ideal place to monitor traffic. The data is already filtered and serialized by the time it gets to the inner loop.

With consciousness occurring at the inner loop, it has direct access to only a small portion of the activation in the full network and examines data after lower levels have produced invariant codes. Attentional filtering limits what information reaches the inner loop. For example, you are currently attending to the words on this page, consciously aware of the data in one degree of the visual field, having little immediate conscious access to the remaining 99% of your visual field that is being activated at the early (V1) stages of the visual system. The inner-loop codes are likely to represent higher level codes providing invariant representations over transformations such as position, location, size, lighting, and orientation.

In this model, consciousness is both after attention and influences what is attended. It is after attention due to its being on the inner loop. What is attended is determined by the map of priority signals coming from the modules. The Wolfe (1994) Guided Search theory provides a model of how multiple priority maps influence what is attended, serially visiting positions in order of the priority level of each location. An attentional controller that has activity and/or priority awareness and message awareness of inner-loop transmissions can enhance the processing of messages related to a target goal state. When a message is transmitted on the inner loop related to the goal state, the controller could determine what modules in the system were activated by the message (i.e., increased their activity/priority shortly after the goal-related message). The controller could then attend to those modules sequentially transmitting each of their messages, determining if the produced inner-loop transmissions related to the goal. If so, the controller would transmit messages from modules activated by the second message. If not, the controller would go back to the next priority message based on the activation effects following the first message. The message awareness of the controller would influence what activations would receive further processing.

To monitor the inner-loop messages, transmissions on the inner loop must activate simplified codes in control regions to allow recognition and categorization of the major messages. The code activated in the controller may be an abstracted symbolic representation of the message transferring on the inner loop. Note the purpose of the code evoked in the controller is not to fully represent the message but to allow categorization of a message sufficient to decide if the message is related to the goals of the controller and thus to determine which messages should be extensively processed. To illustrate, there are many variants of the taste, color, shape, smell, and feel of an apple (see Schooler & Fiore, chapter 12, this volume). Variations in the specific codes in modules throughout the network provide a vast richness of coding. However, if all of this information were to reach the controller, then it would have to deal with hundreds of codes just for apples, increasing the information loading and risking the homunculus trap. We speculate the code evoked in the control region is a schematized code capturing major characteristics of the underlying message. Language provides an analogy to the type of abstract code. Language codes the potentially millions of distinguishable stimuli and events into a vocabulary of typically 10,000 words. The word codes have the disadvantage of providing a crude simplified representation of the represented stimulus but has the advantage that the code is easily stored and transmitted to others.

Dreaming activities may provide illustrations of sequences of controller messages. Both daydreams and sleep dreams can represent specific objects and goals. One can daydream about the taste, shape, and look of "chocolate covered amonds" sufficiently well to nearly uniquely identify the objects. Note the conscious daydream image is a very impoverished representation relative to the actual stimulus. However, that image would be sufficient to allow scanning of the environment to determine if a visual channel contained an object that matched the controller message of the object.

From this perspective, consciousness can now be related to attentional control. In this framework, *consciousness is a module on the inner loop that receives messages that evoke symbolic codes that are utilized to control attention to specific goals.* Consciousness is the message awareness of the controller. Consciousness does not have the richness of the underlying module architecture, but rather works on abstracted codes. This message awareness allows the control decisions to be based on the specific messages transmitted in the network rather than labeled line activity or priority codes from the various modules in the network.

The attentional controller includes *planning and modulation logic*. This logic allows the execution subroutines that involve monitoring messages and module priority/activity reports and modulating the output gain of messages from modules throughout the system (see bottom Fig. 4.1). In the hybrid connectionist control model of learning, Schneider and Oliver (1991) demonstrated how a message aware controller could follow goals and greatly speed up learning. The controller itself was built out of a sequential connectionist net (Elman, 1990). It had the advantage of greatly speeding learning and allowing the network to do task decomposition, learn from instruction, and self-rehearse the solution to a problem. Consciousness is not the same as controlled processing. Controlled processing represents monitoring and modulating transmission throughout the network. For example, switching attention between spatial locations is a controlled operation but not necessarily conscious. The controller is aware of the channels it is processing but not necessarily of the message content of those channels. Of the hundreds of subjects we have run in visual search, never has one commented that they felt they were consciously aware of serially switching attention between memory and display locations even in cases where the switches produce clear evidence of serial processing in the mean and variance data and switching times require hundreds of milliseconds (e.g., Fisk & Schneider, 1984; Schneider & Shiffrin, 1977, Exp. 2). This lack of awareness is reasonable in the current framework because the switching would be occurring within a region and the comparison may be based on an activity report (see Shedden & Schneider, 1990, 1991) without any code being evoked consciously.

A message aware or conscious controller may have extensive within-module processing capacity to allow far more sophisticated control of network traffic. For example, if the consciousness module has local long-term memory, then it could recognize messages that have occurred on the inner loop.

The conscious memory is likely to be separate from the memory in other modules.[3] To the extent that this memory is located in different modules from the other regions, it would represent an independent memory. For example, if the a stimulus failed to evoke a code in the consciousness module or the consciousness module memory failed to "recognize" the code, the controller would not be "aware" that this message had occurred before. However, inde-

---

[3]If the module memory and consciousness share memory, access of either consciousness or the other modules will produce cross talk between messages and limit access of consciousness to the other memory while transmissions are taking place. In contrast, if consciousness has its own memory it can interpret the code evoked in consciousness even while other messages are being transmitted.

pendent connections between modules could still produce effects such as priming as stem completion in the absence of consciousness.

The evidence for independence of recognition and procedural memory (Jacoby et al., 1995) supports the view that consciousness has its own local memory. The inability to recognize previously learned words in a list even when that can be perceived at far lower thresholds (Lightfoot & Shiffrin, 1992) also suggest two memories with very different acquisition and decay rates.

In this framework, consciousness has a pivotal role in the information-processing network. It would be in a position at which it could monitor inner loop message traffic and hence "be aware" of all the critical messages transmitted from all regions. Because this monitoring is for modulation of network traffic, the module would have access to at least control of all the transmissions between regions and perhaps within regions.

The consciousness controller may have evolved a rich set of processing algorithms to speed learning and satisfy goals. Schneider and Oliver (1991) demonstrated how by having the network controller monitor and modulate network traffic it could reduce learning time by orders of magnitude. Giving the controller the ability to track and follow problem space tree hierarchies, as in Newell's (1990) SOAR system, enables the system to perform complex problem-solving operations.

The value of message awareness for satisfying specific goals suggests that conscious processing may be present in many species. The ability of animals to accomplish specific goals (e.g., a dog being told to get the "paper," leaving the room, going outside, and retrieving a specific object) would suggest a "message aware" controller and hence in the current framework a conscious control system.

Consciousness may provide a ready structure on which to evolve a language process. The consciousness monitoring system is already a serial symbolic representation of message traffic that can track complex goal hierarchies. If an input (e.g., word encoding) and output (e.g., pronunciation) were to evolve on top of the consciousness representation, it would allow transfer of conscious codes between individuals allowing rapid transfer of knowledge between individuals and particularly instructed learning (see Schneider & Oliver, 1991).

From this perspective, we will now interpret the work of Shiffrin contrasting automatic/controlled processing to consciousness/unconsciousness, and that of Jacoby in interpreting the independence of conscious and unconscious memory.

SHIFFRIN COMMENTARY

Shiffrin (chapter 3, this volume) assessed the extent to which the relatively un-investigated and ill-defined concepts of consciousness and unconsciousness can be identified with the relatively well-investigated and well-defined concepts of attentive processing and automatic processing (Shiffrin & Schneider, 1977), respectively. Intuitively, these processes provide a potential mapping of consciousness to concepts in cognition. Shiffrin's general strategy for assessing the identity of these conceptual categories is to evaluate the applicability of the criteria that has been traditionally used to distinguish attentive from automatic processing such as capacity, rate of learning, control and depth of processing (Shiffrin, 1988). His general conclusion is that none of the criteria confidently extend to consciousness and thus "the mapping between the two conceptual frameworks is quite poor." The following discussion finds general agreement between Shiffrin's main points and the connectionist-control (hybrid) model of consciousness.

Shiffrin argued that controlled processing is not equivalent to consciousness on the grounds that there may be controlled processes that do not engage consciousness. For instance, in visual search tasks, serial exhaustive search strategies are reliably employed in the absence of the subject's consciousness of using such strategies. These search tasks qualify as controlled processing in that they exhibit the variability with intention and flexibility of processing. For example in character search (Shiffrin & Schneider, 1977, Exp. 4a) subjects can control on a given trial whether they search one diagonal of a square of four characters or search all four items. Subjects can also alter what items are searched. In the hybrid model, low-level interactions between control structures and modules can occur in the absence of conscious involvement (Schneider & Detweiler, 1987; Schneider & Oliver, 1991). Conscious processing works hierarchically, and the higher level goal of detecting a target may implement lower level controlled but unconscious search strategies. In the case of exhaustive search, we would say that consciousness was absent during the search process but became present once completed when an automatic attention response occurred to the target.

We also agree with Shiffrin that there is a general difficulty in teasing apart automatic processing from consciousness in that automatic events often cause passive attention capture and awareness of the event (e.g., in the "pop out" phenomenon in visual search tasks). This is explained in the hybrid model in terms of high priority message vectors being transmitted without interference, and evoking codes in consciousness modules.

Shiffrin observed that unconscious processing can have a wide range of influence on task performance in the absence of attention and consciousness, as in the case of subliminal perception (Shiffrin, Cohen, & Fragassi, 1993). This is consistent with the hybrid model. Automatic transmissions due to high priority codes may not be read by consciousness due to a heavy concurrent processing load. That is, not all highly learned responses may interrupt the consciously directed flow of information. On similar lines, Shiffrin pointed out that unconscious processing can be deep, as reflected in semantic priming effects in subliminal perception (Marcel, 1983). This may be explained in terms of *message bandwidth* according to the hybrid model. The consciousness modules receive and transmit messages of low bandwidth. Modules at the highest levels of processing can communicate information invisible to consciousness, which may exert deeply processed influences on performance. Thus, one can understand low bandwidth conscious information as modulating and routing the processing of higher bandwidth information. In the absence of such modulation, higher bandwidth information can still be processed.

Shiffrin argued that consciousness may operate both serially (e.g., mental calculation), or in parallel (e.g., perception). That consciousness can be a parallel phenomenon finds support in Lightfoot and Shiffrin's (1992) demonstration that consciousness gives rise to impressions of complex holistic configurations of features that are nonetheless processed serially by the attentional mechanism as reflected in steeply rising search times as a function of the number of distracters. Consciousness seems to be able to temporarily integrate feature information. According to our model, consciousness functions to monitor and transmit global messages that are generally received by the whole system serially to avoid the cross-talk problem. To the extent to which compatible codes can be transmitted in the absence of interference, parallel conscious processing is possible. For example, both visual and auditory information can transmit in parallel to evoke a serial code in consciousness about a scene (visual image of two objects colliding and the sound of the collision). There would also be parallel preprocessing that would organize what in attended. This may occur, for instance, when features are organized in a spatial map (e.g., Wolfe, 1994), regions of which may be attended to, giving internal organization to the feature units. Consciousness may also produce a parallel output (e.g., an expectation tuning the sensory and motor systems to seek for and execute escape behaviors). In our view, consciousness is still operating serially but it is modulating parallel input and output.

Lightfoot and Shiffrin (1992) discovered that novel stimuli made up of complex configurations of features would initially be attentionally searched for feature by feature, despite their conscious appearance as single units. With training, however, automatic classification of these stimuli emerged, and they could then be searched for as if they were single units. This may be understood in the hybrid architecture in terms of *chunking*, which is critical in the reduction of network traffic (Schneider & Detweiler, 1987). Chunking involves the formation of new associations between codes in the consciousness module to the message vectors for new configurations. These newly labeled message vectors develop via the attentional mechanism operating on the constituent features of those configurations. These new chunks of several line segments could evoke a single conscious code, reducing the transmission on the inner loop and storage and processing time in the consciousness module. Initially the subject might search for an object that contained the features of chocolate, almonds, and candy coating. With repeated searches, the combined code would be learned so a single high level object could be searched for that was the combination of all of these features, perhaps giving that object an explicit code in conscious and verbal description (e.g., almond M&M candies).

JACOBY'S COMMENTARY

Jacoby's chapter is an attempt to specify the functional relation between conscious and unconscious processes, which he regarded as a critical part of the task of defining the two types of processes. He developed a *process dissociation* paradigm, in which given well-defined assumptions about the relation between conscious and unconscious processing (e.g., independence, redundancy, or mutual exclusivity), the separate contributions of the two types of processing can be estimated. He found that manipulating factors that have "traditionally been identified with the distinction between unconscious (automatic) and conscious processes" results in predicted invariance's of the unconscious component estimate across large variation of the conscious component estimate if one assumes independence, but not if one assumes either redundancy or exclusivity. The factors manipulated included divided versus focused attention, slow versus fast retrieval speed, and old versus young age differences (see also Jacoby, Toth, & Yonelinas, 1993). Jacoby also found similar patterns of results in "remember/know" judgment tasks, in which know judgments (associated with the unconscious component) remain invariant over changes of the remember judgments (associated with the conscious component) under a similar set of performance manipulations (Gardiner & Java, 1991). To supplement these parallel lines of evidence for independence, Jacoby argued convincingly that whereas the conscious component may show cross-modality transfer in a variety of tasks, if this component is controlled for the unconscious component does not (see also Jacoby et al., 1993).

Although we do not object to the inferences that Jacoby drew in his argument for independence, like Shiffrin we disagree with the assumption that unconscious processing is to be identified with automatic processing and consciousness with attentive processing for the reasons already described. This assumption was implicit in Jacoby's argument, and underpinned his choice of task manipulation factors.

Jacoby's data makes a compelling case for consciousness having its own local memory separate from the memory of the other modules. According to the hybrid model, if consciousness has its own memory, then we would expect it to show near independence of memory effects from unconscious processing. Although there may be statistical independence of the memories, we would expect some redundancy of both memory within the consciousness module and memory in other modules contributing to the performance of a task.

CONCLUSIONS

This chapter suggests that consciousness may have evolved as an extension of
attentional control. Consciousness occurs within the attentional control system and involves monitoring the information flow to better attentively modulate the information transfer between regions. We suggest that there are different levels of awareness of the controller, including activity aware, priority
aware, and message aware. We suggest the existence of a consciousness module on the inner loop that it is "message aware," receiving information about
the specific content of the message transmitted on the inner loop. To reduce
information overload on the consciousness module, message information is
assumed to be limited to inner-loop transmissions, which are typically serialized. The coding is of a simplified, symbolic representation of the between region inner loop messages. The controller is assumed to have local memory
and the ability to execute goal hierarchies. The message aware conscious controller provides the organism the ability to accomplish complex specific goal
hierarchies, speed learning, and potentially support language communication
of goal hierarchies. The high utility of such a mechanism and the ability of animals to execute goal hierarchies suggest that many species may have a conscious control system.

This view of consciousness as a message aware control mechanism to modulation cognitive processing is generally in agreement with the Shiffrin and
Jacoby chapters. We agree with Shiffrin that consciousness/unconsciousness
does not map directly to controlled/automatic processing but it is related. We
agree that automatic processing can occur in the absence of consciousness
but can bring information to consciousness and that consciousness may be selective, possibly a lower bandwidth sample of automatic message traffic. We
interpret Shiffrin's evidence of parallel processing and consciousness somewhat differently in that conscious perception is serial but multiple parallel inputs can give rise to a serial code.

Jacoby's evidence makes a provocative and compelling proposal of independent conscious and unconscious memories. The well-supported data for
independence suggests the presence of local memory in the conscious controller, independent from the memory in the modules that the conscious controller is controlling. Such local memory would have the advantage that the
conscious controller could consult its local memory without reducing traffic
on the inner loop and could support cross-modality transfer.

In seeking an interpretation of the mechanisms and purposes of consciousness, it will be beneficial to consider the interrelation between consciousness and attentional control. A controller conscious or message aware
of the information network it is controlling supports faster goal search, learning, and cross-modality transfer, and may even be the basis for evolving language to transfer such messages between individuals.

ACKNOWLEDGMENTS

The writing of this chapter was supported by ARI grant MDA903-92-K-0123 and Office of Naval Research N00014-91-J-1708.

REFERENCES

Briggs, G. E. (1974). On the predictor variable for choice reaction time. *Memory and Cognition, 2,* 575–580.

Broadbent, D. E. (1958). *Perception and communication.* London: Pergamon.

Elman, J. L. (1990). Finding structure in time. *Cognitive Science, 14,* 179–212.

Felleman, D. J., & Van Essen, D. C. (1991). Distributed hieararchial processing in the primate cerebral cortex. *Cerebral Cortex, 1,* 1–46.

Fisk, A. D., & Schneider, W. (1984). Memory as a function of attention, level of processing, and automatization. *Journal of Experimental Psychology: Learning, Memory, and Cognition, 10,* 181–197.

Gardiner, J. M., & Java, R. I. (1991). Forgetting in recognition memory with and without recollective experience. *Memory and Cognition, 19,* 617–623.

Jacoby, L. L., Toth, J. P., & Yonelinas, A. P. (1993). Separating conscious and unconscious influences of memory: Measuring recollection. *Journal of Experimental Psychology: General, 122,* 1–16.

LaBerge, D., Carter, M., & Brown, V. (1992). A network simulation of thalamic circuit operations in selective attention. *Neural Computation, 4,* 318–331.

Lightfoot, N., & Shiffrin, R. M. (1992). On the unitization of novel, complex visual stimuli. In *Proceedings of the 14th Annual Conference of the Cognitive Science Society* (pp. 277–282). Hillsdale, NJ: Lawrence Erlbaum Associates.

Marcel, A. J. (1983). Conscious and unconscious perception: Experiments on visual masking and word recognition. *Cognitive Psychology, 15,* 197–237.

Newell, A. (1990). *Unified theories of cognition.* Cambridge, MA: Harvard University Press.

Olshausen, B. A., Anderson, C. H., & Van Essen, D. C. (1993). A neurobiological model of visual attention and invariant pattern recognition based on dynamic routing of information. *Journal of Neuroscience, 13,* 4700–4719.

Schneider, W., & Detweiler, M. (1987). A connectionist/control architecture for working memory. In G. H. Bower (Ed.), *The psychology of learning and motivation* (Vol. 21, pp. 54–119). New York: Academic Press.

Schneider, W., & Detweiler, M. (1988). The role of practice in dual-task performance: Toward workload modeling in a connectionist/control architecture. *Human Factors, 30,* 539–566.

Schneider, W., & Oliver, W. L. (1991). An instructable connectionist/control architecture: Using rule-based instructions to accomplish connectionist learning in a human time scale. In K. VanLehn (Ed.), *Architectures for intelligence: The 22nd Carnegie Mellon symposium on cognition* (pp. 113–145). Hillsdale, NJ: Lawrence Erlbaum Associates.

Schneider, W., & Shiffrin, R. M. (1977). Controlled and automatic human information processing: I. Detection, search, and attention. *Psychological Review, 84,* 1–66.

Shedden, J. M., & Schneider, W. (1990). A connectionist model of attentional enhancement and signal buffering. In *Proceedings of the 12th Annual Conference of the Cognitive Science Society* (pp. 566–573). Hillsdale, NJ: Lawrence Erlbaum Associates.

Shedden, J. M., & Schneider, W. (1991). A connectionist simulation of attention and vector comparison: The need for serial processing in parallel hardware. In *Proceedings of the 13th An-*

nual Conference of the Cognitive Science Society (pp. 546–551). Hillsdale, NJ: Lawrence Erlbaum Associates.
Shiffrin, R. M. (1984). Automatic and controlled processing revisited. Psychological Review, 91,
   269–276.
Shiffrin, R. M. (1988). Attention. In R. A. Atkinson, R. J. Herrnstein, G. Lindzey, & R. D. Luce
   (Eds.), Stevens' handbook of experimental psychology: Vol. 2. Learning and cognition (pp.
   739–811). New York: Wiley.
Shiffrin, R. M., Cohen, A., & Fragassi, M. (1992). Simultaneous attentive and automatic pro-
   cessing of visually presented characters. Paper presented at 33rd Annual Meeting of the Psychonomic Society, St. Louis.
Shiffrin, R. M., & Schneider, W. (1977). Controlled and automatic human information process-
   ing: II. Perceptual learning, automatic attending, and a general theory. Psychological Review,
   84, 127–190.
Szentagothai, J. (1978). The neuro network of the cerebral cortex: A functional interpretation.
   Proceedings of the Royal Society of London, Britain, 201, 219–248.
Treisman, A. M. (1960). Contextual cues in selective listening. Quarterly Journal of Experi-
   mental Psychology, 12, 242–248.
Treisman, A. (1986). Properties, parts, and objects. In K. R. Boff, L. Kaufmann, & J. P. Thomas
   (Eds.), Handbook of human perception and performance (pp. 35.1–35.70). New York:
   Wiley.
Wolfe, J. M. (1994). Guided Search 2.0: A revised model of visual search. Psychonomic Bulletin
   and Review, 1, 202–238.
Yantis, S., & Johnson, D. N. (1990). Mechanisms of attentional priority. Journal of Experimen-
   tal Psychology: Human Perception and Performance, 16, 812–825.

Subliminal Perception

CHAPTER 5

Do Subliminal Stimuli
Enter the Mind Unnoticed?
Tests With a New Method

Anthony G. Greenwald
Sean C. Draine
University of Washington

In the last decade, there has been a dramatic increase in the acceptability of theoretical interpretations of research findings in terms of unconscious cognition. Part of the shift is in language—many psychologists have become willing to use the word *unconscious* in sentences that, previously, would have been acceptable only by using alternate terms such as "unattended," "automatic," "procedural," or "implicit." However, to characterize this recent change as being just a matter of linguistic style would be to underestimate it severely. There has also been a conceptual and empirical revolution. An important factor in this revolution has been the demonstration of replicability for a class of findings that, until very recently, were widely regarded with great skepticism—findings of *subliminal semantic activation* (see Balota, 1983; Bornstein, 1992; Dagenbach, Carr, & Wilhelmsen, 1989; Fowler, Wolford, Slade, & Tassinary, 1981; Greenwald, Klinger, & Liu, 1989; Groeger, 1988; Hardaway, 1990; Marcel, 1983). Subliminal semantic activation (SSA) can be defined as "indirect evidence for analysis of semantic content of target word stimuli under conditions that limit or prevent awareness of the presence of these words" (Greenwald, 1992, p. 768).[1]

Although at least some types of SSA are now treated by many experts as

---

[1] The term *subliminal* implies a theory of the perceptual threshold (limen) that is no longer justified in the modern era of signal detection theory (Green & Swets, 1966). A more theoretically neutral designation of the class of stimuli with which this chapter is concerned is *marginally perceptible*. The chapter uses subliminal and marginally perceptible as interchangeable designations.

replicable phenomena (see Greenwald, 1992, p. 779), SSA continues to be the focus of controversies. Debate over proper *description* of SSA findings is central to these controversies. Studies of SSA often examine effects of marginally perceptible stimuli on actions the subject is instructed to perform (direct effects), while concurrently observing uninstructed (indirect) effects that are interpreted as likely indicators of unconscious semantic activation. Holender (1986) argued that strategies being used by researchers to assess direct effects of marginally perceptible stimuli in SSA studies were insufficiently sensitive to conscious stimulus effects and, consequently, Holender judged the then-available evidence to be inadequate for assessing crucial details of relations between direct and indirect effects (see also Merikle, 1982; Purcell, Stewart, & Stanovich, 1983).

A HOLY GRAIL OF
SUBLIMINAL ACTIVATION RESEARCH

A *direct effect* of a stimulus is its effect on an instructed response, typically assessed by a measure of accuracy at the instructed task. By contrast, an *indirect effect* is an *un*instructed effect of the task stimulus on behavior, and is often assessed by including an irrelevant or distracting component in the task stimulus, then measuring influences of this distractor on latency or accuracy of instructed responses to it. As illustration, a very well-known indirect effect is the increased latency of response observed in Stroop's (1935) task of naming the color of ink with which a word is printed, caused by (the task-irrelevant stimulus of) that word being the name of a different color.

It has been an elusive goal of SSA research to demonstrate an indirect effect of word stimuli under conditions that preclude any direct effect. Claims to have achieved this long-sought *indirect-without-direct-effect* data pattern have often been met with skeptical appraisals (e.g., the appraisals by Greenwald, 1992; Holender, 1986; Merikle, 1982; Purcell et al., 1983; Reingold & Merikle, 1988).

In order to argue that a set of data demonstrate an indirect effect in the absence of a direct effect, it has been necessary for researchers to claim that a null result has been achieved. In particular, the claimed null result is that the experiment's stimulus presentation conditions had no influence on the direct measure. Such a claim is susceptible to the familiar criticism of inappropriately asserting the truth of a null hypothesis. For this reason, the strongest statistical claim that can be made from existing attempts to demonstrate the indirect-without-direct-effect pattern is that an indirect effect occurred when performance on a direct measure was very likely within some range that included the value of zero. Of course, even stated in that cautious way, the claim may be quite impressive if enough data have been collected to make the statistically credible margin around zero a small one.

REGRESSION METHOD FOR SEEKING THE INDIRECT-WITHOUT-DIRECT-EFFECT PATTERN

The research reported here used a data-analysis strategy that bypasses the usual statistical problems associated with asserting the truth of a null hypothesis. The major innovation in this method was to analyze data using tests of the regression relation between direct and indirect measures. This regression relation can be described by a plot of the equation that relates expected scores on an indirect measure to observed scores on a direct measure. Figure 5.1 illustrates some of the linear functions that might be revealed by regression analysis.

Regression functions such as those in Fig. 5.1 provide answers to questions of the form, "What is the level of performance on the indirect measure that is associated with some specific level of performance on the direct measure?" If, for example, one wishes to know what level of performance on the indirect measure is associated with the *mean* level of performance on the direct measure, the answer will be the mean of the indirect measure. The significance test for the (null) hypothesis associated with this question—that is, the significance test for the hypothesis that level of performance on the indirect measure associated with mean performance on the direct measure = 0—is the usual statistical significance test for the difference of the mean of the indirect measure from zero. In order to test for the indirect-without-direct-effect pattern, the level of performance on the indirect measure associated with the value of *zero* on the direct measure can be tested for the significance of its difference from zero. The sought value in this case is the intercept of the regression equation (the place at which the regression function crosses the vertical axis), and the needed significance test is the usual test of statistical significance (of difference from zero) for this intercept.

This regression analysis strategy entirely reverses the usual difficulty associated with asserting the truth of a null hypothesis. In the context of the regression strategy, researchers who claim that the indirect-without-direct-effect pattern does not exist are the ones left in the position of claiming the truth of a null hypothesis. In particular, they must claim that the regression relation passes through the origin, thereby asserting truth of the null hypothesis that the intercept is equal to zero.

METHOD

The data to be used with the regression-analysis method just described were obtained from three experiments in which subjects' main task was to detect four-letter words accompanied by simultaneous dichoptic masking (see Fig. 5.2). In Experiments 1 and 2, subjects performed at a detection task in which they pressed a key with the right index finger when they judged that a word

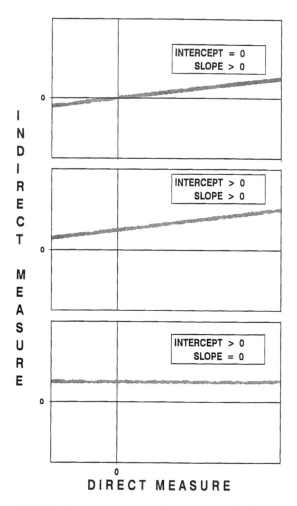

FIG. 5.1. Some expectations of data patterns for linear regression of an indirect measure on a direct measure.

was presented, or a key with the left index finger when they judged that no
word was presented. Experiment 3 also included a detection task; however, for
half of the subjects, the assignment of response keys was reversed so that key-
presses with the left index finger indicated that a word was presented, and
key-presses with the right index finger indicated that no word was presented.
All words were four letters in length. By using effective masking conditions,
performance on the detection task was reduced to a low value for most subjects, but was nevertheless allowed to vary across subjects. Responses to the

stimuli LEFT and RIGH (shortened from RIGHT so that stimulus width was constant for all stimuli at four characters), which were used on a subset of trials in Experiment 1, and on all trials in Experiments 2 and 3, provided the basis for an indirect measure of semantic activation. Specifically, the indirect measures assessed the extent to which the stimuli LEFT and RIGH directed subjects' responses to the left and right response keys, independently of their instructed task.

The direct measure was the signal detection analysis measure of d', which is based on hit and false alarm rates. For purposes of having comparable units, the indirect measure was also computed as a d', by counting presses of the right key in response to RIGH as hits, while counting presses of the right key in response to LEFT as false alarms. (The same value of d' would result from treating left-key responses to LEFT as hits, and left-key responses to RIGH as false alarms.)

Subjects

For the series of three experiments, a total of 881 undergraduate students from lower level psychology courses at University of Washington volunteered in exchange for a modest course credit. Data for 29 subjects were discarded prior to conducting hypothesis tests, either because of equipment malfunction or because they volunteered to the experimenter at the conclusion of the experiment that they had deliberately closed one eye at some time during the experiment. This left analyzable data for 431 subjects in Experiment 1, 175 subjects in Experiment 2, and 246 subjects in Experiment 3.

Apparatus and Masking

Up to three subjects participated concurrently, each in a small (1.5 m × 2.5 m) room containing a table on which was a 33-cm (diagonal) color monitor and keyboard controlled by an IBM/AT-type (80286) computer. Subjects viewed a color (Enhanced Graphics Adapter, EGA) display through a viewing apparatus that presented the images from the left and right halves of the display screen to the left and right eyes, respectively. (The same type of apparatus was used by Cheesman & Merikle, 1986; Greenwald & Klinger, 1990; and Greenwald, Klinger, & Schuh, 1995.)

The apparatus obliged subjects to view the computer's display from a distance of 65 cm, through rotary prisms that were adjusted to superimpose the left-eye and right-eye images. Stimuli (such as instructions) that were presented simultaneously to both halves of the screen were easily viewed with binocular fusion. The placement of the keyboard, on the table that supported this viewing apparatus, allowed the subject to press the "A" key with left forefinger and the "5" key (on the keyboard's numeric keypad) with right forefinger, with these keys being marked with green adhesive dot labels. All responses to the major experimental tasks were made with just these two keys.

Masks were constructed using items in a software-fabricated "character set." Each item in the software character set was composed by blackening selected pixels in the 8 (horizontal) × 14 (vertical) pixel array that comprises a character space for the EGA-interface display. These fabricated characters were constructed so that, with appropriate side-by-side and top-to-bottom juxtapositions, regularly spaced gratings oriented vertically, horizontally, or in either diagonal direction could be constructed. However, rather than using regular gratinglike masks, masks were constructed by randomly selecting, on each trial and with replacement for each position in a 3 row × 15 column rectangular array, elements corresponding to a selected thickness. Sample masks are shown in Fig. 5.2.

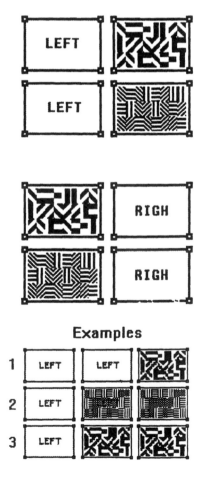

FIG. 5.2. Examples of mask patterns and letter strings used in experiments. These masks are not as wide as those actually used, and frames have been added to facilitate the reader's simulating the effect of dichoptic masking. By de-converging eyes (as if focusing on a more distant object) while looking at a mask + word pair from about 20 cm, the reader should be able to superimpose the two, subjectively seeing three rectangles side by side (as in Examples 1–3 at the bottom). If two of these three rectangles contain a word, as in Example 1, then the eye it names is (at least at the moment) dominant. Example 3 shows the subjective appearance if the right eye is dominant when looking at the topmost mask + word pair. An apparent mixture of the two images in the middle rectangle (as in Example 2) may also occur, and simulates the experience of some subjects in the present research. The topmost mask is made of mask elements 5 pixels thick and the one below it is of elements 2 pixels thick. The reader may find that the mask with 5-thick elements more effectively obscures the word than does the mask with 2-thick elements, when mask and word are superimposed.

Examples

Procedure

Although the detection task is the focus of this chapter, all experiments included at least one additional task that provided an alternative direct measure. For example, an additional task used in all experiments was position discrimination, which required subjects to judge whether a dichoptically masked four-letter word was displayed to the left or right of a fixation point. Critical trials with stimuli LEFT or RIGH were included in the position discrimination task, much as for the detection task.[2]

In all experiments, the first task (detection in Experiment 1, position discrimination in Experiments 2 and 3) included practice at 120 trials of masked displays, and permitted adjustment of masking conditions (usually by making them more difficult) contingent on the subject's performance on a direct measure. Next came two blocks of trials of the experiment's second task, which was position discrimination in Experiment 1. Experiment 1 continued alternating sets of two blocks of trials of its two tasks, until a total of 4 postpractice blocks (56 trials each) had been completed for both tasks.

In Experiment 2, after the 120 trials of position discrimination practice, data were collected for a block of 50 trials of position discrimination, followed by 50 trials of the detection task, 50 trials of a lexical decision task, and 50 trials of an evaluative decision task, and then a second round of 50-trial blocks for each of these four tasks. Each new task was preceded by 10 to 20 practice trials to assure that subjects understood its instructions. The major purpose of the additional tasks in Experiment 2 was to provide alternative direct measures that were used in regression analyses. Results from these alternative analyses are described briefly under the Discussion heading, below.

After practice with the position discrimination task in Experiment 3, data were collected for two 50-trial blocks of position discrimination, after which were two 50-trial blocks of the detection task. After these, there were two more blocks of each task, so that subjects completed 200 trials of each task, not including the initial practice at position discrimination.

In all experiments, all stimulus parameters that could vary across trials within any block of trials (especially side to which the mask was presented and stimulus identity) were varied by an online randomization routine that resulted in a unique sequence of trials for each subject. There were four potentially important differences among procedures for the three experiments.

First was the manner in which stimuli were positioned for the detection task. In Experiment 1, detection stimuli were positioned either to the left or

---

[2]The position discrimination data from Experiment 1, along with those from several other experiments that did not include a detection task, were reported by Greenwald, Klinger, and Schuh (1995). Greenwald et al. reported regression analyses for the position-discrimination task that paralleled those for the detection task reported in this chapter.

right of the fixation point, just as for the position discrimination task. This variable positioning (which subjects knew about from instructions and practice) is unusual for a detection task, and was done in order to use the same types of stimuli for the detection task as for the position discrimination task. Experiments 2 and 3 permitted the potential replication of findings of Experiment 1 using a more standard stimulus presentation for the detection task, with all stimuli centered on the fixation point.

Second, in Experiments 2 and 3, all stimuli were critical stimuli (i.e., only LEFT and RIGH were used as stimuli for the detection task), which permitted more powerful tests of indirect effects with fewer trials overall.

Third, in Experiment 1, data for indirect measures were obtained only on trials using masks to the subject's dominant eye (i.e., the eye to which the mask was observed to be more effective during the practice phase). By contrast, in Experiments 2 and 3 critical trials were presented to both eyes, allowing both direct and indirect measures to be obtained with masking to each eye. The data for Experiments 2 and 3 were analyzed separately for mask to left eye and mask to right eye.

Fourth, whereas all subjects in Experiments 1 and 2 were instructed to press a key with their right finger to indicate word presence and their left finger to indicate word absence, Experiment 3 included a between-subjects manipulation of response key assignment. Thus, approximately half of the data in Experiment 3 were collected using the key assignment of Experiments 1 and 2 (hereafter referred to as the *standard* assignment), and half were collected using the *reverse* key assignment (left-key response to indicate word-presence, and right-key response to indicate word-absence). The manipulation of key assignment was used to control for the possibility of bias in indirect measures that could have been caused by differences in the detectability of the specific stimuli, RIGH and LEFT. Note that indirect effects of semantic activation on detection judgments are shown by relatively more right-key responses when the stimulus is RIGH, rather than LEFT. With the standard key assignment, the same pattern of responses would also be expected if RIGH were more detectable than LEFT. That is, there would be more right-key (i.e., word-present) responses to RIGH. With the reverse key assignment, the effect of RIGH being more detectable than LEFT would be relatively more left-key responses to RIGH. Thus, the variation of key assignment in Experiment 3 permitted the influence of possible differential detectability of RIGH and LEFT to be distinguished from semantic priming.

Each experiment's procedure involved a total of 550 to 600 trials divided among the various tasks. Each experiment required about 50 min to complete, with some variation in session durations resulting from subjects being allowed both to self-initiate trials and to rest ad lib between blocks of trials. Subjects going at the most rapid rate could initiate new trials about 1 sec after response to the prior one.

Analysis Strategies

Because of concern that tests of intercept effects might be sensitive to the presence of outlying scores in either direct or indirect measures, more than 30 regression analyses were conducted for each experiment, using alternative criteria for trimming the predictor and criterion d' measures. The direct measure (the predictor variable) was trimmed on its high accuracy end, based on reasoning that subjects scoring extremely high on this measure were unlikely to show any unconsciously mediated effects on the indirect measure. Two levels of trimming were done on the direct measure, either eliminating scores of d' greater than 3.29 (corresponding to about 95% correct responding) or greater than 2.5 (corresponding to about 90% correct responding). Neither slope nor intercept estimates varied more than slightly between these two levels of trimming.

There were two justifications for trimming the indirect measure (the criterion variable). The first reason was to avoid problems stemming from the possibility that a few subjects had misunderstood the instructions—mistakenly believing that their task was to press the right key if they saw RIGH and the left key if they saw LEFT— and, as a result, had shown unusually high scores on the indirect measure. The second reason was to reduce the variance of the measure, which in turn increased the power of both regression slope and intercept tests. To avoid bias in tests of the intercept effect, subjects were dropped in equal number from both the high and low ends of the distribution. In various tests, between 0% and 4.8% (Experiment 1) or 11.4% (Experiment 2) or .5% (Experiment 3) were trimmed from each tail of the criterion indirect measure. In Experiments 1 and 2, degree of trimming of the indirect measures had no systematic effect of increasing or reducing numerical values of intercepts, but greater trimming did tend to produce higher t values (and, therefore, lower p values of significance tests) because of reduced error variances. In Experiment 3, greater trimming was associated with both larger numerical intercept values and higher t values.

RESULTS

Experiments 1 and 2

Figures 5.3 and 5.4 give results that are representative of the sets of analyses done with varying degrees of trimming on the direct and indirect measures, described just above. Figure 5.3 presents an analysis with an intercept test that is just at the conventionally significant p value of .05, two-tailed. (This was selected as representative because most of the analyses fell near this value, but not consistently either above or below it.) The analyses presented in both

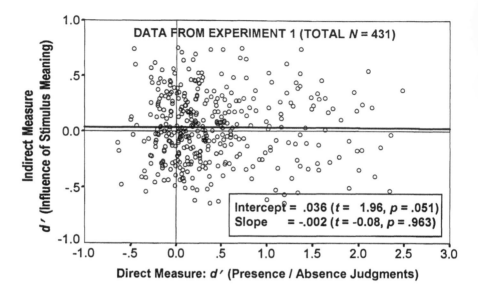

FIG. 5.3. Regression test for Experiment 1 showing indirect measure ($d'_{(i)}$, based on response to position meaning of LEFT or RIGH) as a function of direct measure ($d'_{(d)}$, based on position discrimination accuracy for dichoptically masked left- or right-positioned 4-letter words). The scatter plot is for a representative analysis of the experiment's data (see text).

figures dropped cases with d' scores greater than 2.5 on the direct measure. For Fig. 5.3, the presented analysis of Experiment 1 also dropped the extreme 3.2% (14 cases) from each tail of the indirect measure, and for Fig. 5.4 the presented analyses of Experiment 2 dropped the extreme 8.6% (15 cases) from each tail of the indirect measure. Figure 5.4 gives two analyses for Experiment 2, one based on trials with mask presented to the left eye, the other based on trials with mask to the right eye.

Significance Tests and Effect Sizes. For all regression analyses of Experiments 1 and 2, the value of the intercept was positive. The t values of the intercept test for the three tests were, respectively, 1.96 ($df = 394$, $p = .051$), 1.95 ($df = 133$, $p = .053$), and 2.78 ($df = 135$, $p = .006$), for Experiment 1, Experiment 2 with mask to left eye, and Experiment 2 with mask to right eye. Considered as effect sizes using a **d** statistic (intercept value divided by the untrimmed standard deviation of indirect measure), these intercepts corresponded to **d** values of .095, .134, and .183, approximating the level conventionally regarded as "small" (**d** = 0.2; Cohen, 1977).

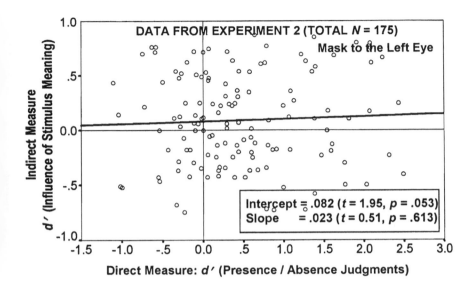

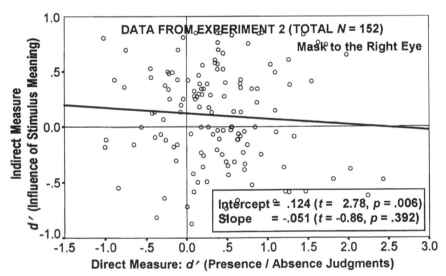

FIG. 5.4. Regression tests for Experiment 2, presented separately for data collected with mask to left eye and mask to right eye. (See Fig. 5.3 caption.)

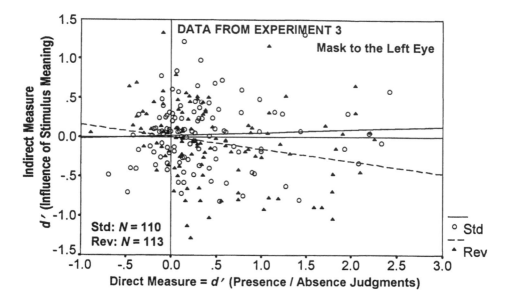

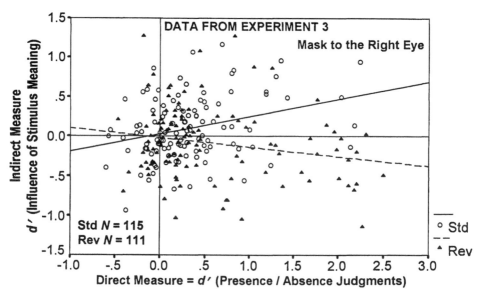

FIG. 5.5.   Regression tests for Experiment 3, presented separately for data
collected with mask to left eye and mask to right eye. Analyses are further broken
down according to response key assignment conditions. The solid and dashed
lines show the best fitting linear regression for the standard and reversed key
assignment conditions, respectively. Data from the scatter plot are shown as
open circles for the standard condition and triangles for the reversed condition.
Significance tests and parameter estimates for the intercepts and slopes are reported in the text.

Experiment 3

Figure 5.5 shows two analyses for Experiment 3—one based on trials with mask presented to the left eye, the other based on trials with mask to the right eye. As with the previous experiments, cases with d' scores greater than 2.5 on the direct measure were dropped from the analysis. In addition, analyses were conducted using various criteria for trimming indirect measures. In contrast to analyses of Experiments 1 and 2, however, greater trimming of indirect measures from Experiment 3 had the systematic effect of increasing the size of intercept parameters. Thus, in the analyses reported just below, only one extreme case from each tail of the indirect measure was dropped from the trials with the mask to the right eye.

Significance Tests and Effect Sizes. Consistent with the previous two experiments, analyses of the standard key assignment condition showed positive intercept values of .025 and .028 with the mask to the left and right eye, respectively. The t values associated with the intercept tests were .55 ($df = 108$, $p = .58$) with left-eye masking and .67 ($df = 109$, $p = .50$) with right-eye masking, neither of which approached statistical significance. The effect sizes corresponding to the intercepts of the standard condition were $d = .058$ and $d = .063$ with masks to the left and right eye, respectively. In contrast, analyses of the reversed condition revealed negative intercept values of $-.001$ with left-eye masking and $-.23$ with right-eye masking, with associated t values of $-.02$ ($df = 113$, $p = .98$) and $-.45$ ($df = 115$, $p = .65$), respectively. The effect sizes for both left and right masks in the reversed condition were also small, with associated d values of $-.002$ and $-.051$.

As can be seen in Fig. 5.5, the slopes of the regression functions in Experiment 3 appeared to differ between the standard condition (positive slopes) and the reversed condition (negative slopes). The standard condition yielded a slope of .035 with the mask to the left eye ($t = .54$, $df = 108$, $p = .59$), and .219 with the mask to the right eye ($t = 3.3$, $df = 109$, $p = .001$). In the reversed condition, the slope was $-.117$ with masking to the left eye ($t = -1.81$, $df = 113$, $p = .07$), and $-.112$ with masking to the right eye ($t = -1.99$, $df = 115$, $p = .048$). This difference in slopes is considered in the Discussion section, just below.

DISCUSSION

The intercept effects obtained in Experiments 1 and 2 conformed to the pattern sought as demonstrating an indirect effect in the absence of a direct effect. If these results validly warrant the conclusion that indirect effects indeed occurred in the absence of direct effects, then a yes answer to the title question of this chapter ("Do subliminal stimuli enter the mind unnoticed?") is justified. However, the strength of the case for that interpretation is moderated by nonreplication of the intercept effect in Experiment 3. Given that the procedures of the standard key assignment condition in Experiment 3 matched those of Experiment 2 in all but a few minor details, no obvious explanation—beyond the uninformative one of Type II error, possibly due to Experiment 3 having the smallest N of the three experiments—is available for its failure to replicate the findings of the previous two experiments.

Implications of Experiment 3

The difference between regression functions across the two key assignment conditions indicated that subjects were more likely to indicate the presence of RIGH than of LEFT. Further, the slopes of these regression functions indicated that as overall detection performance diminished, so did the relative advantage in detectability of RIGH over LEFT. This trend was corroborated by additional analyses, including a series of contrasts between detection performance for RIGH and LEFT based on subject samples with increasingly poorer overall detection performance. For example, *overall* mean detection performance for RIGH was greater than for LEFT with left-eye masking (difference = .050, $t = 1.71$, $df = 246$, $p = .089$) and right-eye masking (difference = .098, $t = 3.53$, $df = 244$, $p = .001$). However, when contrasts were restricted to data points (subjects) with detection performance lower than $d' = .5$, the difference in detection performance between LEFT and RIGH was much reduced for both left-eye (difference = .023, $t = .69$, $df = 156$, $p = .49$) and right-eye (difference = .036, $t = 1.21$, $df = 160$, $p = .229$) masking conditions.

Is it possible that the intercept effects obtained in Experiments 1 and 2 were caused by RIGH being more detectable than LEFT, rather than by the semantic content of these words? Under close scrutiny, this possible conclusion appears both theoretically and empirically implausible. Theoretically, in order to dismiss the intercept effects as artifacts of differences in stimulus detectability, it is necessary to claim that RIGH was more detectable than LEFT when stimulus presentation conditions made the words undetectable overall. That is, when average detection performance for both words was at chance, the word RIGH must nevertheless have been more detectable than the word LEFT. This, however, could only be true if LEFT was less easily detected than no stimulus at all. Given the perplexity of such an argument, a more plausible explanation of the intercept effects of Experiments 1 and 2 is that they were caused by the differing semantic content of the word stimuli. Empirically, the evidence for difference in detectability of LEFT and RIGH in Experiment 3 was statistically inadequate for precisely the range of scores on the direct measure for which this difference would have to be statistically significant in order to provide an alternative interpretation of the Experiment 1 and 2 intercept effects. In other words, differences in detectability of LEFT and RIGH appear capable of explaining slope effects, but not intercept effects, in the regression analysis.

Evidence from the Position Discrimination Task

Because the stimuli LEFT and RIGH were confounded with key assignment in Experiments 1 and 2, it is impossible to separate the effects of differences in detectability from effects of semantic priming for the detection-task data of those experiments. However, the position discrimination tasks in Experiments 1 and 2 did permit differences in stimulus detectability to be tested independently of priming effects. This was possible because the direct measure in the position discrimination task required both left- and right-key presses in responses to the stimulus RIGH, as well to LEFT, depending only on the position of those stimuli. If RIGH was more detectable than LEFT, then position discrimination performance when the stimulus was RIGH should have been superior to when it was LEFT.

Figure 5.6 shows the differences in position discrimination accuracy for

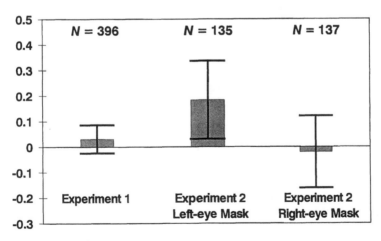

FIG. 5.6. Difference in position discrimination accuracy (computed as d') between trials with RIGH versus LEFT as stimulus. The three tests are based on data from Experiment 1 and Experiment 2. Tests for Experiment 2 are presented separately for data collected with mask to left eye and mask to right eye. The vertical axis represents d' for trials with RIGH as stimulus minus d' for trials with LEFT as stimulus. Error bars represent the 95% confidence interval corresponding to each test.

Experiments 1 and 2 between trials with LEFT as the stimulus and those with RIGH. The results from Experiment 1 showed no significant difference (difference = .032, $t = 1.11$, $df = 395$, $p = .266$) in accuracy of position judgments between the two stimuli, suggesting that differences in stimulus perceptibility did not influence performance in that experiment. Results from Experiment 2 are shown separately by mask side. For trials with masking to the left eye, position discrimination accuracy was significantly greater when RIGH was the stimulus rather than LEFT (difference = .184, $t = 2.36$, $df = 134$, $p = .019$). However, with masking to the right eye, performance with RIGH was slightly poorer than that with LEFT, although this difference was not significant (difference = $-.020$, $t = -.28$, $df = 136$, $p = .783$). The results, particularly those of the left-eye mask condition in Experiment 2, provide some indication that RIGH was more perceptible than LEFT. However, the failure of this pattern to emerge consistently across the three test conditions suggests that the difference in stimulus perceptibility is not robust, and may be influenced by the minor procedural variations existing across these conditions.

In sum, the argument that the intercept effects of Experiments 1 and 2 reflect differences in stimulus detectability is theoretically problematic and, also, difficult to integrate with empirical results from the position discrimination task in Experiments 1 and 2, and from the detection task for subjects whose detection scores were low in Experiment 3. Accordingly, the interpretation that intercept effects in Experiments 1 and 2 indicate unconscious semantic activation continues to appear valid.

**Meta-Analysis of All Dichoptic-Masked
Position Priming Studies**

Meta-analyses of the data from Experiments 1–3 were conducted by computing a weighted mean intercept value, i, with the formula,

$$i = \sum w_j u_j \div k \tag{1}$$

where u_j was the intercept value of the jth data set, k was the total number of data sets, and w_j was a weight for each data set. The weighted mean intercept was tested for statistical significance by transforming the one-tailed p values corresponding to the significance test of each intercept into z scores (see Rosenthal, 1993). The z scores were then combined to yield a single z using the formula,

$$z = \sum w_j z_j \div k^{-2} \tag{2}$$

where z_j was the z score of the jth data set, k was the total number of data sets, and w_j was a weight for each data set. A χ^2 test for heterogeneity of effect among the four data sets was conducted using the formula

$$\chi^2_{(df = k - 1)} = \sum (z_j - z_{wm}) w_j \tag{3}$$

where z_j was the z score of the jth data set, z_{wm} was weighted mean z of the data sets, and w_j was a weight for each data set. In both the tests for significance and for heterogeneity of effect sizes, the weight for each experiment was computed as

$$w_j = kdf_j \div \sum df_j \qquad (4)$$

where df_j was the degrees of freedom for the jth data set, and k was the total number of data sets.

Because data from left- and right-eye masking conditions in Experiments 2 and 3 were collected from the same subject population, z scores from the two masking conditions were first combined, using equation 2, into a single z with corresponding df equal to the average df of the two conditions. The two key assignment conditions in Experiment 3, on the other hand, represented separate subject populations, and were therefore treated as separate data sets in the meta-analysis. Thus, meta-analytic tests of the intercept effects from the detection task were based on four z scores corresponding to the intercept effects of Experiment 1, Experiment 2 (left and right mask conditions combined), Experiment 3 with standard key assignment (left and right mask conditions combined), and Experiment 3 with reversed key assignment (left and right mask conditions combined). The meta-analytic test for significance indicated an overall statistically significant intercept effect for the detection task ($i = .039$, $z = 3.38$, $p = .0004$). The nonsignificant test for heterogeneity among the intercept effects suggested that these effects were homogeneous ($\chi^2 = 4.93$, $df = 3$, $p = .18$).[3]

The present results from the detection task supplement the larger set of data that tested semantic activation in the position discrimination task, reported by Greenwald et al. (1995). Together, these two data sets include well over 2,000 subjects, each participating in one of five versions of the position discrimination task, and thus provide a very powerful test of SSA effects obtained from position priming experiments using dichoptic masking. Using the meta-analytic techniques already described, the combined test of significance for the intercept effects of the position discrimination tasks of the two data sets indicated a statistically significant aggregate effect ($i = .020$, $z = 3.85$, $p = .00006$), but with some heterogeneity of effect sizes ($\chi^2 = 10.88$, $df = 4$, $p = .020$).

Meta-analytic combination of the intercept effects from both the detection and position discrimination tasks indicated a statistically significant aggregate effect ($i = .025$, $z = 5.14$, $p = .0000001$) and, again, significant hetero-

---

[3]A second meta-analysis was conducted in which left- and right-eye masking conditions were treated as separate data sets. The resulting significance tests were consistent with those of the former procedure, yielding a weighted mean intercept of .040 ($z = 3.52$, $p = .0002$) with $\chi^2 = 8.15$ ($df = 5$, $p = .23$), indicating statistically acceptable homogeneity of findings across experiments.

geneity of effect sizes ($\chi^2 = 15.52$, $df = 8$, $p = .05$). The mild heterogeneity of effects remains unexplained (see next paragraph). However, the meta-analytic tests for significance of the combined intercept effects demonstrate that those intercept effects, although small, are almost certainly not Type I errors.

Cautious Conclusion.   The conclusion that the intercept effects in Experiments 1 and 2 demonstrate semantic activation must be stated cautiously. There is not yet a clear explanation for the failure of Experiment 3 to replicate the intercept effects observed in Experiments 1 and 2, and there was a similar unexplained failure to find an intercept effect in a portion of the position discrimination data of Greenwald et al. (1995). At the same time, as previously demonstrated, examinations of all available statistical tests of intercept effects—including the few nonsignificant effects—provide overwhelming support for the statistical significance of the intercept effect in the full combined set of relevant data.

Evaluation of Assumptions Underlying the Regression Analysis Strategy

Four assumptions underlie the regression method used here to conclude that indirect effects occur in the absence of direct effects of marginally perceptible stimuli. First, both the direct and indirect measures must have rational zero points. That is, zero values on both measures must indicate absence of their respective effects. Second, the relation between direct and indirect measures is assumed to be linear. Third, the predictor variable (in this case, detection accuracy) is assumed to be measured without error. Fourth, the logical analysis underlying use of the regression method to infer existence of unconscious cognition assumes that the direct measure is at least as sensitive as the indirect measure to consciously perceivable stimulus effects. Possible criticisms of the claim to have demonstrated the indirect-without-direct-effect data pattern and, with it, the existence of unconscious cognition, follow from possible error of these four assumptions. Consider the possible failure of each assumption.

Rational Zero Points.   The assumption of rational zero points is easily met, because the theory underlying the d' measure (i.e., signal detection theory) provides this property. At the same time, it is reassuring that other measures that have rational zero values also produced the same pattern of positive intercept effects. In particular, the present analyses were repeated using both gamma and the very simple measure of hit rate minus false alarm rate, both of which also have zero values that indicate absence of stimulus effects. The analyses based on these two alternative measures yielded the same pattern of significant and nonsignificant results that were obtained with d', and in some cases yielded even stronger evidence of statistical significance than did the d' measure.

Linear Relation Between Direct and Indirect Measures. The assumption of linearity may well be wrong. However, because the regression method is readily extended to nonlinear functions, incorrectness of the linear relation assumption need not be damaging. In order to examine the possibility that a nonlinear function might provide a superior fit to the data, the regression method was used to test several forms of nonlinear functions, especially the quadratic (U-shape) function that was observed in a subset of the data reported by Greenwald et al. (1995). In the present data, nonlinear effects were generally not apparent. In any case, analyses that tested the fit of nonlinear functions did not alter the statistical significance of intercept effects, nor did they alter magnitudes of intercept effects more than slightly.

No Measurement Error in the Predictor Variable. The assumption of error-free measurement of the predictor is very clearly invalid. Evidence concerning reliability of the predictor was obtained in Experiment 1 by computing the direct measure separately for trials on which the stimuli were LEFT or RIGH, and ones on which the stimuli were other four-letter words. Although the reliability correlation between these two measures was high ($r = .907$), it was clearly less than perfect. Evidence for measurement error in the predictor is important because there are circumstances under which such error can cause a statistically significant intercept to materialize when the true underlying regression function passes through the origin. This possibility is illustrated in Fig. 5.7. However, a spurious intercept effect of the type illustrated in Fig. 5.7 can appear only when the true regression function has a markedly positive slope, and the mean of the predictor is substantially above zero. For those regression analyses yielding significant intercept effects, shown in Figs. 5.3 and 5.4, it can be seen that regression slopes were essentially flat, and the means of the predictor variables were not much above zero. As a result, measurement error in the predictor is not plausibly responsible for the statistically significant intercept effects of Experiments 1 and 2.

Direct Measures Are at Least as Sensitive to Conscious Stimulus Effects as Indirect Measures. As others (especially, Holender, 1986; and Reingold & Merikle, 1988) have noted, the translation of data patterns involving indirect and direct effects into assertions about unconscious cognition critically depends on one's assumptions about how conscious and unconscious cognition map onto direct and indirect measures. Figure 5.8 (left panel) shows the assumptions made by Holender (1986; as analyzed by Reingold & Merikle, 1988) in arriving at a skeptical conclusion about the existence of unconscious cognition. Holender assumed that in order to draw conclusions direct measures must be sensitive to all conscious effects of task stimuli, and must reflect only conscious effects. With these (exhaustiveness and exclusiveness) assumptions, the demonstration of an indirect effect in the absence of a direct effect provides unambiguous evidence for unconscious effects, as well as indicating that unconscious effects are dissociated from conscious cognition. The relevance of intercept effect findings such as those in the present experiments to conclusions about dissociation has been discussed in detail by Greenwald et al. (1995).

Although Holender's exclusiveness and exhaustiveness assumptions sim-

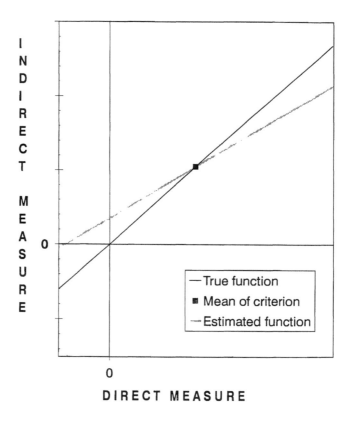

FIG. 5.7.   Possible effect of unreliability of the regression predictor variable on the estimate of the regression intercept (intersection with Y axis). The predictor's unreliability causes the regression slope, but not the mean on the criterion variable—through which the function passes—to be slightly underestimated. This results in an overestimate of the intercept to the extent that the predictor mean is greater than zero *and* the regression slope is positive. An overestimate of the intercept should *not* have occurred in the present research, because observed regression slopes were approximately flat (see plotted slopes in Figs. 5.3 and 5.4).

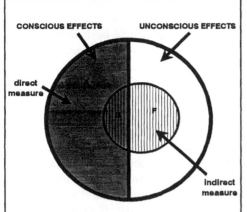

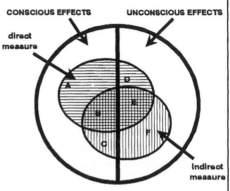

STRONG ASSUMPTIONS FOR DIRECT
MEASURE (Holender, 1986)

Exclusiveness: Direct measure is influenced
only by conscious stimulus effects.

Exhaustiveness: Direct measure is sensitive to
all conscious stimulus effects.

Therefore: Finding indirect effect ((B + F) > 0)
in the absence of direct effect ((A + B) = 0)
indicates unconscious cognition (F > 0)

MINIMAL ASSUMPTIONS FOR DIRECT
MEASURE (Reingold & Merikle, 1988)

Inclusiveness: Direct measure can include both
conscious and unconscious stimulus effects

Relative sensitivity: Conscious contribution to
direct measure is at least as great as
conscious contribution to indirect measure
(A ≥ C)

Therefore: Finding indirect effect (B + C + E + F)
greater than direct effect (A + B + D + E)
indicates unconscious cognition (F > 0)

FIG. 5.8. Alternative assumptions about use of direct and indirect measures
as indicators of conscious and unconscious cognition. Both panels use an inclusiveness assumption for the indirect measure—that is, the indirect measure
can be sensitive to both conscious and unconscious stimulus effects. For the
right panel, not only the indirect-greater-than-direct effect (as diagramed) but
also the indirect-without-direct-effect finding yields the conclusion of unconscious cognition (F > 0), as follows: When the direct effect (A + B + D + E) is
zero, then A, B, and E must all be zero; C = 0 then follows from the relative sensitivity assumption (A ≥ C); and F > 0 then follows from the indirect effect (B +
C + E + F) being greater than zero, given that B, C, and E have been demonstrated to equal 0. (Areas represent magnitudes of stimulus effects on direct
and indirect measures, and cannot be negative.)

plified the problem of empirically defining unconscious cognition, those assumptions also sparked controversy. The controversy was well articulated by Reingold and Merikle (1988), who considered implausible both that unconscious stimulus effects would have no influence on direct measures, and that direct measures would generally be sensitive to all conscious stimulus effects. Accordingly, Reingold and Merikle suggested that subsequent analyses of unconscious cognition be based on the more cautious assumption that direct measures (like indirect measures) might include both conscious and unconscious contributions (see also Jacoby, Lindsay, & Toth, 1992) and, further, that direct measures need not be sensitive to all conscious stimulus effects.

Replacing Holender's exclusiveness assumption for the direct measure with the inclusiveness assumption shown in Fig. 5.8 (right panel) had the undesired side effect of making it impossible—in the absence of any other changes of assumptions—to interpret any patterns of direct and indirect effects in terms of unconscious cognition. Reingold and Merikle responded to this difficulty by introducing what they described as a minimal assumption to enable conclusions about unconscious cognition. Their additional assumption was that direct measures were at least as sensitive to conscious stimulus effects as were comparable indirect measures. In the right panel of Fig. 5.8, this *relative sensitivity* assumption is interpreted as assuming that the region labeled A is at least as large as that labeled C.

Importantly, Reingold and Merikle's (1988) analysis did not logically exclude interpretation of the indirect-without-direct-effect data pattern, such as the one evidenced by a significantly positive intercept effect using the present regression method. Examination of their assumptions, as shown in the right panel of Fig. 5.8, reveals that the indirect-without-direct-effect data pattern yields a conclusion of demonstrating unconscious cognition (i.e., region F > 0; see Fig. 5.8's caption). Based on this analysis, the significant intercept effects found in Experiments 1 and 2 provide evidence for existence of unconscious cognition.

In the present research, two strategies addressed the possibility that the relative sensitivity assumption was not met (i.e., the possibility that indirect measures were relatively more sensitive to conscious stimulus effects than were direct measures). First, the experiments provided two other direct measures, a position-discrimination measure based on discriminating the left versus right spatial position of four-letter word stimuli (in both experiments) and a lexical decision task based on judging whether four-letter stimulus words were displayed forward or backward (in Experiment 2). When the detection measure was replaced by either of these other direct measures in the regression analysis, the same positive intercept effects were obtained. The second strategy followed from the use of a detection task to provide the major direct measure. Because of the nature of the detection task, any consciously perceived stimulus attribute, whatever the stimulus, should have led subjects to respond by indicating that a stimulus was present. Thus, to the extent that subjects consciously perceived any information that might have produced above-zero scores on the indirect measure, their scores on the direct measure should also have been (at least as much) above zero.

Dissociation Interpretation of Regression Functions

Debates in the recent literature concerning the nature of unconscious cognition encompass three competing views: unconscious cognition does not exist (nonexistence); unconscious cognition exists, but only in association with conscious cognition (association); and unconscious cognition exists and is independent of conscious cognition (dissociation). As discussed earlier, the intercept effects of the present study, in conjunction with the relative sensitivity assumption, reject the nonexistence view. However, decisive interpretation of the intercept effect findings as supporting the strong conclusion of dissociation (rather than the weaker conclusion of association) requires the exhaustiveness assumption employed by Holender (see left panel of Fig. 5.8) —the assumption that the direct measure is sensitive to all conscious stimulus effects. Although the exhaustiveness assumption is clearly not generally valid, it may nevertheless be valid for the direct measure provided by the detection task, which should have been sensitive to the effects of any consciously perceived stimulus attributes. To the extent that the exhaustiveness assumption holds for the detection task, the present findings can be seen as supporting the dissociation view. (See Greenwald et al., 1995, for detailed consideration of the plausibility of the exhaustiveness assumption for the direct measures provided by detection and position discrimination measures in the present series of experiments.)

CONCLUSION

Using a data analysis strategy based on regression of indirect on direct measures, the present research found evidence for an indirect-without-direct-effect pattern in the form of positive intercept effects. Validity of the resulting conclusion that indirect effects can occur in the absence of direct effects depended on appraisal of four assumptions underlying the regression analysis method. A detailed analysis of the possibilities for failure of each assumption provided no basis for revising the conclusion. The present data, further, support the conclusion that unconscious effects can occur in the absence of conscious effects (dissociation), if—as is plausible, but not definitively established —the direct measures used in the present research are accepted as providing exhaustive measures of conscious stimulus effects. The dissociation interpretation constitutes an affirmative answer to the title question; it amounts to the conclusion that subliminal stimuli can enter the mind unnoticed.

Theoretically, evidence for dissociation is important because it is incompatible with the large class of models that assume sequential processing of information through an ordered series of increasingly complex levels (or stages) of analysis. Such models generally assume that some processing occurs at a stage prior to focal attention. Unconscious cognition has often been identified with this preattentive processing stage. In these information-processing models, any stimulus that is processed preattentively should be capable of achieving focal attention (i.e., conscious awareness) when it is goal relevant, as it was in the detection task of the present experiments. Interpreted in terms of such models, the present findings indicated that stimuli that must have been preattentively processed (as indicated by their producing indirect effects) were not focally attended (as indicated by their failing to produce effects on the direct-effect measure of detection). Such findings are inconsistent with models that treat outputs of preattentive processing as being routinely available for focal attention.

Practically, evidence for dissociation is important because it implies the possibility of cognitive influences that, because they are produced by undetectable stimuli, cannot be consciously defended against. Subliminal techniques of the sort used now in laboratory research could possibly be developed for use in mass media to produce significant influences on behavior. Importantly, such influences have not yet been compellingly demonstrated in research. Nevertheless, in a recent ruling, a Nevada state court suggested that evidence for subliminal influence could justify an exclusion of subliminal messages from the constitutional protection of free speech afforded by the First Amendment of the United States Constitution (*Vance v. Judas Priest*, 1990).

Because of the theoretical and practical importance of the indirect-without-direct-effect data pattern, and the long history of skeptical regard for claims to have obtained that finding, it is unlikely that those who have been skeptical about previous claims to have found this long-sought pattern will be thoroughly persuaded by the present findings. At the same time, the present methods avoided problems for which previous claims have been criticized and, therefore—together with the similar findings of Greenwald et al. (1995)—provide substantially stronger support for the indirect-without-direct-effect pattern than has been available previously. The strong claim for the present findings (that they constitute evidence that subliminal activation is producible by stimuli that entirely escape conscious detection) no doubt invites, even provokes, further skeptical reaction. This is as it should be. Only by continued findings of data patterns such as those in the present research will the conclusion survive skeptical criticism and become strongly established.

ACKNOWLEDGMENTS

Research reported in this chapter was partially supported by grants from National Institute of Mental Health (MH-41328) and National Science Foundation (DBC-9205890). A report based on this research was presented at the meetings of the Psychonomic Society in Washington, DC, November 5, 1993. Portions of the introduction section of this chapter are based on material presented in Greenwald, Klinger, and Schuh (1995), to which the topic of this report is closely related.

REFERENCES

Balota. D. A. (1983). Automatic semantic activation and episodic memory encoding. *Journal of Verbal Learning and Verbal Behavior, 22,* 88–104.

Bornstein. R. F. (1992). Subliminal mere exposure effects. In R. F Bornstein & T S. Pittman (Eds.), *Perception without awareness: Cognitive, clinical, and social perspectives* (pp. 191–210). New York: Guilford.

Cheesman. J., & Merikle, P. M. (1986). Distinguishing conscious from unconscious perceptual processes. *Canadian Journal of Psychology, 40,* 343–367.

Cohen. J. (1977). *Statistical power analysis for the behavioral sciences* (rev. ed.). New York: Academic Press.

Dagenbach. D., Carr. T. H., & Wilhelmsen. A. (1989). Task-induced strategies and near-threshold priming: Conscious influences on unconscious perception. *Journal of Memory and Language, 28,* 412–443.

Fowler. C. A., Wolford. G., Slade. R., & Tassinary. L. (1981). Lexical access with and without awareness. *Journal of Experimental Psychology: General, 110,* 341–362.

Green. D. M., & Swets. J. A. (1966). *Signal detection theory and psychophysics.* New York: Wiley.

Greenwald. A. G. (1992). New Look 3: Unconscious cognition reclaimed. *American Psychologist, 47,* 766–779.

Greenwald. A. G., & Klinger. M. R. (1990). Visual masking and unconscious processing: Differences between backward and simultaneous masking? *Memory and Cognition, 18,* 430–435.

Greenwald. A. G., Klinger. M. R., & Liu. T. J. (1989). Unconscious processing of dichoptically masked words. *Memory and Cognition, 17,* 35–47.

Greenwald. A. G., Klinger. M. R., & Schuh. E. S. (1995). Activation by marginally perceptible ("subliminal") stimuli: Dissociation of unconscious from conscious cognition. *Journal of Experimental Psychology: General, 124,* 22–42.

Groeger. J. A. (1988). Qualitatively different effects of undetected and unidentified auditory primes. *Quarterly Journal of Experimental Psychology, 40A,* 323–329.

Hardaway. R. A. (1990). Subliminally activated symbiotic fantasies: Facts and artifacts. *Psychological Bulletin, 107,* 177–195.

Holender. D. (1986). Semantic activation without conscious identification in dichotic listening, parafoveal vision, and visual masking: A survey and appraisal. *Behavioral and Brain Sciences, 9,* 1–23.

Jacoby. L. L., Lindsay. D. S., & Toth. J. P. (1992). Unconscious influences revealed: Attention, awareness, and control. *American Psychologist, 47,* 802–809.

Marcel, A. J. (1983). Conscious and unconscious perception: Experiments on visual masking and word recognition. *Cognitive Psychology, 15,* 197–237.

Merikle, P. M. (1982). Unconscious processing revisited. *Perception and Psychophysics, 31,* 298–301.

Purcell, D. G., Stewart, A. L., & Stanovich, K. E. (1983). Another look at semantic priming without awareness. *Perception and Psychophysics, 34,* 65–71.

Reingold, E. M., & Merikle, P. M. (1988). Using direct and indirect measures to study perception without awareness. *Perception and Psychophysics. 44,* 563–575.

Rosenthal, R. (1993). Cumulating evidence. In G. Keren & C. Lewis (Eds.), *A handbook for data analysis in the behavioral sciences: Methodological issues* (pp. 519–559). Hillsdale, NJ: Lawrence Erlbaum Associates.

Stroop, J. R. (1935). Studies of interference in serial verbal reactions. *Journal of Experimental Psychology, 18,* 643–662.

Vance v. Judas Priest, CBS, Inc., et al. (1990). Nos. 86-5844, 86-3939, 1990 WL 130920 (Nev. Dist. Ct. Aug. 24, 1990).

CHAPTER

Measuring
Unconscious Influences

Philip M. Merikle
Steve Joordens
University of Waterloo, Ontario

Experimental studies of perception without awareness can be traced to the very beginnings of psychology in North America (see Merikle, 1992). In fact, Kihlstrom, Barnhardt, and Tataryn (1992) suggested that the first published study from the first psychological laboratory in North America at Johns Hopkins University was Peirce and Jastrow's (1884) demonstration that it is possible to perceive small differences in pressure on the skin without any subjective awareness of different sensations. However, despite its relatively long history, research on perception without awareness has been plagued by continual controversy.

Much of the controversy over perception without awareness stems directly from the logic underlying the dissociation paradigm, which is the predominate approach used in experimental studies. With the dissociation paradigm, perception without awareness is demonstrated whenever it can be shown that stimulus information that is completely unavailable to conscious awareness is nevertheless perceived. Thus, before it is possible to demonstrate unconscious perception, it is necessary to find a satisfactory measure of conscious perception and to demonstrate that perception occurs even when this measure exhibits null sensitivity. If the selected measure of awareness cannot be shown to exhibit null sensitivity, then it is always possible that some relevant conscious information was perceived.

In this chapter, we first review the problems that arise whenever one attempts to demonstrate perception without awareness based on a dissociation between a measure of awareness and a second measure of perception assumed to reflect unconsciously perceived information. The conclusion reached on the basis of this review is that traditional approaches based on the dissociation paradigm are probably doomed to continual controversy because of the inherent difficulties in demonstrating that any behavioral measure actually exhibits null sensitivity to all relevant conscious information. For this reason, we suggest an alternative approach. It is based on a behavioral measure that is assumed to reflect the magnitude of unconscious influences relative to conscious influences. This measure predicts qualitative differences in performance such that stimuli perceived without awareness lead to different consequences than stimuli perceived with awareness. In addition, quantitative predictions can also be made. This alternative approach has considerable potential because it provides a methodology for studying perception without awareness that does not require a convincing demonstration of a null effect.

THE TRADITIONAL APPROACH: MEASURING THE ABSENCE OF CONSCIOUS INFLUENCES

In the vast majority of studies directed at demonstrating perception without awareness, the goal has been to demonstrate a dissociation between two different measures of perception. One measure (C) is assumed to indicate whether any relevant stimulus information is consciously perceived, while the second measure ($C + U$) is assumed to be sensitive to both consciously and unconsciously perceived information. To demonstrate perception without awareness, it must be shown that stimulus information that is completely unavailable to conscious awareness is nevertheless perceived. Thus, using this approach, perception without awareness is demonstrated whenever C shows no sensitivity and $C + U$ shows some sensitivity to the same perceptual information.

Although the logic underlying the dissociation paradigm is straightforward, it has proven difficult to design experiments that provide compelling, non-controversial evidence for perception without awareness. This difficulty stems from both a critical assumption that must be made concerning the measure C and the methodological requirement that this measure must exhibit null sensitivity. As noted previously (e.g., Merikle, 1992; Merikle & Reingold, 1992; Reingold & Merikle, 1988, 1990), the measure C must be assumed to be sensitive to all relevant conscious experience. If this assumption cannot be justified, then any dissociation between measures may simply indicate that C was insensitive to a least some potentially relevant conscious information. In other words, any sensitivity exhibited by $C + U$ in the absence of sensitivity for C may simply reflect consciously perceived information that C failed to detect. Thus, before the dissociation paradigm can be used successfully to study perception without awareness, it is necessary to find an exhaustive measure of all relevant, conscious experience and to demonstrate that this measure exhibits null sensitivity.

In efforts to find adequate exhaustive measures of conscious experience, two general classes of measures have been used: *subjective* and *objective*. Subjective measures are characterized by the fact that conscious awareness is indexed by subjects' self-reports of their perceptual experiences. In many instances, these self-reports indicate the subjects' confidence that perceived information was useful for guiding their decisions (e.g., Cheesman & Merikle, 1986; Peirce & Jastrow, 1884). When these subjective reports show that subjects do not believe any useful stimulus information was perceived, it is assumed that subjects did not have any relevant conscious experiences. With objective measures, on the other hand, conscious experiences are assumed to be indexed by some measure of a subject's discriminative capabilities. Examples of objective measures used in recent studies are forced-choice, presence–absence decisions (e.g., Kemp-Wheeler & Hill, 1988) and forced-choice decisions among a small set of stimulus alternatives (e.g., Dagenbach, Carr, & Wilhelmsen, 1989). When objective measures are used to index conscious awareness, null sensitivity (e.g., $d' = 0$) is assumed to reflect the complete absence of relevant, consciously perceived information.

The problem with all assumed measures of conscious awareness—subjective and objective—is that it is impossible to demonstrate in a completely convincing fashion that null sensitivity necessarily indicates no relevant information was consciously perceived. Many of the continuing debates can be characterized as Investigator A claiming to demonstrate perception without awareness because Measure X exhibits null sensitivity and Investigator B claiming either (a) that the methodology was inadequate to establish null sensitivity, or (b) that Measure X was not a satisfactory exhaustive measure of relevant conscious information. An example of the debates over methodology is Marcel's (1983) claim that he demonstrated perception in the absence of stimulus detection, and Merikle's (1982) counterclaim that the apparent null stimulus detection probably reflected the small number of trials Marcel used to establish a detection threshold. An example of the debates over the adequacy of measures of awareness is Cheesman and Merikle's (1986) claim that they demonstrated perception in the absence of relevant subjective experiences, and Holender's (1986) counterclaim that Cheesman and Merikle's measure of subjective awareness did not rule out the possibility that partial stimulus information was consciously perceived and used to guide decisions.

All of these claims and counterclaims have made investigators very cautious when interpreting their data. It is probably for this reason that the phrase "near-threshold priming" is sometimes used (e.g., Dagenbach et al., 1989). But "near-threshold" is not satisfactory because it does not allow one to meet the critical assumption underlying the dissociation paradigm. If the assumed exhaustive measure of relevant conscious experience is not shown to exhibit null sensitivity in a convincing manner, then there is no basis for concluding that an observed dissociation between the measures C and $C + U$ necessarily reflects an unconscious influence.

What all of this controversy over the results of experiments based on the dissociation paradigm illustrates is that null effects are almost always open to alternative interpretations. In studies of perception without awareness, these alternative interpretations usually call into question either the methodology used to establish null sensitivity or the assumption that the chosen measure of C is an exhaustive measure of all relevant conscious experience. Although it may be possible to design better experiments (cf. Miller, 1991), given how difficult it is to demonstrate null sensitivity for any measure of perception (see Macmillan, 1986), it may not be possible to ever obtain compelling, noncontroversial evidence for perception without awareness, as long as the experimental logic requires a convincing demonstration that an exhaustive measure of awareness exhibits null sensitivity.

AN ALTERNATIVE APPROACH: MEASURING THE RELATIVE MAGNITUDE OF UNCONSCIOUS INFLUENCES

As an alternative to demonstrating dissociations between a measure of awareness and another measure of perception, a better strategy for studying perception without awareness may be to find a measure of the relative magnitude of unconscious influences and to use this measure to predict performance. At first consideration, it may seem that finding a satisfactory measure of unconscious influences may be as difficult a task as finding a satisfactory measure of conscious perceptual experience. However, this is not necessarily the case. The reason it is so difficult to find a satisfactory measure of conscious perceptual experience is that the logic of the dissociation paradigm requires an exhaustive measure of all relevant conscious experiences. The logical requirements are not nearly as stringent if one attempts to find a task capable of reflecting the relative magnitude of unconscious influences. Given that most measures of perception are probably affected by both conscious and unconscious influences (cf. Jacoby, 1991; Reingold & Merikle, 1988, 1990), any task that is differentially affected by unconscious as opposed to conscious influences meets the requirements of this approach. Thus, with this approach, there is no need to assume that any task exhaustively measures either conscious or unconscious influences. Rather, all that needs to be assumed is that the task is sensitive to the conscious and unconscious influences of the same subset of stimulus characteristics.

One measure that appears to have considerable promise as a measure of the relative magnitude of unconscious influences is the exclusion task developed by Jacoby and his colleagues (e.g., Debner & Jacoby, 1994; Jacoby, 1991; Jacoby, Toth, & Yonelinas, 1993). The distinguishing characteristic of the exclusion task is that subjects are instructed not to use particular responses when performing the task. As an example of an exclusion task, consider the task used recently by Debner and Jacoby (1994) to study perception without awareness. In their studies, a single, five-letter word (e.g., table) was presented and masked on each trial. Immediately following the presentation of each word, a three-letter word stem (e.g., tab__) was presented and the subjects were instructed to complete the stem with any word that came to mind except the word that had just been presented. For present purposes, the important result is that the subjects were unable to exclude the immediately preceding word on a significant proportion of the trials when the stimulus onset asynchrony (SOA) between the word and the mask was relatively short (i.e., 50 ms).

The fact that the immediately preceding words were used as responses despite the instructions not to use these words suggests that the masked words were perceived without awareness, at least on some proportion of the trials. This interpretation follows from two critical assumptions. First, conscious perception of the immediately preceding word leads subjects not to use it to complete the stem. This assumption is consistent with the belief that an important characteristic of conscious awareness is that it enables one to act on the world and to produce effects on the world (cf. Searle, 1992). Second, responses are controlled by conscious influences whenever either conscious influences alone or both conscious and unconscious influences are present. Unless this assumption is made, there is no reason to expect the exclusion instructions to cause subjects not to complete the stems with consciously perceived words.

Whereas Jacoby and his colleagues have used exclusion tasks in conjunction with inclusion tasks and an underlying model of how conscious and unconscious processes are related to study the separate contributions of conscious and unconscious process to task performance (e.g., Jacoby et al., 1993), our goals were much more modest. We simply wanted to establish if an exclusion task can be used to estimate the relative magnitude of the unconscious influences. The logic underlying using the exclusion task in this manner is quite straightforward; the more frequently subjects fail to exclude words they are instructed not to use, the greater are the unconscious influences relative to the conscious influences.

Defining Critical Stimulus Durations. The first goal was to establish whether the exclusion task can be used to define the critical stimulus duration at which unconscious influences are maximal. The procedure was as follows: On each trial, a single, four-to-six-letter target word was presented. To control stimulus duration, the presentation of each target word was both preceded and followed by a row of 10 random letters. Immediately following the offset of the row of random letters that followed each target word, the first three letters of the target word were displayed. The task for the subjects was simply to complete the word stem with any word they could think of except the word just presented. The SOA between the onset of a target word and the onset of the letter row that followed each target word was 0, 29, 43, 57, 71, or 214 ms. These six SOAs were randomly intermixed across a series of 180 trials so that a total of 30 words was presented for each stimulus duration.

Figure 6.1 shows the mean proportion of responses across 16 subjects that matched the immediately preceding target word at each stimulus duration. There are three important aspects of these results. First, baseline performance, as indicated by the mean proportion of stems completed with target words at the 0 ms duration, was .14. In other words, the word stems were completed with words we identified as target words on 14% of the trials even when no word was presented. Second, at the 214 ms stimulus duration, the subjects were able to successfully exclude target words from their responses. The mean proportion of matching responses at this duration was .07, which was significantly lower than baseline performance. This finding demonstrates that the subjects were able to follow the task instructions when they were given a suf-

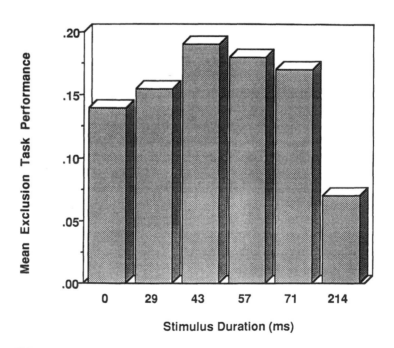

FIG. 6.1.   Mean proportions of responses that matched a target word at each stimulus duration.

ficiently long exposure to a target word before the onset of the mask. However, as shown in Fig. 6.1, at the four intermediate stimulus durations (29, 43, 57, and 71 ms) the mean proportion of matching responses was higher than baseline. More specifically, the mean proportions were .16, .19, .18, and .17 at the 29 ms, 43 ms, 57 ms, and 71 ms durations, respectively. Thus, the third and most critical aspect of the data shown in Fig. 6.1 is that the subjects were unable to exclude the immediately preceding target words from their responses at the intermediate stimulus durations. Statistical analysis of the group data shown in Fig. 6.1 indicated that mean exclusion task performance at the 43 ms and 57 ms stimulus durations was significantly higher than baseline.

Given the logic underlying the experiment, the data suggest that on some proportion of the trials, the subjects perceived words without any subjective awareness of these words. If this interpretation is correct, then it should be possible to use the exclusion task to determine for individual subjects the stimulus duration at which perception without the subjective experience of perception occurs most frequently. The real advantage of the exclusion task relative to other methods that have been used to identify critical stimulus durations is that unconscious perception is inferred on the basis of a positive deviation from baseline performance rather than on the basis of a null difference. By using a positive effect rather than a null effect to define the conditions under which unconscious perception is maximal relative to conscious perception, the problems associated with attempting to demonstrate null sensitivity are avoided.

Qualitative Differences in Performance.   Although the exclusion task offers a potential solution to the problems associated with using null findings to establish the conditions under which unconscious processes may dominate conscious processes, the following question needs to be addressed: How can it be established that the exclusion task is in fact a valid indicator of unconscious processes? Clearly, converging evidence is needed to support the conclusion that the inability to exclude the target words as completions for the word stems reflects an unconscious influence and, conversely, that the ability to exclude the target words reflects a conscious influence.

Our search for converging evidence followed from previous suggestions that the best way to validate a potential measure of awareness is to demonstrate that the selected measure predicts a qualitative difference in performance across conscious and unconscious perceptual states (e.g., Joordens & Merikle, 1992; Merikle, 1992; Merikle & Cheesman, 1986; Merikle & Reingold, 1990). The importance of demonstrating qualitative differences cannot be overemphasized. Not only can qualitative differences validate a task as a measure of unconscious perception by constraining alternative interpretations, but qualitative differences also serve to move the study of unconscious influences forward by showing how conscious and unconscious processes differ. In fact, it can even be argued that the distinction between conscious and unconscious processes is of questionable value if conscious and unconscious processes do not lead to qualitatively different consequences (e.g., Merikle, 1992; Reingold & Merikle, 1990).

To establish whether or not the exclusion task can predict a qualitative difference in performance, we decided to model an experiment on a study reported previously by Merikle and Cheesman (1987). The experiment was based on a variant of the Stroop task in which two color words—RED or GREEN—are used to prime responses to two target colors (also red or green). An interesting characteristic of this two-color variant of the Stroop task is that subjects name the colors faster on incongruent (e.g., GREEN-red) than on congruent (e.g., RED-red) trials whenever incongruent trials are more probable than congruent trials (cf. Logan, Zbrodoff, & Williamson, 1984). In other words, the typical Stroop effect is reversed when the proportion of incongruent trials significantly exceeds the proportion of congruent trials (e.g., 75% incongruent trials and 25% congruent trials). The straightforward explanation of this reversed Stroop effect is that subjects capitalize on the predictive information provided by the primes. Given that there are only two possible colors, the intelligent strategy is to expect that the target color on each trial will be the color not named by the word. Such a strategy would facilitate performance on the incongruent trials and slow performance on the congruent trials.

For present purposes, the important aspect of Merikle and Cheesman's (1987) results is that subjects only adopted the predictive strategy in the two-color variant of the Stroop task when the primes were consciously perceived. This conclusion was supported by comparing each subject's performance when the primes were presented below an individually determined threshold for subjective awareness (cf. Cheesman & Merikle, 1986) to performance when the primes were clearly visible. Primes presented below the subjective threshold led to a typical Stroop effect in that responses were faster on congruent than on incongruent trials. However, when the primes were presented above the subjective threshold, the typical Stroop effect was reversed in that responses were faster on incongruent than on congruent trials. This crossover in performance led Merikle and Cheesman to conclude that perception without awareness leads to qualitatively different consequences than perception with awareness.

If the exclusion task does provide a measure of when subjects perceive without subjective awareness, then it should be possible to use the exclusion task to predict a qualitative difference in performance on the Stroop task that closely parallels the one found by Merikle and Cheesman when they presented primes above and below a subjectively defined threshold. To evaluate whether such a prediction can in fact be based on exclusion task performance, we tested 16 subjects in a two-part experiment. The first part of the experiment was very similar to our initial experiment in which we showed that the number of responses matching the target words was higher than baseline at intermediate stimulus durations and lower than baseline at the longest stimulus duration (see Fig. 6.1). For each subject, we found the stimulus duration associated with the highest proportion of stems completed with the target words. Across the 16 subjects, this critical stimulus duration ranged from 29 ms to 71 ms, and the median critical duration was 57 ms. Once the critical duration was determined, each subject was tested immediately in the second part of the experiment which consisted of a Stroop task that was very similar to the one used by Merikle and Cheesman (1987). Basically, there were two blocks of 256 trials. In one block of trials, the color-word primes were presented for the critical duration, whereas in the other block of trials, the primes were presented for 214 ms. In both blocks of trials, the primes were masked by a row of seven ampersands that changed color to either red or green 300 ms after the onset of each prime. To encourage subjects to adopt a predictive strategy, the prime-target pairings were incongruent on 75% of the trials in each block. We expected the subjects to exhibit the predictive strategy when the primes were presented for 214 ms, that is, faster RTs on incongruent than congruent trials. However, if unconscious influences are greater than conscious influences at the critical duration, then performance at this duration should reveal a typical Stroop effect, that is, faster RTs on congruent than on incongruent trials.

Figure 6.2 shows the mean RTs to respond to the colors in each condition of the experiment. In general, the pattern of results is very similar to the one found by Merikle and Cheesman (1987). At the critical duration, a typical Stroop effect was found in that subjects responded faster on congruent than incongruent trials. However, when the primes were presented for 214 ms, the subjects responded faster on incongruent than congruent trials. This crossover in performance provides evidence that the consequences of perceiving the primes at the critical duration and the 214 ms duration were qualitatively different. As such, these data provide converging evidence that the critical duration as defined by exclusion-task performance reflects a condition in which unconscious influences were greater than conscious influences.

Quantitative Differences in Performance.   In the preceding experiment, there were two measures of unconscious influences: the exclusion task and the Stroop effect. Both measures were assumed to reflect the relative magnitude of the unconscious influences of the color-word primes. More specifically, it was assumed that both the number of failures to exclude target words as completions for the stems and the size of the Stroop effect as indicated by the difference in RT between incongruent minus congruent trials measured the relative magnitude of the unconscious influence of the primes. If both of these measures reflect similar unconscious influences, then they should be

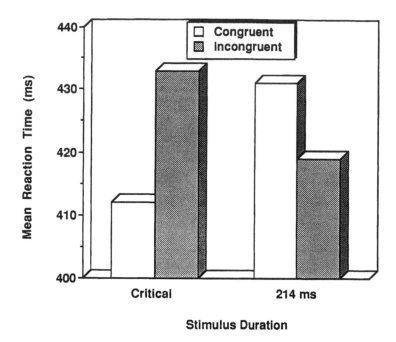

FIG. 6.2.   Mean reaction times to the colors.

positively correlated. In other words, it should be possible to predict the size of the Stroop effect on the basis of performance on the exclusion task.

To establish whether exclusion-task performance and the Stroop effect are in fact correlated, we compared performance on each of these measures. Figure 6.3 shows a scatterplot of the performance on these measures at the critical duration by the 16 subjects in the preceding study. As suggested by Fig. 6.3, there was a modest positive correlation between the measures, $r(14) = .44$, $p < .10$, indicating that higher performance on the exclusion task (i.e., more failures to exclude) was associated with larger Stroop effects. Thus, these data provide suggestive evidence that both measures do in fact index overlapping unconscious influences.

Given these suggestive findings, we decided to conduct a larger scale study of the possible quantitative relation between exclusion-task performance and the magnitude of the Stroop effect. The exclusion task and the Stroop task with 75% incongruent trials were administered in the same way as in the preceding study. However, rather than attempting to define a critical stimulus duration for each subject, we simply tested different groups of six subjects each on both tasks at one of six different stimulus durations, that is, 29 ms, 43 ms, 57 ms, 71 ms, 85 ms, and 100 ms. For all subjects, the exclusion task was administered prior to the Stroop task. The results are shown in Fig. 6.4. Once again, performance across these tasks was positively correlated, $r\,(34) = .39$, $p <$ .02. Thus, these data show that it is possible to make a quantitative prediction regarding the size of the observed Stroop effect on the basis of performance on the exclusion task. These findings suggest that both measures do in fact index the relative magnitude of the influence of primes perceived without awareness.

CONCLUDING COMMENTS

On the basis of the findings described here, it appears that the exclusion task provides a satisfactory measure of the relative magnitude of unconscious influences. The results are consistent with the assumption that unconscious influences are greater than conscious influences following short exposure durations but that conscious influences are greater than unconscious influences following long exposure durations. Thus, the subjects were able to follow the exclusion task instructions and not use the target words to complete the stems

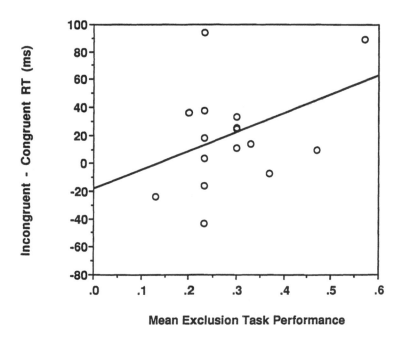

FIG. 6.3.   Scatterplot of the correlation between exclusion task performance and the difference in RT between incongruent and congruent trials at the critical stimulus duration.

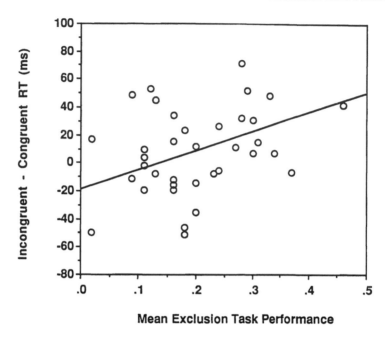

FIG. 6.4.   Scatterplot of the correlation between exclusion task performance
and the difference in RT between incongruent and congruent trials as measured at six different stimulus durations.

when the words were presented for 214 ms. However, at shorter stimulus durations, the subjects were unable to follow the exclusion task instructions and
the target words were used to complete the stems despite instructions to the
contrary. A straightforward interpretation of these different patterns of performance at long and short stimulus durations is that they reflect how often
the target words were either consciously or unconsciously perceived. At the
214-ms stimulus duration, most target words were consciously perceived, and
the resulting awareness of the target words allowed the subjects to censor or
exclude the target words as possible completions for the stems. On the other
hand, at shorter stimulus durations, unconscious perception was probably
much more frequent than conscious perception. For this reason, many target
words were used to complete the word stems as the subjects were unable to exclude target words they never consciously perceived.

   This interpretation of performance on the exclusion task is strengthened
considerably by the converging evidence found when the exclusion task was
used to predict performance on the Stroop priming task. When primes were
presented for 214 ms, subjects were able to use the primes in a strategic manner to anticipate the color of the targets. However, when primes were presented for the critical stimulus duration, the primes were perceived but the

subjects could not use the primes in a strategic manner. This qualitative difference in performance across long and short stimulus durations is entirely consistent with earlier findings (e.g., Cheesman & Merikle, 1986; Merikle & Cheesman, 1987) showing that primes presented either above or below a subjectively defined threshold for awareness are perceived, but only consciously perceived primes presented above the awareness threshold can be used in a strategic manner.

Given that the exclusion task measures the relative magnitude of unconscious influences, it is important to consider how conscious and unconscious influences affect performance on this task. Following the lead of Jacoby and his colleagues (e.g., Debner & Jacoby, 1994; Jacoby, 1991; Jacoby et al., 1993), a critical assumption underlying our interpretation is that performance on the exclusion task is always controlled by conscious influences whenever either conscious influences alone or both conscious and unconscious influences are present. In other words, performance on any given trial is always controlled exclusively by conscious influences if such influences are present. Thus, performance is never affected simultaneously by both conscious and unconscious influences, and unconscious influences only affect performance when there are no conscious influences. However, even though performance on any given trial is affected exclusively by either conscious or unconscious influences, when performance is averaged across trials, the average performance gives an indication of the relative dominance of conscious or unconscious influences. In this way, the exclusion task provides both a discrete measure of unconscious influences on each trial and a continuous measure of the relative magnitude of unconscious influences across trials.

The exclusion task implies a considerably different concept of how different stimulus contexts are related to conscious and unconscious perception than is implied in more traditional studies directed at establishing thresholds for awareness. The threshold concept implies that it is possible to define stimulus contexts that either reflect conscious influences exclusively (i.e., above the awareness threshold) or reflect unconscious influences exclusively (i.e., below the awareness threshold). Thus, whenever an attempt is made to establish a threshold for awareness, the objective is to define stimulus contexts that either always lead to unconscious perception or always lead to conscious perception. With the benefit of considerable hindsight, it is now clear that it is probably impossible to ever find stimulus contexts that always lead to unconscious perception or even always lead to conscious perception (cf. Jacoby, 1991; Reingold & Merikle, 1988, 1990). The best that can probably be done is to find stimulus contexts that vary as to whether conscious or unconscious influences are dominant.

One of the more important advantages of using the exclusion task to establish stimulus contexts in which unconscious influences are greater than conscious influences is that such contexts are defined in terms of deviations from baseline performance. Thus, with the exclusion task, unconscious influences are defined in terms of statistically significant effects. In contrast, with more traditional approaches based awareness thresholds, unconscious influences are defined in terms of null or statistically nonsignificant effects. By not having to establish a null effect before it is possible to demonstrate unconscious influences, the exclusion task offers a method for studying unconscious influences that avoids the critical methodological issue that has plagued attempts to study perception without awareness for more than 100 years. For this reason, the exclusion task should find wide application in the study on unconscious influences in perception and memory.

ACKNOWLEDGMENTS

Preparation of this chapter was facilitated by a grant from the Natural Sciences and Engineering Research Council of Canada to Philip M. Merikle and a scholarship from the Natural Sciences and Engineering Research Council of Canada to Steve Joordens.

REFERENCES

Cheesman, J., & Merikle, P. M. (1986). Distinguishing conscious from unconscious perceptual processes. *Canadian Journal of Psychology, 40,* 343–367.

Dagenbach, D., Carr, T. H., & Wilhelmsen, A. (1989). Task-induced strategies and near-threshold priming: Conscious influences on unconscious perception. *Journal of Memory and Language, 28,* 412–443.

Debner, J. A., & Jacoby, L. L. (1994). Unconscious perception: Attention, awareness, and control. *Journal of Experimental Psychology: Learning, Memory, and Cognition, 20,* 304–317.

Holender, D. (1986). Semantic activation without conscious identification in dichotic listening, parafoveal vision, and visual masking. *Behavioral and Brain Sciences, 9,* 1–66.

Jacoby, L. L. (1991). A process dissociation framework: Separating automatic from intentional uses of memory. *Journal of Memory and Language, 30,* 513–541.

Jacoby, L. L., Toth, J. P., & Yonelinas, A. P. (1993). Separating conscious and unconscious influences of memory: Measuring recollection. *Journal of Experimental Psychology: General, 122,* 139–154.

Joordens, S., & Merikle, P. M. (1992). False recognition and perception without awareness. *Memory & Cognition, 20,* 151–159.

Kemp-Wheeler, S. M., & Hill, A. B. (1988). Semantic priming without awareness: Some methodological considerations and replications. *Quarterly Journal of Experimental Psychology, 40A,* 671–692.

Kihlstrom, J. F., Barnhardt, T. M., & Tataryn, D. J. (1992). Implicit perception. In R. F. Bornstein & T. S. Pitman (Eds.), *Perception without awareness: Cognitive, clinical, and social perspectives* (pp. 17–54). New York: Guilford.

Logan, G. D., Zbrodoff, N. J., & Williamson, J. (1984). Strategies in the color-word Stroop task. *Bulletin of the Psychonomic Society, 22,* 135–138.

Macmillan, N. A. (1986). The psychophysics of subliminal perception. *Behavioral and Brain Sciences, 9,* 38–39.

Marcel, A. J. (1983). Conscious and unconscious perception: Experiments on visual masking and word recognition. *Cognitive Psychology, 15,* 197–237.

Merikle, P. M. (1982). Unconscious perception revisited. *Perception & Psychophysics, 31,* 298–301.

Merikle, P. M. (1992). Perception without awareness: Critical issues. *American Psychologist, 47,* 792–795.

Merikle, P. M., & Cheesman, J. (1986). Consciousness is a "subjective" state. *Behavioral and Brain Sciences, 9,* 42–43.

Merikle, P. M., & Cheesman, J. (1987). Current status of research on subliminal perception. In M. Wallendorf & P. F. Anderson (Eds.), *Advances in Consumer Research, Vol. XIV,* (pp. 298–302). Provo, UT: Association for Consumer Research.

Merikle, P. M., & Reingold, E. M. (1990). Recognition and lexical decision without detection: Unconscious perception? *Journal of Experimental Psychology: Human Perception and Performance, 16,* 574–583.

Merikle, P. M., & Reingold, E. M. (1992). Measuring unconscious perceptual processes. In R. F. Bornstein & T. S. Pitman (Eds.), *Perception without awareness: Cognitive, clinical, and social perspectives* (pp. 55–80). New York: Guilford.

Miller, J. (1991). Threshold variability in subliminal perception experiments: Fixed threshold estimates reduce power to detect subliminal effects. *Journal of Experimental Psychology: Human Perception and Performance, 17,* 841–851.

Peirce, C. S., & Jastrow, J. (1884). On small differences in sensation. *Memoirs of the National Academy of Sciences, 3,* 73–83.

Reingold, E. M., & Merikle, P. M. (1988). Using direct and indirect measures to study perception without awareness. *Perception & Psychophysics, 44,* 563–575.

Reingold, E. M., & Merikle, P. M. (1990). On the inter-relatedness of theory and measurement in the study of unconscious processes. *Mind & Language, 5,* 9–28.

Searle, J. R. (1992). *The rediscovery of the mind.* Cambridge, MA: MIT Press.

CHAPTER 7

Subliminal Perception:
Nothing Special, Cognitively Speaking

Lynne M. Reder
Jolene Scully Gordon
Carnegie Mellon University

Intrigue with the suggestion that humans have subliminal perception is evidenced by the tremendous literature and continuing controversy on the topic. Merikle and Joordens (chapter 6, this volume) and Greenwald and Draine (chapter 5, this volume) have made significant contributions to this literature and to methodologies aimed at gaining further understanding of subliminal perception or subliminal semantic activation.

The basis for the persistent interest in establishing evidence for subliminal perception ranges from theoretical interest of psychologists in the elusive nature of consciousness to the pragmatic interest of those who would hasten to capitalize on the phenomenon were the evidence unequivocal. Yet the theme of this symposium gives us a more general explanation of the interest in the phenomenon of perception without awareness: Cognitive psychology strives to understand the underlying mechanisms giving rise to cognition—to what extent can we tool our instruments and methodology to observe effects on behavior of stimuli about which we are unaware? And if it can be established that behavior can be influenced by visual presentations that elude awareness, then is it reasonable to presume that other sources of unconscious influence also affect behavior? Finally, can the influence of unconscious cognitive processes be explained within the same general framework one adopts to explain conscious processes, or will tapping into processes unaware to the perceiver require architectural assumptions or mechanisms heretofore unnecessary?

METHODOLOGICAL ISSUES

Central to the current issues surrounding this topic is the question of how to determine when a stimulus is truly subliminal. This is the quagmire that Merikle and Joordens and Greenwald and Draine have addressed, and one that has kept the research area of unconscious perception in a state of controversy for decades. After an extensive review of the extant literature in three basic paradigms (visual masking, dichotic listening, and parafoveal vision), Holender (1986) concluded that existing methodology was simply not sufficient to demonstrate semantic activation without conscious identification. Trying to demonstrate null sensitivity on a direct detection measure in order to prove that an effect on a second, indirect measure was caused by an unconscious process is, in essence, like trying to prove the null hypothesis: There is always the possibility that some amount of stimulus information was detected on at least some occasions that eluded detection by conventional subjective or objective measures.

Merikle and Joordens cogently argued that it is virtually impossible to conceive of an exhaustive measure of unconscious influence, and to convincingly demonstrate null sensitivity on that measure. Both groups of researchers abandoned trying to demonstrate subliminal perception by the traditional methods critiqued by Holender and took great strides to move the methodology forward. In many ways their solutions share the same spirit in the way they finesse the problem statistically and methodologically.

Merikle and Joordens developed an elegant technique of using Jacoby's exclusion task (Debner & Jacoby, 1994; Jacoby, 1991) to obtain clearly different patterns of stem completion conditional on exposure duration. Jacoby's exclusion task involves instructing subjects to complete a word stem, such as "tab__," with any word that comes to mind other than the preceding prime word ("table"). Jacoby found that at short stimulus onset asynchronies (SOAs of 50 ms), subjects were unable to exclude the prime in their response. By manipulating the SOA between the onset of the prime word, and the onset of the mask, over a range of 0 to 214 ms, Merikle and Joordens were successful in determining the critical stimulus duration at which the unconscious influence of the prime exceeded the conscious influence of the prime (to comply with instructions to exclude the prime word). Furthermore, they have successfully used this crossover point (the critical stimulus duration) at which the relative magnitude of unconscious influence of a percept exceeds conscious influence to predict the critical exposure duration necessary for unconscious perception to influence responses in a completely different task, the two-color Stroop task. In addition, Merikle and Joordens found significant (albeit modest in magnitude) correlations between performance on the exclusion task and the size of the Stroop effect. Thus, Merikle and Joordens demonstrated that both qualitative and quantitative predictions can be made on the basis of a measure of the relative influences of conscious and unconscious influences. They concluded that the exclusion task is a satisfactory measure of these relative influences. Merikle and Joordens succeeded in moving this area of research forward in both methodological development and the use of statistical techniques to establish converging evidence between task domains.

Like Merikle and Joordens, Greenwald and Draine moved beyond attempting to prove there is no conscious component in subliminal perception, to a method that analyzes the regression relation between direct and indirect measures of responses to near-threshold stimuli. Conditions were designed such that subjects could perform on a continuum from less than chance to greater than chance accuracy on direct measures of conscious perception (d'), as measured on a lexical detection task of four-letter stimuli presented under conditions of dichoptic masking. An important manipulation was the occasional flash of the word (LEFT) or the non-word (RIGH). Tendencies to respond with the left index finger (intended for nonword responses) to the word *left* or with the right index finger to the nonword *righ* were viewed as unconscious influences.

Collecting data from over 2,000 subjects in 20 experiments, Greenwald and Draine used two values of d'. Objective accuracy at discriminating words from nonwords was the direct measure. The indirect measure involved computing d' differently: Erroneous responses in which subjects pressed the right key (intended for words) for R-I-G-H were treated as hits, and those same presses for the word LEFT were treated as false alarms. They regressed performance on the indirect measure against the direct measure (as the predictor). The critical result was to find an intercept that was significantly greater than zero. Although Greenwald and Draine are to be commended on taking a fresh approach to their investigation of unconscious influences, their statistical methods rely on assumptions that are not fully met by the data.[1] Greenwald and Draine argued that the critical result for demonstrating unconscious influences is a positive value of the intercept measuring the stimulus word influence (left, righ) on position response. Although the positive intercept is consistently found in their regression models, the use of the regression techniques in this instance seems problematic despite their claims to the contrary. Nevertheless, Greenwald and Draine's technique offers promise as yet another method for surpassing the limitations of earlier methodology in the investigation of nonconscious perception.

Where do we go from here? These data from Merikle and Greenwald's laboratories with their improved methods for investigating the phenomenon of subliminal perception are convincing evidence that such a phenomenon exists. Now we are in a position to ask how and why they occur. It is important to

---

[1] Particularly problematic is the necessary assumption that there is no measurement error in their predictor variable.

explore how we can understand these effects in terms of a general, theoretical framework.

THEORETICAL ISSUES

Holender (1986) went beyond the issue of criticizing the extant methodologies' ability to demonstrate subliminal perception: He decried the absence of any theoretical framework in which such phenomena could be explained. Morton (1986), however, asserted that Holender's position was not justified. Morton pointed out that subliminal effects could be accommodated within a variety of existing information-processing models (including very early versions of Morton's own logogen model; e.g., Morton, 1986), without concern about the specific nature or definition of "consciousness."

In the spirit of Morton's commentary, the viewpoint we would like to propose is that *subliminal perception is nothing special, cognitively speaking.* It can be explained and subsumed by a general cognitive architecture, the mechanisms of which produce comparable results in many other task domains. The remainder of this commentary focuses on how to understand phenomena such as subliminal perception as merely a specific instance in a general cognitive architecture.

ASSUMPTIONS

The assumptions come from the SAC model of memory (Reder & Schunn, 1996; Schunn, Reder, Nhouyvanisvong, Richards, & Stroffolino, in press). This model bears similarity to a class of frameworks (e.g., ACT*, Anderson, 1983; and CAPS, Just & Carpenter, 1992). The differences among these theories or frameworks is probably not of importance for accounting for these phenomena. The following are the most essential assumptions in this context:

1. Memory is organized into a perceptual and semantic network of connected ideas, with each concept node in memory varying in base strength, transitory strength, and short-term activation, all as a function of environmental or internal (self-activation) exposure.

2. The base strength of a concept can be thought of as its resting level of activation. The strength of a connection between two concepts is independent of the strength of the concepts that are connected by it.
   **For a concept to be in conscious awareness, its activation must be above threshold. Magnitude of activation is partly a function of the exposure duration of the stimulus.
   **Availability of a concept is a function of its current level of activation.

That is, the ease with which a concept may pass over threshold depends on its resting level of activation.

3. In addition to a declarative memory that sends out activation effortlessly, there is a more controlled processing mechanism that represents strategic decisions or rules, such as "whenever you see a red traffic light ahead of you, stop your car." The execution of these rules vary in their time to "fire" (execute an action). Actions are not necessarily motoric. They may include goal setting or activating something else in declarative memory.
**The speed with which a production rule can fire is partly a function of the level of activation of the elements it tries to match in its condition clauses.

With these few assumptions, we can explain a number of phenomena as well as subliminal perception (see Reder & Schunn, 1996, for a fuller explanation), and show that subliminal perception can be understood as a result of routine cognitive functioning, and does not really behave in an exceptional fashion.

Exclusion Task

In light of the assumptions just enumerated, consider first the exclusion task performed by Merikle and Joorden. As in Jacoby's experiments, the subject is subliminally primed with "table" and then asked to complete the word stem, "tab___." When presented for a very short duration, "table" gets only partially activated. As a result, when asked to complete "tab___," "table" becomes available more easily than other words.

Whether presented subliminally or for a longer duration, table is more available as a completion if it was so primed. In addition, there is the explicit rule (special purpose production that is set up based on instructions): *If the word you think of was just seen, do not use it for the completion.* For that rule to be followed subjects must of course be consciously aware of the previous presentation. Therefore, when the word "table" is presented for a longer duration it exceeds the threshold for conscious awareness, and a context tag is generated, from another production rule: *If I recognize the word that is flashed, generate a context tag for it so that I do not use it for the completion task.* "Table" is still activated first when presented with the "tab___" stem, but it is not explicitly used as the completion because it has been tagged as having been seen and thus the condition for the Exclusion rule is fulfilled—that is, the action is to respond with any word other than "table."

Note that this explanation predicts that subjects will take *longer* to complete the stem with any other word than with table.[2] That is because they think of table in both cases.

---

[2]Unfortunately, those data were not provided.

Stroop Task

The experiment described by Merikle and Joordens as a two-alternative, red/green, forced-choice Stroop task, was sometimes structured so that the probability of an incongruent prime-target pair was three times as probable as a congruent prime-target pair. When the prime "RED" preceded the opposite color (green) target 75% of the time, the subject learned to expect that a color word meant the opposite color target was more likely to occur. In this situation, a supraliminal color cue had the opposite effect of a subliminal color cue: If the color word prime was not consciously detected, the congruent color was facilitated in a choice reaction time; however, when the prime was consciously detected, there was facilitation for the opposite color, and corresponding inhibition for the matching color.

Our explanation is that subjects develop an adaptive rule (in our model, a production rule) that states: If "red" is flashed, prime (activate) "green" and dampen (inhibit) "red." A complementary production for "green" is also created as the subjects learn the contingencies.

The key point is that these productions cannot fire in time if the color word is only briefly flashed. There is not enough stimulus energy to get the production to fire and thus to prime the opposite word, before the color patch is presented. The speed with which productions fire is a function of the production's strength and the activation level of the nodes that match its condition elements. Other productions were also created—for example, when the ampersands are colored red, press the left key that is labeled "red." The subliminal prime of the word *red* will raise the activation level of the "red" term in the production to push the red button, although it is not sufficient to actually get the production to fire—nor should it, because primes are not always accurate, and thus there would be too many errors.

It is important to note the explanatory connection between the pattern of results and the subliminal flash. The basic notion is that the subliminal flash raises the activation of the corresponding element, but not enough to reach threshold. Productions cannot fire without the elements passing threshold (hence the production to "prime" or activate the opposite color cannot fire), but a subliminal flash can raise the current activation level of an element enough to make it easier to fire a production later.

In other words, under conditions of below-threshold activation of the color word there is enough activation to raise the level of the word element "red" so that when the color red is seen, the production will apply faster, enabling the subject to press the red button faster. In contrast, the production that enables anticipation (the priming) of "green" will not fire unless the elements reach threshold—the opposite color cannot have its activation raised (nor inhibit the congruent color) unless a production fires.

Note that this type of explanation does not depend on external, subliminal perception. Rather, it depends on below-threshold activation (that could arise from internal stimulation) and the concomitant inability of the production to fire. The same result would occur if for some other reason the production could not fire fast enough to inhibit the congruent response. Indeed, a classic result of Neely (1977) involving priming in a lexical decision task can be explained with similar assumptions.

LEXICAL DECISION

In Neely's experiments, subjects were instructed to make lexical decision judgments for word and nonword targets preceded by a prime. The prime could either be semantically and categorically related to the word (e.g., BIRD-sparrow) or unrelated (e.g., BIRD-popsicle). In addition, subjects were explicitly instructed that a nonsemantic, experimental relation had been constructed for some of the primes. Specifically, when the prime "BODY" was presented, a building part, such as "door," was likely to be given as a target to be judged; likewise, when subjects saw "BUILDING," they were likely to see a body part, such as "arm" as the target. Lexical items from the category of body parts were not always preceded with the prime "BUILDING," nor were building words always preceded with the prime "BODY"; however, the proportion was the same as in the case of the opposite cuing of Green to Red described earlier.

Whether there was facilitation in lexical decisions (faster decision times compared with conditions involving a neutral prime of xxxx preceding the target) depended both on the relation of the prime to the target and the SOA (lag from prime to target) between them: Semantically related prime-target pairs produced facilitation at the shortest SOA (250 ms), but inhibition at the longest SOA (750 ms) if they expected the other category of words (i.e., in the "shift" condition); unrelated, but expected, pairs such as "BODY-door" on the other hand, produced the complementary results: Facilitation increased with increasing SOA. Thus, as in Merikle and Joordens' data, at short SOAs, automatic processes (spreading activation between semantic associates) dominate strategic processes (the firing of a production rule created by instructions); long SOAs, however, provide time for the production rule to fire that states "when I see the word *BODY* think of *building parts,*" and to shift activation from the automatically primed word to the new focus of attention.

The important point, from our perspective, is that the aforementioned results mirror those for the Merikle and Joorden results in that SOA maps onto duration of the flash. With short SOAs or brief exposures, there is not enough time or activation for the strategic production to fire. Only automatic activation (not caused by special-purpose, experiment-specific productions) can occur in this context.

It is also important to mention that Merikle and Joordens actually confounded SOA and duration of the flash. We have been supposing that their results occurred because the duration of the flash was not long enough to
achieve sufficient activation; however, because the offset of the prime was
confounded with the onset of the colored ampersands, it is conceivable that
there simply was not enough *time* to fire the production that would prime the
other color. In other words, the comparison with the Neely study may be even
more direct.

The result that performance can be affected by manipulations about which
the subject is *unaware* have made people feel that there was something magical about subliminal perception. However, such results occur in many arenas
besides subliminal perception, and not just in those described earlier. Mis-
attributions of familiarity occur in a wide variety of situations (see Jacoby,
Bjork, & Kelly, 1994; Kamas & Reder, 1994, for fuller discussions).

Savings in re-learning is another example of a phenomenon that can be
thought of in these terms, although it has not heretofore been characterized
in this way. Specifically, savings in re-learning is a situation where a person
learns (relearns) information encountered previously that could not be recollected or recognized and from the subject's perspective might well never have
been presented prior to the new learning situation. In this paradigm, the information might be a list of words that when tested 6 months later seems to
have left no memory "trace." Nevertheless, this previously presented list can
be learned faster than a new one. This can be explained in terms of residual
activation of the connection from the word units to an experimental node (experimental context) that has fallen below the threshold for awareness, yet remains at a level sufficient to be strengthened and thus facilitates relearning.

Merikle and Joorden's Correlations

The correlations between performance on the exclusion task and the size of
the Stroop effect found by Merikle and Joordens were modest. One reason for
the low correlations may be the size of the Stroop effects they obtained. Although statistically significant for different SOAs, the mean Stroop effect was
only approximately 20 ms at the critical stimulus duration, and 12 ms at the
longest stimulus duration. Thus, the magnitude of the correlations between
the Stroop effect and performance on the exclusion task was restricted by the
range of these Stroop effects.

Other factors too may have influenced the relatively weak correlations. Although individual differences in ability to perceive the words should increase
the correlation across the two tasks, other variables that differ among individuals are not shared by the Stroop and exclusion tasks. For example, people
probably vary in the strength of their productions to prime the incongruent
color in the Stroop task. The strength of those productions will influence re-

sponse time in the corresponding conditions. Likewise, people probably differ in their baseline strength or resting activation level of words that are subliminally flashed in the exclusion task. Those differences in resting activation levels will translate into performance differences, but only in the exclusion task. Differences in baseline activation levels and differences in strength of productions are probably uncorrelated among individuals, thus reducing the size of the correlation between the two tasks.

CONCLUSIONS

This commentary claims that the same mechanisms are operating in many arenas and show the same general pattern. Through discussion of data from several different experimental paradigms it has been illustrated that subliminal perception is really only an instance of a more general mechanism. If stimulus energy is insufficient to pass threshold, the words (elements) cannot be in the focus of attention. Partially activated elements that fail to capture attention are also insufficient to activate strategic processes controlled by the firing of production rules created by instructional manipulations. A production can also fail to fire *in time* and therefore have no impact on behavior when the interval between the exposure of an element that is above threshold and the target word (SOA) is very short.

In summary, we believe that research in subliminal perception, with its recent progress in methodology, provides a rich source of fascinating results that are nicely accommodated by current cognitive processing models such as SAC.

ACKNOWLEDGMENTS

Preparation of this chapter was supported by grant BNS-8908030 from the National Science Foundation to Lynne M. Reder and also by the Neural Processes in Cognition postdoctoral training fellowship (NSF Grant BIR-901437) to Jolene Scully Gordon.

REFERENCES

Anderson, J. R. (1983). *The architecture of cognition.* Cambridge, MA: Harvard University Press.

Debner, J. A., & Jacoby, L. L. (1994). Unconscious perception: Attention, awareness, and control. *Journal of Experimental Psychology: Learning, Memory, and Cognition, 20,* 304–317.

Holender, D. (1986). Semantic activation without conscious identification in dichotic listening, parafoveal vision, and visual masking: A survey and appraisal. *Behavioral and Brain Sciences, 9,* 1–66.

Jacoby, L. L. (1991). A process dissociation framework: Separating automatic from intentional uses of memory. *Journal of Memory and Language, 30,* 513–541.

Jacoby, L. L., Bjork, R., & Kelly, C. (1994). Illusions of comprehension and competence. In D. Druckman & R. A. Bjork (Eds.), *Learning, remembering, and believing: Enhancing performance.* Washington, DC: National Academy Press.

Just, M. A., & Carpenter, P. N. (1992). A capacity theory of comprehension: Individual differences in working memory. *Psychological Review, 99,* 122–149.

Kamas, E., & Reder, L. M. (1994). The role of familiarity in cognitive processing. In E. O'Brien & R. Lorch (Eds.), *Sources of coherence in text comprehension: A festschrift in honor of Jerome L. Meyers.* Hillsdale, NJ: Lawrence Erlbaum Associates.

Morton, J. (1986). What do you mean by conscious? *Behavioral and Brain Sciences, 9,* 43.

Neely, J. H. (1977). Semantic priming and retrieval from lexical memory: Roles of inhibitionless spreading activation and limited-capacity attention. *Journal of Experimental Psychology: General, 106,* 226–254.

Schunn, C. D., Reder, L. M., Nhouyvanisvong, A., Richards, D. R. Stroffolino, P. J. (in press). A spreading activation model of feeling-of-knowing and strategy selection. *Journal of Experimental Psychology: Learning, Memory, and Cognition.*

Reder, L. M., & Schunn, C. D. (1996). Metacognition does not imply awareness: Strategy choice is governed by implicit learning and memory. In L. M. Reder (Ed.), *Implicit memory and metacognition.* Mahwah, NJ: Lawrence Erlbaum Associates.

Implicit Learning and Memory

IV

CHAPTER

How to Differentiate
Implicit and Explicit Modes
of Acquisition

Arthur S. Reber
Brooklyn College
The City University of New York

The term *implicit learning* was first coined back in the 1960s, and used to refer to the process whereby reasonably complex knowledge about structured stimulus displays is acquired largely independent of awareness of both the process and the products of acquisition. After an extended period of dormancy, the 1980s witnessed a renewed focus on the study of implicit learning, stimulated largely by a revival of interest in several other processes whose basic functions also operate largely independently of consciousness—specifically, implicit (or subliminal) perception and implicit memory. Indeed, implicit learning has become at least a "warm" topic in cognitive psychology as an increasing number of laboratories have undertaken systematic examination of the phenomenon. For those who had been laboring in relative (and cool) obscurity during the previous decades this interest in the topic was most welcome. It had the desired effect of stimulating an array of perspectives and approaches that differed in interesting ways from the agendas of the original researchers. Consequently, we are slowly coming to understand what implicit learning is and is not, which conditions do and do not allow for the emergence of dissociations between the implicit and the explicit, which kinds of theoretical models give reasonable accounts of the database and which are lacking in explanatory power, and how implicit and explicit cognitive processes fit into an overall characterization of cognitive function.

The picture that is gradually emerging is not a simple one; implicit perception, learning, and memory, which make up what Kihlstrom dubbed the *cognitive unconscious*, are complex processes whose properties are delicately intertwined with those of more familiar processes that operate largely within the control parameters of consciousness. As interest in implicit learning has grown, an important role has been played by those who have questioned many of the initial findings. Dulany, Perruchet, Shanks, and colleagues (Dulany, 1991; Dulany, Carlson, & Dewey, 1984; Perruchet, Gallego, & Savy, 1992; Perruchet & Pacteau, 1990; Shanks & St. John, 1994) have, for example, challenged the notion that the knowledge acquired in implicit learning studies is held independent of conscious retrieval and questioned whether this so-called tacit knowledge base can truly be characterized as abstract. This is as it should be. Novel proposals properly attract intense scrutiny and only if they hold up under it should they be regarded as legitimate topics for further scientific examination.

However, it will probably come as no surprise for me to tell you that I do not concur with recent proposals (Dulany, 1991; Haider, 1992; Shanks & St. John, 1994) that the entire endeavor to understand the cognitive unconscious is misguided; that all is artifact; that there is no evidence for the acquisition and representation of complex knowledge independent of consciousness. The purpose of this chapter is to rebut these arguments, because whereas some of them are interesting, indeed provocative, they are nevertheless simply wrong. In short, there is a pressing need to identify those factors that permit us to distinguish between implicit and explicit cognitive process.

Note, by arguing that the critics of implicit learning are wrong, I am not claiming that the present characterization of implicit cognitive processes is entirely correct. Future work will most certainly allow us to see that some of the current notions are in error, but it will also show that implicit perception, learning, and memory are not methodological artifacts, as some have maintained, but rather foundational components of mind. To make this case, there are four lines of attack on these issues. First, the discussion speculates about our philosophical heritage and why so many find the notion of a sophisticated cognitive unconscious so difficult to accept. Second, it explains why the verbalizability criterion for implicitness that so many have focused on is but a large red herring that swam by a couple of years ago and has caused no end of confusion. Third, it presents evidence from diverse groups of studies that clearly implicate a dissociation between implicit and explicit modes of acquisition independent of the verbalizability criterion. And, finally, it shows that when the cognitive unconscious is put into an evolutionary framework, distinctions between the implicit and the explicit turn out to be anything but controversial; indeed, they are natural consequences of basic evolutionary biology.

SOME PHILOSOPHIC MUTTERINGS

Why has the proposal that implicit learning involves unconscious abstraction attracted such vigorous, even hostile, reactions from various quarters of the psychological community? Are there deeper issues here lurking behind the straightforward empirical questions about levels of awareness and degrees of abstractness in mental representation? Put another way, are these attempts on the part of so many to show that the claims made are false diagnostic of particular philosophical leanings? Let me read you some examples to give you a sense of the tone of the discussion: Shanks and St. John (1994) concluded that "proponents of implicit learning . . . have failed to demonstrate that it describes a class of human learning abilities. On the contrary, human learning is almost invariably accompanied by conscious awareness." Perruchet and Pacteau (1990), stated unambiguously that "we do not question human abstraction ability, no more than we question the existence of unconscious processes. What we do question is the joint possibility of unconscious abstraction." And finally, Dulany (1991) concluded that "psychologists of the next century may even look back on the metatheory of a cognitive unconscious as a kind of shared madness, a folie à deux milles or so."[1]

But, how can cognitive psychologists be so convinced about the nonexistence of a cluster of processes whose ontological status is not only empirically secure but utterly obvious? What do psychologists think is going on when a child acquires a natural language or becomes socialized and inculcated with the mores of society? With language development the case is quite clear. Formal instruction is essentially irrelevant, explicit cognitive processes are absent, learning is essentially unintentional, individual differences in the basic skill are minimal, language users have virtually no access to the rules of their language, and the end product of the acquisition process is a rich, complex, and abstract representation that mirrors that of the structure of the linguistic corpus. A similar picture is easily painted for the processes of socialization and acculturation. Why, when we have such obvious real-world examples of complex knowledge acquired and memorially represented independent of consciousness do so many still strive to demonstrate that it does not occur within the controlled settings of the cognitive scientist's laboratory? To understand the source of the resistance here, we need to do a little historical and philosophical meandering.

There are two broad philosophical traditions that dominate Western psychological thought: the Cartesian and the Lockean. The former emphasizes rationalistic approaches with attendant emphases on intuitionism and nativism; the latter advocates empiricist orientations with attendant emphasis on experimentalism and acquired knowledge. These great traditions have been the intellectual forces behind the exploration of consciousness and awareness within both philosophy and psychology. From the Cartesian perspective, in-

---

[1]Note. significantly, that exactly the same critical attitude tends to emerge when the topic of subliminal (or implicit) perception is broached (see Holender, 1986). In fact, both Dulany (1991) and Shanks and St. John (1994) subjected the possibility of implicit perception to the same vigorous criticism as they do implicit learning.

tuition and rational considerations became the primary basis for achieving understanding; from the Lockean point of view, consciousness and mind became equated and introspection became the primary basis for discerning epistemic truth. Despite the vast differences between Locke and Descartes (for they did not agree on much), their approaches to epistemology share this common theme. In both, consciousness is held to be equivalent to mind; that which is mental is that which is conscious. For Descartes, the indubitable that cemented his existence was the capacity to recognize and be cognizant of his own mental action. For Locke, the mind was transparent to itself, mental acts took place within the full light of consciousness; as Dennett (1987) put it, from Locke's perspective "unconscious thinking or perceiving was . . . dismissed as incoherent, self-contradictory nonsense."

This common thread in our two dominant philosophical traditions suggests that the current attacks on the claim that implicit learning embodies unconscious abstraction are motivated, at least in part, by deep philosophical considerations about what is, in principle, cognitively possible. It is likely that those whose thinking embodies either of these orientations neither classifies nor recognizes themselves as such. However, I suspect that the passion behind recent criticisms of implicit learning derives from the lingering vestiges of both Lockean and Cartesian thought.

Ultimately, of course, resolution of the issues that have risen in the study of the cognitive unconscious will rest in the hands of the experimentalists and truth will out no matter what philosophic framework we work within; it is an empirical question whether unconscious abstraction can take place. So, whether we be Lockeans, Cartesians, or agnostics, it would be best to return to methodological and experimental issues.

THE VERBALIZABILITY CRITERION:
A RED HERRING

Much of the criticism of the work in implicit learning has turned on the question of whether or not subjects can verbalize the knowledge they acquired during the experiments. Proponents claim that subjects cannot readily communicate held knowledge; critics argue that subjects were not asked the right questions and when these are asked, evidence of awareness of mental content emerges. In the early artificial grammar (AG) learning studies (Reber, 1965, 1967), I argued that the knowledge subjects took from a learning session was not held consciously simply because when subjects were asked to tell what rules they were using they were unable to do so. It seemed quite sensible at the time; if they could not tell you what they knew, then they did not know what they knew—hence the learning was implicit and the knowledge held, tacit. In recent years, many questions have been raised about the use of this verbalizability criterion in the determination of unconscious learning. Several studies have reported that when subjects are questioned in various, more sophisticated ways, they often provide evidence of having some knowledge of the structure that characterized the stimulus display (Dulany, Carlson, & Dewey, 1984; Mathews, Buss, Stanley, Blanchard-Fields, Cho, & Druhan, 1989). These findings have had the effect of focusing a good deal of attention on the verbalizability criterion as the basis for determining the implicit or explicit status of mental representations. The point that needs to be made here is that the verbalizability criterion is a red herring—a small odoriferous cousin to the sardine that when dragged across one's path, disturbs the scent and diverts one's attention away from the main issues. Answering the following questions will make the point:

How Do We Measure the Contents of Mind?

The determination of which aspects of mental life are conscious and/or unconscious is a complex measurement problem. Some mental content acquired in the typical experiment on implicit learning may indeed be available to consciousness, but determining this is, of course, dependent on the procedures used and, unfortunately, all those currently in use are flawed—whether or not they depend on a subject's ability to give verbal expression to held knowledge. There are two classes of measures here, α and β, where α consists of measures of the knowledge held outside the reach of consciousness and β of knowledge available to consciousness. Implicit cognitive processes will be said to have occurred whenever $\alpha > \beta$; explicit whenever $\alpha \leq \beta$. It is easily shown that for each measure of α or β that has been proposed the assessment is indeterminant. Consider a few of the more widely cited techniques.

An oft-cited technique is that of Dulany et al. (1984), which was developed to show that subjects were in fact aware of knowledge acquired during the acquisition phase of an AG experiment. The study used the standard procedure where subjects first memorize strings of letters generated by an AG and then make well-formedness judgments about novel grammatical and nongrammatical letter strings. Dulany et al.'s innovation was to have subjects indicate their judgments, not a simple yes/no but by marking each string. When they thought a string was nongrammatical, they crossed out the perceived offending letter or letters; whey they judged a string grammatical, they underlined the letter or letters deemed important in making that string acceptable. Dulany et al. argued that these marked elements represented a valid measure of subjects' held knowledge because when projected into the full set of items that made up the well-formedness task they accounted for their subjects' actual discrimination performance without residual; that is, $\alpha = 0$, all is β. This finding appears robust and has been replicated by Turner (1993).

A closer look at this procedure, however, yields a possibly different characterization. First, it is not clear that subjects carrying out this task are giving expression to consciously held knowledge. It is plausible that subjects simply underline letter groups that yield vague, familiar feelings and cross out those that trigger an implicit sense of something being wrong. If such essentially implicit judgments are made with any measure of consistency, they will produce validity scores that are high and yield the exact same result. Moreover, even if subjects are able to verbalize knowledge about the marked chunks, it does not mean they were aware of them during the well-formedness judgment phase; they could have become conscious of these aspects of the strings only after making the identification responses. Put simply, it is not clear that knowledge that Dulany et al. assigned to β belongs there.

Perruchet and Pacteau (1990) followed up on Dulany et al.'s suggestion that subjects learn small letter groups. They trained one group of subjects in an AG learning experiment using only the bigrams that the grammar can generate and compared their performance on the well-formedness task with that of subjects who were exposed to full letter strings. They found that the bigram-only subjects were able to distinguish novel grammatical from nongrammatical letter strings and were, moreover, aware of the letter groups they were taught. Perruchet and Pacteau argued that these results undermine not only the implicit element but also the abstract representation argument. But again, there are indeterminacies present that considerably soften Perruchet and Pacteau's conclusions. First, whereas the bigram-only group did perform above chance, their performance was significantly poorer than subjects who trained with full letter strings. Second, it turns out that these bigram-only subjects are very different from subjects who learn using complete strings. Gomez and Schvaneveldt (1994) and Manza and Reber (in press) replicated the finding that bigram-only subjects can indeed discriminate between grammatical and nongrammatical strings. However, both also reported that the bigram-only subjects are significantly poorer at making well-formedness judgments of novel strings than control subjects and, importantly, both found that these subjects could only make such judgments when the novel strings are instantiated using the same letter set. That is, bigram-only subjects do not show transfer and cannot make well-formedness judgments about strings instantiated with a novel letter set, an ability that subjects who learned from full letter strings possess. Perruchet and Pacteau's measures of the knowledge subjects hold consciously and what they reflect in terms of their memorial representations turn out to be incomplete.

The other procedure that has frequently been used to consciously assess held knowledge is the generate/predict technique introduced by Nissen and her colleagues in studies of implicit sequence learning. In these experiments, subjects make speeded reactions to the location of a stimulus appearing on a monitor. Two variations on this procedure have been used: sequences with a repeating structure (Cohen, Ivry, & Keele, 1990; Nissen & Bullemer, 1987;

Willingham, Nissen, & Bullemer, 1989) and sequences that follow a complex set of rules (Cleeremans & McClelland, 1991; Lewicki, Czyzewska, & Hoffman, 1987). The standard finding in these experiments is that RTs to structured and repeating sequences are faster than those to sequences that lack structure. Generally, subjects reveal little or no conscious knowledge of the underlying structure of these sequences and hence, these findings have been taken as evidence for implicit learning, as measures of α. Nissen and her coworkers (e.g., Hartman, Knopman, & Nissen, 1989) were the first to suggest that subjects' conscious knowledge of the sequence (β) could be ascertained by requiring subjects to predict the next location of each event prior to its occurrence. Shanks and St. John (1994), who also favored using this technique, put the argument neatly: "If subjects are . . . able to do (this) with above-chance accuracy, then this is evidence of conscious knowledge since their predictions must be based on conscious expectancies."

However, it is far from clear that the prediction task is an explicit task. Knowing explicitly where a stimulus will occur is certainly not the same thing as explicitly knowing the underlying rule that determines its location. Someone reading this sentence could predict each successive word at rates far better than chance without conscious knowledge of the generative rules of English. Are prediction tasks measures α or β? There is a good bit of evidence that under conditions like these, subjects' increased facility in generating or predicting sequences with a rich underlying structure actually reflects implicit processing—that is, it is a measure of α.

Some years ago, a number of studies (e.g., Reber & Millward, 1971) of prediction behavior were carried out using variations on the probability learning technique where subjects predict which of two events would occur on each trial. Rather than simply use Bernoulli sequences with events of differing probabilities, the subjects worked with sequences where successive events were determined by complex rules. Subjects learned to make above chance predictions of event sequences of considerable complexity. In one study, subjects learned to track events that occurred with a shifting probabilistic structure with a period of 50 trials (Reber & Millward, 1971). In another, subjects were able to predict the target event on each Trial n when it was made stochastically dependent on whatever event had occurred on an earlier Trial n-j, where j ranged from 1 to 7 across sessions (Millward & Reber, 1972). This latter study is of particular interest because an earlier experiment had indicated that subjects lack reliable explicit memory for events more remote than five trials (Millward & Reber, 1968). In addition to these studies, recent work (Kushner, Cleeremans, & Reber, 1991) has shown that subjects learn to make above chance predictions of the location of a target event determined by a complex, biconditional rule. This finding is particularly interesting because the rule used is deterministic; hence, explicit knowledge would be reflected by perfect performance. In short, these prediction and generation tasks are much more complex than they are typically thought to be; subjects' explicit responses cannot be taken as evidence for explicitly held knowledge.

Finally, similar mismeasures of α and β have also been made in the study of implicit memory. The problems stem from the fact that techniques used to measure explicit memorial content turn out to be contaminated by hidden contributions from implicit representations. For example, in one version of Jacoby's *task dissociation* technique (Jacoby, 1991, 1992; Jacoby, Lindsay, & Toth, 1992) subjects are given a standard stem-completion task but are instructed not to use any words in the original list. The argument is simple: If free recall can be taken as an accurate assessment of explicit memory (as the verbalizability criterion suggests it should be), then subjects should be able to carry out this task and never use an item from the original stimulus set. However, in a series of experiments, Jacoby and his colleagues showed that subjects cannot perform this task properly. The problem is that they have implicit representations for stimuli that they cannot recall; hence items from the original list leak in at rates well above the known base rate. The implication is that even free recall, the optimum verbalizability condition, yields overestimates of β because implicit memories slip by undetected by consciousness.

Measuring mental content is a difficult task. When we are required further to determine whether our measures are assessing material held tacitly (α) or consciously (β), the problems are greatly compounded. Our best tack here is to eschew any singular procedure and to try ensure that whatever devices we use are valid and are measuring what we think they are.

Why Should We Have Access to Held Knowledge at All?

The simple fact that subjects are not naturally introspectively aware of either the process or the products of implicit learning is, in and of itself, interesting and important. Although there may be circumstances under which one can extract some communicable information from subjects, doing so has a distinctly unnatural quality to it. As the scientist/philosopher Polanyi argued (1961), under the appropriate conditions implicit acquisition of knowledge is the natural process; the verbal explication of knowledge gained is the labored and unnatural one. This pattern is seen most clearly in Mathews et al. (1989), where subjects were stopped periodically during the well-formedness phase of an AG experiment and asked to provide detailed statements about what they knew and how they were making decisions for an "unseen partner" who would use this information later. Subjects worked at this task over 4 days; the protocols from each day were then given to groups of yoked subjects who used them to make well-formedness decisions about the same stimuli. The overall performance of the yoked subjects was above chance but below that of the experimental subjects, suggesting that some knowledge was held consciously and could be communicated and some was not. However, and this is the interesting finding, there was a group × day interaction caused by an substantial increase in performance of yoked subjects on Day 4. Indeed, by the last day, the yoked subjects were making well-formedness decisions nearly as well as the experimental subjects—who themselves had ceased improving in their decision-making ability after the second day. The message is clear: The capacity to examine and report held knowledge improved with practice—but much more slowly and laboriously than the acquisition process.

Generally speaking, it is more remarkable that we can give verbal form to our cognitive processes than the reverse and I find the argument that all cognitive processes are transparent to consciousness utterly implausible. A series of arguments based on some standard principles of evolutionary biology that will give substance to this claim are discussed later. It will, I hope, become clear that by virtue of the hierarchical properties of evolutionary systems the implicit takes primacy over the explicit and the processes that operate independently of consciousness should be viewed as forming the epistemic foundations of mind. Finally, to put to rest the verbalizability issue, consider a third question:

What Is the Real Relation Between Implicit and Explicit Systems?

This point turns on appreciating that the implicit–explicit distinction is not between two isolated cognitive modules but between two poles on a continuum. At one pole are processes that are palpably conscious; at the other there are those that are manifestly unconscious. There is a problem because those that are conscious take on a phenomenological poignancy that is hard to escape. And, of course, because we are not aware of the implicit (or only become so after labored effort) the cognitive unconscious plays little phenomenological role. But it is quite clear that human behavior runs the full gamut and the fingerprints of a top-down consciousness are found in many places and in varying degrees. Solving the Tower of Hanoi, ascertaining the correct rule in the classic concept formation studies of Bruner, Goodnow, and Austin (1956), rotating a mental image, giving directions to someone, doing a crossword puzzle are all palpably conscious operations. No doubt much of what we do takes place in the full light of a modulating, controlling consciousness that, as Baars (1988) argued, displays a rather rich array of functions. On the other hand, acquiring one's natural language, becoming socialized, being able to discern that a new piece of music is indeed Mozart, vaguely sensing in the middle of a chess game that there may be something to be gained by a novel line of attack, having an intuitive sense about a productive line of research, coming to be able to anticipate the appropriate location of a stimulus in a sequence learning experiment, learning to distinguish acceptable sequences of letters from those containing violations all take place largely in the absence of awareness of the epistemic contents of mind that accompany them. Indeed, if we had access to these kinds of knowledge and could readily communicate them, linguists, anthropologists, and musicologists would have been out of work long ago. Let us not forget that one of the reasons Titchener's Structuralism failed was because his observers were not able to verbalize reliably the contents of their own minds, even when confronted with relatively simple psychophysical displays.[2]

This relation between the implicit and explicit, between the conscious and the unconscious mental domains, can be made sense out of when these functions are viewed against the backdrop of evolutionary biology. The controlling and modulating functions of consciousness differentially interpenetrate various cognitive processes; the degree to which we can become aware of the operations of these processes is given by their evolutionary histories. It is senseless to attempt to identify any particular criterion as dividing the implicit from explicit or to use any particular method to determine the contents of α and β.

To summarize, the proper approach here is to use an individual's relative inability to provide verbal description of mental content as a kind of common sense marker for increasing the likelihood that implicit processes are present. It has heuristic value but it should not be treated as a diagnostic marker. Moreover, we need to take care that methods that may appear to be tapping consciousness are not actually allowing implicit functions to contaminate the measure. We need to look not for cases where $\beta = 0$, but only for those where $\alpha > \beta$.

DISSOCIATIONS BETWEEN IMPLICIT AND EXPLICIT PROCESSING

With the red herring back in the creel, the following is a review of the small but growing literature that points to a dissociation between the implicit and the explicit without drawing on the verbalizability criterion.

Differential Robustness of Implicit and Explicit Cognition

Put simply, implicit memory and learning display a robustness and resiliency to neurological and psychiatric insult that is not characteristic of explicit processes. Amnesics who show chance performance on explicit memory tests,

---

[2]Titchener, of course, was an unabashed Lockean and apropos my earlier philosophical speculations, his entire empirical program was predicated on the assumption that mental processes were open to introspection by consciousness. However, even Titchener acknowledged that reporting the contents of mind was not a natural process and that his observers had to undergo special training in introspection in order to be able to carry out the task effectively.

such as recognition and recall, display virtually normal abilities on implicit memory tasks as repetition priming or word-stem completion (see, e.g., Graf & Schacter, 1985; Schacter, 1987; Schacter & Graf, 1986; Warrington & Weiskrantz, 1968, 1974). Patients with prosopagnosia so extreme that they do not recognize the faces of members of their own family show virtually normal implicit facial memory when indirect tasks are used (De Haan, Young, & Newcombe, 1987; Newcombe, Young, & De Haan, 1989; Young & DeHaan, 1988). In cases of blindsight, patients with scotomas and other visual disorders give clear evidence of tacit knowledge of stimuli presented in the damaged sectors of the visual field (Weiskrantz, 1986; Weiskrantz, Warrington, Sanders, & Marshall, 1974). Hemineglect patients with impaired ability to report items presented in the neglected field show implicit memories for such stimuli (Volpe, LeDoux, & Gazzaniga, 1979). Patients with acquired dyslexia reveal virtually normal lexical knowledge of presented words provided they were presented at rates too rapid to be decoded consciously (Coslett, 1986; Shallice & Saffran, 1986). Wernicke's aphasics show essentially normal patterns of response latencies on lexical decision tasks despite showing chance performance on explicit knowledge of the meanings of the very same words (Blumstein, Milberg, & Shrier, 1982; Milberg & Blumstein, 1981).

Psychotics and chronic alcoholics who were incapable of discovering relatively simple letter-to-number rules when presented as explicit problems nevertheless showed intact implicit learning of the underlying structure of an AG (Abrams & Reber, 1988). Amnesics also display normal AG learning when using standard experimental procedures but below normal performance when instructed to make their judgments based on explicit comparisons with the learning stimuli (Knowlton, Ramus, & Squire, 1992; Knowlton & Squire, 1994). Alzheimer's patients performed normally on an implicit sequence learning task but poorly on tasks that required reflection and conscious control of cognition (Nissen & Bullemer, 1987). Amnesics also develop preferences for melodies based on Korean melodic patterns that they are unaware they have heard before (Johnson, Kim, & Risse, 1985). In an extended case study, an amnesic patient learned to operate a computer, including carrying out data entry, using disk storage and retrieval procedures, and writing and editing simple programs (Glisky & Schacter, 1989; Glisky, Schacter, & Tulving, 1986).

There is much more here but these citations will do. Implicit and explicit processes are distinguished from each other on the basis of differential robustness to insult and injury.

Individual Differences in Implicit and Explicit Processes

Although there are but a few studies, the evidence suggests that performance on implicit tasks shows fewer individual differences than explicit tasks. In the one study that directly explored this issue, sample variance in performance on an explicit task was over four times as great as it was for an equivalently difficult implicit task (Reber, Walkenfeld, & Hernstadt, 1991).

Aaronson and Scarborough (1977) reported a similar pattern of results in a serial reading task where words are presented one at a time. Two independent processes emerged in this study; one involved reading individual words and organizing them into phrase units, the other was based on the integrative process of encoding sentential information. Standard psycholinguistic analysis treats the former as a conscious, overt process and the latter as an automatic, unconscious one. Large individual differences were found on the explicit process with slow readers taking nearly three times as long as normal readers, but no differences were observed on the implicit task where slow readers were indistinguishable from normal readers.

An interesting corollary to this issue is the relation between IQ and implicit and explicit learning. I know of only two studies to date that have examined this issue and both have found no significant correlations between performance on an implicit learning task and IQ. In one of these experiments (Reber et al., 1991) this lack of correlation was accompanied by a positive significant correlation between IQ and performance on a parallel explicit task, in the other (Knowlton et al., 1992) this factor was not explored.

Direct Versus Indirect Tests of Held Knowledge

In a recent experiment, Reed and Johnson (1994) looked at a variety of direct and indirect methods of assessing mental content using the sequence learning procedure. Following standard conventions, implicit knowledge was evidenced by decreases in RTs on repeating sequences. Explicit knowledge was measured using three direct measures: recognition, where subjects were asked whether or not two- and three-element subsequences were part of the original sequences; generation, where they were asked to produce the sequences; and verbalization, where they were asked for overt descriptions of sequences. Reed and Johnson found that none of these direct measures supported the notion that subjects held knowledge of the sequences in a memorial form available to consciousness. In addition, no significant correlations were found between the slopes of individual subject's learning curves and the direct measures of held knowledge; this result is in agreement with Perruchet and Amorim's (1992) argument that if knowledge has truly been acquired implicitly, then performance on direct and indirect measures should show no concordance.

I suspect that the expectation that implicit and explicit systems show independence, although supported in these experiments, will not ultimately be borne out. To date, there are but three studies that have tested for it. In one

(Reed & Johnson, 1994) no concordance was found; however, in the other two (Curran & Keele, 1993; Reber et al., 1991), positive correlations between implicit and explicit performances were reported but, owing largely to small n's, they fell short of significance. From the evolutionary perspective developed later, there are good biological reasons for viewing conscious, explicit processes as functioning as top-down modulating systems that interpenetrate the foundational implicit systems. The degree to which this cognitive penetration occurs should, in principle, depend on the evolutionary history of the subsystems involved. Put simply, the systems of greatest evolutionary antiquity (e.g., basic physiological functions) will operate almost completely outside of conscious control; those of more recent vintage (e.g., the capacity for solving complex problems) will be more open to awareness. In a sense these two systems operate, to use Mathews' term, *synergistically,* and the degree of mutual interaction between them will likely be shown to map onto the relations between neuroanatomical structures and their evolutionary histories.

Speeded Tests

Turner and Fischler (1993) introduced the "speeded test" technique to the standard AG learning procedure. In these studies, subjects were given biasing instructions to encourage them to engage in either implicit or explicit processing and well-formedness judgments were made under either lax or strict time controls. In two separate experiments they found that the performance of explicitly instructed subjects was differentially impaired by the speeded tests relative to the implicit subjects who were unaffected by time constraints. Interestingly, in another experiment where subjects were asked to make similarity judgments rather than the traditional well-formedness judgments, differential effect of the speeded test disappeared. Turner and Fischler argued that the results support the existence of distinct and separate modes of processing.

Differential Effects of Distractor Tasks

Curran and Keele (1993) carried out a series of experiments investigating the roles of attention and awareness in sequence learning. In the signature study, they compared the performance of one group of subjects in a wholly explicit mode (accomplished by the simple device of teaching them the exact sequence to be used) with two other groups who were more or less aware of the sequence during learning. These latter two groups were determined ad hoc based on a questionnaire administered after learning was complete. The three groups differed from each other in a simple sequence learning task with more efficient performance associated with more awareness. However, when they were transferred to a dual task condition where they had to keep track of the frequency with which a particular tone was sounded while engaging in the sequence task the differences between the groups was no longer displayed. Several additional experiments yielded comparable results, leading Curran and Keele to conclude that there were parallel but independent forms of learning operating here. While they pointed out that there are reasons for suspecting that the attentional and nonattentional varieties of acquisition identified here may not be directly equivalent to the implicit/explicit distinction, these findings support the general argument that differences in expressed awareness accompany dissociations between situations. This kind of interaction is important because it makes moot the question of whether some awareness remains in cases that some may have dubbed implicit, because these variations in awareness likely have no effect on performance.

Verbalization May Actually Compromise Implicit Processing

There has been a good bit of work on this issue lately, most of it by Schooler and his colleagues. The typical finding is that requiring subjects to give verbal expression to mental content disrupts performance relative to having them engage in unrelated activity (Schooler & Engstler-Schooler, 1990; Schooler, Olhsson, & Brooks, in press). The one recent study of direct relevance here is Fallshore and Schooler's (1993) demonstration that requiring subjects in an AG learning to give explanations of what they know prior to the well-formedness test significantly lowers their performance compared with subjects who are left to operate in the implicit mode. Incidentally, this result suggests that the differences between the regular subjects and the yoked controls in Mathews et al. (1989) may actually be greater than reported because the required verbalization would have likely lowered overall performance levels.

The Role of Affective States

Rathus, Reber, Manza, and Kushner (1994) found that implicit and explicit modes of processing become dissociated under conditions of high anxiety. They reported that whereas high levels of anxiety disrupt explicit processing, they had no impact on implicit processes in an AG learning experiment. Interestingly, Rathus et al. also found that subclinical levels of depression do not show this dissociation; neither explicit nor implicit processes were significantly impaired. The relation between affect and the implicit–explicit distinction will likely turn out to be complex.

This section can be summarized rather succinctly: There is ample evidence for the dissociation between implicit and explicit modes of cognitive processing independent of the verbalizability criterion.

EVOLUTION AND CONSCIOUSNESS

The point here is that the processes of implicit perception, learning, and memory are neither unlikely nor surprising cognitive processes. If we step back and take an evolutionary biology perspective it becomes immediately apparent that cognitive processes that operate independently of consciousness have evolutionary primacy. Consciousness, at least the compelling top-down, modulating cluster of operations that is so phenomenologically compelling, is a late arrival on the evolutionary scene and was preceded by a considerable gap of phylogenetic time by processes that operated without this kind of awareness. Operations of perception, acquisition, and memorial representation of the complex stimulus displays of the environment must, therefore, have been carried out by mechanisms that operated independently of this ineffable sense of introspective awareness.

As argued elsewhere (Reber, 1992a, 1992b, 1993), recognition of the phyletic primacy of implicit systems carries with it a number of intriguing entailments—some that have already received empirical support and others yet to attract sufficient attention to know whether they will be borne out. Specifically, phylogenetically older structures and forms and the processes that they control will, by virtue of the hierarchical nature of evolution (Simon, 1962), display various properties relative to structures and forms of more recent vintage. Among these properties are two of those cited earlier as differentiating the implicit from the explicit, specifically: Evolutionarily older systems tend to be more robust than those more recently evolved and they tend to show less individual-to-individual variability. The evolutionary model also predicts a variety of other modes of dissociation between the implicit and the explicit that have, to date, received relatively little attention, specifically: Evolutionarily older systems should operate relatively independently of developmental level in childhood and be relatively immune to the effects of aging and implicit systems ought to show phylogenetic continuity. There are some hints that these are also borne out (see Reber, 1993) but we have a lot of work to do before we can feel confident in these predictions.

SUMMARY AND SOME SPECULATIVE EXTENSIONS

So, where are we today on the matters of implicit learning, the cognitive unconscious generally, and the vexing questions invited by the very existence of protoplasm-based organisms who are both self-aware and self-reflecting? The quick review of the literature included here strongly supports the notion of a hierarchically structured mind/brain with a foundational implicit acquisition and representation system that operates largely independently of consciousness hovered over by an explicit cognizing system whose operations are intimately tied to consciousness. These two systems, although interactive in ways that we are only beginning to understand, are demonstratively dependent on neurological structures with different evolutionary histories and easily seen to be functionally dissociable.

Indeed, the argument I have tried to develop here has been, as noted earlier, that of the *primacy of the implicit.* That is, the operations that make up the cognitive unconscious represent a cluster of related processes that are subserved by neuroanatomical systems of considerable antiquity. These implicit systems, when operating in an acquisitive mode, yield a knowledge base that allows an individual to function in a variety of domains in which decisions can be made, problems can be solved, and choices arrived at without conscious recourse to mental content. The form of representation of the tacit knowledge base that is induced here is in large measure functionally determined by the context within which it was acquired and by constraints placed on the circumstances of its use. Hence, it may be highly abstract or sharply tuned to the physical form of the initial inputs (Manza & Reber, in press; Whittlesea & Dorken, 1993). In short, the cognitive unconscious needs be viewed in the larger context of functionalism. Given this overall framework, I hope it is now clear that we cannot allow the issue of verbalizability of held knowledge to confound issues. The extent to which an individual is or is not capable of explicating held knowledge is a pragmatic issue.

Implicit acquisition systems are those that have a long evolutionary history and can be found operating in virtually every species of even modest neurological complexity. Indeed, they are phylogenetically ubiquitous for the most obvious of reasons; their singular capacity to encode and represent the patterns of covariations among events enabled the organisms that evolved them to function effectively.

Some Speculations

Assuming that this quick-and-dirty overview has the virtues of capturing the existing database reasonably well and making sense, one still needs to examine a host of related questions, specifically what indeed is consciousness and what kinds of relations exist between its processes and functions and those of the phyletically primary implicit system. To deal with these issues, I'd like to entertain a few loose thoughts about the nature of consciousness and its relation to those processes that make up the cognitive unconscious.

First, it seems to me that there are at least two distinguishable kinds of cognitive systems that we can consider as capturing the notion of consciousness, call them C_1 and C_2. C_1 is the evolutionarily older one and C_2 is our familiar, top-down, modulating consciousness, the one we typically think of when we use the term *consciousness* and that we tend to associate with our species. C_1, however, is an interesting cognitive system; I think of it as a kind of consciousness, the kind that enables an organism to be aware of itself as an entity and to appreciate the distinction between self and not-self. C_1 is the kind of system that enables an organism to do many things: to detect covariations between events and thus display the capacity for classical conditioning and to make volitional behavioral decisions; to "decide" to move a limb, turn a head, or select several alternative sources of food; in short, to engage in operant behavior.

To push this point, turn to an early passage in James' *The Principles of Psychology* (1890, Vol. 1, pp. 14–18) where he presents the now-classic gloss on the relation between the various levels of the central nervous system and their corresponding behavioral functions. James used the technique of making a variety of cuts at higher and higher levels in the CNS of a laboratory frog. A normal, intact frog has, as he noted, a reasonably rich behavioral repertoire, although most of its actions are tied to fairly specific stimulus events. A pinch on the bottom provokes hopping, stroking the frog's side yields leg twitches, and moving objects with the proper size and flight patterns evoke directed tongue flicks, and so on. As we all know from high school biology, progressive cuts at increasingly higher levels along the CNS leave more and more of the frog's behavioral repertoire intact. When the very highest cut is made, the one that removes only the higher brain centers, the changes in the frog's behavior are subtle. Indeed, virtually the entire range of behaviors are still present; pinches still elicit hopping, strokes evoke leg twitches, and flies tongue extensions. But, as James noted, to an astute observer the frog "feels" very different; it no longer seems to have any volitional elements to its behaviors. The raw behaviors that are in large part mechanistic, reflexive, and wedded to particular lower brain structures are still there, but the operant aspects of action are not. To the observer the frog seems, in so many words, to have lost its "frogness." Although it certainly does frog-appropriate things, it just does not seem like a proper frog.

What the decorticate frog has lost is its C_1, that cognitive capacity to differentiate "me from thee," to distinguish between self and not-self, to emit operant behaviors. Surely this does not mean that frogs have anything approximating human consciousness; the point is simply that there are good reasons for assuming that a viable and significant form of consciousness, C_1 is likely emergent in even fairly primitive amphibian forms.

However, C_1 is not the kind of consciousness that, to scholars such as Dulany (this volume), lies at the heart of the arguments raised in the context of implicit learning theory. That epistemic entity, C_2, is a cognitive system of a rather different sort. C_2 functions not merely to differentiate self from other. It also incorporates a large number of cognitive functions that allow us to modulate and refine the actions of self (see Baars, this volume, for some of the various possible functions of consciousness). It is C_2, not C_1, that has attracted so much philosophical and psychological interest over the millennia for the simple reason that it is C_2 that is so introspectively compelling.

The question now practically asks itself: How can the term *consciousness*, in either of these senses, be profitably applied to the issue of implicit learning? The answer, clearly affirmative, is entailed by a simple, common finding: In our experiments on implicit learning, our subjects are conscious of the fact that they have learned something. They are aware that cognitive change has taken place during the learning phase of the experiment; they know that they know something they did not know before. They have a "feeling of knowing" and when making decisions or solving problems, confidence ratings correlate with performance (Manza & Reber, in press). But the subjects do not, by and large, know what it is they know. Hence, they have a kind of consciousness that emerges in these experiments, but it is not a proper C_2 kind of consciousness; it is something cognitively less well developed. It is epistemically closer to the "cognitions" (using that term loosely) of James' frog with an intact cortex; there is an awareness without the self-reflective, modulating functions linked with true C_2-type consciousness.

So, my first major speculative conclusion is that implicit learning requires no more in the way of consciousness than the kinds of mentations that accompany C_1 processes. This point of view, of course, fits with the evolutionary framework developed earlier.

Second, what do we know about the nature of the relation between these two forms of consciousness? A natural gloss here is that C_1 is a bottom-up system, one that detects covariations among events, whereas C_2 is a top-down system that imposes structure and modulates other, ongoing processes.[3] One of C_2's modes of operation is penetrating all of those operations and functions that lie within its scope to either modulate their action or bring them within the full spotlight of awareness. C_2's scope, however, is limited and those limits are determined in large measure by evolutionary considerations. That is, the evolutionary antiquity of the processes it is attempting to modulate dictate the extent to which such modulation is successful. Cognitive processes, such as those involved in solving puzzles like the Hobbits and Orcs or in rotating a mental image or reading a map or doing a crossword puzzle, are all dependent on recently evolved structures. These are not only "open" to introspection by C_2, their very existence as processes are intimately linked with it. But older, more primitive processes, such as those within the domain of C_1, lie largely outside of the modulating influences of C_2. Hence, implicitly acquired knowl-

---

[3] Interestingly, and unfortunately way beyond the scope of this chapter, is the intriguing possibility that the processes of C_1 can be captured by connectionist architectures and those of C_2 by production systems or other symbol manipulation systems like SOAR. If perhaps my speculations about how C_1 and C_2 interrelate seem unconvincing to some, then I gain some solace from the fact that no one has yet developed a cognitive architecture that integrates both kinds of systems.

edge is not easily available to examination by consciousness. At the other ex-treme are the vegetative functions that provide the physiological foundation for our most basic operations; these are completely opaque to consciousness. No one knows what their liver is doing.

As noted earlier, this conceptualization of the relation between that which is conscious and that which is not puts C_1 and C_2 on a continuum. That is, they should not be seen as separate or encapsulated modules and the relation be-tween them is best thought of as interactive or synergistic. Indeed, conscious and unconscious cognitive processes should be viewed as lying at separate poles on a continuum of penetrability. Consciousness did not evolve as a sep-arate faculty of mind; it emerged gradually over time as a process that allowed its possessor to gain control over other cognitive functions and that is its pri-mary adaptive feature.

Finally, consider one last speculation on the phylogenetic continuity of im-plicit learning entailed by the previous arguments: The evolutionary biology framework implies that the implicit cognitive functions that members of our species display are of a kind with those found in, say, frogs. But surely this can-not be right; frogs may be capable of differential classical conditioning but they cannot learn artificial grammars. My view on this issue is that the under-lying process of detection of covariations is the same in all cases but the kinds of covariations that the system is sensitive to change with increasing phylo-genetic complexity. In the case of relatively uncomplicated creatures (e.g., frogs), the covariations among the several events to be learned must be pre-sented with virtually perfect fidelity. CS_1 needs to be followed UCS_1 and CS_2 by UCS_2 on all trials. Stochastic displays will not be learned or learned only with considerable struggle. As we move up the phylogenetic scale, we find that the degree of covariation among the several events can be lessened and learn-ing still takes place. In the case of humans, we find a species whose members can acquire exceedingly tenuous connections between events that are remote, highly complex, and multiconditional. The heart of the process, detecting co-variations, remains the same; it is the kinds of covariations that can be readily detected that changes.

To summarize, implicit learning is a foundational process that lies at the very core of the functioning of virtually every complex species. It is dependent on evolutionarily old structures that are robust, coherent, and display rela-tively little variation both ontogenetically and phylogenetically. Its essential features can be seen in behaviors that range from simple classical condition-ing to the acquisition of natural languages and the processes of socialization. And although it is clear that implicit and explicit cognitive modes are disso-ciable from each other, it is equally clear that there is continuity between them. The patterns of dissociation that we see are not those driven by a clean demarcation between systems; rather, they are ones that reflect continuity.

ACKNOWLEDGMENT

Preparation of this chapter was supported by Grant BNS-89-07946 from the
National Science Foundation.

REFERENCES

Aaronson, D., & Scarborough, H. S. (1977). Performance theories for sentence coding: Some
   quantitative models. *Journal of Verbal Learning and Verbal Behavior, 16*, 277–303.
Abrams, M., & Reber, A. S. (1988). Implicit learning: Robustness in the face of psychiatric dis-
   orders. *Journal of Psycholinguistic Research, 17*, 425–439.
Baars, B. J. (1988). *A cognitive theory of consciousness.* New York: Cambridge University Press.
Blumstein, S. E., Milberg, W,. & Shrier, R. (1982). Semantic processing in aphasia: Evidence from
   an auditory lexical decision task. *Brain and Language, 17*, 301–315.
Bruner, J. S., Goodnow, J., & Austin, G. (1956). *A study of thinking.* New York: Wiley.
Cleeremans, A., & McClelland, J. L. (1991). Learning the structure of event sequences. *Journal
   of Experimental Psychology: General, 120*, 235–253.
Cohen, N. J., Ivry, R., & Keele, S. W. (1990). Attention and structure in sequence learning. *Jour-
   nal of Experimental Psychology: Learning, Memory, and Cognition, 16*, 17–30.
Coslett, H. B. (1986). *Preservation of lexical access in alexia without agraphia.* Paper pre-
   sented at the Ninth European Conference of the International Neuropsychological Society,
   Veldhoven, The Netherlands.
Curran, T., & Keele, S. W. (1993). Attentional and nonattentional forms of sequence learning.
   Journal of Experimental Psychology: Learning, Memory, and Cognition, 19, 189–202.
De Haan, E. H. F., Young, A. W., & Newcombe, F. (1987). Face recognition without awareness.
   Cognitive Neuropsychology, 4, 385–415.
Dennett, D. C. (1987). Consciousness. In G. L. Gregory (Ed.), *The Oxford companion to the
   mind* (pp. 161–164). New York: Oxford University Press.
Dulany, D. E. (1991). Conscious representation and thought systems. In R. S. Wyer, Jr. & T. K.
   Srull (Eds.). *Advances in social cognition* (Vol. 4). Hillsdale, NJ: Lawrence Erlbaum Associ-
   ates.
Dulany, D. E., Carlson, R. A., & Dewey, G. I. (1984). A case of syntactical learning and judgment:
   How conscious and how abstract? *Journal of Experimental Psychology: General, 113*, 541–
   555.
Dulany, D. E., Carlson, R. A., & Dewey, G. I. (1985). On consciousness in syntactical learning and
   judgment: A reply to Reber, Allen, and Regan. *Journal of Experimental Psychology: General,
   114*, 25–32.
Fallshore, M. & Schooler, J. W. (1993). [Unpublished data].
Glisky, E. L., & Schacter, D. L. (1989). Extending the limits of complex learning in organic am-
   nesia: Computer training in a vocational domain. *Neuropsychologia, 27*, 107–120.
Glisky, E. L., Schacter, D. L., & Tulving, E. (1986). Computer learning by memory-impaired patients: Acquisition and retention of complex knowledge. *Neuropsychologia, 24*, 313–328.
Gomez, R. L., & Schvaneveldt, R. W. (1994). What is learned from artificial grammars? Transfer
   tests of simple association. *Journal of Experimental Psychology: Learning, Memory, and
   Cognition, 20*, 396–410.
Graf, P., & Schacter, D. L. (1985). Implicit and explicit memory for new associations in normal
   and amnesic subjects. *Journal of Experimental Psychology: Learning, Memory, and Cognition, 11*, 501–518.

Haider, H. (1992). Implicit knowledge and learning: An artifact. *Zeitschrift fur experimentelle und angewandte psychologie, 39,* 68–100.

Hartman, M., Knopman, D. S., & Nissen, M. J., (1989). Implicit learning of new verbal associations. *Journal of Experimental Psychology: Learning, Memory, and Cognition, 15,* 1070–1082.

Holender, D. (1986). Semantic activation without conscious identification in dichotic listening, parafoveal vision, and visual masking: A survey and appraisal. *The Behavioral and Brain Sciences, 9,* 1–66.

Jacoby, L. L. (1991). A process dissociation framework: Separating automatic from intentional uses of memory. *Journal of Language and Memory, 30,* 513–541.

Jacoby, L. L. (1992, November). *Influences of memory: Attention, awareness, and control.* Paper presented at the meeting of the Psychonomic Society, St. Louis, MO.

Jacoby, L. L., Lindsay, D. S., & Toth, J. P. (1992). Unconscious influences revealed: Attention, awareness, and control. *American Psychologist, 47,* 802–809.

James, W. (1890). *Principles of psychology* (Vol. 1). New York: Holt.

Johnson, M. K., Kim, J. K., & Risse, G. (1985). Do alcoholic Korsakoff's syndrome patients acquire affective reactions?. *Journal of Experimental Psychology: Learning, Memory, and Cognition, 11,* 27–36.

Knowlton, B. J., Ramus, S. J., & Squire, L. R. (1992). Intact artificial grammar learning in amnesia: Dissociation of abstract knowledge and memory for specific instances. *Psychological Science, 3,* 172–179.

Knowlton, B. J., & Squire, L. R. (1994). The information acquired during artificial grammar learning. *Journal of Experimental Psychology: Learning, Memory, and Cognition, 20,* 79–91.

Kushner, M., Cleeremans, A., & Reber, A. S. (1991). Implicit detection of event interdependencies and a PDP model of the process. In *Proceedings of the 13th Annual Conference of the Cognitive Science Society.* Hillsdale, NJ: Lawrence Erlbaum Associates.

Lewicki, P., Czyzewska, M., & Hoffman, H. (1987). Unconscious acquisition of complex procedural knowledge. *Journal of Experimental Psychology: Learning, Memory, and Cognition, 13,* 523–530.

Manza, L., & Reber, A. S. (in press). *Inter- and intra-modal transfer of an implicitly acquired rule system.* In D. Berry (Ed.), *How implicit is implicit learning?* London: Oxford University Press.

Mathews, R. C., Buss, R. R., Stanley, W. B., Blanchard-Fields, F., Cho, J.-R., & Druhan, B. (1989). The role of implicit and explicit processes in learning from examples: A synergistic effect. *Journal of Experimental Psychology: Learning, Memory, and Cognition, 15,* 1083–1100.

Milberg, W., & Blumstein, S. E. (1981). Lexical decision and aphasia: Evidence for semantic processing. *Brain and Language, 14,* 371–385.

Millward, R. B., & Reber, A. S. (1968). Event-recall in probability learning. *Journal of Verbal Learning and Verbal Behavior, 7,* 980-989.

Millward, R. B., & Reber, A. S. (1972). Probability learning: Contingent-event sequences with lags. *American Journal of Psychology, 85,* 81-98.

Newcombe, F., Young, A. W., & De Haan, E. H. F. (1989). Prosopagnosia and object agnosia without covert recognition. *Neuropsychologia, 27,* 179–191.

Nissen, M. J., & Bullemer, P. (1987). Attentional requirements of learning: Evidence from performance measures. *Cognitive Psychology, 19,* 1–32.

Perruchet, P., & Amorim, M. (1992). Conscious knowledge and changes in performance in sequence learning: Evidence against dissociation. *Journal of Experimental Psychology: Learning, Memory and Cognition, 18,* 785–800.

Perruchet, P., Gallego, J., & Savy, I. (1990). A critical reappraisal of the evidence for unconscious abstraction of deterministic rules in complex experimental situation. *Cognitive Psychology, 22,* 493–516.

Perruchet, P., & Pacteau, C. (1990). Synthetic grammar learning: Implicit rule abstraction or explicit fragmentary knowledge? *Journal of Experimental Psychology: General, 119,* 264–275.

Polanyi, M. (1961). Knowing and being. *Mind, 70,* 458–470.

Rathus, J., Reber, A. S., & Manza, L, & Kushner, M. (1994). Implicit and explicit learning: Differential effects of affective states. *Perceptual and Motor Skills, 79,* 163–184.

Reber, A. S. (1965). *Implicit learning of artificial grammars.* Unpublished master's thesis, Brown University.

Reber, A. S. (1967). Implicit learning of artificial grammars. *Journal of Verbal Learning and Verbal Behavior, 77,* 317–327.

Reber, A. S. (1992a). The cognitive unconscious: An evolutionary perspective. *Consciousness and Cognition, 1.*

Reber, A. S. (1992b). An evolutionary context for the cognitive unconscious. *Philosophical Psychology, 5,* 33–51.

Reber, A. S. (1993). *Implicit learning and tacit knowledge: An essay on the cognitive unconscious.* New York: Oxford University Press.

Reber, A. S., & Millward, R. B. (1971). Event tracking in probability learning. *American Journal of Psychology, 84,* 85–99.

Reber, A. S., Walkenfeld, F. F., & Hernstadt, R. (1991). Implicit and explicit learning: Individual differences and IQ. *Journal of Experimental Psychology: Learning, Memory, and Cognition, 17.* 888–896.

Reed, J., & Johnson, P. (1994). Assessing implicit learning with indirect tests: Determining what is learning about sequence structure. *Journal of Experimental Psychology: Learning, Memory, and Cognition, 20,* 585–594.

Schacter, D. L. (1987). Implicit memory: History and current status. *Journal of Experimental Psychology: Learning, Memory, and Cognition, 13,* 501–518.

Schacter, D. L., & Graf, P. (1986). Preserved memory in amnesic patients: Perspectives from research on direct priming. *Journal of Clinical and Experimental Neuropsychology, 8,* 727–743.

Schooler, J. W., & Engstler-Schooler, T. Y. (1990). Verbal overshadowing of visual memories: Some things are better left unsaid. *Cognitive Psychology, 17,* 36–71.

Schooler, J. W., Ohlsson, S., & Brooks, K. (1993). Thoughts beyond words: When language overshadows insight. *Journal of Experimental Psychology: General, 22,* 166–183.

Shallice, T., & Saffran, E. (1986). Lexical processing in the absence of explicit word identification: Evidence from a letter-by-letter reader. *Cognitive Neuropsychology, 3,* 429–458.

Shanks, D. R., & St. John, M. F. (1994). Characteristics of dissociable human learning systems. *Behavioral and Brain Sciences, 17,* 367–447.

Simon, H. A. (1962). The architecture of complexity. *Proceedings of the American Philosophical Society, 106,* 467–482.

Turner, C. W. (1993). *Classification and explicit memory for instances: Separate and unequal?* Unpublished manuscript.

Turner, C. W., & Fischler, I. S. (1993). Speeded tests of implicit knowledge. *Journal of Experimental Psychology: Learning, Memory, and Cognition, 19,* 1165–1177.

Volpe, B. T., LeDoux, J. E., & Gazzaniga, M. S. (1979). Information processing of visual stimuli in an "extinguished" field. *Nature, 282,* 722–724.

Warrington, E. K., & Weiskrantz, L. (1968). New method of testing long-term retention with special reference to amnesic patients. *Nature, 217,* 972–974.

Warrington, E. K., & Weiskrantz, L. (1974). The effect of prior learning on subsequent retention in amnesic patients. *Neuropsychologia, 12,* 419–428.

Weiskrantz, L. (1986). *Blindsight.* New York: Oxford University Press.

Weiskrantz, L., Warrington, E. K., Sanders, M. D., & Marshall, J. (1974). Visual capacity in the hemianopic field following a restricted occipital ablation. *Brain, 97,* 709–728.

Whittlesea, B. W. A., & Dorken, M. D. (1993) Incidentally, things in general are particularly determined: An episodic-processing account of implicit learning. *Journal of Experimental Psychology: General, 122,* 227–248.

Willingham, D. B., Nissen, M. J., & Bullemer, P. (1989). On the development of procedural knowledge. *Journal of Experimental Psychology: Learning, Memory and Cognition, 15,* 1047–1060.

Young, A. W., & DeHaan, E. H. F. (1988). Boundaries of covert recognition in prosopagnosia. *Cognitive Neuropsychology, 5,* 317–336.

CHAPTER 9

Cognitive Mechanisms for Acquiring "Experience": The Dissociation Between Conscious and Nonconscious Cognition

Pawel Lewicki
University of Tulsa

Maria Czyzewska
Southwest Texas State University

Thomas Hill
University of Tulsa

Perhaps one of the least controversial and most commonsense conclusions from psychology research is that expertise requires experience, and in many areas of expertise, acquisition of declarative, articulable knowledge is not sufficient to become a truly proficient expert (Chi, Glaser, & Farr, 1988; Ericsson & Smith, 1991). For example, many researchers agree that a minimum of 10 years of "active experience" is necessary to achieve true mastery in an area (the "10-year rule"; see Chase & Simon, 1973) and this specific minimum amount of experience applies to a surprisingly wide variety of domains, such as chess, musical composition, livestock judgment, mathematics, tennis, science, and interpretation of X-ray pictures (e.g., Chase & Simon, 1973; Chi et al., 1988; Gustin, 1985; Hayes, 1981; Krogius, 1976; Lehman, 1953; Lesgold, 1984; Sosniak, 1985).

As is demonstrated later, many cognitive skills important for human adjustment and performance represent knowledge that cannot be articulated by the person. Moreover, often the person who exhibits a skill clearly has no access to its nature (i.e., to the rules and principles on which the skill is based); thus, the inability to articulate that knowledge represents not merely a difficulty with verbalizing (i.e., "putting into words") something that a person intuitively "knows" or "feels," but rather a fundamental lack of access to the content (meaning) of the relevant rules and principles.

For example, experienced, professional comedians do not *know* how to be funny, they just "do" or "say" things that appear funny more often to their audiences than things said by other people. Moreover, they have no means of communicating to others how to improvise or "make up" jokes. For example, Johnny Carson could not make another "Johnny Carson" by "sharing the secret" of his success with someone else; in one of his interviews Carson said that he just does not know what it is that makes him funny, and he does not have any specific method and does not know any prescription that he could sell or patent.

Experienced, good managers are not those who have "memorized really well" some specific "rules for being a good manager"; they just do their job and their actions tend to consistently produce desirable outcomes. If a corporation needs to "groom" new managers, they typically do not attempt to accomplish this by having the potential candidates learn specific "rules for being a successful leader"—instead, the candidates are given the opportunity to "gain experience" (Chi et al., 1988).

Experienced radiologists cannot easily communicate their skills (allowing them to accurately detect abnormalities in fuzzy X-ray pictures) to others. It takes years of experience to acquire the skills (Lesgold, 1984), and, in fact, nobody knows what precisely those skills entail; attempts to develop computer programs to aid in even the "preliminary scanning" of X-ray pictures have not been successful (physicians do not know what specifically the computer should be programmed to look for).

If the knowledge structures responsible for such expertise of a person were acquired from experience but are not fully accessible to the person's conscious awareness, then they must be acquired at least partially through implicit (nonconscious) learning.[1] This chapter reviews some of the recent research on the mechanisms of nonconscious acquisition of knowledge that illustrate how nonconscious processes of encoding participate in the development of "expertise that requires experience."

ENCODING OF STIMULI, ENCODING ALGORITHMS

The starting point for the present reasoning is the widely accepted notion that memories are the joint product of external and internal events, and there is no straightforward relation between external stimuli and the "mind's impres-

---

[1]This reasoning is not intended to imply that acquiring knowledge in a consciously controlled manner is unimportant in the development of expertise. It is, obviously, not only very important as an independent component of many advanced cognitive abilities, but it also provides the necessary foundations for the efficient acquisition of this type of experience that is not mediated by conscious processes (and which is the focus of this chapter). To give a simple example, radiologists need explicit and very sophisticated knowledge about the types of "fuzzy" shapes on an X-ray picture that represent the normal parts of the anatomy and that should be ignored, so that they can efficiently focus on those that are potentially suspect and should be examined very carefully. However, they still are usually unable to articulate what specific attributes of the "suspect" shapes are decisive for making the final determination about a potential abnormality.

sion of those stimuli, much less our memory for them" (Johnson & Raye, 1981, p. 68; see also Cofer, 1973; Hochberg, 1978; Kaufman, 1974; R. Lachman, J. L. Lachman, & Butterfield, 1979; Neisser, 1967). Human perceptual processes—in particular the interpretation of stimuli—depend to a large extent on learned encoding algorithms that impose preexisting categories or prototypes on encountered stimuli. In other words, the actual (subjective) meaning of a stimulus for a subject, and the subjective experience of encountering the stimulus, depends not only on the "objective" characteristics of the stimulus, but also on the preexisting encoding algorithms used to "translate" the stimulus into subjectively meaningful terms.

In perception of social stimuli, the crucial mediating role of these encoding algorithms is obvious because social stimuli are especially ambiguous in nature, and thus open to alternative interpretations (Cantor & Mischel, 1979; Lewicki, 1984). However, these algorithms play an equally important role in the interpretation of many other (even simple visual) stimuli. For example, they are responsible for the recognition of shapes of objects or for the determination of distances between locations in three-dimensional space (Hochberg, 1978; Kaufman, 1974; Rock, 1975). A good example of the operation of such algorithms in visual encoding is face recognition in nearsighted persons. They are often unable to recognize from a distance a familiar face until the other person says something, which, according to the nearsighted perceiver, instantaneously causes the face to appear sharper to the point of being recognizable (Lewicki, 1986a). In this case, recognition of the person's voice activates the appropriate visual encoding algorithm that allows the perceiver to recognize and "see" what could not be seen before.

Regardless of whether it is a visual shape or spoken words that are part of a conversation, every stimulus that is about to be stored in memory and "become an experience" for a person, first has to be encoded according to some preexisting inferential (encoding) algorithms. Those encoding algorithms translate objective stimuli (e.g., a specific pattern of sensations in the retina of the eye) into subjectively meaningful impressions and interpretations (e.g., recognizing a friendly looking face). Therefore, encoding algorithms function as the major interpreter of stimuli that directly determines the final shape and the scope of an individual's experiences.

The concept of *encoding algorithms* (which is a more convenient way of referring to inferential rules followed by the individual in the processes of categorization and identification) is closely related to the concept of *schema* used in cognitive as well as cognitively oriented personality and social psychology (Brewer & Nakamura, 1984; Higgins & King, 1981; Markus, 1977; Rumelhart, 1984; Wyer & Srull, 1981). Using the terms proposed by Rumelhart and Ortony (1977), one might say that inferential encoding algorithms are responsible for the "instantiation of schemas," that is, they perform the task of assignment of variables (Rumelhart & Ortony, 1977) to external or internal events. Thus, the term *encoding algorithms* denotes the basic element of schemata: algorithms that are followed when deciding whether to impose specific schemata on perceived events.

Encoding processes operate with such speed that they are practically unnoticeable to the perceiver who experiences only the final outcome: a subjective impression (Hochberg, 1978; R. Lachman et al., 1979). Moreover, the specific inferential rules involved in encoding are not available to the perceiver's conscious awareness, and this fact constitutes a ubiquitous feature of encoding of a variety of classes of stimuli (from elementary pattern recognition to social perception).

For example, perceivers have no access to the processing rules used to encode shapes or relative locations of objects in three-dimensional space (Hochberg, 1978). Most people are not (and have never been) aware of the semantic and syntactic rules of their native language (Chomsky, 1980), despite the fact that at the same time these rules are constantly utilized by the mechanisms of speech production and comprehension (Reber, 1989). Numerous studies indicate that people lack access to the inferential encoding rules involved in their person perception and interpretation of social stimuli (Lewicki, 1986a; Nisbett & Wilson, 1977).

It is worth noting at this point that this lack of awareness regarding the contents of inferential encoding algorithms is not as obvious as the trivial fact that one is not aware of the internal biochemical workings of one's brain (e.g., one does not feel the physical locations of information in the cortex or changes in states of synapses). *Encoding algorithms* denote purely psychological and not physiological phenomena. They are based on knowledge about reality (stored in memory), they clearly involve inferential operations on information, and, potentially, they could be made accessible to one's awareness to assume the same status as consciously controlled inferential rules. (Of course, this is not to say that there are no corresponding physiological phenomena and mechanisms; however, to borrow from the computer metaphor, this research is concerned with the software architecture rather than the hardware design.)

ACQUISITION OF ENCODING ALGORITHMS

Inferential encoding algorithms represent knowledge about covariations or contingencies (e.g., certain facial features *coincide* with the friendliness of a perceived person). The knowledge about a covariation between two features A and B allows the perceiver to infer (i.e., identify) B on the basis of A (e.g., one can infer friendliness of individuals on the basis of particular aspects of their facial expression). Although perceivers cannot articulate many of the inferential encoding algorithms that they use, there is ample evidence to suggest that those encoding rules result from some form of learning. Namely, the human cognitive system is capable of detecting (and storing in memory) information about contingencies and covariations between events or features present in the environment even if the attention of the perceiver is not directed at those covariations (e.g., Lewicki, 1986a, 1986b; Lewicki, Czyzewska, & Hoffman, 1987; Musen & Squire, 1993). The knowledge acquired in this way is later used to encode new stimuli, even if the particular encoding rule was learned without any conscious awareness, that is, perceivers may not know that they use or even "know anything about" this knowledge.

The independence of this process from subjects' awareness has been demonstrated in a number of studies focusing on the question of consciousness. For example, the phenomenon of nonconscious learning of information about covariations (color–word associations) was demonstrated in a recent study with patients with amnesia who did not have any conscious knowledge about the pattern manipulated in the stimulus material (Musen & Squire, 1993).

In another study, subjects nonconsciously acquired information about a complex pattern of locations of a target on a computer screen and this knowledge specifically facilitated their performance in the subsequent search task (e.g., it was found that when the pattern was changed, subjects' performance deteriorated). The entire sample of subjects tested in this study consisted of professors from a university psychology department who knew the experimenters investigated nonconscious processes (Lewicki, Hill, & Bizot, 1988). These subjects tried very hard to figure out what was being manipulated, and especially what caused their performance to suddenly deteriorate at one point during the task. However, none of the subjects came close to identifying the true nature of what was manipulated and investigated in the study, specifically, the nonconsciously acquired information about the nature of the pattern of target locations (Perruchet, Gallego, & Savy, 1990, demonstrated that the information about the pattern acquired in this study can be represented in terms of differential frequencies of certain types of target movements). Most subjects suspected that the experimenters used "subliminal stimuli," and at one point, the content of these stimuli somehow interfered with task performance. In another study, subjects who nonconsciously acquired information about a complex covariation could not report any explicit knowledge about what in fact they had learned (as was evident from their response latency patterns) even when promised a $100 cash reward (Lewicki, Czyzewska, & Hoffman, 1987).

It should be noted at this point that there is a clear difference between the acquisition of information from subliminal exposures (Greenwald, Klinger, & Liu, 1987; Greenwald, Klinger, & Schuh, 1995), and nonconscious processing of covariations. The nonconsciously acquired information about covariations is unavailable to subjects' awareness not because of any physical characteristics of the stimuli (e.g., exposure time), but rather because of the limited capacity and efficiency of the human consciously controlled cognitive system. Although usually the perceivers can see all relevant features of each item (stimulus), they naturally do not look for consistencies or covariations between features across the items unless they are instructed to deliberately focus their attention on such relations. However, even if so instructed, research indicates that perceivers experience great difficulty with consciously controlled covariation detection tasks (Nisbett & Ross, 1980) unless the covariation is very strong and salient (Crocker, 1981; see also Alloy & Tabachnik, 1984). Nevertheless, these covariations are still processed (outside of the perceiver's conscious attention) and stored in long-term memory, and the information about these nonconsciously acquired covariations is used to encode new stimuli.

Moreover, recent research indicates that knowledge about relations (between features or events) acquired outside of conscious awareness can be much more complex than the knowledge that the person is able to detect or even comprehend on the level of consciously controlled cognition (Lewicki et al., 1987; see the section on complex covariations, below).

These laboratory results are consistent with real-life observations that working (procedural) knowledge necessary for normal human cognition is more complex than what the average user of this knowledge is able to understand on the level of conscious awareness. For example, most fluent speakers of a language are not aware of the complex syntactic rules that they follow (e.g. most of us know that "a big, red barn" sounds better than "a red, big barn," but very few of us could say why), or most perceivers do not understand the complex geometric transformations that are in fact constantly performed by their cognitive system in order to recognize patterns or shapes of objects in perspective.

The process of acquisition of encoding algorithms discussed so far is based directly on the information that is objectively present and correctly acquired from the environment; thus, it should result in more or less objective generalizations of experience. In other words, this process can account for learning of true relations between features or events, that is, for the acquisition of those inferential encoding algorithms that reflect the true nature of reality and are similar in most perceivers (e.g., the procedures for interpreting spatial relations between objects in perspective, the semantic and syntactic rules of a language, or the rules for encoding some common types of social cues such as *smiling* $\Rightarrow$ *happy*).

However, this mechanism cannot explain how people acquire those subjective encoding rules that clearly do not reflect the objective nature of the environment. At the same time, the common existence of such biased encoding algorithms implies that the processes of acquisition of encoding algorithms suffer from some deficiencies. These deficiencies sometimes lead to the development of interpretive rules that are relatively independent of (or even inconsistent with) the objective nature of the environment encountered by the individual.

Recent research has uncovered one such general deficiency, which, in a sense, happens to be a by-product of the normal encoding process. Specifically, human memory tends to self-perpetuate existing encoding rules when objective, clear evidence is not available.

SELF-PERPETUATION OF ENCODING ALGORITHMS

Encoding processes impose on stimuli preexisting categories (or prototypes) even if the stimuli do not match well with the categories. This process of imposing such imperfectly fitting interpretive categories has been shown in a number of studies on pattern recognition and prototype abstraction (e.g., Posner, Goldsmith, & Welton, 1967; Reed, 1972), person perception (e.g., Cantor & Mischel, 1979; Higgins & King, 1981), and, more recently, in research on the nonconscious acquisition of information about covariations and its influence on subsequent encoding processes (Lewicki, 1986a, 1986b; Lewicki et al., 1987; Lewicki et al., 1988). Moreover, there is evidence indicating that biased encoding of stimuli (as more consistent with the inferential encoding rules than they actually are) may occur even when subjects are not aware that they are making any inferences. Instead, subjects may believe that they actually "see" the features that are in fact only inferred (Lewicki, 1986a, 1986b; Lewicki, Hill, & Sasaki, 1989).

Consider an encoding rule that implies a relation between features A and B. If the perceiver encounters a stimulus that is clearly A but ambiguous (i.e., uninformative) regarding B, then the stimulus is likely to be encoded as being both A and B (i.e., the ambiguity whether the stimulus is B or ¬B will be resolved and encoded in favor of B). For example, if the encoding rule implies a positive covariation between having short hair and being kind (Lewicki, 1986b), then ambiguous behavior of a newly encountered person with short hair is likely to be encoded as kind rather than unkind.

Therefore, as a result of the inferential processes involved in encoding, the final encoded (and stored in memory) representation of a stimulus consists of both the *objective* features of the stimulus (i.e., those actually present in the external world and directly perceived) as well as the *subjective* features that were not present (or could not be directly perceived) but rather inferred by the perceiver.

There is evidence to suggest that these two types of information (*directly perceived* and *inferred*) do not assume differential status in human memory, and once they are stored (i.e., become part of the memory representation of the stimulus) they are indistinguishable (Lewicki et al., 1989). In other words, the experience of perceiving some stimulus as being both A and B is stored in memory in the same way regardless of whether both features (A and B) were directly perceived by the perceiver or only feature A was perceived, whereas the stimulus was ambiguous regarding feature B and thus its presence was only inferred based on some preexisting encoding correlational rule stating that if A is present, then also B is present.

This proposition has far-reaching implications for the development of encoding algorithms. Namely, if the information about a stimulus possessing a set of characteristics is stored in the same way regardless of whether all the characteristics were directly perceived in the outside world or only inferred in the process of encoding, then it would mean inferential rules that control the encoding processes are capable of fabricating self-supportive evidence. Considering the fact that encoding algorithms develop and become stronger as a function of the amount of evidence (stimuli) interpreted as consistent with the rules (Hochberg, 1978; Lewicki, 1986a, 1986b; Lewicki et al., 1988; Lewicki et al., 1989), encoding algorithms may gradually develop and become stronger in a self-perpetuating manner.

In other words, ambiguous stimuli (that are objectively neither inconsistent with nor supportive of a given inferential encoding algorithm) would be encoded and stored in memory as consistent and, thus, contribute to the cumulative evidence supportive of the encoding algorithm. For instance, after encountering a number of short-haired persons whose behavior was objectively ambiguous regarding kindness (see the previous example), the encoding algorithm relating kindness to short hair would still be reinforced because the (in fact ambiguous) behavior of the persons with short hair is more likely to be interpreted as kind than similar behavior of persons without short hair.

Obviously, the self-perpetuation process cannot start until there is something to be self-perpetuated. In other words, an initial encoding rule (at least a weak one or one that is in an early stage of development) needs to exist before it can develop further via self-perpetuation. Thus, the question arises at this point as to where these initial erroneous rules come from.

Recent research has demonstrated that such initial encoding rules can be acquired incidentally from an even extremely limited amount of consistent evidence: Encountering just a few instances that are consistent with a particular rule is sufficient to cause a respective encoding tendency to appear (Lewicki, 1985, 1986b). Such encoding biases are initiated outside of conscious control, thus the individual can neither prevent them nor discard them as invalid. However, initially they are very weak (i.e., their influence on encoding is small relative to other factors that can potentially affect the final impression). Thus, due to the statistical nature of reality (i.e., the low probability to repeatedly encounter, by chance, instances of a factually nonexistent covariation), it is reasonable to assume that most of the self-perpetuation processes are stopped or reversed by unambiguous evidence encountered by the perceiver before strong biases develop.

However, if such unequivocal inconsistent (with the initial bias) evidence is not present in one's environment, and if at the same time ambiguous evidence (that is open to biased encoding) is encountered, then the bias will start to self-perpetuate. This may be particularly common in social cognition because social cues are ambiguous by nature and thus open to alternative interpretations. Eventually, the bias might become so strong that it is practically impervious to falsification through objective inconsistencies (as is, for example, the case in neurotic patients who cannot control their dysfunctional emotional responses).

The existence of the phenomenon of self-perpetuation has been demonstrated using a wide variety of experimental procedures and stimulus materials containing the hidden manipulated covariations (e.g., Hill, Lewicki, Czyzewska, & Boss, 1989; Hill, Lewicki, Czyzewska, & Schuller, 1990; Hill, Lewicki, & Neubauer, 1991; Lewicki & Hill, 1989; Lewicki et al., 1989). For example, subjects developed in a self-perpetuating manner new encoding dispositions in the process of searching for a target digit in matrices of digits; searching for a target digit in matrices of letters; learning to "predict movements of dynamic targets"; evaluating relative lengths of geometric figures; evaluating weights of objects based on their sizes; acquiring "an intuitive knowledge" of a new ("Polynesian") language; learning to "judge intelligence of a patient based on brain scans"; learning to "interpret personality characteristics of stimulus persons based on their silhouettes"; learning to estimate fairness of a university professor based on hidden proportions of his schematic face; learning to detect "hidden sadness" of stimulus persons based on videotaped episodes.

DEVELOPMENT OF COMPLEX (CONDITIONAL) ENCODING ALGORITHMS

It has been shown in several studies that covariations of considerable complexity can be implicitly detected and processed. Subjects in those experiments nonconsciously acquired procedural knowledge about formal structures of the material that were not only too complex and too confusing to be consciously noticed, but even exceeded the complexity level of knowledge that one can use in consciously controlled inferences (e.g., they were four-way interactions; Lewicki et al., 1987). Because the development of real-world experience and expertise must involve acquisition of very complex procedural knowledge, this finding is of importance because it demonstrates how such knowledge can be acquired.[2]

---

[2]Ceci and Liker (1986) analyzed patterns of reasoning applied by successful horse race handicappers and concluded that their subjects showed signs of consciously employing models involving higher order interactions. However, this interpretation is yet to be supported by controlled

In one of the experiments, subjects performed a search task in which they were supposed to quickly search for locations of a target character hidden in matrices of distractor characters on a computer screen—the so-called matrix scanning paradigm.[3] The pattern of target locations manipulated in the experiment was extremely complex; specifically, it followed a sequence determined by a fourth-order interaction involving five (random) variables. Nevertheless, the analysis of subjects' performance on this task revealed that every subject tested in this study eventually acquired procedural knowledge of this complex pattern. Moreover, after the experiment, subjects, who were college students, were offered a sizable cash award ($100) for identifying any systematic feature of the material. Some subjects spent several hours trying to find a systematic pattern, but none of them succeeded; yet all of these subjects had earlier detected and learned (nonconsciously) at least some aspects of the pattern, and had successfully used that knowledge, as evidenced by their improved performance in the search task (response latency patterns; for a replication and extension of that study, see Stadler, 1989).

A series of studies on the process of acquisition of such complex encoding algorithms suggests that subjects learn those complex knowledge structures via a process of "conditional elimination": An encoding algorithm based on a simple covariation (between two features or events) can be abandoned (and replaced by a new one) when it does not fit the current stimuli well, but the

---

laboratory studies. Moreover, there is convergent evidence indicating that even if subjects are instructed to do so, they experience great difficulty with consciously controlled covariation detection tasks even when simple covariations are manipulated (Nisbett & Ross, 1980) unless the covariation is very strong and salient (Crocker, 1981; see also Alloy & Tabachnik, 1984).

[3]The *matrix scanning procedure* was used in numerous studies on nonconscious learning (e.g., Lewicki, 1986a; Lewicki et al., 1987; Stadler, 1989) and self-perpetuation (e.g., Hill et al., 1989, Exp. 1; Lewicki et al., 1989, Exps. 3a and 3b). In the learning phase of the task, subjects participate in a search task in which they speed-search a matrix of distractor characters (usually digits or letters) for a target character (usually digit 6 or letter *K* or *H*). The manipulated covariations are built into the relations between target locations and features of the background matrix of distractors. In the testing ("guessing") segment, the matrices are exposed for a very short time (ostensibly subliminally), and the subjects are asked to "guess" the location of the target "following their intuition." Because in this phase, the matrices are hardly visible, subjects cannot find out that these matrices in fact do not contain the target. Thus, if they have a "feeling of seeing" the target (and they see it in locations consistent with the pattern manipulated in the learning phase), it can only be due to "inferences" they make in the process of encoding by applying the encoding categories (covariations) learned in the training phase. The guesses of subjects in the matrix scanning tasks indicate that the subjects follow the encoding rules acquired in the learning phase and that the consistency of the responses with the rule gradually increases over time in the self-perpetuating manner (i.e., it increases despite the lack of any supportive evidence present in the testing phase material). This procedure also allows for the use of sensitive response-latency indices capable of detecting relatively early stages of the development of specific procedural knowledge (affecting the encoding). For example, in some of those studies, during the regular learning phase, the manipulated covariation was suddenly changed (typically "reversed"), causing an immediate increase of subjects' response latencies (indicative of the expected effect of learning).

abandoned algorithm is not lost entirely but only deactivated (i.e., conditionally eliminated), and it can be reactivated and used again when stimulus material consistent with the old algorithm is encountered (Lewicki, 1986a). If the crucial feature of the material that determines which encoding algorithm should be used is detected, then a higher order encoding algorithm begins to develop. This research also revealed that there are individual differences in cognitive styles between subjects in terms of the degree to which the perceivers can flexibly switch from one encoding algorithm to another (in the process of "conditional elimination").

INDIRECT INFERENCES IN
THE DEVELOPMENT OF IMPLICIT KNOWLEDGE

There is evidence indicating that the processes of acquisition of information about covariations (reviewed in the previous sections) may prompt a spontaneous development of new relations between concepts or features (Lewicki, Hill, & Czyzewska, 1994). Specifically, new relations can emerge between variables that have not been found to be connected in the environment and whose relation could only be inferred "indirectly" (i.e., by applying the rule of transitivity). In other words, if an individual acquires information about a covariation between features A and B, and independently, between features B and C, then this may result in the spontaneous development of an expectation that A and C are also related (i.e., a new encoding algorithm would emerge representing the implicit knowledge that objects that are A, are also C).

This phenomenon of nonconscious indirect inferences—generating new encoding algorithms by applying the rule of transitivity to connect existing algorithms—was recently demonstrated with different stimulus materials. Most of the studies used modified versions of procedures tested in previous research (matrix scanning tasks, schematic pictures of stimulus persons, etc.).

One of the experiments used the *kinematics procedure*[4] (Lewicki et al.,

---

[4]The *kinematics procedure*, based on an experimental paradigm introduced by Runeson and Frykholm, 1983, was used in a study on self-perpetuation of nonconsciously learned encoding biases (Hill et al., 1989, Exp. 2) and in a recent series of studies on nonconscious indirect inference (Lewicki, Hill, & Czyzewska, 1992). One of its advantages is that subjects find it interesting (or even entertaining) and thus they attend very well to the material even if it is long. In the learning phase, labeled "training designed to put you [subjects] in a particular mode of thinking about other people," subjects are exposed to a series of videotaped episodes schematically presenting people performing a variety of simple operations (drinking a canned beverage, lifting an object, etc.). During each episode, subjects are informed (via a prerecorded soundtrack synchronized with the video presentation) whether the particular person shown in the video possesses a particular characteristic (e.g., likability as determined in case studies). The material contains hidden covariations between the respective characteristic (e.g., likability) and some nonsalient details of the videotaped pictures (e.g., a proportion of the stimulus person's body: in one version

1994) in which subjects were exposed to videotapes presenting abstracted movements of selected points on bodies of invisible stimulus persons engaged in various activities. Information about the personality of each stimulus person was provided. In the first phase of the study, subjects nonconsciously learned a covariation between the personality information (A) and a subtle variation of distances between the dots identifying the legs of the stimulus persons (B). In the second phase, dots on subjects' arms were introduced (C) and their distances covaried with distances between the dots on legs (B), however, no information about the personality of the stimulus persons was provided (A). In the testing phase, only dots on arms were shown (C) and subjects were asked to "make intuitive judgments of personality" (A) of the persons "based on the dynamics of their body language."

Consistent with expectations, subjects' judgments (of feature A) were found to be affected by the distances between dots on arms (feature C), although in this arrangement of the stimulus material the relation between the two features (A and C) could be established by subjects only indirectly (i.e., "via" feature B). As usual, tests of participants' awareness revealed no trace of their knowledge about any relations between the crucial features manipulated in the stimulus material. Moreover, not a single subject noticed any variation of the (in fact systematically varied) distances between dots marking the limbs of the stimulus persons.

NONCONSCIOUS DEVELOPMENT OF "META-KNOWLEDGE" ABOUT RELATIONS BETWEEN VARIABLES

Convergent evidence from research on the nature of expertise suggests that mental models of experts are not just more accurate, and containing more specific details than corresponding models of novices (Rouse & Morris, 1986). Instead, the novice–expert shift requires a more conceptual change involving a movement from representational (i.e., more concrete) to abstract models of the relevant areas (Larkin, 1983). Expertise appears to entail global, "pattern-oriented representations" (Chase & Simon, 1973), which must involve gen-

---

of this procedure the videotape was prepared such that the background was completely white, while the only visible elements of the stimulus persons' bodies were black spots marking their ankles and joints). During the testing phase, the personality information is not available; subjects are asked to rate the (previously manipulated) characteristics of additional (different) stimulus persons "based on intuition." (In the previous implementation of this paradigm, in order to increase the believability of the instructions. each participant was handed a copy of Runeson & Frykholm (1983), published in the *Journal of Experimental Psychology: General*, providing "scientific evidence" that one can accurately make such "intuitive" judgments based on "kinematics.")

eral concepts of crucial dimensions indicative of the qualities or outcomes that need to be properly identified or predicted by the expert.

Taking into account the fact that experts cannot articulate significant aspects of their expertise and this expertise must be acquired through implicit learning processes, it was expected that the cognitive system is capable of extending the knowledge about implicitly learned specific covariations into abstractions (or at least more general representations) that go beyond the specifics of the situation from which the covariation was originally acquired. It was hypothesized that the transfer of knowledge from one situation to another (relevant) situation is not limited to nonabstract dimensions associated with concrete instances (e.g., density of small components of a specific type of background), but it also involves more abstract (possibly even semantic) concepts, such as "the density of elements" (independent of what these elements are and what role they play in the display).

In addition to the general functional argument for expecting that the transfer involves abstraction (i.e., the argument that it is required for effective development of expertise, as mentioned earlier), this expectation can also be derived from theories of semantic memory and previous research on semantic concept formation as a result of nonconscious acquisition of information about covariations (Lewicki, 1986a, Exps. 4.3, 4.4, 4.5, and 4.6). The results of that previous research indicate that newly (nonconsciously) acquired information about a covariation between variables A and B does specifically influence the respective semantic memory representations of concepts A and B. The changes in semantic memory (registered via reaction time measures in the semantic judgment paradigm) indicated that as a result of acquisition of information about the covariation, the memory representations of the respective concepts A and B change (or "develop"; see Lewicki, 1986a, pp. 71–75), thus facilitating abstractions of knowledge acquired from the covariation. In a sense, the hypothesized process of abstraction does not require a "designated" mechanism; instead, one can say that the expected process may represent a natural consequence of the organization of knowledge. (This reasoning is consistent with both the computational and pre-storage types of theories of semantic memory; see Lewicki, 1986a, chap. 4.)

In functional terms, such generalization of implicitly acquired knowledge about a covariation between variables A and B (which may involve processes similar to those studied in classical research on learning; Fields & Verhave, 1987) would lead to retention of information about global characteristics of variables A and B. This, in turn, would result in the development of meta-knowledge about the covariation, helpful in identifying similar systematicities in other environments. For example, a person who has acquired information about a covariation between A and B would be more sensitive to "A-" and "B-type variables" in other relevant environments in which similar variables may also be related, and thus that person may be better equipped to encode information about those new situations. In most simple terms, this process can be described as implicit learning not only about a specific covariation but also about what general kinds of variables covary and thus "where to look" for systematicities in other similar environments.

This expectation was confirmed in a recent experiment in which participants were exposed to two segments of stimulus material, containing two nonidentical but structurally similar covariations. The first segment contained only a learning phase. The second segment consisted of a learning and a testing phase. The results suggest that subjects' acquisition of information about the second covariation (measured by the degree to which their performance in the testing phase of the second segment was consistent with the pattern manipulated in the learning phase of the second segment) was facilitated by the fact that they had previously acquired the procedural knowledge of the covariation in the first segment. Specifically, in the control group where no consistent covariation was embedded into the first segment of the material (thus the participants could not acquire any relevant knowledge before they were exposed to the second segment material), the acquisition of the covariation in the second segment was slower than in the experimental groups where a consistent covariation was present in the first segment. The matrix scanning procedure (described before) was used. The target character was "6" and the background characters were letters in the first segment and numbers in the second segment; thus, even though the two covariations were different and the knowledge about the first one could not directly facilitate subjects' performance in the second segment, the two covariations were structurally similar because they both involved knowledge about the relation between "the character composition of the background" and "the location of the target." It was found that subjects who learned about the covariation involving letters in the first segment learned the covariation about the numbers (in the second segment) more quickly than those who were exposed to the control version of the first segment with no systematic covariation present.

The ability to acquire meta-knowledge such as demonstrated in this matrix scanning experiment appears to be an indispensable, core component of the process of "gaining experience" with particular areas of reality (especially experience that is difficult to articulate; see the first section of this chapter).

CONCLUDING COMMENTS

This chapter has reviewed research on a class of cognitive mechanisms that contribute to nonconscious acquisition of procedural knowledge capable of influencing encoding processes. The results of this acquisition are (potentially complex) structures of information necessary for the advanced levels of performance in a variety of areas, such as those investigated in studies on mental models of experts. It should be noted, however, that the generality and applicability of the mechanisms discussed here is not limited to the areas traditionally associated with advanced or unusual levels of performance, such as chess mastery or the ability to interpret X-ray pictures. Most likely, the same or a comparable level of formal complexity of nonconscious acquisition of knowledge is necessary in order to acquire common skills such as linguistic abilities, which most of us happen to possess, that is, they apply to skills not commonly thought of as requiring advanced levels of performance. The human cognitive system is equipped to nonconsciously acquire complex encoding skills; some of them are necessary to maintain the normal level of adjustment (e.g., speaking, thinking, problem solving); others require, among other factors, extended interaction (experience) with some specific areas of reality and allow for the development of expert level performance in the respective specialized areas.

ACKNOWLEDGMENT

This research was supported by National Science Foundation Grant BNS-89-20726 and National Institute of Mental Health Grant MH42715.

REFERENCES

Alloy, L. B., & Tabachnik, N. (1984). Assessment of covariation by humans and animals: Joint influence of prior expectations and current situational information. *Psychological Review, 91*, 112–149.

Brewer, W. F., & Nakamura, G. V. (1984). The nature and functions of schemas. In R. S. Wyer, Jr. & T. K. Srull (Eds.), *Handbook of social cognition* (Vol. 1, pp. 119–160). Hillsdale, NJ: Lawrence Erlbaum Associates.

Cantor. N., & Mischel, W. (1979). Prototypes in person perception. In L. Berkowitz (Ed.), *Advances in experimental social psychology* (Vol. 12, pp. 3–52). New York: Academic Press.

Ceci, S. J., & Liker, J. K. (1986). A day at the races: A study of IQ, expertise, and cognitive complexity. *Journal of Experimental Psychology: General, 115*, 255–266.

Chase, W. G., & Simon, H. A. (1973). The mind's eye in chess. In W. G. Chase (Ed.), *Visual information processing* (pp. 215–281). New York: Academic Press.

Chi, M. T., Glaser, R., & Farr, M. J. (Eds.). (1988). *The nature of expertise.* Hillsdale, NJ: Lawrence Erlbaum Associates.

Chomsky, N. (1980). Language and unconscious knowledge. In N. Chomsky (Ed.), *Rules and representations* (pp. 217–254). New York: Columbia University Press.

Cofer, C. N. (1973). Constructive processes in memory. *American Scientist, 61*, 537–543.

Crocker, J. (1981). Judgment of covariation by social perceivers. *Psychological Bulletin, 90*, 272–292.

Ericsson, K. A., & Smith, J. (1991). *Toward a general theory of expertise: Prospects and limits.* Cambridge, England: Cambridge University Press.

Fields, L., & Verhave, T. (1987). The structure of equivalence classes. *Journal of the Experimental Analysis of Behavior, 48*, 317–332.

Greenwald, A. G., Klinger, M. R., & Liu, T. J. (1987). *Unconscious processing of word meaning.* Unpublished manuscript.

Greenwald, A. G., Klinger, M. R., & Schuh, E. S. (1995). Activation of marginally perceptible ("subliminal") stimuli: Dissociation of unconscious from conscious cognition. *Journal of Experimental Psychology: General, 124,* 22–42.

Gustin, W. C. (1985). The development of exceptional research mathematicians. In B.S. Bollm (Ed.), *Developing talent in young people* (pp. 270–331). New York: Ballantine.

Hayes, J. R. (1981). *The complete problem solver.* Philadelphia: Franklin Institute Press.

Higgins, E. T., & King, G. A. (1981). Accessibility of social constructs: Information processing consequences of individual and contextual variability. In N. Cantor & J. Kihlstrom (Eds.), *Personality, cognition and social interaction* (pp. 69–121). Hillsdale, NJ: Lawrence Erlbaum Associates.

Hill, T., Lewicki, P., Czyzewska, M., & Boss, A. (1989). Self-perpetuating development of encoding biases in person perception. *Journal of Personality and Social Psychology, 57,* 373–387.

Hill, T., Lewicki, P., Czyzewska, M., & Schuller, G. (1990). The role of learned inferential encoding rules in the perception of faces: Effects of nonconscious self-perpetuation of a bias. *Journal of Experimental Social Psychology, 26,* 350–371.

Hill, T., Lewicki, P., & Neubauer, R. M. (1991). The development of depressive dispositions: A case of self-perpetuation of encoding biases. *Journal of Experimental Social Psychology, 27,* 392–409.

Hochberg, J. (1978). *Perception.* Englewood Cliffs, NJ: Prentice-Hall.

Johnson, M. K., & Raye, C. L. (1981). Reality monitoring. *Psychological Review, 88,* 67–85.

Kaufman, L. (1974). *Sight and mind: An introduction to visual perception.* New York: Oxford University Press.

Kihlstrom, J. (1987). The cognitive unconscious. *Science, 237,* 1445–1452.

Krogius, N. (1976). *Psychology in chess.* New York: RHM Press.

Lachman, R., Lachman, J. L., & Butterfield, E. C. (1979). *Cognitive psychology and information processing: An introduction.* Hillsdale, NJ: Lawrence Erlbaum Associates.

Lehman, H. C. (1953). *Age and achievement.* Princeton, NJ: Princeton University Press.

Lesgold, A. M. (1984). Acquiring expertise. In J. R. Anderson & S. M. Kosslyn (Eds.), *Tutorials in learning and memory: Essays in honor of Gordon Bower* (pp. 31–60). New York: Freeman.

Larkin, J. H. (1983). The problem of knowledge representation in physics. In D. Gentner & A. L. Stevens (Eds.), *Mental models* (pp. 75–98). Hillsdale, NJ: Lawrence Erlbaum Associates.

Lewicki, P. (1984). Self-schema and social information processing. *Journal of Personality and Social Psychology, 47,* 1177–1190.

Lewicki, P. (1985). Nonconscious biasing effects of single instances on subsequent judgments. *Journal of Personality and Social Psychology, 48,* 563–574.

Lewicki, P. (1986a). *Nonconscious social information processing.* New York: Academic Press.

Lewicki, P. (1986b). Processing information about covariations that cannot be articulated. *Journal of Experimental Psychology: Learning, Memory, and Cognition, 12,* 135–146.

Lewicki, P., Czyzewska, M., & Hoffman, H. (1987). Unconscious acquisitions of complex procedural knowledge. *Journal of Experimental Psychology: Learning, Memory, and Cognition, 13,* 523–530.

Lewicki, P., & Hill, T. (1989). On the status of nonconscious processes in human cognition: Comment on Reber. *Journal of Experimental Psychology: General, 118,* 239–241.

Lewicki, P., Hill, T., & Bizot, E. (1988). Acquisition of procedural knowledge about a pattern of stimuli that cannot be articulated. *Cognitive Psychology, 20,* 24–37.

Lewicki, P., Hill, T., & Czyzewska, M. (1992). Nonconscious acquisition of information. *American Psychologist, 47,* 796–801.

Lewicki, P., Hill, T., & Czyzewska, M. (1994). Nonconscious indirect inferences in encoding. *Journal of Experimental Psychology: General, 123,* 257–263.

Lewicki, P., Hill, T., & Sasaki, I. (1989). Self-perpetuating development of encoding biases. *Journal of Experimental Psychology: General, 118,* 323–337.

Markus, H. (1977). Self-schemata and processing information about the self. *Journal of Personality and Social Psychology, 35,* 63–78.

Musen, G., & Squire, L. R. (1993). Implicit learning of color-word associations using a Stroop paradigm. *Journal of Experimental Psychology: Learning, Memory, and Cognition, 19,* 789–798.

Neisser, U. (1967). *Cognitive psychology.* New York: Appleton-Century-Crofts.

Nisbett, R. E., & Ross, L. (1980). *Human inference: Strategies and shortcomings of social judgment.* Englewood Cliffs, NJ: Prentice-Hall.

Nisbett, R. E., & Wilson, T. D. (1977). Telling more than we can know: Verbal reports on mental processes. *Psychological Review, 84,* 231–259.

Perruchet, P., Gallego, J., & Savy, I. (1990). A critical reappraisal of the evidence for unconscious abstraction of deterministic rules in complex experimental situations. *Cognitive Psychology, 22,* 493–516.

Posner, M. I., Goldsmith, R., & Welton, K. E., Jr. (1967). Perceived distance and the classification of distorted patterns. *Journal of Experimental Psychology, 73,* 28–38.

Reber, A. S. (1989). Implicit learning and tacit knowledge. *Journal of Experimental Psychology: General, 118,* 219–235.

Reed, S. K. (1972). Pattern recognition and categorization. *Cognitive Psychology, 3,* 382–407.

Rock, I. (1975). *Introduction to perception.* New York: Macmillan.

Rouse, W. B., & Morris, N. M. (1986). On looking into the black box: Prospects and limits in the search for mental models. *Psychological Bulletin, 100,* 349–363.

Rumelhart, D. E. (1984). Schemata and the cognitive system. In R. S. Wyer, Jr. & T. K. Srull (Eds.), *Handbook of social cognition* (Vol. 1, pp. 161–188). Hillsdale, NJ: Lawrence Erlbaum Associates.

Rumelhart, D. E., & Ortony, A. (1977). The representation of knowledge in memory. In R. C. Anderson, R. J. Spiro, & W. E. Montague (Eds.), *Schooling and the acquisition of knowledge* (pp. 99–135). Hillsdale, NJ: Lawrence Erlbaum Associates.

Runeson, S., & Frykholm, G. (1983). Kinematic specification of dynamics as an informational basis for person-and-action perception: Expectation, gender, recognition, and deceptive intention. *Journal of Experimental Psychology: General, 112,* 585–615.

Sosniak, L. A. (1985). Learning to be a concert pianist. In B. S. Bloom (Ed.), *Developing talent in young people* (pp. 19–67). New York: Ballantine.

Stadler, M. (1989). On learning complex procedural knowledge. *Journal of Experimental Psychology: Learning, Memory, and Cognition, 15,* 1061–1069.

Wyer, R. S., & Srull, T. K. (1981). Category accessibility: Some theoretical and empirical issues concerning the processing of social stimulus information. In E. T. Higgins, C. P. Herman, & M. P. Zanna (Eds.), *Social cognition: The Ontario symposium* (pp. 161–197). Hillsdale, NJ: Lawrence Erlbaum Associates.

C H A P T E R

Consciousness in
the Explicit (Deliberative)
and Implicit (Evocative)

Donelson E. Dulany
University of Illinois at Urbana-Champaign

In an instant we can consciously symbolize a moment in the past, part of a distant present, or even some remote and only possible future. Obviously, too, we do somehow deal with the not here and not now. Nevertheless, there is a deep and unusually difficult question for this discipline and beyond. How do we carry the symbolic representations we actually use? Consciously or unconsciously, or both? The aim of this chapter is to sketch more deeply contrasting views of consciousness, show their expression in contrasting theories of implicit and explicit learning, summarize some illustrative evidence, and consider a few broader implications and possible responses. In this domain, we ask about the conscious or unconscious representation of what is learned and used in learning and judging.

Central to my proposal is the idea that the explicit and implicit are distinguished, not by one being conscious and the other unconscious, as is generally assumed, but by different forms of nonconscious operations on different forms of conscious contents that carry symbolic representation. We should identify the *explicit* with *deliberative* mental episodes and the *implicit* with *evocative* mental episodes (associative-activational), but one is no more conscious or nonconscious than the other.

A STANDARD COGNITIVE VIEW

From the emergence of cognitive psychology, consideration of consciousness has largely been an attempt to locate it in alien architectures, a strategy oddly

179

needed to legitimize the topic, but one with entailments that are questionable (Dulany, 1984, 1991).

One is the assumption that a limited place for consciousness leaves abundant room for *unconscious symbolic representation*. On the *computational view* of mind, what is sometimes called cognitivism in artificial intelligence (AI) and philosophy, consciousness is only an occasional, incidental emergent of a level of cognitive representation where all computation, all real mental activity, is said to take place (Haugeland, 1978; Jackendoff, 1987; Velmans, 1991). Consider Gardner's (1987) comment: "To my mind, the major accomplishment of cognitive science has been the clear demonstration of the validity of positing a level of mental representation . . . but this form of representation does not involve processes of which the organism is in any way conscious or aware" (pp. 383–384). And, according to Dennett (1987), "so far as cognitive science is concerned, the important phenomena are the explicit unconscious mental representations" (p. 218). On the *information-processing view,* consciousness is identified with a short-term or working memory or focus of attention, with unconscious representations briefly registered in a sensory memory, plentifully stored in long-term memory, and even actively involved in a working memory just outside a limited focus of attention (e.g., Bower, 1975; Mandler, 1975; Posner, 1978). Summing up that tradition, R. Lachman, J. Lachman, and Butterfield (1979) concisely put consciousness in its place: "Most of what we do goes on unconsciously. It is the exception, not the rule, when thinking is conscious" (p. 207). On a related *global-distributed view,* consciousness is a globally active coordinating system for countless specialized and unconscious processors, each taking orders and reporting back, each operating with a significant degree of independent and symbolic intelligence (Baars, 1988). Johnson-Laird (1988) summed up this view: "I have presented a theory of mental architecture. The conscious mind depends on the serial processing of explicitly structured symbols. The unconscious mind depends on the parallel processing of distributed symbolic representations" (p. 382). What has recently been heralded as a "rediscovery (or reclaiming) of the unconscious," or even a "paradigm shift," could be better described as a late celebration of what has been a foundational commitment throughout the cognitive movement.

Connectionist architectures have been exceptions, of course, with no one promoting a habitat for consciousness. It is typically viewed as only an emergent of a settled and stable network that is itself characterized as nonsymbolic (Rumelhart, Smolensky, McClelland, & Hinton, 1986, p. 39). Indeed, the question raised here cannot be addressed until there is more systematic effort to identify units with the world of symbolic representations.

A second entailment of standard architectures is what I have called the *separate systems assumption* (Dulany, 1991). If what is conscious and what not are assigned to separate systems, they can operate independently, and hence they can be said to have different, if overlapping, roles carried out in different ways: Unconscious processing is usually said to be fast, parallel, effortless, and efficient; and consciousness is said to be slow, serial, effortful, and inefficient (Baars, 1988; Mandler, 1985, chap. 3). With processing theoretically divided, the routine of our lives is said to be governed by intelligent but insensible agencies, whereas consciousness itself shows up for a modest try when there is novelty to be handled, problems to be solved, or troubles to be shot.

From these assumptions, it is but a short step to the third: the postulation of a *personal unconscious,* or cognitive unconscious, as it is more often called now (Kihlstrom, 1987; Reber, 1993). What we are imperfectly aware of doing as persons—searching, recognizing, comparing, deciding, inferring—is assigned to another system, which thereby gains personhood in just that significant sense: as a system of those mental episodes. Indeed, mainstream theorists have filled an unconscious memory with just those abstract and pictorial symbols that populate our own conscious remembrances; the one is deeply analogous to the other. It becomes natural, then, to think that we can learn and use knowledge within an unconscious if we can do those things within consciousness—but in ways less fettered by its commonly assumed limitations.

Generally, too, within the standard tradition we find a *nonanalytic posture* with respect to consciousness. There is relatively little analysis of modes and contents of consciousness, with the consequence that it is difficult in that language even to speak clearly of just what is and is not conscious. Claims for "unconscious learning" (or for that matter, "unconscious perception" or "unconscious memory") lack weight with little clarity beyond the prefix.

AN ALTERNATIVE MENTALISTIC VIEW

As standard claims for unconscious processing are challenged in various literatures (e.g., Dulany, 1991; Holender, 1986; Shanks & St. John, 1994), we need a clearer alternative to that standard cognitive metatheory. The essence of this mentalistic view is the strong hypothesis that symbolic representation is carried exclusively by conscious contents, and their representational power, in turn, is yielded by their role in the mental activity, deliberative and evocative mental episodes, that permits coping in varying degree with a world beyond. Indeed, I would suggest that this is the principal adaptive significance of consciousness. With consciousness, we represent and come to terms with our external and internal worlds, real and imagined, with varying and imperfect— but generally adaptive—success. For a start, our commonsense meanings of "symbolic representation" and "consciousness" should be clear and common enough for the last three sentences to communicate, but the meaning of both terms is necessarily elaborated in the development that follows.

Analysis of the Intentionality of Consciousness.   From Brentano (1874/ 1973) onward, analyses of intentionality have varied considerably (see Dennett, 1987; Dretske, 1988; Fodor, 1994; Putnam, 1988; Searle, 1983, 1990, 1992), but a central thread is the effort to say what it is about mental activity that permits it to represent symbolically, to refer or to stand for—to be, in the traditional term, *about* something. Although philosophers have claimed the topic, we should develop the analysis of intentionality that can work best for us in this science. If we do, it may help us gain a better understanding of how we symbolically represent with consciousness, although it is an empirical question as to whether only conscious states possess intentionality. In this context, the core meaning of "intentionality" is semantic, and not, on a common misunderstanding, volitional—although volitional intentions themselves may have semantic value.

As I see it, the key principle is this: An intentional state lies at the intersection of several kinds of central variables for the science—agency, mode, and content, in each case a variable that may be defined over conscious states. These are the variables on which we may theoretically position conscious states within the mental episodes that can give those contents their symbolic representational power. Consider, for example, the thought: "I believe that implicit learning is a reality."

1. *Agency* is a sense of "I" varying in frequency and intensity, a sense that appears in that thought. It is a challenge certainly for experimentalists, but well known to clinicians who encourage neurotics to own their feelings and see psychotics disown some of their beliefs—and then attribute them to the TV or voices. Agency is centrally involved in learning and maintaining a coherent sense of self.

2. *Content* is infinitely varied, as is the world it represents. But there is a theoretically significant distinction to be made between contents that are propositional and those that are not. With a propositional content, there is a value on a subject variable (its "argument" in the predicate calculus): "implicit learning" rather than "explicit learning." And something is predicated of it: "is a reality" rather than "is not a reality." A propositional content takes values on both subject and predicate variables, whereas a nonpropositional content takes a value on only one variable. In ordinary language, for example, we speak of a belief *that* ____, where the blank would be filled with a propositional content, but we refer to a perception or image *of* ____, where the blank would be filled with a nonpropositional content.

Contents of awareness also come in different kinds of *codes*—a common enough cognitive concept. *Literal* (or "perceptual") codes are first the outputs of sensory transduction, the code that is called "analogue" when later it is a content of remembrance or creative imagery. *Semantic* codes identify an object as such and may stand apart in abstract representations. Furthermore, content at the moment has a *focus* and a *surround*, with that context usually appearing in literal code, but nevertheless capable of influencing the semantic interpretation of the focus. Conscious contents may also be *correlated,* given co-occurrences of the world events they symbolically represent. Subjects in a "covariation" learning experiment, for example, may be unaware that it was long and short hair that predicted certain personal traits—significant contents for the experimenter—but still do rather well on their correlated awareness that it was short and long necks (Dulany & Poldrack, 1991).

3. *Modes* of awareness carry those contents and come in familiar types and subtypes. Sometimes the content they carry is propositional, and so those modes have traditionally been called "propositional attitudes"—for example, a belief, perception, or intention *that* ____. But sometimes the mode carries a nonpropositional content and might reasonably be called a "nonpropositional attitude"—only a perception, feeling, or intuition *of* ____. Modes provide variables, some very familiar, such as degree of belief or strength of feeling or brightness.

4. But what about *aboutness*—the traditional essence of intentionality? Sometimes it is said (e.g., Palmer, 1978; Rumelhart & Norman, 1988) that symbolic representations are about something else in the sense that there can be a mapping of events in a "represented world" to a "representing world." Sometimes, too, it is said (e.g., Haugeland, 1985; Pylyshyn, 1984) that human mental activity, like the running of a computer program, symbolically represents just by virtue of the syntax of its computation, although the view is often defended by noting that the syntax of computation can sometimes be mapped to the syntax of external events: Adding three to three others produces six both computationally and physically. A difficulty for the mapping criterion, however, is that mental contents function with considerable independence of whatever they seem to be about—the "intentional inexistents," in Brentano's (1973) term. This is seen most clearly in all sorts of symbolic imaginings and imperfect ideas of remote pasts and only possible futures that are simply unmappable to any represented world.

How then might mental contents get their symbolic power? They get it in two ways that yield useful theoretical and experimental criteria of symbolic representation. In one, we *predicate* something of a mental content, and we can then use that propositional form in deliberative mental episodes. If you believe that snow is slushy (the predication), and you want dry feet, decide to wear your boots—the conclusion of a deliberative mental episode. In another way, mental contents gain symbolic power by *evoking* other symbols. The idea of snow evokes images of sleds, downhill racers, or just getting stuck—in each case, an evocative mental episode. Because transducers link intervening mental activity to stimulus energies and movement, both kinds of mental episodes permit us to cope in some degree, in this case with the sometimes snowy world beyond, thereby conveying aboutness on the intervening mental contents. As they must, these general criteria of use can apply alike to symbols of the real and unreal, of the present, past, or future. Furthermore, developmentalists (DeLoache, 1989; Perner, 1991) are already studying how the child forms the belief that one mental content symbolizes another—on this account, the meta- awareness that is the traditional *sense of aboutness*.

5. What is the *domain* for symbolic representation in consciousness? The simple and programmatic answer is this: between the outputs of sensory transducers and the inputs to motor transducers. Sensory transducers transform stimulus energies yielding a first symbolic code: literal awareness of color, brightness, edges, pitch, amplitude, movement, depth, and so on. And motor transducers transform a last symbolic code in conscious intentions and percepts, yielding control down through Gallistel's (1981) hierarchy of muscle groups, single muscles, and muscle fibers. Within those transducers is the domain of the nonsymbolic: the "informationally encapsulated" (Fodor, 1983) and "cognitively impenetrable" (Pylyshyn, 1984). Between those transducers is the standard domain of symbolic representation (Pylyshyn, 1984). But we need to examine the strong hypothesis underlying this view by seeing empirically how far and consistently toward those boundaries conscious representation goes, to see what the evidence will show and how much that is now thought to be unconscious representation might be better explained otherwise.

A simple notational device—Agency[Mode(content)]—can help us see what constitutes a mental event with intentionality. We as agents represent the world in one of many modes by one of many contents—and on the view here, we do that only *in consciousness*.

Unified System for the Conscious and Nonconscious. I suggest a rather grand principle: Within the domain of symbolic representation, mental episodes consist of nonconscious operation upon conscious intentional states, yielding other conscious intentional states. These operations and states are what James (1890, p. 243) so evocatively called the "flights and perchings" of thought.

1. Consider a general form for a mental episode, with a conscious state i at moment $n + 1$ yielded by operation on other conscious states, j and k, at prior moments:

$$S_{in+1} \leftarrow \text{Operation upon } (S_{jn}, \ldots, S_{kn-m}).$$

This looks like a production, but it has this identity not in inactive procedural memory but only in use with all contents in awareness. It could also be thought of as an n-tuple within a function describing a class of mental episodes, such as a kind of inference or decision:

$$S_{in+1} = f(S_{jn}, \ldots, S_{kn-m}).$$

For the theorist of mental episodes, variables are defined over conscious states, and a function describes a class of operations within a class of mental episodes.

2. What are those *operations?* An operation is not a state, and so by its nature cannot be a conscious state. In mentalistic language, an operation is a transformation of conscious states, and in neural language, it is what in the brain moves us from state to state. But a nonconscious operation has no other "cognitive" existence—though we sometimes speak harmlessly enough of "mental" or "cognitive" operations. "Cognition" as the name of something other than a conscious or neural state labels no more than "the ether" did.

3. Consider two *types* of mental episodes, each embodying one of the criteria of symbolic representation, episodes that will also provide us with an analysis of the explicit and the implicit: In *deliberative* mental episodes, the contents are propositional (predicative), and the operations are the deliberative operations of inference, decision, and judgment. In our work on causal reasoning, for example, subjects believe in some degree that a suspect is guilty and inferentially revise that belief on belief in asserted evidence (Carlson & Dulany, 1988). All are conscious modes with propositional contents that we assess with reports, and their interrelations specify the deliberative operation. The work illustrates some of the characteristic properties of deliberative episodes: use of novelty, range of abstraction, rapidity of acquisition or reversal, and relative effortfulness. On any trial, for example, clues can be fully asserted of any suspect in novel pairings, and under specifiable conditions, belief may move immediately (or gradually) to certainty of guilt or innocence—regardless of prior belief or novelty of coupling. Deliberative warrant is not equivalent to association strength, and hence these characteristics of deliberative episodes should make their connectionist modeling awkward.

In *evocative* mental episodes, the contents are nonpropositional, nonpredicational—a sense *of* _____, as we would say in ordinary language. A sense of one or more contents evokes a sense of another by an activation operation. Awareness of a red light evokes a conscious sense of stop. Or awareness of letters in a word fragment evokes awareness of its completion, or literal awareness of a word's form evokes semantic awareness of its identity. These are associative relations that are gradually strengthened, resistant to reversal, require little effort, and are relatively specific—specific under literal similarity or featural overlap, as we would expect of what is automatic (Logan, 1990). With units identified with nonpredicative conscious contents, the evocative should be a natural domain for connectionist models.

4. Deliberative and evocative episodes are *interrelated,* not isolated mental activities. We often move from something propositional in a deliberative episode to its analogue in an evocative episode. From repeatedly entertaining the propositional form, Belief that [Red means stop], comes the full blown episode, ([Red] activates [stop]). For conscious contents, we might put the learning principle this way: Predication produces activation. Indeed, this

probably explains the effectiveness of elaborative processing, an inherently predicative activity. How might we move from the evocative to its propositional expression? By remembering, an evocative episode may be recoded as a propositional content of awareness. From ([Red] activates [stop]) comes (Remembrance that [Red means stop]). Furthermore, most complex activities contain both kinds of mental episodes.

5. *Meta-awareness* comes from recursive mental episodes; they should provide an understanding of introspective access, personal theories of mind, and conceptions of self. Within a remembering episode, we may be aware of prior modes and contents of awareness:

$$\text{Aware}_2(\text{Aware}_1[S_{11}]).$$

By a nonconscious remembering operation, those prior conscious states, like any other natural event, may become objects of a symbolic awareness. This amounts to a remembrance theory of second order (reflective) awareness, an alternative to the appendage, mental eye, and intrinsic theories so thoroughly examined by Natsoulas (1993a, 1993b). We can also rather successfully map that awareness into experimental report variables when we meet well-established verbal and memorial conditions of report validity (Ericsson & Simon, 1993). Furthermore, by one or more remembering episodes, together with inference, we may be aware in some degree that a mental operation or mental episode occurred,

$$\text{Aware}_{n-2}(S_{in-1} \leftarrow \text{Operation upon}[S_{jn}, \ldots, S_{kn-m}])$$

In the same way, we may even be imperfectly aware of the form of a class of episodes:

$$\text{Aware}_{n+2}(S_{in+1} = f[S_{jn}, \ldots, S_{kn-m}]).$$

Nevertheless, the *forms* of mental operations and episodes are often beyond the limits of memory and inference. In fact, all this together explains why our subjects validly report at values on content and modal variables that are interrelated in ways we closely describe by equations that none of them knows (Carlson & Dulany, 1988; Dulany, 1968; Dulany & Carlson, 1983). But scientific theory is our business, not theirs—even though they may validly access their own inadequate causal theories that could become the subject of investigation in a study of the self and naive explanation. This also explains the obvious fact that we are often unaware *that* we are learning or how, and it shows that this is "unconscious learning" in a misleading and relatively trivial sense (e.g., Greenwald, 1992, p. 771), rather than in the fundamental sense of our operating on unconscious symbolic representations. That is what a cognitive unconscious is said to do.

6. What of the *memory* traces themselves? It is only conscious remembrance in mental episodes that functions symbolically, not traces in a non-

conscious memory. Traces should be specific to their particular learning episodes and to the episodes they construct in remembering, with action a by-product when motor transducers are also activated. Although our ordinary and theoretical languages can easily mislead us by speaking of dormant "beliefs" and "intentions" and "propositions," I suggest that memory traces are nonsymbolic and lack intentionality in this fundamental sense: They don't do what symbolic representations do in consciousness, namely, participate in predications within deliberative episodes and activate other symbols within evocative episodes. The nonconscious memory traces themselves are only *dispositional* within their activating networks.

Personal Level Description.   To speak of a person is to speak of an entity capable of fully formed mental episodes of the kind that we as persons know about—episodes of searching, inferring, deciding, perceiving, recognizing, and so on. Now, on this mentalistic view, fundamental aspects of our mental lives are inseparable: the conscious and the nonconscious in mental episodes, and intentionality and mental episodes. Conscious modes and contents are transformed by nonconscious operations. Furthermore, the intentionality of consciousness, in the broader sense, provides the variables for those mental episodes that are either deliberative or evocative—the two ways contents gain the power to represent symbolically, the "aboutness" that is traditionally the essence of intentionality.

If this should be correct, there would be two fundamental implications: There could be no personlike entity capable of fully formed mental episodes outside awareness, and that is the nature of a personal or cognitive unconscious worthy of the name. And, for the science, conscious states (the mode, content, and agency they carry) would naturally become the coin of the realm: values on the variables we assess and specify theoretically.

THEORY OF EXPLICIT AND IMPLICIT LEARNING

Assumptions this deep are usually safely remote from the laboratory, but they have recently been exposed in several experimental arenas, perhaps most prominently in the study of implicit and explicit learning. Indeed, they give us reason to be interested in the odd little learning tasks of the domain. The standard theory of the learning process follows directly from the standard metatheory.

The Standard Theory.   In this literature, *explicit* learning is usually a bother to be minimized, but it is typically characterized for contrast in much the same way (e.g., Reber, 1989): (a) It is intentional learning. (b) Hypotheses are formed and tested on the evidence, all well represented in awareness.

(c) What is learned is usually rules, accessible to consciousness for control of action. (d) And it is a process, given the limitations of consciousness, best suited to domains with relatively simple and salient regularities. Explicit learning is conscious learning.

Implicit learning, in contrast, is said to be (a) incidental learning, either in the sense of proceeding without any intention to learn or without intention to learn the complex structure of a domain. (b) It consists of a "nonconscious abstraction system" (Reber, 1976) that can count frequencies and abstract rules. (c) It therefore produces implicit knowledge, in the sense that subjects "are not consciously aware of aspects of the stimuli which lead them to their decision" (Reber & Allen, 1978, p. 218) and may have "no access whatsoever to it" (Lewicki, 1986a). (d) Furthermore, the process is said to be aroused when the regularities to be learned are complex or nonsalient enough to evade a limited consciousness (Hayes & Broadbent, 1988; Reber, 1993). When the going gets tough, the unconscious gets going. Implicit learning is unconscious learning, with judgment usually said to reveal unconscious implicit memory.

Central assumptions of the standard metatheory are fully embodied. What is learned—the regularities—is symbolically represented outside consciousness. Explicit and implicit learning are assigned to separate systems for conscious and unconscious processing. And implicit learning is assigned to a cognitive unconscious (Reber, 1989, 1993) capable of person-level mental episodes of abstracting, retrieving, and deciding. Indeed, according to Lewicki and Hill (1989), the ability "to nonconsciously acquire complex knowledge structures . . . constitutes one of the necessary metatheoretical assumptions of contemporary cognitive psychology" (p. 239).

A Mentalistic Theory.   Can the alternative metatheory be as productive as the standard metatheory in yielding a theory of explicit and implicit learning?

Explicit learning consists of *deliberative* mental episodes. In different domains, it takes different forms, consistent with the general description of deliberative mental episodes as consisting of nonconscious deliberative operations on propositional conscious contents. In studies from this literature, subjects (a) venture propositional hypotheses in awareness, hypotheses that often have the form of rules—in one experimental example (Dulany & Poldrack, 1991), "Persons with long hair (subject) are kind" (predicate). (b) They consciously represent evidence propositionally as well· the assertion that co-occurrence did or did not occur. (c) Then, by a nonconscious operation of inference, they confirm and disconfirm hypotheses, a kind of elaborative processing that leaves traces of confirmed hypotheses. (d) At a test of memory, subjects again observe evidence, and the nonsymbolic trace is manifest as a propositional remembrance, an explicit remembrance, that represents the learned rule in awareness. Explicit learning is deliberative learning.

Implicit learning, on the other hand, should be identified with *evocative* mental episodes. It consists of the establishment and use of evocative relations among nonpropositional but fully conscious contents—perceptions, images, thoughts of _____, for example, long hair or kindliness. (a) By co-occurrence, successively or simultaneously, activational associations between nonpropositional contents are established and progressively strengthened. A somewhat Hebbian learning principle is suggested: Co-activation produces mutual activation of nonpropositional conscious contents, thereby leaving a trace of the full evocative mental episode (Dulany, 1991). What operates on the contents, then, is activation. (b) At a test of memory, the nonsymbolic trace participates in construction of the evocative mental episode. In this way, contents of awareness that have been associated tend to directly activate their associates. Thus, this is implicit remembering but not "unconscious memory," as it is usually termed. Implicit memory is no more "remembering without awareness" than it is "awareness without remembering;" it is evocative remembering and nonpropositional awareness apart from deliberative remembering and propositional awareness. It is also not unconsciously remembering something not consciously remembered (an item or task), but the specific evocation in consciousness of what was primed in consciousness.

There are inherent relations between the explicit and implicit, and in most tasks, some of both kinds of learning should occur. (a) By another learning principle (Dulany, 1991)—predication produces activation—explicit learning should also tend to establish evocative relations, especially in advanced stages. As a consequence, explicit learning should also be revealed in implicit remembering. (b) Evocative episodes should also provide evidence for explicit hypothesis testing. (c) Furthermore, by explicitly remembering an evocative episode, the episode can be symbolically represented as a rule to be acted on. This is implicit learning revealed in explicit remembering. For example, the episode, ([Long hair] activates [kind]), may be remembered as the rule, [Long hair is described as kind]. (d) All traces are specific to what is experienced, but rules can be used in deliberative episodes that provide for flexibility and abstraction. As a result, what is learned can be concrete or abstract in varying degrees. It depends on whether the learning or the remembering is evocative or deliberative, and on the range of events over which the derived rules have dominion, from 1 to N. (e) Nevertheless, the evocative is inherently simpler than the deliberative and probably more robust, but the deliberative should better be able to benefit from additional time or elaborative information.

We see then that the mentalistic theory embodies central assumptions of the mentalistic metatheory. Symbolic representation of what is learned is carried exclusively by conscious contents, propositional in explicit learning and nonpropositional in implicit learning. The knowledge acquired in both cases is neither explicit nor implicit; it is a system of neural traces produced and used in explicit (deliberative) or implicit (evocative) mental episodes. In both cases, too, what is conscious (the modes and contents) and what is not (the operations) constitute a united system of mental episodes. The explicit and implicit differ then, not by one being conscious and the other not, but by the forms of conscious states and nonconscious operations. It is these differences that should account for differences across standard explicit and implicit conditions as we vary other factors or populations. The standard and alternative views are contrasted in Table 10.1.

EXPERIMENTAL PARADIGMS

Paradigms for studying implicit and explicit learning are especially interesting because they have been designed to reveal unconscious symbolic representation in both inactive memory and a working system. The tasks are governed by rules too complex or obscure for the grasp of any ordinary consciousness. So as the standard rationale goes, if ordinary subjects learn, they must have laid up tacit knowledge of the rules, representations that yield the tacitly working knowledge that controls judgment on each trial. What is learned, but out of awareness, would have to be used evocatively or deliberatively, and so would earn its spurs as a symbolic representation. It would have to evoke other symbols and/or be the subject of predications, and so would meet our criteria of symbolic representation.

In wide ranging reviews, Berry and Dienes (1993), Seger (1994), and Shanks and St. John (1994) have disagreed, with the former two concluding that learning without the critical awareness has been satisfactorily established, and the latter concluding that it has not. This account is much more selective, but will go beyond their reviews in summarizing our recent work, drawing implications, and addressing possible responses to this particular view. How well do the standard and alternative theories fare in the light of recent experimental re-examination of earlier claims? And is the intuition of unconscious implicit learning better reformulated as evocative implicit learning?

Hidden Covariation. In introducing this paradigm, Lewicki (1986b) presented his subjects a relatively obscure contingency between hair length and personality of six pictured women: hair that was long or short and one paragraph descriptions as kind or capable. They then evaluated eight new pictures, again with long or short hair, answering the questions Kind? or Capable? When the "correct" answer should have been Yes, RT was longer, as expected on the Glucksberg and McCloskey (1981) model holding that No answers would be quicker when there is no information to report. Subjects also answered Yes correctly to a significant degree. What gives these findings their dramatic impact, however, is the report that when questioned, "Not one subject mentioned haircut or anything connected with hair" (p. 13). That is evi-

TABLE 10.1
The Implicit and Explicit

Standard View: Correlated with Consciousness and the Unconscious

	Unconscious Processing	Conscious Processing
Explicit		X
Implicit	X	

Alternative View: Orthogonal to Consciousness and the Nonconscious

	Nonconscious Operations	Conscious Contents and States
Explicit	Deliberative	Propositional
Implicit	Evocative	Nonpropositional

dence for unconscious symbolic representation of what is learned, the hidden covariation.

Duplicating the reported procedures closely, Dulany and Poldrack (1991) were unable to replicate either of those reported effects. We did, however, find that we could closely recreate the RT effects by perfectly biasing the sequence of test pictures, presenting the four for which the correct answer was Yes first, while exactly observing the five detailed constraints on the sequence specified by Lewicki (1986b, pp. 137–138). The effect was a simple response bias: Early responses take longer than later responses. We also found that we could replicate the effect on correctness of judgment by supplying an awareness hint just before the learning trials: "Try to figure out the relationship between the physical appearance and the personality of each person." In order to examine the role of rules in awareness, we first empirically determined the validity of each rule reported, the degree to which contents in the subjects' awareness were correlated with that in the experimenter's awareness when scoring correct. We asked each subject to sort the pictures using the rule reported, and scored its validity, the correctness of the sort into pictures of long and short hair. The study then found random judgment when reported rules had zero validity, and a linear relationship between the validity of the reported rules and the frequency of correct judgments. In short, we found clear evidence for conscious, not unconscious, representation of what was learned. This was not a task, however, in which the deliberative explicit could be separated from the evocative implicit.

We can, of course, only describe what we found necessary to replicate the effects; we cannot know the full set of conditions under which the original experiments were run in laboratories and dormitories at the University of Warsaw. Our experiments also re-examined only one variety of the paradigm. Nevertheless, providing their subjects with neither of the aids ours found so useful, De Houwer, Hendrickx, Baeyens, and Van Avermaet (1993) were unable to obtain reliable effects in six variants of the hidden covariation paradigm.

Dynamic Systems.   In the best-known variant of this paradigm, Hayes and Broadbent's (1988) subjects learned to keep a "person in a computer" friendly by making overtures on a scale of personal warmth, obtaining responses varying around a point two steps cooler, then in a reversal phase, two steps warmer. In a "current mode" the response was governed by the current overture, and in a "lagged mode" by the preceding overture—a complex contingency the authors believed would evoke unconscious implicit learning. Their two most frequently cited results appear to support the standard theory. Introduction of a secondary task, number or letter generation, impaired performance only in the Current mode of control, leaving presumably unconscious learning in the Lagged mode unimpaired. Moreover, postlearning reports of conscious rules were unrelated to performance in the Lagged mode and in both modes when there was a secondary task.

These results are questionable, however, for two main reasons: With 9 or 10 subjects per condition, evidence for selective impairment may not be stable. And subjects cannot reasonably be expected to validly report a single rule that will predict mean performance over two phases of the experiment, not only the last 10 trials of reversal, but also an unmarked last 10 trials of acquisition, 40 to 60 trials earlier and governed by different rules.

In a re-examination, Dulany and Wilson (1994) reproduced their procedures closely but with two principal changes. In order to have a sensitive test for selective impairment, we ran 24 subjects per condition. We also obtained reports either just before or just after a last block of 10 trials in one or the other of each phase, *trials providing the performance measures to which those reports should predict.* (a) Testing for it in several ways, this study failed to replicate selective impairment; the secondary task impaired performance equally in the two modes of control (a finding also recently reported by Green & Shanks, 1993). (b) With reported rules scaled on their validities, the degree they could produce a "friendly" response if acted on, the reports predicted equally well in both modes of control, and with and without a secondary task. Most important, there was no underprediction of performance from rules in awareness. (c) Furthermore, a quantitative model of deliberative learning, one borrowing from theories of control and of hypothesis revision (Dulany & Carlson, 1983), fitted performance with increasing accuracy over trials in all conditions. Subjects used the deliberative strategy more often in the Current than in the Lagged mode, but equally well when they did. (d) Overall, use of the deliberative strategy predicted validities of final rules reported, demystifying those reports to a degree, but with a residual that left room for implicit learning, not as unconscious learning, but as a direct evocative association of conscious contents. (e) Finally, separating out those trials in which subjects used the deliberative strategy, we found evidence of learning over the remaining trials that was also selectively spared by demand placed on the subject—direct evidence of implicit evocative learning. A sense of "polite" comes to evoke a sense of "friendly," a mental episode that can be remembered as a conscious rule to act upon: "A polite overture gets a friendly response."

There have been reports of unconscious implicit learning in other variants of the dynamic systems paradigm (Berry & D. E. Broadbent, 1984; D. E. Broadbent, Fitzgerald, & M. H. P. Broadbent, 1986), but these interpretations, too, have been challenged on the evidence (Marescaux, Luc, & Karnas, 1989; Sanderson, 1989; Stanley, Mathews, Buss, & Kotler-Cope, 1989).

Finite State Grammar (FSG). In this paradigm (Mathews, Buss, Stanley, Blanchard- Fields, Cho, & Druhan, 1989; Reber 1989, 1993), subjects inspect strings generated by an FSG, with explicit instructions to find rules or implicit instructions simply to memorize or select a string. Clearly they learn, as shown by successfully classifying old and/or new strings as grammatical or nongrammatical. On the standard interpretation, these studies (especially in the implicit conditions) provide evidence for learning by, or in, a cognitive unconscious—for unconscious symbolic representation of what is learned and used, a set of rules that may be abstract or as concrete as individual exemplars. For several reasons, however, none of these earlier or later studies satisfactorily supports that interpretation: They omit any attempt to assess the relevant contents of awareness (e.g., Reber, 1976; Reber, Walkenfield, & Hernstadt, 1991). Or, they obtain only qualitative protocols that yield no metric for determining whether a set of correct and correlated rules does or does not account for the judgments made (e.g., Abrams & Reber, 1988; Reber & Allen, 1978). Or, their report procedures introduce unacceptable error, especially in remembering contents to be reported (e.g., Mathews et al., 1989). "Yoked partners," who see no strings at learning, do judge less well than do real subjects providing them with briefings every 10 trials. Nevertheless, real subjects will often fail to remember up to 10 rules, a source of error compounded by their yoked partners' failure to assimilate, remember, and comply.

In the study that opened the experimental re-examination of implicit learning, Dulany, Carlson, and Dewey (1984) chose a task thought to "keep those explicit processes at a minimum" (Reber & Allen, 1974, p. 195). After learning, however, subjects not only classified strings by underlining the grammatical and crossing out the ungrammatical, but they did so by simultaneously marking features in the strings that suggested to them that classification. They reported rules in awareness, rules in which a grammatical classification is predicated of features. The rules are readily arrayed on a validity metric inherent in the task, the degree to which the rules if acted on would yield a correct classification. In short, we found that these rules predicted the judgments without significant residual: Proportion correct judgments increased with validity of rules by a slope of .99 and intercept of .01, and error of prediction scattered around a mean of .005 with individual subjects acceptably within sampling limits of zero. Each rule was of limited scope, and most of imperfect validity, but in aggregate they were adequate to explain the imperfect levels of judgment found.

We should first reject the continuing characterization of these reports as "guesses" (Reber, 1989, p. 230; 1993, p. 66), a saving assumption clearly and strongly refuted by the evidence we presented earlier. It is an oddity of the FSG task that validities of guessed rules will correlate with proportion correct judgments—and hence the confusion—but no oddity when Pearson r alone is inadequately diagnostic (Anderson, 1982). Extensive computer simulations (Dulany & Carlson, 1985; Dulany, Carlson, & Dewey, 1985, pp. 29–30) confirmed what was explained in the original report (Dulany et al., 1984, p. 553). If subjects responded and then guessed the rules, there would be two consequences: The mean validities of each subject's rules would progressively underpredict correct judgments as number of correct judgments increased over subjects, producing a *slope* greater than one for number correct as a function of rule validity, not the .99 we found. And, overall *mean difference,* proportion correct minus mean rule validity, would be a significant positive value, not the approximately zero we found. In fact, our 100 simulations of our 50 subjects guessing 100 rules revealed a distribution of slopes exclusively greater than one and mean differences exclusively of positive value, in both cases with the quite small variance expected of so many means of so many judgments. We therefore were able to determine that if subjects guessed their rule reports, the obtained slope "is a random event to be expected about once in 10 billion occasions," and the obtained mean difference would be expected with a "vanishingly small likelihood of 1.38×10^{-24}" (Dulany et al., 1985, pp. 29–30). These are unattractive odds, and the functions and means we found do follow gracefully from a theory that subjects acted on reported rules in awareness.

Nevertheless, it is a fascinating possibility that studies such as Reber, Walkenfield, and Hernstadt (1991) and Abrams and Reber (1988) have tapped a kind of implicit learning that has great significance in being less variable over persons, more robust in pathology, and perhaps more primitive evolutionarily (Reber, 1992): the evocative, but no less conscious, implicit learning described earlier. Indeed, this may be the greater significance of that work. It is also a fascinating possibility that evocative implicit learning occurred to some degree in Dulany et al. (1984). Rules may have come not only by inference from perceiving a test string and explicitly remembering learning strings, the process described (p. 552), but also in some degree from the more immediate evocation of a sense of grammaticality from features in test strings, features that had been in learning strings. That is the evocative implicit process we isolated in the computer person task. It is the process Dulany, Pritchard, Greenberg, and Krause (1993) more directly controlled in the FSG task in an effort to examine its relative specificity of transfer and trial-by- trial representation in awareness.

In order to produce a predominance of *evocative implicit learning,* we presented letter strings at the computer and instructed subjects to respond with the first category that came to mind, either W (for well formed) or N (for not). We allowed only 1 sec to view each letter string and respond, then after the first block another .5 sec for the experimenter's feedback of W or N. To produce a predominance of *deliberative explicit learning,* we presented instructions to find rules that would suggest whether the string was W or N. We then allowed 2 full seconds for viewing and responding, and after the first block, another second to inform them of the true category.

After 10 blocks of learning trials, all subjects classified strings in an 11th block. Memory conditions for the implicit evocative or explicit deliberative were now like the learning condition of that type, except for the omission of the experimenter's classification and the addition of novel strings generated by the same grammar. In a 12th block, we presented the same strings as in the 11th, but now subjects not only classified a string but also reported a rule asserting what feature of the string suggested the classification. Theoretically, subjects reported a rule that was either the direct outcome of a deliberative explicit process or the remembrance of an evocative episode in an implicit process. The theoretically interesting results were of two kinds: transfer in the 11th block and the relation of rules to judgment in the 12th block.

In all four combinations of learning and memory conditions, there was a shift to better performance on the old strings, from around 50% correct to 65% to 70% correct. Subjects learned in both of the learning modes and in ways that both memory modes would reveal. But only when there was *explicit* learning *explicitly* tested was there a significant transfer to the *novel* strings in the 11th block. Thus, with implicit conditions at *either* learning or remembering, nothing but classification of the old strings showed the benefit of learning. On the alternative theory, learning and remembering in the evocative implicit mode should be specific—with the effects of the specific trace revealed only when there is considerable similarity or feature overlap.

In addition, validities of reported rules predicted correct judgments without significant residual over all conditions. A slope not greater than 1 rules out guessing when reporting, and there was no significant underprediction of proportion correct from mean rule validity for any of the subjects. If subjects were aware of features directly evoking a sense of grammaticality, and that mental episode was immediately remembered as a rule for action, the validities of those rules should predict proportion correct judgments as well as they do when deliberatively formed. We found that they did. This is direct evidence for evocative episodes with their contents in awareness. Even in the implicit case, without time for explicit remembrance, subjects were aware of what it is about a string that guides their decisions.

Other studies have also challenged the view that subjects acquire an unconscious representation of the abstract structure of an FSG task by showing at least one of the following: The effects on judgment could be explained by more specific contents of awareness reported (Dienes, Broadbent, & Berry, 1991). What is learned is specific features, contents that could be in awareness during judgment (Perruchet & Pacteau, 1990; Servan-Schreiber & Anderson, 1990). What is learned is specific items, also contents that could be in awareness during judgment (Vokey & Brooks, 1992). Or, the apparent evidence for unconscious abstraction, as in Reber and Lewis (1977), is only a statistical artifact (Perruchet, Gallego, & Pacteau, 1992).

Sequences. In this paradigm, subjects must respond as quickly as possible to the position of a figure, usually in one of four quadrants, and the figure moves by some complex sequential rule (e.g., Willingham, Nissen, & Bullemer, 1989) or by a complex set of simpler rules (e.g., Lewicki, Hill, & Bizot, 1988). Unconscious implicit learning is inferred when RT is shorter for the rule constrained sequences than for control sequences, either within or between subjects, and subjects by some index are unable to report those rules. So far it has not been possible in this paradigm to frame the issues in the quantitative manner we have in the other three paradigms because the usual performance measure, RT, is not on the same scale as available report measures.

Nevertheless, at this point claims for unconscious symbolic representation of sequential rules fail for three principal reasons: In some of the studies, the learning effects are most directly attributable to response biases driven by event frequencies that are correlated with the sequential rules. In Lewicki et al.'s (1988) task, Perruchet, Gallego, and Savy (1990) showed that a relative speeding of responses during a rule-constrained sequence was attibutable to a relative *slowing* of responses during the unconstrained sequence—a slowing produced by relatively infrequent backward and horizontal movements in the unconstrained sequence. In the Willingham et al. (1989) task, too, Shanks, Green, and Kolodny (1994) examined subjects who failed to report any knowledge of the experimental sequence, DBCACBDCBA, and found another kind of response bias. Those experimental subjects no longer showed a learning effect when compared with the proper controls whose pseudo-random sequence matched the unequal and biasing frequencies of the experimental events—the 3 B&C to 2 A&D ratio. For other tasks (Cohen, Ivry, & Keele, 1990; Nissen & Bullemer, 1977), Perruchet and Amorim (1992) also presented evidence that subjects have significant awareness of the rules prior to appreciable changes in RT. In fact, correlation of mean recognition of four-trial sequences with mean RT for final position was −.82. Perhaps most interestingly, subjects after learning are significantly able to predict where a sequence will go next—at least, if the prediction task is constructed so as to maximize the set of cues available in the original learning (e.g., Cleeremans & McClelland, 1991, Exp. 2; Perruchet et al., 1990; Perruchet & Amorim, 1992).

What is the significance of work in the sequence learning task for the alternative theory of implicit and explicit learning? Reporting where a figure will appear next obviously expresses a conscious content: a conscious expectancy of what comes next. Clearly, too, recognition of sequences is a conscious content. Furthermore, it is at least plausible that the response biases found are in turn caused by just these conscious expectancies and recognitions. Nevertheless, a theory of conscious evocative learning leads one to wonder about the sources of those propositional recognitions and expectancies that "This comes next." To what degree are they the explicit remembrance of an evocative episode, one in which awareness of a stimulus in sequence has come to activate a awareness of the next? And to what degree are they inferred from explicit remembrance that "This did come next," a rule deliberatively learned over the prior series of trials? The interesting question now is this: How much of a consciously represented and controlling sequence is the evocative implicit and how much the deliberative explicit?

Other Related Paradigms.   Other paradigms have also been designed to demonstrate learning while keeping logically necessary information out of awareness.

In reviewing studies of *subliminal mere exposure,* Bornstein (1992) described eight in which repeated brief exposures of a figure, usually masked, produced preference for the exposed figure over another, despite inability to recognize which of the two figures had been presented earlier. The effect is usually said to be "unconscious learning" of what is "unconsciously perceived" or "perceived without awareness." These eight studies were also certified as sound because subjects showed greater responsiveness of an "indirect measure" (preference) than of a "direct measure" (recognition awareness) on the logic elaborated by Reingold and Merikle (1988). The logic is inadequately satisfied, however, when the indirect judgment can be based on some content of awareness that escapes the one direct measure the investigators choose. In this paradigm, the correlated and useful awareness could easily be a greater (and conscious) sense of brightness of the exposed stimuli, as Mandler, Nakamura, and Van Zandt (1987) found, or a greater (and conscious) sense of familiarity of those stimuli, as Bonnano and Stillings (1986) found— or for that matter, a greater (and conscious) positive affect, as Zajonc (1968) originally thought. A more analytic mentalistic view can help us to recognize a range of conscious contents and reject the fundamental conceptual error running through this literature (and beyond, for example, into studies of prosopagnosia; Young & De Haan, 1992): What has systematic effects and is not consciously recognized must be unconsciously recognized. Consciousness is richer than that; it permits inferences from contents on other dimensions of awareness, in this case dimensions correlated with frequency of exposure.

Kihlstrom, Schacter, Cork, Hurt, and Behr (1990) reported that subjects can learn under *anesthesia,* as revealed not in explicit but in implicit memory tests. This is now generally cited as evidence of learning with the logically necessary information represented outside consciousness, although that possibility is not developed beyond the authors' first paragraph. With a different anesthetic, a subsequent study (Cork, Kihlstrom, & Schacter, 1992) failed to reveal learning in either type of test. Others have also failed to find the effect (Eich, Reeves, & Katz, 1985). Before spending too much time on the relation between type of anesthetic and implicit memory, however, we might look for the more direct relation between type of anesthetic and emergence of awareness. With awareness of the prime, implicit memory effects are a commonplace. And because safety calls for minimum dosages, emergence of awareness under anesthesia is commonplace enough for anesthesiologists to devote conferences to examination of the problem (Bonke, Fitch, & Millar, 1990; Rosen & Lunn, 1987).

Of *natural sleep,* suffice it to say here that a strong recent study found no evidence of learning revealed in either implicit or explicit memory tests (Wood, Bootzin, Kihlstrom, & Schacter, 1992). This is consistent with what was found with reasonably alert sophomores: concept learning only when the necessary featural information was consciously identified during learning (Carlson & Dulany, 1985). The results of Johnson, Peterson, Yap, and Rose (1989) and Hawley and Johnston (1991) converge.

IMPLICATIONS

Our own experiments within the covariation, dynamic systems, and FSG paradigms found (a) no learning in the absence of correct or correlated rules in awareness, (b) a linear relation between proportion correct judgments and the validities ("correlation") of those rules, (c) prediction without significant residual in the sense that proportions correct are no greater than would be expected from action on those rules in awareness. This has been more than a test for effects with "zero awareness," as Baars (1988) and others have called it, and it certainly is not an attempt to make sure that we have "removed the possibility that some flickering residue of awareness remains in the subject's mind" (Reber, 1993, p. 87). The "indirect measures" of learning fail to exceed the "direct measures" of awareness at every level of awareness. Furthermore, the overall network of results provides not only a degree of construct validation for the assessments, but also competitive support for the theory. Others' critical findings converge. In the sequence paradigm, others' findings are now reasonably explained by awareness revealed in recognition and prediction tests. On a best interpretation at this point, I believe, awareness is the medium for symbolically representing what is learned and used in these paradigms: awareness of rules, concrete to abstract, that are the direct outcome of deliberative explicit learning or the indirect outcome of evocative implicit learning. Consciousness is essentially and systematically involved in both.

On the mentalistic theory, we also see how conditions and populations may selectively impair or benefit either the deliberative explicit or evocative implicit, consistent with what is learned and used being represented in awareness in both cases. For example, Turner and Fischler (1993, Exp. 1) found that rule discovery subjects did less well with a short than with a longer deadline for judgment in a FSG task, but that memorization subjects did equally well. But rule discovery subjects selectively benefited from a longer deadline when there was training in using the rules (Exp. 3). Furthermore, if amnesia selectively impairs deliberative learning and remembering, and a task can be evocatively learned, it would be consistent with the theory for amnesiacs to be equal to normals in later judgment while inferior in explicit deliberative memory for the earlier learning task (Knowlton, Ramus, & Squire, 1992). Because reports of the relevant awareness predict without significant residual in other studies, dissociations in the absence of those reports are better interpreted now as dissociations, not of the explicit with awareness and the implicit without, but of the deliberative and evocative with awareness in both.

This is a literature that is controversial but not uninterpretable. It has a revealing overall trajectory and asymmetry of methodological weaknesses. Critical challenges repeatedly fail to replicate some small effect or point to response biases, statistical artifacts, or improper control conditions in the target studies. Some of the challenged studies have left the critical awareness unassessed. In others, the assessments are unclear, or presented after an unreasonable time lag, or after a confusing change of contingencies. Or the possible role of correlated conscious contents is disregarded despite a body of analyses revealing their quantitative predictive values (Carlson & Dulany, 1985; Dulany, 1961, 1962, 1968; Dulany et al., 1984). As those weaknesses are removed, learning is found only with symbolic representation in awareness. Indeed, if we really can, and readily do, use working memory to lay up enduring but tacit representations of complex domains, we should expect their effects to be robustly revealed in these experiments. This is not a body of evidence that can carry the standard metatheoretical freight.

POSSIBLE RESPONSES

1. Before troubling ourselves further, however, should we recognize, as some suggest, that where there is "conscious learning" there must also be "uncon-

scious learning"? This is also said of conscious and unconscious perception (e.g., Merikle & Reingold, 1992). If reports are sensitive enough to explain all of the performance effects, the reasoning goes, they may or even must be reflecting not only conscious but unconscious learning (or perception). If we think about the matter more concretely, however, we can reject this kind of explanation on two grounds. It lacks antecedent credibility to say that when subjects report that they are aware that X covaries with Y that they are also reporting that they are *unaware* that X covaries with Y. It also removes the hypothesis of unconscious representation from vulnerability to the data, as we can see with this analogy: Where strong experiments show that reports of sensory perception predict stimulus effects without significant residual, we could just as well say that sensory perception is always—and untestably—accompanied by extrasensory perception. To selectively support the claim for unconscious representation of what is learned, it is necessary to show in some way that it has independent and separable effects. That test fails when reports of awareness predict without significant residual.

2. Should we, as some also suggest, now look for *qualitative differences* between learning with and without awareness—a logic recently introduced in attempts to demonstrate unconscious perception (e.g., Jacoby & Whitehouse, 1989; Merikle & Reingold, 1992)? For converging reasons, the logic is defective. It does not, as claimed, bypass a difficult determination of absence of the relevant awareness, but merely presupposes that absence under some type of impoverished stimulus condition. Furthermore, experimental demonstrations invalidate the logic by corroborating our common sense (Bernstein & Welch, 1991; Joordens & Merikle, 1992): Strong and weak stimulation may have qualitatively different effects while we are aware of what produces both kinds. In the Jacoby–Whitehouse paradigm, for example, an impoverished word may transduce a literal awareness that activates a conscious sense of familiarity (or "perceptual fluency") of the recurrent word, whereas a clear stimulus produces conscious identification as such—different contents of awareness that can be used in different ways, to say that the word did or did not occur before.

3. Should we, as some suggest, try to place learning with and without awareness in *opposition* (Jacoby, 1991)? If an effect overcomes some conscious intention to behave otherwise, the reasoning goes, it must be an unconscious effect (Jacoby & Kelley, 1992). This logic, too, is defective because it presupposes that what is uncontrollably automatic is unconscious—an identity disconfirmed every time we consciously but uncontrollably see, hear, and feel the world around us in one way rather than another. Literal awareness of physical properties usually compels awareness of identity. Despite the loose association of "automatic" and "unconscious" in ordinary language, theoretical conceptualization must be precise enough to encompass the necessary distinctions. Nevertheless, placing controls in opposition can be, and I believe

has been (for example, Jacoby & Kelley, 1992), extremely useful for separating what is primarily evocative (and automatic) from what is deliberative (and intentional). In fact, I believe this is its significance.

4. Would a longer series of trials in these learning paradigms produce a kind of *automaticity* that requires unconscious symbolic representation? The most relevant evidence so far comes from Cleeremans and McClelland's (1991) finding within a blending of the FSG and sequence tasks: After 62,000 trials, subjects were influenced by only three elements of prior context, a distance consistent with recurrent awareness of those prior elements evoking a conscious anticipation of their successor. Furthermore, with a shorter series of trials, Cohen et al. (1990) found that a secondary task eliminated the influence of all but the preceding trial, presumably by preventing that recurrent awareness and evocative episode.

But does automaticity more broadly challenge a mentalistic view? As many have seen it, and as Klatzky (1984) summarized it, judgment and action in automaticity are much like earlier conscious processing, just faster, less effortful—and unconscious, of course. So let me say that if advanced automaticity of perceptual recognition and action should require unconscious symbolization, so be it. There would still be an immense domain for the investigation of conscious symbolic representation, even if only at the top nodes of hierarchical structures, leaving the lower nodes to a more molecular and simpler form of unconscious representation.

The automatization literature, however, shows convergence on two principles by which the process greatly changes, principles that challenge that common view. Most fundamental are various descriptions of a change from something deliberative to something more *direct*—and I would say, evocative—as in J. R. Anderson's (1983) concept of proceduralization, in Schneider's (1985) shift from attentional processing to the strengthening of direct associations, and in Logan's (1988) algorithm to memory-based processing. Indeed, Cohen, Servan-Schreiber, and McClelland's (1992) connectionist model describes strengthening of direct associations in the Stroop paradigm—associations, I would think, between literal awareness of words or colors and their conscious identities as such. Another is *unitization,* revealed in experimental work from LaBerge (1981) to Czerwinski, Lightfoot, and Shiffrin (1992), and expressed in J. R. Anderson's (1983) theoretical concept of composition. The hierarchy flattens. Common to both principles is something that is very important: *process deletion.* In automatization, symbolic representations drop out, not down, to an unconscious.

A mentalistic program for automaticity would build on that foundation: In an automatized perceptual unit, literal awareness of form, the output of sensory transducers, directly activates semantic awareness of a word or object's identity in an evocative episode. In an automatized action unit, a conscious intention directly activates inputs to motor transducers. In neither case would we need specialized processors computing and unconsciously representing the intermediate levels of a hierarchy. I have also suggested two processes for getting to that stage: predication produces activation, and co-activation produces mutual activation. On this view, explicit and implicit learning are both aspects of automatization, and automaticity consists of strongly evocative mental episodes. In the especially challenging domain of language as well, theory on this program would describe nonconscious operations on conscious contents yielding others and action: on literal forms to be comprehended and thoughts to be expressed. We need alternatives to the traditional assumption that rules in the consciousness of the sophisticated linguist are in the unconscious of the naive speaker-hearer.

The exciting possibility for automaticity theory is that skilled perception, comprehension, and performance are faster and less effortful, not because so much more is done unconsciously, but because so much less is done. In this way, we would see the conscious and nonconscious as orthogonal to, rather than correlated with, the automatic (evocative) and nonautomatic (deliberative). They both require nonconscious operation on conscious contents.

Concepts Used.   What in the general mentalistic view most helps us to understand what is going on in this domain? There are several answers: the need to speak more analytically about consciousness and symbolization in order to frame the questions more clearly; a failure of meta-awareness as consistent with operating on conscious contents; correlated contents of awareness as explanatory; meeting memory conditions as necessary for reported awareness to be introspectively valid; literal as well as semantic (identity) codes in awareness; and, perhaps most important in this domain, consciousness in the evocative (implicit) as well as in deliberative (explicit).

THE QUESTIONS

Each author was given a list of questions and asked to "let your hair down" in answering. My answers flow in part from the mentalistic view sketched.

How Should Consciousness Be Defined?   *Consciousness,* like the label for any rich and interesting domain, is not well captured by a definition. We inevitably start with our commonsense conceptions of a domain we call consciousness, something we all live with. We should then refine and systematize those conceptions in metatheoretical and theoretical forms, as I and others try to do, then move on to the laboratory to examine competing formulations in the usual ways. What we understand consciousness to be is inherent in what we learn it does, perhaps uniquely. That is not a matter of prior announcement, but something that should evolve throughout inquiry as our commonsense conceptions are refined within meta-theory and theory, then supported or revised. If analysis of intentionality should be theoretically productive, and in the long run empirically supportable for consciousness but not for an unconscious, we would have learned something very significant about the nature of consciousness.

Is It Possible to Operationalize Consciousness in the Context of Your Research?   The answer is Yes, if "operationalize" only means assess or manipulate states of awareness—a common and relatively harmless misuse of the term. But the answer is No, if it means "operational definition" in the proper sense. On one kind, the explicit definition (Bergmann, 1951), the operational definiens is unique, complete, and stipulative. Among other limitations, the defined term thereby refers only to the operations, and so the formal representation of consciousness as a phenomenon is abandoned. On the other kind, reduction sentences (Carnap, 1937), multiple operations only partially define terms that can also refer to theoretical entities. But the relation of any theoretical entity to its mappings is inherently hypothetical and not definitional. Referring to "operational definition" is too often simply an attempt to bootleg validity of a mapping that is not otherwise justified ("But that's the way I operationally defined it")—with a common result in this literature being a spurious finding of "unconscious" learning or perception. We should venture measures and manipulations, observing established conditions for validity (the Bayesian a priori), and gain further confidence in their validity as they are found to behave as theory of what they map says they should and not otherwise (the Bayesian diagnosticity.) An examination of operationally defining contents of awareness, and alternatives, can be found in Dulany (1968, pp. 374–382).

What Is the Relation Between Conscious and Nonconscious Thought? Central to what I have sketched is the idea that mental activity consists of nonconscious operations on conscious states (modes, contents, and senses of agency). Contrary to occasional claims (e.g., Greenwald, 1992, p. 778), there is nothing circular about this. Mental operations, on the one hand, and states, on the other, are distinguished by their different descriptions and separate roles within mental episodes. The assignment of the nonconscious and conscious to those roles is then subject to empirical inquiry.

Is Consciousness Equivalent to Attention?   This common and toocomfortable identification has been misleading. Literal awareness precedes and goes well beyond what is ordinarily called an *attentional focus*—some smaller focus where we consciously identify objects, persons, and ideas as such. Consider focusing on an oboe while hearing much more of a symphony in the conscious surround. Or, think of the wraith of a voice that hangs in lit-

eral awareness before we wrench attention from reading and grant it the semantic awareness of identification. Or, for that matter, think of that surplus in
the span of consciousness that produces the partial report advantage in the
Wundt–Sperling paradigm. Furthermore, attentional operations, like all mental operations, are inherently outside consciousness (Carr, 1992; Treisman,
Vieira, & Hayes, 1992). Whether object-based or place-based, they should operate to select what in literal awareness will be activated into conscious identification at the focus.

*Does What is Supraliminal Versus Subliminal Separate the Conscious
and Nonconscious?*   The question arises in the learning literature because
"unconscious perception" would have to be the basis for learning that used information outside awareness. But no available threshold (or "limen") specified by stimulus value can reliably separate the conscious and nonconscious
because the sensitivities of consciousness and reporting are inevitably variable over many factors.

One of Cheesman and Merikle's (1986) conceptions of a *subjective threshold* is the stimulus value at which subjects fail to report its presence when
questioned. This cannot reliably distinguish the conscious and unconscious
for reasons central to signal detection theory and known long before its rise:
One can say No on a high criterion for Yes for any of many reasons, despite
discriminability that could be based on semantic awareness, literal awareness,
partial awareness, or correlated awareness. Their other conception is the
highest stimulus value below which subjects are unable to report covariation
of stimulus and response presence and absence. Belief that we are discriminating a weak stimulus value, however, is a content of meta-awareness requiring a metacognitive feat roughly equivalent to computing a chi-square in the
head, one beyond most of us despite sometimes being aware of the stimulus.

Neither can an *objective threshold*—the highest stimulus value with random association of response presence and absence—reliably separate the
conscious and nonconscious. A chosen response, such as recognition, may reveal only one of several possibly effective contents of awareness, as we saw in
consideration of the mere exposure effect. Moreover, in signal detection theory, a $d' = 0$ simply indicates equivalent means for uncorrelated distributions
of noise and signal-plus-noise; and values within the latter could therefore randomly exceed values within the former sufficiently to activate a usable state of
awareness. In SDT there is no rigorously specifiable "low threshold" below
which stimuli are confidently outside awareness for a "crucial experiment."
We should therefore make the better interpretation of a recurrent pattern of
results for experiments in aggregate. With stimulus values close to that for
$d' = 0$ for detection, the most sensitive and inclusive report, positive effects
not only vanish with removal of discoverable artifacts, they have been small
and nonrobust at best (Holender, 1986).

Greenwald (1991), for example, reported that when indirect measures of semantic priming were plotted against the objective measure of detection for over 1,800 subjects in several experiments, the function passed through the origin and the slope did not exceed 1. There were no effects beyond those explainable by detection. On a new look at his data, Greenwald (1993) later reported an intercept greater than zero in two experimental tasks, but the effect revealingly failed to replicate in two closely similar and five somewhat similar tasks.

Is Consciousness Unitary? It presents itself in many modes, varying in degree, and in many contents, varying in form. It achieves unity momentarily in the mental episodes that nonconscious operations yield. And it achieves unity over time, an enduring theory of self (Harré, 1984), from a recurrent sense of agency and from remembering prior states, contents, and mental episodes. *I* as agent recurrently remember *my* private experiences and no one else's.

Is Consciousness Only an Emergent Property—an Epiphenomenon—of the Information Processing System as a Whole? The popularity of this view is largely due to a commitment to information-processing systems developed while consciousness was, in Norman's (1980, p. 16) phrase, "the peculiar stepchild of our discipline," predictably found emerging from one of those alien systems. Indeed, it is intrinsic to the computational theory of mind that consciousness is merely an epiphenomenon. The view is then sustained by a mistaken belief that attributing causality to conscious states and contents entails dualism. Dualism is an ultimate ontological assumption that we are unable to examine in our scientific inquiry as we now know it. Saying consciousness is causal, however, is a theoretical claim that we address in the usual empirical ways. It is equivalent to asserting that conscious states and contents are part of a theory that embodies a causal model, one that generates effects (Harré & Madden, 1975), and the claim at any time is as good as the competitive support for that theory. When theory of this kind becomes more advanced, we may even begin to ask whether "working memory" has been only our own emergent abstraction from a richer domain of intentionality and mental episodes.

For now, too, we should recognize that if consciousness is only an emergent of independent cognitive processes, what it emerges from should be shown to operate with systematic and robust independence described in some cognitive theory. Otherwise an emergence claim reduces to the assertion that consciousness has a neural substrate, a matter not at issue here. Indeed, we can look at the literature of unconscious learning (and unconscious perception) and say that if what is so commonly theorized were correct, if we routinely use such unconscious symbolic representations, their independent experimental effects should be too strong and robust for these literatures even to be controversial.

FINAL COMMENT

What difference might a broader mentalistic program make? For one thing, a greater use of constrained introspective reports could provide the richer data needed for stronger competitive support of theories and perhaps reveal a domain of unusually strong orderliness. Consciousness would become the central topic of the discipline, or as Miller (1980) put it, "the constitutive problem of psychology" (p. 146). We would also be provided another source of plausibility constraints for theorizing. Listen again to James (1890): Belief in an unconscious is "the sovereign means for believing what one likes in psychology, and of turning what might become a science into a tumbling ground for whimsies" (p. 163). In the present climate, the half-life of too many models is about two-thirds the journal lag. And listen to Freud (1915/1957), who for one shining moment may have had it right: "The psychoanalytic assumption of unconscious mental activity appears to us, on the one hand, as a further expansion of the primitive animism which caused us to see copies of our own consciousness all around us" (p. 171). We would move beyond the explanatory primitivism that has dominated much of the discipline throughout the cognitive movement—the practice of attributing the poorly understood to a resident humanoid that secretly does what persons do. On the mentalistic view expressed, there's nobody home but us.

ACKNOWLEDGMENTS

I am indebted to David Adams, Richard Carlson, Gregory Murphy, Russell Poldrack, and Evan Pritchard for their helpful comments on an earlier version of this chapter.

REFERENCES

Abrams, M., & Reber, A. S. (1988). Implicit learning: Robustness in the face of psychiatric disorders. *Journal of Psycholinguistic Research, 17*, 425–439.
Anderson, J. R. (1982). *The architecture of cognition.* Cambridge, MA: Harvard University Press.
Baars, B. J. (1988). *A cognitive theory of consciousness.* New York: Cambridge University Press.
Bernstein, I. H., & Welch, K. R. (1991). Awareness, false recognition, and the Jacoby-Whitehouse effect. *Journal of Experimental Psychology: General, 120*, 324–328.
Berry, D. C., & Broadbent, D. E. (1984). On the relationship between task performance and associated verbalizable knowledge. *Quarterly Journal of Experimental Psychology, 36*, 209–231.
Berry, D., & Dienes, Z. (1993). *Implicit learning: Theoretical and empirical issues.* Hillsdale, NJ: Lawrence Erlbaum Associates.
Bergmann, G. (1951). The logic of psychological concepts. *Philosophy of Science, 18*, 93–110.

Bonke, W., Fitch, W., & Millar, K. (Eds.). (1990). *Memory and awareness in anaesthesia.* Lisse/ Amsterdam: Swets & Zeitlinger.

Bonnano, G. A., & Stillings, N. A. (1986). Preference, familiarity, and recognition after repeated brief exposures to random geometric shapes. *American Journal of Psychology, 99,* 403-415.

Bornstein, R. F. (1992). Subliminal mere exposure effects. In R. F. Bornstein & T. S. Pittman (Eds.), *Perception without awareness: Cognitive, clinical, and social perspectives* (pp. 191–210). New York: Guilford.

Bower, G. H. (1975). *Cognitive Psychology: An introduction.* In W. K. Estes (Ed.), *Handbook of learning and cognitive processes* (Vol. 1, pp. 25–80). Hillsdale, NJ: Lawrence Erlbaum Associates.

Brentano, F. (1973). *Psychology from an empirical standpoint.* London: Routledge & Kegan Paul. (Original work published 1874)

Broadbent, D. E., Fitzgerald, P., & Broadbent, M.H.P. (1986). Implicit and explicit knowledge in the control of complex systems. *British Journal of Psychology, 77,* 33–50.

Carlson, R. A., & Dulany, D. E. (1985). Conscious attention and abstraction in concept learning. *Journal of Experimental Psychology: Learning, Memory, and Cognition, 11,* 45–58.

Carlson, R. A., & Dulany, D. E. (1988). Diagnostic reasoning with circumstantial evidence. *Cognitive Psychology, 20,* 463–492.

Carnap, R. (1937). Testability and meaning. *Philosophy of Science, 4,* 1–40.

Carr, T. (1992). Automaticity and cognitive anatomy: Is word recognition "automatic"? *American Journal of Psychology, 105,* 201–238.

Cheesman, J., & Merikle, P. M. (1986). Distinguishing conscious from unconscious perceptual processes. *Canadian Journal of Psychology, 40,* 343–367.

Cohen, A., Ivry, R. I., & Keele, S. W. (1990). Attention and structure in sequence learning. *Journal of Experimental Psychology: Learning, Memory and Cognition, 16,* 17–30.

Cohen, J., Servan-Schreiber, D., & McClelland, J. (1992). A parallel distributed processing approach to automaticity. *American Journal of Psychology, 105,* 239–269.

Cork, R. C., Kihlstrom, J. F., & Schacter, D. (1992). Absence of explicit or implicit memory in patients anesthetized with sufentanil/ nitrous oxide. *Anesthesiology, 76,* 892–898.

Cleeremans, A., & McClelland, J. L. (1991). Learning the structure of event sequences. *Journal of Experimental Psychology: General, 120,* 235–253.

Czerwinski, M., Lightfoot, N., & Shiffrin, R. M. (1992). Automaticity and training in visual search. *American Journal of Psychology, 105,* 271–316.

De Houwer, J., Hendrickx, H., Baeyens, F., & Van Avermaet, E. (1993). *Hidden covariation detection might be very hidden indeed.* Manuscript submitted for publication.

Dennett, D. (1987). *The intentional stance.* Cambridge, MA: MIT Press.

DeLoache, J. S. (1989). The development of representation in young children. In W. H. Reese (Ed.), *Advances in child development and behavior* (Vol. 22, pp. 1–39). New York: Academic Press.

Dienes, Z., Broadbent, D., & Berry, D. (1991). Implicit and explicit knowledge bases in artificial grammar learning. *Journal of Experimental Psychology: Learning, Memory, and Cognition, 17,* 875–887.

Dretske, F. (1988). *Explaining behavior: Reasons in a world of causes.* Cambridge, MA: MIT Press.

Dulany, D. E. (1961). Hypotheses and habits in verbal "operant conditioning." *Journal of Abnormal and Social Psychology, 63,* 251–263.

Dulany, D. E. (1962). The place of hypotheses and intentions: An analysis of verbal control in verbal conditioning. In C. E. Eriksen (Ed.), *Behavior and awareness* (pp. 102–129). Durham, NC: Duke University Press.

Dulany, D. E. (1968). Awareness, rules, and propositional control: A confrontation with S-R behavior theory. In T. Dixon, & D. Horton (Eds.), *Verbal behavior and general behavior theory* (pp. 340–387). Englewood Cliffs, NJ: Prentice-Hall.

Dulany, D. E. (1984, November) *A strategy for investigating consciousness.* Paper presented at the Psychonomic Society, San Antonio, TX.

Dulany, D. E. (1991). Conscious representation and thought systems. In R. S. Wyer, Jr., and T. K. Srull (Eds.), *Advances in social cognition* (Vol. 4, pp. 97–120). Hillsdale, NJ: Lawrence Erlbaum Associates.

Dulany, D. E., & Carlson, R. A. (1983, November). *Consciousness in the structure of causal reasoning.* Paper presented at the meeting of the Psychonomic Society, San Diego, CA.

Dulany, D. E., & Carlson, R. A. (1985). Syntactical judgment and rules: A computer simulation for the separation of guessing of rules from control by rules. *Psychological Documents, 15,* 19.

Dulany, D. E., Carlson, R. A., & Dewey, G. I. (1984). A case of syntactical learning and judgment: How conscious and how abstract? *Journal of Experimental Psychology: General, 113,* 541–555.

Dulany, D. E., Carlson, R. A., & Dewey, G. I. (1985). On consciousness in syntactic learning and judgment: A reply to Reber, Allen, and Regan. *Journal of Experimental Psychology: General, 114,* 27–34.

Dulany, D. E., & Poldrack, R. A. (1991, November). *Learned covariation: Conscious or unconscious representation?* Paper read at Psychonomic Society, San Francisco, CA.

Dulany, D. E., Pritchard, E., Greenberg, N., & Krause, K. (1993, November). *Awareness and novelty in the explicit and implicit.* Paper presented at the Psychonomic Society, Washington, DC.

Dulany, D. E., & Wilson, T. L (1994). *Learning to control a person in a computer: Consciousness in the explicit and implicit.* Manuscript submitted for publication.

Eich, E., Reeves, J. L., & Katz, R. L. (1985). Anesthesia, amnesia, and the memory/awareness distinction. *Anesthesia and analgesia, 64,* 1143–1148.

Ericsson, K. A., & Simon, H. (1993). *Protocol analysis: Verbal reports as data.* MIT Press.

Fodor, J. (1983). *The modularity of mind.* Cambridge, MA: MIT Press.

Fodor, J. (1994). *The elm and the expert: Mentalese and its semantics.* Cambridge, MA: MIT Press.

Freud, S. (1957). The unconscious. In J. Strachey (Ed. and Trans.) *The standard edition of the complete psychological works of Sigmund Freud* (Vol. 14, pp. 59–215). London: Horarth Press. (Original work published 1915)

Gallistel, C.R. (1981). Précis of Gallistel's *The organization of action: A new synthesis. The Behavioral and Brain Sciences, 4,* 609–650.

Gardner, H. (1987). *The mind's new science: A history of the cognitive revolution.* New York: Basic Books.

Glucksberg, S., & McCloskey, M. (1981). Decisions about ignorance: Knowing that you don't know. *Journal of Experimental Psychology: Human Learning and Memory, 7,* 311–325.

Green, R.E.A., & Shanks, D. R. (1993). On the existence of independent learning systems: An examination of some evidence. *Memory and Cognition, 21,* 304–317.

Greenwald, A. G. (1991, November). *Subliminal semantic activation between objective and subjective thresholds.* Paper presented at the Psychonomic Society, San Francisco.

Greenwald, A. G. (1992). New look 3: Unconscious cognition reclaimed. *American Psychologist, 47,* 766–779.

Greenwald, A. G. (1993). *Do subliminal stimuli enter the mind unnoticed?* Paper presented at the 25th Carnegie Symposium on Cognition, Pittsburgh, PA.

Haugeland, J. (1978). The nature and plausibility of cognitivism. *Behavioral and Brain Sciences, 2,* 215–260.

Haugeland, J. (1985). *Artificial intelligence: The very idea.* Cambridge, MA: MIT Press.

Hawley, K. J., & Johnston, W. A. (1991). Long-term perceptual memory for briefly exposed words as a function of awareness and attention. *Journal of Experimental Psychology: Human Perception and Performance, 17*, 807–815.

Hayes, N. A., & Broadbent, D. E. (1988). Two modes of learning for interactive tasks. *Cognition, 28*, 249–276.

Harré, R. (1984). *Personal being*. Cambridge, MA: Harvard University Press.

Harré, R., & Madden, E. H. (1975). *Causal powers*. Totowa, NJ: Rowman & Littlefield.

Holender, D. (1986). Semantic activation without conscious identification in dichotic listening, parafoveal vision and visual masking. *Behavioral and Brain Sciences, 9*, 1–23.

Jackendoff, R. (1987). *Consciousness and the computational mind*. Cambridge, MA: MIT Press.

Jacoby, L. L. (1991). A process dissociation framework: Separating automatic from intentional uses of memory. *Journal of Memory and Language, 30*, 513–541.

Jacoby, L. L., & Whitehouse, K. (1989). An illusion of memory: False recognition influenced by unconscious perception. *Journal of Experimental Psychology: General, 118*, 126–135.

Jacoby, L. L., & Kelley, C. M. (1992). A process-dissociation framework for investigating unconscious influences: Freudian slips, projective tests, subliminal perception, and signal detection theory. *Current Directions in Psychological Science, 1*, 174–178.

James, W. (1890). *The principles of psychology*. New York: Holt.

Johnson-Laird, P. N. (1988). *The computer and the mind: An introduction to cognitive science*. Cambridge, MA: Harvard University Press.

Johnson, M. K., Peterson, M. A., Yap, E. C., & Rose, P. M. (1989). Frequency judgments: The problem of defining a perceptual event. *Journal of Experimental Psychology: Learning, Memory, and Cognition, 15*, 126–136.

Joordens, S., & Merikle, P. M. (1992). False recognition and perception without awareness. *Memory and Cognition, 20*, 151–159.

Kihlstrom, J. F. (1987). The cognitive unconscious. *Science, 237*, 1445–1452.

Kihlstrom, J. F., Schacter, D. L., Cork, R. C., Hurt, C. A., & Behr, S. E. (1990). Implicit and explicit memory following surgical anesthesia. *Psychological Science, 1*, 303–306.

Klatzky, R. (1984). *Memory and awareness*. San Francisco: Freeman.

Knowlton, B. J., Ramus, S. J., & Squire, L. R. (1992). Intact artificial grammar learning in amnesia: Dissociation of classification learning and explicit memory for specific instances. *Psychological Science, 3*, 172–179.

LaBerge, D. (1981). Unitization and automaticity in perception. In H. Howe & J. Flowers (Eds.), *Nebraska symposium on motivation* (pp. 43–71). Lincoln: University of Nebraska Press.

Lachman, R., Lachman, J. L., & Butterfield, E. C. (1979). *Cognitive psychology and information processing: An introduction*. Hillsdale, NJ: Lawrence Erlbaum Associates.

Lewicki, P. (1986a). *Nonconscious social information processing*. New York: Academic Press.

Lewicki, P. (1986b). Processing information about covariation that cannot be articulated. *Journal of Experimental Psychology: Learning, Memory, and Cognition, 12*, 135–146.

Lewicki, P., & Hill, T. (1989). On the status of nonconscious processes in human cognition: Comment on Reber. *Journal of Experimental Psychology: General, 1989, 118*, 239–241.

Lewicki, P., Hill, T., & Bizot, E. (1988). Acquisition of procedural knowledge about a pattern of stimuli that cannot be articulated. *Cognitive Psychology, 20*, 24–37.

Logan, G. D. (1988). Toward an instance theory of automatization. *Psychological Review, 95*, 492–527.

Logan, G. D. (1990). Repetition priming and automaticity: Common underlying mechanisms? *Cognitive Psychology, 22*, 1–35.

Mandler, G. (1975). Consciousness: Respectable, useful, and probably necessary. In R. Solso (Ed.), *Information processing and cognition: The Loyola symposium* (pp. 229–254). Hillsdale, NJ: Lawrence Erlbaum Associates.

Mandler, G. (1985). *Cognitive psychology: An essay in cognitive science*. Hillsdale, NJ: Lawrence Erlbaum Associates.

Mandler, G., Nakamura, Y., & Van Zandt, B.J.S. (1987). Nonspecific effects of exposure to stimuli that cannot be recognized. *Journal of Experimental Psychology: Learning, Memory, and Cognition, 13,* 646–648.

Marescaux, P-J., Luc, F., & Karnas, G. (1989). Modes d'apprentissage sélectif et nonsélectif et connaissances acquises au controle d'un processus: Evaluation d'un modèle simulé. *Cahiers de Psychologie Cognitive, 9,* 239–264.

Mathews, R. C., Buss, R. R., Stanley, W. B., Blanchard-Fields, F., Cho, J. R., & Druhan, B. (1989). Role of implicit and explicit process in learning from examples: A synergistic effect. *Journal of Experimental Psychology: Learning, Memory, and Cognition, 15,* 1083- 1100.

Merikle, P. M., & Reingold, E. M. (1992). Measuring unconscious perceptual processes. In R. F. Bornstein & T. S. Pittman (Eds.), *Perception without awareness* (pp. 55–80). New York: Guilford.

Miller, G. A. (1980). Computation, consciousness and cognition. *Behavioral and Brain Sciences, 3,* 146.

Natsoulas, T. (1993a). Consciousness₄: Varieties of intrinsic theory. *Journal of Mind and Behavior, 14,* 107–132.

Natsoulas, T. (1993b). What is wrong with an appendage theory of consciousness. *Philosophical Psychology, 6,* 137–154.

Nissen, M. J., & Bullemer, P. (1987). Attentional requirements of learning: Evidence from performance measures. *Cognitive Psychology, 19,* 1–32.

Norman, D. (1980). Twelve problems for cognitive science. *Cognitive Science, 4,* 1–32.

Palmer, S. (1978). Fundamental aspects of cognitive representation. In E. Rosch & B. B. Lloyd (Eds.), *Cognition and categorization* (pp. 262–300). Hillsdale, NJ: Lawrence Erlbaum Associates.

Perner, J. (1991). *Understanding the representational mind.* Cambridge, MA: MIT Press.

Perruchet, P., & Amorim, M-A. (1992). Conscious knowledge and changes in performance in sequence learning: Evidence against a dissociation. *Journal of Experimental Psychology: Learning, Memory, and Cognition, 18,* 785–800.

Perruchet, P., & Pacteau, C. (1990). Synthetic grammar learning: Implicit rule abstraction or explicit fragmentary knowledge? *Journal of Experimental Psychology: General, 119,* 264- 275.

Perruchet, P., Gallego, J., & Pacteau, C. (1992). A reinterpretation of some earlier evidence for abstractness of implicitly acquired knowledge. *Quarterly Journal of Experimental Psychology, 44A,* 193–210.

Perruchet, P., Gallego, J., & Savy, I. (1990). A critical reappraisal of the evidence for unconscious abstraction of deterministic rules in complex experimental situations. *Cognitive Psychology, 22,* 493–516.

Posner, M. (1978). *Chronometric explorations of mind.* Hillsdale, NJ: Lawrence Erlbaum Associates.

Putnam, H. (1988). *Representation and reality.* Cambridge, MA: MIT Press.

Pylyshyn, Z. W. (1984). *Computation and cognition: Toward a foundation for cognitive science.* Cambridge, MA: MIT Press.

Reber, A. S. (1976). Implicit learning of synthetic languages: The role of instructional set. *Journal of Experimental Psychology: Human Learning and Memory, 2,* 88–94.

Reber, A. S. (1989). Implicit learning and tacit knowledge. *Journal of Experimental Psychology: General, 118,* 219–235.

Reber, A. S. (1992). The cognitive unconscious: An evolutionary perspective. *Consciousness and cognition, 1,* 93–133.

Reber, A. S. (1993). *Implicit learning: An essay on the cognitive unconscious.* New York: Oxford University Press.

Reber, A. S., & Allen, R. (1978). Analogic and abstraction strategies in synthetic grammar learning: A functionalist interpretation. *Cognition, 118,* 189–221.

Reber, A. S., & Lewis, S. (1977). Implicit learning: An analysis of the form and structure of a body of tacit knowledge. *Cognition, 5*, 333–361.

Reber, A. S., Walkenfeld, F. F., & Hernstadt, R. (1991). Implicit and explicit learning: Individual differences and IQ. *Journal of Experimental Psychology: Learning, Memory, and Cognition, 17*, 888–896.

Reingold. E. M., & Merikle, P. M. (1988). Using direct and indirect measures to study perception without awareness. *Perception & Psychophysics, 44*, 563–575.

Rosen, M., & Lunn, J. N. (1987). *Consciousness, awareness, and pain in general anaesthesia.* London: Butterworth.

Rumelhart, D., & Norman, D. A. (1988). Representation in memory. In R. Atkinson, D. Luce, & G. Lindzey (Eds.), *Revised Stevens handbook of experimental psychology* (pp. 511–587). New York: Wiley.

Rumelhart, D. E., Smolensky, P, McClelland, J. L., & Hinton, G. E. (1986). Schemata and sequential thought processes in PDP models. In J. L. McClelland & D. E. Rumelhart (Eds.), *Parallel distributed processing* (Vol. 2, pp. 8–57). Cambridge, MA: MIT Press.

Sanderson, P. M. (1989). Verbalizable knowledge and skilled task performance: Association, dissociation and mental models. *Journal of Experimental Psychology: Learning, Memory, and Cognition, 15*, 729–747.

Schneider, W. (1985). Toward a model of attention and the development of automatic processing. In M. I. Posner & O.S.M. Marin (Eds.), *Attention and performance XI* (pp. 475–492). Hillsdale, NJ: Lawrence Erlbaum Associates.

Searle, J. (1983). *Intentionality: An essay in the philosophy of mind.* New York: Cambridge University Press.

Searle, J. (1990). Consciousness, explanatory inversion, and cognitive science. *Behavioral and Brain Sciences, 13*, 585–586.

Searle, J. (1992). *The rediscovery of mind.* Cambridge, MA: MIT Press.

Seger, C. A. (1994). Implicit learning. *Psychological Bulletin, 115*, 163–196.

Servan-Schreiber, E., & Anderson, J. R. (1990). Learning artificial grammars with competitive chunking. *Journal of Experimental Psychology: Learning, Memory, and Cognition, 16*, 592–608.

Shanks, D. R., & St. John, M. F. (1994). Characteristics of dissociable human learning systems. *Behavioral and Brain Sciences, 17*, 367–447.

Shanks, D. R., Green. R.E.A., & Kolodny, J. (1994). A critical examination of the evidence for nonconscious (implicit) learning (837–861). In C. Umilta & M. Moscovitch (Eds.), *Attention and Performance XV: Conscious and nonconscious information processing* (pp. 837–861). Cambridge, MA: MIT Press.

Stanley. W. B., Mathews, R. C., Buss. R. R., & Kotler-Cope, S. (1989). An analysis of the interaction of procedural and declarative knowledge in a simulated process control task. *Quarterly Journal of Experimental Psychology, 41*, 553–577.

Treisman, A., Vieira, A., & Hayes, A. (1992). Automaticity and preattentive processing. *American Journal of Psychology, 105*, 341–362.

Turner, C. W., & Fischler, I. S. (1993). Speeded tests of implicit knowledge. *Journal of Experimental Psychology: Learning, Memory, and Cognition, 19*, 1165–1177.

Velmans, M. (1991). Is human information processing conscious? *Brain and Behavior Sciences, 14*, 651–726.

Vokey, J. R., & Brooks, L. R. (1992). The salience of item knowledge in learning artificial grammars. *Journal of Experimental Psychology: Learning, Memory, and Cognition, 18*, 328–344.

Wood, J. M., Bootzin, R. R., Kihlstrom, J. F., & Schacter, D. L. (1992). Implicit and explicit memory for verbal information presented during sleep. *Psychological Science, 3*, 236–239.

Willingham, D. B., Nissen, M. J., & Bullemer, P. (1989). On the development of procedural knowledge. *Journal of Experimental Psychology: Learning, Memory, and Cognition, 15,* 1047–1060.

Young, A. W., & De Haan, E.H.F. (1992). Face recognition and awareness after brain injury. In A. D. Milner & M. D. Rugg (Eds.), *The neuropsychology of consciousness* (pp. 69–90). San Diego: Academic Press.

Zajonc, R. B. (1968). Attitudinal effects of mere exposure. *Journal of Personality and Social Psychology Monograph, 9,* 2–27.

CHAPTER 11

Remembering and Knowing as States of Consciousness During Retrieval

Suparna Rajaram
State University of New York at Stony Brook

Henry L. Roediger III
Washington University, St. Louis

Consciousness is the most difficult topic that psychologists investigate. Philosophers have debated the meaning of consciousness for thousands of years and experimental psychologists have tried, with mixed success, to study the topic objectively in the last hundred years. The term has many distinct meanings and senses. As Miller put it over 30 years ago, "Consciousness is a word worn smooth by a million tongues" (1962, p. 25). In trying to gain a firmer grasp on the topic, we were driven to the *Macmillan Dictionary of Psychology,* edited by Stuart Sutherland (1991), where we found the following definition:

> **Consciousness.** The having of perceptions, thoughts, and feelings; awareness. The term is impossible to define except in terms that are unintelligible without a grasp of what consciousness means. Many fall into the trap of equating consciousness with self consciousness—to be conscious it is only necessary to be aware of the external world. Consciousness is a fascinating but elusive phenomenon: It is impossible to specify what it is, what it does, or why it evolved. Nothing worth reading has been written on it. (p. 90)

Many psychologists perhaps share Sutherland's gloomy conclusion, although certainly other authors in this volume represent exceptions to the general rule. However, why do most of us perceive that psychologists have made little headway beyond the ruminations of philosophers in understanding consciousness? Early psychologists in the structural school assiduously applied the method of analytic introspection to gain understanding of conscious awareness of the world. As is pointed out in every history of psychology textbook, current judgment is that the great effort they put into such introspective study came to naught, at least in terms of enhancing our understanding of consciousness. One fundamental problem is that the introspectionists' methods proved unreliable from laboratory to laboratory (and even within people in the same laboratory). Give two people what seems to be exactly the same experience and their list of the basic attributes (the sensations, the images, and the affections, as Titchener would have described them) might be rather different in the two cases. In short, one overarching difficulty in the study of consciousness is that researchers have had great difficulty in producing reliable methods for distinguishing states of awareness.

This chapter aims to help revive a study of states of awareness that accompany the act of retrieval. Essentially, the method described here involves having people determine the nature of conscious experience while they are retrieving previous experiences. Following Tulving (1985) and Gardiner (1988), we believe that subjects can reliably and usefully distinguish at least two conscious states of awareness during recollection: *remembering and knowing*. To try to illustrate the distinction, think back to your most recent trip to a scientific conference. Try to retrieve as much as possible about your travel experience. In performing this exercise, you can retrieve many details of the adventure and (in a sense) mentally reexperience the event from beginning to end, recalling many details along the way. This ability to revive the experience, or to mentally relive it, is referred to as remembering; remembering is this feeling of reexperiencing and of recollecting many details that authenticate the memory.

Next, try to retrieve your travel experience to a scientific conference held, say, 20 years ago. To pick an example, the second author is certain that he attended the Midwestern Psychological Association meeting in 1975 and in fact that he drove from West Lafayette, Indiana, to the conference in Chicago. However, despite knowing that he attended the conference and drove to it, he cannot remember the travel. He can remember no details whatsoever—the weather, the traffic conditions, or even the companions (if any). This latter experience represents knowing an event: We are confident that it happened, but we do not remember it.

The distinction between remembering and knowing in the senses just described was introduced by Tulving (1985). He argued that remembering is a reflection of the episodic memory system (our memory for personal happenings, or autobiographical memory as some call it), whereas knowing reflects output from the semantic memory system (our repository of impersonal, ahistorical knowledge). This distinction is a bit tricky, however, because in the travel example, and in the experiments described later, subjects make judgments of knowing about events in their personal past. The general argument is that we can know about episodes of our lives in an impersonal way, just as amnesic patients may know something but not be able to remember it. In short, the second author's 1975 trip to the meetings of MPA has the same ahistorical and impersonal character as his knowledge that Abraham Lincoln was the 16th president of the United States. He knows both statements to refer to true facts about the world, but has no personal recollection of the occurrence of either event. Although Tulving (1985) originally introduced knowing this way, more recently others (e.g., Jacoby, Yonelinas, & Jennings, chapter 2, this volume) have conceived the process in terms of familiarity, as one of the two components in recognition memory (as proposed by Mandler, 1980, 1989). This chapter considers other ideas about what the Know state may signal.

REMEMBER AND KNOW JUDGMENTS

Tulving (1985) introduced the distinction between remembering and knowing the past. In his experiment he presented subjects with a categorized list of words and then tested them by various means, with increasing power of retrieval cues across tests. Subjects studied category name–instance pairs (e.g., musical instrument–VIOLA) and then they participated in three successive recall tests. The power of the cues increased at each test from a free recall test to a category cued recall test (e.g., musical instrument–_____), to a category name and letter cued recall test (e.g., musical instrument–V____). After subjects recalled an item, they made a Remember or Know judgment to it.

The results showed that the proportion of Remember responses declined as the cues provided at test increased in their power. That is, the proportion of Remember responses given to recalled items was greatest in free recall and least in the test with category name and letter cues. Correspondingly, Know responses increased with stronger cues. Furthermore, in a recognition memory experiment, Remember responses declined more with retention interval over an 8-day period relative to overall recognition performance.

Following Tulving's original description of the Remember/Know judgment technique to analyze states of consciousness during a memory test, John Gardiner and his colleagues have employed the test in an extensive series of experiments. Others have also used the paradigm to good effect. Gardiner and Java (1993) have recently reviewed this body of work in an excellent chapter so we will not attempt to review exactly the same evidence. This chapter reviews the representative evidence to date from use of the technique and describes a new approach to explaining these results in light of further evidence carried out mostly in the first author's laboratory. To achieve this goal, the chapter is organized in the following way. First, it delineates the instructions, remarks on the nature of these judgments and the relation between them, and discusses the relations between Remember/Know judgments and the conceptual/perceptual components of recognition and of priming. Second, it describes the influence of conceptual and perceptual manipulations on Remember/Know judgments. This section documents three types of dissociations—that is, the influence of the independent variables on Remember but not Know judgments, the opposite influence of the independent variables on Remember and Know judgments, and the influence of the independent variable on Know but not Remember judgments—that initially supported the substantial degree to which conceptual/perceptual processing distinction mapped on to Remember/Know distinction. Third, it reviews evidence that is problematic for the conceptual/perceptual processing distinction and proposes a new framework based on distinctiveness and fluency that accounts for most of the evidence to date. The last section evaluates the Remember/Know paradigm, its usefulness in studying the states of consciousness that accompany retrieval, and comments on the current alternate approaches to the study of conscious recollective states.

REMEMBER/KNOW JUDGMENTS: INSTRUCTIONS AND RELATED REALMS OF RESEARCH

In the experiments reported by Gardiner and his colleagues and by Rajaram, subjects were instructed that if production of a word was "accompanied by a conscious recollection of its prior occurrence in the study list," then they should write *R* for Remember beside the word. Subjects were further told:

> Remember is the ability to become consciously aware again of some aspect or aspects of what happened or what was experienced at the time the word was presented (e.g., aspects of the physical appearance of the word, or something that happened in the room, or of what you were thinking or doing at the time). In other words, the Remember words should bring back to mind a particular association, image, or something more personal from the time of study, or something about its appearance or position (i.e., what came before or after the word). On the other hand, Know responses should be made when you recognize that the word was in the study list but you cannot consciously recollect anything about its actual occurrence or what happened or what was experienced at the time of its occurrence.

These instructions were taken from those used by Rajaram (1993, p. 102), but they were modeled on previous instructions used by Gardiner (1988). It is important to emphasize that instructions to the subject are absolutely critical and in some cases researchers require subjects to repeat back the instructions in their own words to make sure they understand them before the experiment proceeds. Subjects must be told precisely what judgments are to be made, and must be clear that they can separate the two states of consciousness reliably. Although the technique seems fraught with difficulties, its careful use has produced a reasonably consistent body of evidence.

For example, in one experiment Gardiner (1988, Exp. 1) manipulated levels of processing while subjects studied a long list of words. During study subjects were either led to think about the phonemic property of words, or to make a semantic judgment about the words. This levels of processing manipulation has powerful effects on recall and recognition in most explicit tests (Craik & Lockhart, 1972). Gardiner was interested in whether the effect would be seen in Remember judgments or Know judgments, or both. He gave subjects a recognition test and asked them, after judging a word to be old, to say whether they remembered that the word had appeared in the list or whether they simply knew that it had. A powerful levels of processing effect was obtained in overall recognition. When the effect was decomposed into subjects' Remember and Know judgments, Remember judgments determined the levels of processing effect, whereas judgments of Know were equivalent for the two levels of processing.

Note that Gardiner (1988) followed the convention introduced by Tulving in considering overall recognition to be composed of two types of recollective experiences: Remember and Know. Subjects are asked to judge recognized items on this basis, and the experimenter decomposes overall recognition into the two components. This practice has been criticized by others (Jacoby et al., chapter 2, this volume) for assuming that there is an exclusive relation between the two states of knowledge: Items are either remembered or known, without any possibility of a mix, as would be assumed if the underlying processes were thought to be independent.

This chapter also follows the convention of decomposing recognition responses into the two entities of interest, Remember and Know. We are interested in a first-person account of subjects' states of awareness, and so leave it to them to determine this quality for each item. This is not to say that the processes involved may not somehow overlap, but we leave it to the subject to determine whether recognized items should be judged as Remember or Know based on the predominance of information on which the retrieval experience is based. As is true in all the first-person accounts of states of awareness, we assume the subject is capable of making these judgments and we prefer to trust the categorization of behavior provided by the subject rather than to apply statistical techniques that, at least in the theorist's mind, provides a truer picture of the subject's states of awareness. After all, the hallmark of first-person accounts of the study of consciousness, of which Remember/Know is a clear instance (see Gardiner, 1991), is to rely on the subject's reports. These of course may be faulty in some instances (e.g., Nisbett & Wilson, 1977), but in matters of recollection—an inherently private experience—we must rely on the subject's reports. Therefore, this chapter does so, although others would prefer to make different assumptions and apply different models to structure the data. This issue is considered again later.

Several other variables have an effect similar to levels of processing on Remember and Know judgments, as reviewed by Gardiner and Java (1993). That is, manipulation of an experimental variable often affects Remember responses, leaving Know responses unaffected. One worry in the face of such a pattern is that Know responses are simply a residual category—just what subjects put down when they think that something might have occurred but cannot really remember it—but do not reflect a special state of consciousness. Although a valid concern, the feeling of familiarity that accompanies the retrieval of many events in the absence of clear and vivid recollection is a common everyday experience. It may also be that Know judgments are simply insensitive to any experimental variation. At least this second charge can be shown to be untrue, because as becomes evident later, experiments have permitted people to dissociate Remember and Know judgments by experimental variables. That is, manipulation of a variable can have opposite effects on Remember and Know judgments, or in some cases have an effect on Know judgments but not on Remember judgments (Rajaram, 1993).

Another potential worry is that Remember/Know judgments simply reflect different degrees of confidence, with Remember responses reflecting high confidence and Know responses reflecting low confidence. If so, Remember/Know judgments would not represent states of consciousness any different from those represented by confidence judgments. Although it is true that Remember judgments are often accompanied by high degrees of confidence (Tulving, 1985), Gardiner and Java (1990), Parkin and Walter (1992), and Rajaram (1993) have shown that these responses are not one and the same. In these studies, independent variables produced different patterns of interactions with Remember/Know judgments compared to those produced with confidence judgments ("sure"/"unsure").

Gardiner's results (already presented), and many other results too, fit comfortably with two other bodies of evidence. This fact has caused researchers to conceive of Know judgments in quite a different way from Tulving's (1985) initial conceptualizations of these as reflecting impersonal knowledge from semantic memory. First, several other researchers have proposed that there are two bases of recognition memory. Mandler (1979, 1980) termed these components integration and elaboration, where integration refers to processing elements of an individual item and elaboration refers to relating an item to others. Jacoby (1983a, 1983b; Jacoby & Dallas, 1981) similarly distinguished between perceptual and conceptual bases of recognition. In some cases, we recognize an event from our past because of perceptual fluency—it "jumps off the page at us"—whereas in other cases we recognize that a test event has a similar meaning to a prior study event. The ideas of Mandler and Jacoby might map on to Remember and Know judgments in a natural way. Remember judgments may reflect elaborative or conceptual processing in those theorists' terms, whereas Know judgments might reflect perceptual fluency or intraitem integration. Putting the same ideas into Jacoby et al.'s (chapter 2, this volume) terms, Remember judgments may reflect conscious recollection, whereas Know judgments might reflect the automatic influence of past events that bias current experience. However, note that for Jacoby's and Mandler's conceptualizations, Know items provide a warm feeling of familiarity, whereas this attribute seemed absent from Tulving's original conceptualization of Know responses as involving retrieval from semantic memory.

The second realm of research that may be related to Remember/Know judgments is perceptual priming in implicit memory tests (Jacoby, 1983b; Roediger, 1990). These authors have distinguished between perceptual and conceptual bases of priming. In many experiments, such as Gardiner's, a variable affecting subjects' strategies has large effects on Remember judgments but leaves Know judgments unaffected. This same pattern occurs in many perceptual implicit memory tests, with variables such as levels of processing having little or no effect on tests such as perceptual identification (Jacoby & Dallas, 1981), word stem completion (Graf, Mandler, & Haden, 1982), and word fragment completion (Roediger, Weldon, Stadler, & Riegler, 1992). One plausible idea is that the same sorts of factors that affect perceptual priming also drive Know judgments in the Remember/Know paradigm, because Know judgments are also little affected by levels of processing. On the other hand, Remember judgments appear to increase with manipulation of variables, such as levels of processing, that also improve performance on the conceptual explicit memory tasks such as free recall (see Roediger, Weldon, & Challis, 1989). This similarity is explored later, too. However, to presage the conclusion, neither of the simple stories worked out previously turn out to accommodate all the data.

REMEMBER/KNOW JUDGMENTS: EFFECTS OF PERCEPTUAL AND CONCEPTUAL MANIPULATIONS

A selective review of the evidence is presented in this section to demonstrate the similarities and differences in the pattern of results between Remember/Know judgments and explicit/implicit memory tasks. (Note that in this chapter, all discussions of explicit memory tasks pertain specifically to explicit tasks that are conceptual in nature. Similarly, all discussions of implicit memory tasks are only with reference to perceptual priming tasks.) As becomes evident, the initial set of Remember/Know data described here maps onto the conceptual and perceptual processing distinctions in recognition memory very nicely. In addition, our selection of studies also illustrates the three different types of dissociations that validate the Remember/Know distinction, namely, dissociations in which a given variable affects only Remember judgments, affects both Remember and Know judgments in opposite ways, or affects only Know judgments.

The previous section described the results that Gardiner (1988, Exp. 1) ob-

tained by having subjects produce rhyme associates or semantic associates for
the study items. Rajaram (1993, Exp. 1) also used this variable in order to
replicate Gardiner's findings. These data are presented in Fig. 11.1. Rajaram
(1993) replicated the basic levels of processing finding for the overall recognition data and for the Remember judgments. In addition, she found that subjects gave more Know responses to items for which rhyme associates rather
than semantic associates were produced at study. Although this reversed levels of processing effect for Know judgments supports the hypothesis that perceptual factors mediate Know judgments, this pattern has not typically been
reported with perceptual implicit memory tasks (but see Luo, 1993). As mentioned earlier, the levels of processing manipulation is known to have little effect on implicit memory tasks under most conditions (Roediger & McDermott, 1993). Of course, Gardiner (1988) did not find the reversed levels of
processing effect, so further research is warranted.

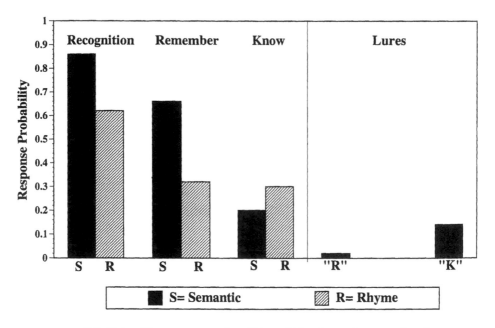

FIG. 11.1.   The mean proportion of hits and false alarms shown as a function
of the levels of processing manipulation in Rajaram (1993, Exp. 1). Similar to
Gardiner's finding, a levels of processing effect—that is, superior performance
for semantically processed items compared to phonetically processed items—
was obtained in the overall recognition and in Remember judgments. This effect was magnified for the Remember judgments and reversed for the Know
judgments. These data support the notion that Remember judgments are enhanced by conceptual processing during study whereas Know judgments increase as a function of perceptual processing. From Rajaram (1993). Copyright
1993 by Psychonomic Society. Adapted with permission.

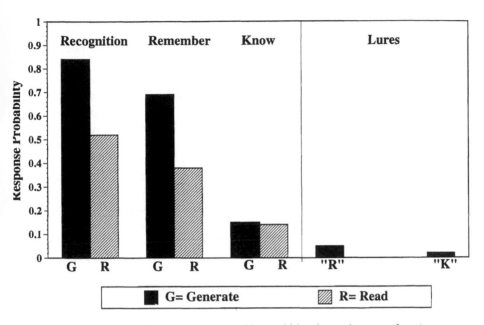

FIG. 11.2. The mean proportion of hits and false alarms shown as a function of the generate–read manipulation at a 1-hour retention interval in Gardiner (1988, Exp. 2). The generation effect, that is, better memory for items generated than read at study, was obtained for recognition and Remember judgments. This variable had no effect on the Know judgments. These data bear a resemblance to the effects of the generate–read manipulation on conceptual explicit memory and perceptual implicit memory tasks respectively, with one difference. In the perceptual implicit memory tasks, the generation effect is typically reversed, unlike the null effects obtained here for the Know judgments. From Gardiner (1988). Copyright 1988 by Psychonomic Society. Adapted with permission.

The generate–read manipulation also produces a similar dissociation between Remember and Know judgments as the one found between explicit and implicit tasks, with one difference. When subjects *generate* the target words at study in response to semantic cues (e.g., generating antonyms given the semantic cue, hot–???) as opposed to simply reading them with context (hot–cold) or without any context (e.g., xxx–cold), performance in explicit memory tasks such as recognition is better for generated items compared to items that were read (Jacoby, 1978; Slamecka & Graf, 1978). This pattern reverses in implicit memory tasks when the items in the *read* condition are presented without the context (i.e., xxx–cold): reading generally produces more priming on perceptual implicit memory tasks than does generating (see Roediger & McDermott, 1993, for a review, and Masson & MacLeod, 1992, for an exception). As shown in Fig. 11.2, when Gardiner (1988, Exp. 2) had subjects generate or read (with context, i.e., hot–cold) target items and later make Remember and Know judgments to recognized items, subjects gave more Remember responses to generated than to read items. This manipulation had no effect on Know judgments. Furthermore, Java (1994) found that even when study items in the Read condition were presented without context (i.e., xxx–cold, a condition in which the generation effect reverses in the perceptual priming tasks), the generate/read manipulation had no effect on Know judgments. Thus, once again we find that the manipulated variable affects Know judgments and perceptual priming tasks differently. However, with the levels of processing variable, the effect of the independent variable is observed on Know judgments (at least in Rajaram's experiment) but not obtained on perceptual priming tasks, whereas the generate–read manipulation is known to affect perceptual priming but had no effect on Know judgments.

Another variable of interest in this regard is processing of items under full versus divided attention conditions. This manipulation has been reported to have differential effects on explicit and implicit memory tasks (e.g., Jacoby, Woloshyn, & Kelley, 1989; Parkin & Russo, 1989). Explicit memory tasks are adversely affected for information studied under conditions of divided attention compared to the full attention condition. This variable does not seem to influence performance on perceptual implicit memory tasks. Similar dissociations were reported between Remember and Know judgments by Gardiner and Parkin (1990). When subjects studied words under divided attention conditions in which they had to perform a secondary task of tone monitoring, Remember responses declined compared to the full attention condition in which no such secondary task was performed. Know judgments however, remained unaffected. In this study, the pattern of dissociation between Remember and Know judgments bears a close resemblance to that observed between explicit and implicit memory tasks.

One of the concerns in evaluating the psychological distinction between Remember and Know judgments is that single dissociations in which a variable affects Remember but not Know judgments do not establish Know judgments as an independently manipulable entity. To consider Know judgments as different from other forms of memories, opposite effects of an independent variable on Remember and Know judgments should be demonstrated. In addition to the levels of processing manipulation (Rajaram, 1993, Exp. 1) described earlier, two other independent variables have influenced Remember and Know judgments in opposite ways. In one experiment, Rajaram (1993) had subjects study words and pictures and later participate in a recognition memory task in which all the studied and nonstudied items were presented in verbal form. Thus, the perceptual format between study and test varied for half the items (i.e., pictures at study and words at test), whereas it was held constant for the other half (words at study and at test). Sample materials and the design are displayed in Fig. 11.3.

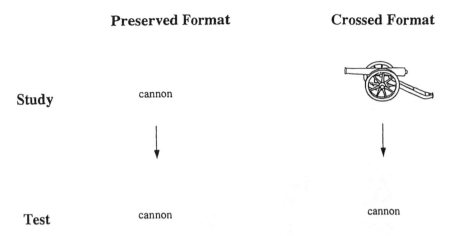

FIG. 11.3.   The design and a sample of stimuli used in Rajaram (1993, Exp. 2).

Based on the well-documented findings of the picture superiority effect—
better memory for pictures than for words—in most explicit memory tasks
(e.g., Madigan, 1983), Rajaram (1993) predicted that a higher proportion of
Remember judgments would be given for studied pictures than for studied
words. The predictions for Know judgments were derived from Weldon and
Roediger's (1987) report that the picture superiority effect reverses in perceptual implicit memory tasks such as word fragment completion. Thus,
priming on the word fragment completion task is greater following the study
of words (i.e., the names of pictures) rather than following the study of the pictorial counterpart of these items. This is because perceptual implicit memory
tasks such as word fragment completion benefit from the perceptual match
between the study and test materials, according to the principle of transfer appropriate processing (Roediger et al., 1989). Based on these findings, Rajaram (1993) reasoned that Know judgments should be higher for items in the
same format condition compared to items in the different format condition
(see Fig. 11.3).

The predictions for Remember and Know judgments were borne out very
nicely, as shown in Fig. 11.4. Rajaram (1993) found that a picture superiority
effect was obtained for the overall recognition in that the hit rate for pictures
was .90 and that for words was .69. When the data were decomposed into Remember and Know, the picture superiority effect was enhanced for Remember judgments (.81 for pictures and .51 for words). Conversely, significantly
more Know judgments were given for studied words (.18) than for studied pictures (.09). Once again we find that the patterns of dissociations are very similar between Remember/Know judgments on one hand, and (conceptual) ex-
plicit/(perceptual) implicit memory tasks on the other.

Another variable that produces opposite effects on Remember and Know

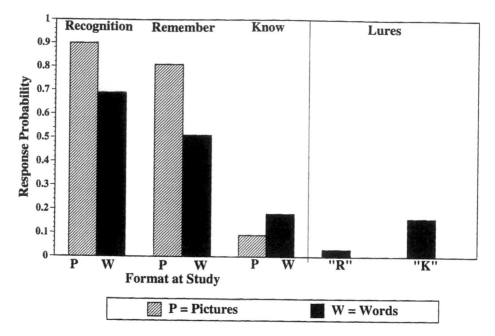

FIG. 11.4.   The mean proportion of hits and false alarms obtained by Rajaram
(1993, Exp. 2) for studied words and pictures in a word recognition memory
task. Subjects studied words and pictures and then took a recognition test requiring recognition of studied and nonstudied items presented in word form.
Items studied as pictures were recognized better than those studied as words.
This picture superiority effect was magnified when only Remember responses
are considered, indicating greater conscious recollection for pictures than for
words. However, Know responses were greater for words than for pictures, presumably because Know judgments are driven by the same factors as produce
priming on perceptual implicit memory tasks (i.e., the match in perceptual processes between study and test.) From Rajaram (1993). Copyright 1993 by Psychonomic Society. Adapted with permission.

judgments is study and testing of words and nonwords. Gardiner and Java
(1990, Exp. 2) found that after studying words and nonwords, subjects gave
more Know responses to studied nonwords than to studied words, whereas
more Remember judgments were given to studied words than nonwords. Gardiner and Java reasoned that subjects likely engage more in the perceptual
analyses of the stimuli that are nonwords (because no conceptual information
is available for these items), thereby giving more Know responses to these
stimuli. On the other hand, word stimuli are more amenable to conceptual
processing, thereby receiving more Remember responses. Like most of the
data described so far, this explanation fits well with the conceptual and perceptual bases of recognition proposed by Jacoby and Mandler.

In the studies described so far, variables that influenced Know judgments

TABLE 11.1
The Design and Results for Proportion of Hits and False Alarms
for the Masked Repetition Manipulation

Study items (target)—table, plate

	Targets		Lures	
	Masked Repetition	Unrelated Prime	Masked Repetition	Unrelated Prime
Mask	XXXXX	XXXXX	XXXXX	XXXXX
Prime	table	scale	glass	chalk
Target	TABLE	PLATE	GLASS	SHIRT
Response Required	"Yes"	"Yes"	"No"	"No"
Recognition Response	"Yes"	"Yes"	"Yes" (FA)	"Yes" (FA)
	.67	.60	.23	.18
Remember Response	.43	.42	.05	.05
Know Response	.24	.18	.18	.13

Note: From Rajaram (1993). Copyright 1993 by Psychonomic Society. Reprinted with permission.

also had an effect on Remember judgments. Rajaram (1993, Exp. 3) reasoned that if Know judgments are dissociable from Remember judgments, they should be *selectively* influenced by independent variables that increase the perceptual fluency with which items are processed. This prediction was based on the findings from previous experiments suggesting that Know judgments are sensitive to variables that enhance perceptual processing of the to-be-recognized stimuli.

Jacoby and Whitehouse (1989) argued that masked repetition of tested items increases the perceptual fluency with which these items are processed. This increased perceptual fluency gives rise to a feeling of familiarity, which in turn improves recognition performance for the studied items and increases false alarms for nonstudied items in a recognition memory task. To test the hypothesis that perceptual factors increase Know responses, Rajaram (1993) presented some of the studied and nonstudied words in the recognition test twice, such that the first presentation of the stimulus was not available for conscious report[1] because it was preceded by a forward mask and followed by its second presentation in uppercase (see Table 11.1 for an example). For the other half of the studied and nonstudied words, the preceding masked words

---

[1] It should be noted that we are not claiming that these items were presented subliminally. We consider that the masked items were not available for conscious report only in as much as subjects claimed not to see these words and were unable to report them.

were different from the targets. This design is illustrated with an example in the top half of Table 11.1. Based on Jacoby and Whitehouse's (1989) findings, the prediction was that the masked repetition manipulation should only increase Know judgments while leaving the proportion of Remember judgments unaffected, if indeed Know judgments are selectively affected by changes in the perceptual processing of items. The results obtained by Rajaram supported this perceptual fluency hypothesis for Know judgments, as seen in the data presented in the bottom half of Table 11.1.

The results reported here present a generally consistent story with reference to the two comparisons mentioned earlier. The first comparison is drawn between the conceptual explicit and perceptual implicit memory dissociations on one hand, and the Remember and Know dissociations on the other. As has been noted, although there are some differences observed in the results with Remember/Know judgments, it seems that, by and large, Remember judgments capture the processes affecting conceptual explicit memory tests whereas Know judgments are influenced by processes that also determine performance on perceptual implicit memory tasks.

The second comparison was drawn between Remember/Know judgments and conceptual/perceptual bases of recognition memory, respectively. When subjects encode the conceptual, elaborative, and meaningful aspects of the to-be-recognized materials, they are more likely to actually "remember" the event later. Thus, variables such as semantic encoding of the stimuli in the levels of processing manipulation, generating words in the generate–read manipulation, studying pictures versus words, and studying words versus nonwords all benefit recollective processes that preserve the conscious and vivid aspects of memories. Conversely, if the to-be-recognized materials are processed more for their perceptual (surface) characteristics, subjects "know" that the event was encountered before but cannot "remember" the actual occurrence of the event. Thus, phonemic processing of stimuli in the levels of processing manipulation, reading the stimuli in the generate–read manipulation, the perceptual overlap between studied and tested stimuli in the picture–word manipulation, and the presumed enhanced perceptual processing of nonwords relative to words all increase the likelihood of "knowing" that the stimulus was encountered before. In addition, increasing the perceptual fluency of processing by masked repetition of an item selectively increases Know judgments. Some other studies have examined the effects of manipulating certain subject variables and have obtained results consistent with the findings described so far.

Individual Differences in Remember/Know judgments

Remember and Know judgments have been reported to dissociate as a function of normal aging and neurological impairments as well. For instance,

Parkin and Walter (1992) predicted that older adults should produce fewer Remember responses than do young adults because older adults likely benefit less from contextual information (typically associated with episodic memory) compared to younger adults (Craik, L. W. Morris, R. G. Morris, & Loewen, 1990). Further, older adults may rely more on the perceptual fluency of the studied stimuli to recognize them (if conceptual factors are impoverished), which should yield a higher proportion of Know judgments from these subjects compared to younger adults. This was exactly the pattern of results obtained by Parkin and Walter (1992). Once again, this pattern resembles the dissociation obtained between perceptual implicit memory tasks and conceptual explicit memory tasks (e.g., Light & Singh, 1987) as a function of age, because performance on explicit memory tests is found to be impaired in older adults whereas their perceptual priming is comparable to that of young adults.

In another study, Blaxton (in press) tested normal subjects and temporal lobe epileptics (TLEs) for their recognition memory for novel line drawings. TLEs exhibit memory impairments as a function of their neurological condition. In one experiment, normals and TLEs with left temporal lobe damage gave mostly Know judgments to the recognized line drawings, suggesting that these novel figures were presumably processed for perceptual features in the right hemisphere. That is, because the left temporal lobe was damaged, TLEs presumably relied on the right hemisphere processes to support recognition and the right hemisphere is assumed to be largely responsible for perceptual processes involved in object recognition. On the other hand, TLEs with damage to the right temporal lobe gave mostly Remember judgments to correctly recognized stimuli. Based on these results and the fact that left TLEs show deficits in learning meaningful verbal materials, Blaxton (in press) reasoned that the left TLEs presumably show deficits in conceptual processing of information, whereas the right TLEs show perceptual processing deficits. To test this idea, Blaxton had normals, left TLEs, and right TLEs perform semantic or surface encoding of the same novel figures in another experiment. Normal subjects gave more Remember judgments following semantic encoding of the nonverbal materials and more Know judgments following an analysis of the surface features. The left TLEs showed a deficit in conceptual transfer, as indicated by a high proportion of Know judgments even after the semantic encoding of the nonverbal stimuli. On the other hand, the right TLEs produced a high proportion of Remember responses even when the study instructions specified a perceptual analysis of the materials, exhibiting a deficit in perceptual transfer.

Blaxton's (in press) work supports the conceptual/perceptual distinction drawn between Remember and Know judgments, and Parkin and Walter's (1992) study with young and aging adults provides yet another instance of the similarity between explicit and implicit memory tasks and Remember and Know judgments. Thus, based on the evidence presented in this and the previous sections, it appears that Remember and Know judgments capture the two bases of recognition memory proposed by Jacoby and Dallas (1981) and Mandler (1980). As clear-cut and consistent as these findings may seem, new evidence collected by the first author complicates these conclusions and is discussed next.

EVIDENCE INCONSISTENT WITH
PERCEPTUAL BASES OF KNOW JUDGMENTS

There are two different patterns of results where the dissociations between Remember and Know judgments do not seem to map onto conceptual and perceptual factors, respectively. The first set of results documents the failure to obtain an increase in Know judgments in conditions that presumably facilitate perceptual processing. The second set of results demonstrates the effects of perceptual manipulations on Remember judgments. Each of these patterns is discussed in turn.

The first set of results was obtained by manipulating the modality in which the study and test items were presented to the subjects. There are two reports on the effects of preserved or changed modality between study and test on Remember/Know judgments. In one study, Rajaram (1993) had subjects either read words or listen to words before participating in a visual word recognition task. The expected benefit for Know judgments in the perceptual match condition (visual presentation at study and at test) relative to the perceptual mismatch condition (auditory presentation at study and visual presentation at test) was not obtained. The results are shown in Fig. 11.5. One possible reason why Rajaram (1993) failed to obtain an effect of modality match for Know judgments might be that there was no effect of preserved modality on the overall recognition memory performance.

However, the lack of a modality effect in recognition is also problematic for two-component theories of recognition. Modality has substantial effects on perceptual priming tests, with words presented visually producing greater priming than words presented auditorily on visual priming tests (fragment completion, stem completion, etc.; see Roediger & McDermott, 1993, for a review). If recognition memory is partly driven by the same factors that determine perceptual priming, then one would expect an effect of modality on recognition memory. Although some have reported a small effect (Kirsner & Smith, 1974), most researchers have not obtained this effect. For example, in a particularly powerful experiment, Challis et al. (1993) failed to find any modality effect in recognition.

Gregg and Gardiner (1991) also failed to find perceptual effects of modality match on Know judgments. In their study, subjects read words either silently or aloud at study and at test. When the modality of presentation was

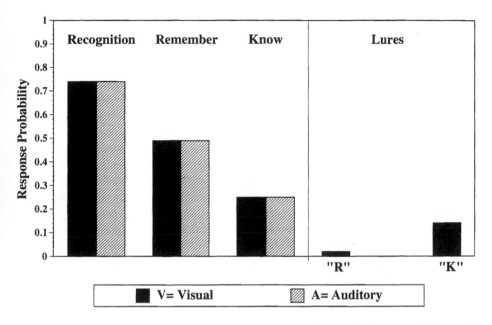

FIG. 11.5.   The mean proportion of hits and false alarms obtained by Rajaram (1993, Exp. 1) for items studied in the visual or auditory modality and tested on a recognition task in the visual modality. There was no effect of modality of presentation on any of the response type (Recognition, Remember, or Know). Although a modality match is expected to enhance Recognition and Know judgments, there are no consistent reports of this modality effect reported in the literature for the recognition memory task. From Rajaram (1993). Copyright 1993 by Psychonomic Society. Adapted with permission.

held constant (e.g., reading an item silently at both study and test, or aloud at both study and test), the perceptual match was maximized as opposed to when the modality of presentation was changed (i.e., read silently at study and aloud at test, or vice versa). Based on the idea that perceptual factors drive Know judgments, one would predict that greater proportions of Know judgments would be obtained in the perceptual match rather than the perceptual mismatch conditions. Conversely, more Remember judgments should be obtained in the modality mismatch conditions, because subjects have to rely on information that is relatively conceptual to recognize an item presented in a different modality. In a recent study Gregg and Gardiner (1994) found the effect of modality match on Know judgments using a powerful manipulation. However, in an earlier study (1991), they found that Know judgments did not increase in the modality match conditions, and similar to Rajaram's findings (1993), they also failed to find an effect of modality match on overall recognition. These puzzling findings may be due to the problem that, in general, the modality manipulation has not yielded consistent patterns in conceptual explicit tasks in the literature. In any case, it is not clear to us why the modality match did not increase Know judgments in Gregg and Gardiner's (1991) and Rajaram's (1993) work. Further work (or a better theory) is needed to resolve this issue.

The second set of findings that is difficult to reconcile with the hypothesis that Remember judgments reflect conceptual processes and Know judgments reflect perceptual processes are those showing effects of perceptual variables on Remember judgments instead of on Know judgments. The first such finding was reported by Gardiner and Java (1990, Exp. 1), who assessed subjects' recognition memory for high and low frequency words. A typical finding in recognition memory tasks is that subjects recognize significantly more low frequency words compared to high frequency words. This superior recognition memory for low frequency words has been explained by assuming that low frequency words are processed with increased perceptual fluency (Jacoby & Dallas, 1981) or familiarity (Mandler, 1980) relative to high frequency words, when they are processed on the test (i.e., a second time.) A straightforward prediction from this reasoning is that better recognition memory for low frequency words should be captured in Know judgments. However, Gardiner and Java (1990) obtained the opposite pattern of results. Subjects gave significantly more Remember judgments to low frequency words than to high frequency words, whereas Know judgments for these two types of words did not differ.

Rajaram (1996; Rajaram & Coslett, 1992, 1993) reported a similar finding with both verbal and pictorial stimuli. In one experiment, subjects studied pictures and words, and were given a recognition task in which all the studied and nonstudied items were presented in the pictorial form. Note that this experiment is the complement to the one reported by Rajaram (1993, Exp. 2), where the study phase was exactly the same as described here but at test, all items were presented in the word form (see Fig. 11.3). In Rajaram's (1993) experiment, the maximal perceptual match was present for items studied and tested in word form and, accordingly, Know judgments in this same format condition were higher than in the different format condition (that is, pictures at study and words at test). Based on these prior results, a higher proportion of Know judgments would be expected for items studied and tested in the pictorial format in Rajaram's (1996) experiment, because the perceptual overlap is greater in this condition than in the different format condition (that is, words at study and pictures at test). However, the results did not bear out this prediction, as shown in Fig. 11.6. Not surprisingly, studied pictures were recognized better than were studied words, and this difference was magnified for the Remember responses. However, rather than showing the same pattern for Know responses, as predicted, the pattern actually reversed. That is, items studied as words received more Know judgments following their judgment of "old" on picture recognition than did items studied as pictures. This outcome is inconsistent with the idea that Know judgments are driven mostly by perceptual fluency.

Rajaram (1996) manipulated the size and the reflection of the studied and tested pictorial stimuli to determine the effects of such changes in surface features on Remember and Know judgments. In one experiment, subjects studied pictures presented in small or large sizes. At test, these pictures were either presented in the same size as at study (perceptual match condition) or in a different size from study (perceptual mismatch condition). In addition, half of the nonstudied pictures were small and the other half were large in size. The results showed a slight but statistically significant effect of size on overall recognition, with same size pictures at study and test being recognized better than pictures that differed in size on the two occasions. Given that size had an

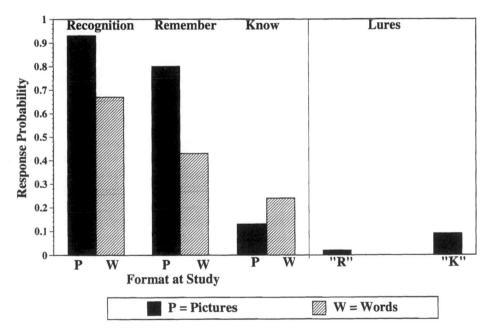

FIG. 11.6. The mean proportion of hits and false alarms obtained by Rajaram (1996) when subjects studied words and pictures and were tested on a picture recognition memory task. Overall recognition was better for items studied and tested in the pictorial form compared to the items studied as words and tested in the pictorial form. This effect was magnified for Remember judgments, whereas a greater proportion of Know judgments were given to items studied as words and tested in the pictorial form compared to the same format condition. These data are problematic for the notion that greater perceptual match increases Know judgments. From Rajaram (1996). Copyright 1996 by American Psychological Association. Adapted with permission.

effect in overall recognition, one might expect that such a perceptual factor would influence Know judgments but not Remember judgments. However, exactly the opposite happened. The effect of size was entirely on the Remember judgments and indeed there was actually a reversal (that approached significance) in Know judgments, such that different sized pictures at study and test produced a greater proportion of Know responses than did same size pictures at study and test.

Rajaram (1996) manipulated another perceptual attribute to determine its effect on Remember and Know judgments: the reflection of pictorial stimuli across study and test. Half the objects presented at study faced right whereas the other half faced left. At test, the reflection for half of studied items was preserved with respect to the study phase and was changed for the other half (such that objects facing right at study now faced left, and vice versa.) Furthermore, half the lures faced left and the other half right. It is reasonable to assume that the perceptual match in the same reflection condition would be greater than the changed reflection condition. Therefore, overall recognition should be greater in the same reflection condition compared to the changed reflection condition and the difference should appear in Know judgments (due to enhanced perceptual fluency). On the other hand, same or changed reflection should not affect Remember judgments. However, the results did not bear out any of these predictions. The effect of preserved or changed reflection was not obtained either for the overall recognition judgments or for the Know judgments. Surprisingly, a significantly greater proportion of Remember judgments were obtained in the same reflection condition relative to the changed reflection condition. Therefore, Rajaram's (1996) results were directly opposite to those predicted by the idea that perceptual fluency drives Know responses.

These effects of size and reflection changes on Remember/Know judgments are clearly problematic for the idea that the conceptual/perceptual processing distinction underlies Remember and Know judgments. It is worth noting that the problematic results obtained in the experiments with size and reflection were conducted under the same conditions, with the same instructions, and with the same general type of subjects that led to the positive results found in the earlier experiments. However, in the context of other experiments, these findings do not seem too surprising, because they are in accord with the evidence reported by Biederman and E. Cooper (1992) and L. Cooper, Schacter, Ballesteros, and Moore (1992) for explicit and implicit memory tasks. Biederman and E. Cooper (1992) manipulated the size of line drawings of common objects in an object naming task (presumably a perceptual implicit task) and a recognition task (presumably a conceptual explicit task). L. Cooper et al. (1992) manipulated size and reflection of novel line drawings (in separate experiments) in an object decision task (presumably a perceptual implicit task) and a recognition task (presumably a more conceptual explicit task). The performance in the object naming task (Biederman & E. Cooper, 1992) and the object decision task (L. Cooper et al., 1992) was unaffected by changes in the size and reflection of stimuli even though these tasks are presumed to rely on perceptual operations. On the other hand, recognition performance was adversely affected by size and reflection changes across study and test phases in both these studies. That is, recognition was better for objects that were tested in the same size or reflection given at study rather than in a different size or reflection.

In the light of these findings, Rajaram's (1996) results with Remember and Know judgments may not be surprising. Specifically, if Know judgments are presumed to be sensitive to variables that affect perceptual implicit tasks and if size and reflection (for whatever reasons) do not affect priming, then it is no surprise that these variables do not affect Know judgments. In fact, increased Know judgments were produced with size or reflection changes although they were rather small. More importantly, the effects of size and reflection transformations in Rajaram's experiments on Remember judgments were similar to the effects obtained in recognition memory in Biederman and E. Cooper's (1992) and L. Cooper et al.'s (1992) studies. Regardless of this consistency in the pattern of results, the size and reflection effects and the other perceptual effects on Remember judgments reported in this section cannot be accommodated within the currently favored explanation that only conceptual factors affect conscious and vivid aspects of the recollective processes.

How can these findings be explained and what implications do they have for the nature of recollective processes? One obvious conclusion to draw from all these findings is that the conceptual/perceptual processing distinction, although useful up to a point, does not encompass the gamut of variables that influence the two different ways of accessing memories for prior events. Indeed, these data seem highly problematic for dual process theories of recognition.

One way to understand the overall picture painted by these apparently conflicting patterns of data would be to revisit the instructions that subjects are given in the Remember/Know paradigm. The instructions given for making Remember judgments specify not only the associations from the study phase that come to mind (i.e., the conceptual attributes), but also memory for the "aspects of the physical appearance of the words [pictures], . . . or something about its appearance." Further, subjects were told that, "the 'remembered' word [picture] should bring back to mind a particular association, image, or something more personal from the time of study, or something about its appearance or position." Thus, conscious recollective processes can be influenced by certain perceptual attributes as well (Hunt & Toth, 1990). Then, why were perceptual effects obtained selectively for Know judgments in the early studies?

Perhaps a distinction is necessary between factors that induce fluency of processing and factors that provide salient or distinctive information about the studied stimulus (Rajaram, 1996). The basic idea is that either fluent processing or encoding of highly distinctive information influences recognition memory. Indeed, many others have proposed that distinctiveness of events greatly affects their recognition (see Hunt & McDaniel, 1993). A further assumption is that whereas fluency affects Know responses, distinctiveness of information against a background of relatively uniform information, as in the von Restorff effect (von Restorff, 1933; also see Wallace, 1965) and related phenomena, aids Remember judgments. Based on these ideas, several specific predictions can be made. For instance, if the independent variable increases the perceptual fluency with which an item is processed, its effects should be observed on Know judgments. ("Fluency" could be gauged by how the variable affects implicit memory tests.) However, if the independent variable increases the salience of the stimulus or consists of stimuli with distinctive attributes, this manipulation would affect the Remember judgments. L. Cooper et al. (1992) also suggested that distinctive information influences episodic memory judgments. This hypothesis about how fluency and distinctiveness affect Remember/Know judgments can be applied not only to perceptual attributes, but to conceptual, temporal, contextual, and spatial attributes as well. For example, inducing *conceptual* fluency of processing should influence Know judgments, and not Remember judgments. These predictions will be tested in future experiments.

The problematic data obtained with the Remember/Know paradigm can be accommodated within the distinctiveness/fluency of processing framework with some reasonable assumptions. For example, low frequency words are likely more distinctive than high frequency words (Gardiner & Java, 1990), leading to better recognition and greater Remember judgments for low frequency words compared to high frequency words. Similarly, studied pictures may be considered more distinctive than studied words, regardless of the test format (Rajaram, 1993, 1996). These assumptions would help explain the results obtained by Rajaram (1993, 1996) for studied pictures and words that were tested in either verbal or pictorial format. Furthermore, if we consider size and reflection of stimuli as attributes that are relevant and distinctive for episodic memory judgments, then Rajaram's (1994) data can also be accommodated within this framework. In fact, L. Cooper et al. (1992) speculated that size and reflection likely constitute distinctive properties of events used by the episodic memory system. Clearly, this section has provided a post hoc analysis of the inconsistent data. However, the explanation outlined here, although a bit tentative, can accommodate both the early and recent evidence (with the exception of modality effects, for which we have no explanation) on the issue, and also provide a framework for conducting future research to test the proposed ideas. In addition, the idea that fluency of processing increases recognition is supported by considerable additional evidence (e.g., Luo, 1993), although not uniformly so (Watkins & Gibson, 1988). Of course, the same is true for the concept of distinctiveness, because considerable research also shows the usefulness of this construct in explaining a variety of memory phenomena (see Hunt & McDaniel, 1993).

EVALUATION OF THE REMEMBER/KNOW PARADIGM AND CONCLUDING REMARKS

The Remember/Know distinction introduced by Tulving (1985) is a meta-memorial judgment, one of a large number of judgments that people are able to make about their own memories. This field of metamemory has grown rapidly since around 1970 (see Nelson, 1992, for a collection of important papers on this topic). Subjects have provided feeling-of-knowing judgments (they know if they can recognize information on a multiple choice test that they could not retrieve on a cued recall test); judgments of a tip-of-the-tongue state (missing information is on the tip of one's tongue or not); judgments of reality monitoring (judging whether something actually occurred in the external world or was imagined by the subject); judgments of confidence of responses (sure or unsure); judgments of learning (whether or not studied information will be recalled later). (References for all these examples can be found in Nelson, 1992.) This list can be extended, but the point would remain the same: The Remember/Know judgment is another in a long series of interesting judgments that psychologists have asked people to make about properties of their own memorial experience. In each of these other cases, the simplest form of the response involves a dichotomous rating (sure or unsure, know or don't know, etc.), although various types of rating scales and other procedures can also be used (Nelson, 1984). The relation of Remember/Know judgments to these other types of judgments awaits future research, but we firmly believe that the Remember/Know judgment is an interesting, reliable, and psychologically meaningful judgment for subjects to make, as attested by the growing body of literature showing generally encouraging and reproducible results. As mentioned previously, the Remember/Know judgment requires even more careful instructions to subjects than is usual, but the dividends seem worth the effort.

As stated earlier, the Remember/Know paradigm provides a viable framework to study the states of awareness that accompany different memories. Remember judgments provide an index of conscious experience as this judgment is given by the subjects to memories for which they are aware how it is they know that they had encountered them on a previous occasion. Alternately, for Know judgments, even though the subjects are fully aware that the memory belongs to their personal past, they are unable to determine the basis of this conscious experience. Of course, the nature of Know judgments is more complicated because this judgment could either be based on the sort of experience just described or some nonconscious processes may also be articulated as Knowing as long as they give rise to a strong feeling of familiarity. In any case, these two judgments provide a valid measure of the conscious experience during retrieval and take researchers one step further in their attempts to study at least one aspect of consciousness, that is, related to the retrieval process, within a scientific framework.

Our analysis of Remember/Know judgments followed the prior work of Tulving, Gardiner, and others by assuming that overall recognition hit rate could be decomposed into two meaningful entities based on judgments by the subjects. The assumption is that the subjects themselves can divide the recognized items into Remember and Know categories. Jacoby et al. (chapter 2, this volume) have criticized the assumption of an exclusive relation between Remember and Know categories and have advocated treating these responses as statistically independent and applying the same equations as in the process dissociation procedure (Jacoby, 1991). The assumption that subjects can make mutually exclusive responses does not necessarily mean, however, that information driving both types of processes could not be present in the same test event. We prefer to leave it to our subjects (rather than to a mathematical model) to decide which is the predominant source of information that guides their response. In addition, the logic for considering Remember and Know as mutually exclusive would seem to be the same as in most of the other metamemory judgments, in which subjects essentially make dichotomous and mutually exclusive responses (know or don't know, tip of the tongue or not, sure or unsure, etc.). Although information on which these judgments are made may be continuous and overlapping, subjects are always asked to gauge the information by whatever means they have and produce appropriate responses.

Jacoby's (1991) process-dissociation procedure also aims to provide a decomposition of performance into components reflecting conscious recollection and those reflecting a more automatic basis. The logic of opposition on which the procedure is based and the resulting equations represent a mixture of a first-person and a third-person account of conscious experience. This means that subjects' abilities to follow the exclusion and inclusion instructions rely on their first-person judgments; they decide what to include or to exclude. However, the use of the technique to determine how responding is based on conscious recollection and on some more automatic basis resembles more a third-person account, because the data are subjected to a model to determine the outcome. This outcome of the procedure tells the experimenter what subjects were able to consciously control and what influences were automatic, rather than letting the subjects make direct claims for themselves on these matters. Certainly there is no necessary conflict between the application of the Remember/Know procedure and the use of the process-dissociation procedure. Both would seem to have their benefits and weaknesses. As we see it, the benefit of the Remember/Know distinction is to permit a first-person account in the determination of the *nature* of memories for the retrieved items from a study episode. Jacoby et al. (this volume) argue that the Remember/Know responses should be considered independent and therefore subjected to the process-dissociation procedure. Indeed, they find good agreement between some results produced from the process-dissociation procedure and others from the Remember/Know procedure when it is assumed that Remember and Know judgments are independent. Others have suggested that it makes more sense to consider Remember and Know judgments as redundant; for example, all items recognized might be considered known by the subject, but remembered items have some special additional features that permit them to be consciously recollected. If one assumes a redundancy relation between the measures, then the equations used to arrive at estimates of conscious recollection and automatic influences are different and so are the likely outcomes (Joordens & Merikle, 1993). It will doubtless take considerable theoretical and experimental effort to unravel these complicated issues, but in the meantime we prefer to stick with subjects' judgments, as is the custom in the study of other metamemory phenomena. Nonetheless, it may be that some of the problematic data reviewed toward the end of the empirical part of this chapter would change if the process-dissociation procedure or other measurement assumptions were employed. (On the other hand, some of the supportive data for dual process recognition theories might also evaporate).

One aspect that the Remember/Know procedure and the process-dissociation procedure share is to argue that even explicit memory measures such as recall and recognition involve a mixture of processes. Recognition memory seems to have more than one basis, however it is to be conceptualized (e.g., Jacoby, 1991; Mandler, 1980, 1989). One important offshoot of this point is that a standard application of signal detection theory to recognition memory data is inappropriate, because it assumes that subjects make judgments along a single dimension of strength or familiarity. Because recognition memory has at least two (and maybe multiple) bases, not just judgments of strength, it seems pointless to correct overall recognition data using signal detection to gain estimates of d' and β. Signal detection theory might still be quite useful in analyzing judgments of familiarity, as Jacoby et al. (this volume) advocate. However, the standard use of signal detection theory to analyze recognition memory, at least in some instances, may be inappropriate.

One can conceive of the Remember/Know procedure as a method to "purify" recognition scores (or cued recall, or free recall, for that matter) into two components. The Remember component more accurately reflects the output of an episodic memory system, according to Tulving (1985). A problem for future research is the phenomenological status of Know responses. Tulving (1985) argued that these were output of a semantic memory system that could also be changed by recent events. Others (Gardiner, 1988; Jacoby et al., chapter 2, this volume; Rajaram, 1993) have shown that Know responses are driven by perceptual fluency, or familiarity, or some other automatic process. The important issue for future research is to address whether Know responses reflect anything more than a residual category: Perhaps subjects respond positively to test items but then cannot recollect their actual occurrence in the study list and so simply say, "Know." This inarticulateness or blankness of conscious experience associated with Know responses would be consistent with the idea that they are driven by some automatic process that is not under conscious control.

Tulving's (1985) Remember/Know paradigm has produced a considerable body of evidence, showing that it is useful and perhaps telling us important facts about subjective experience during recollection. This section has tried to point to some future directions that might be profitable. Research using this paradigm is just beginning and the unanswered questions seem to loom larger (or at least represent a longer list) than the answered questions. So, future research in this area should pay important dividends in the study of conscious experience.

ACKNOWLEDGMENTS

The authors thank Jonathan Cohen, John Gardiner, Kathleen McDermott, Jonathan Schooler, and Endel Tulving for comments on an earlier draft of this chapter. Preparation of this chapter was supported by the AFOSR Grant F49620-92-J-0437 to Henry L. Roediger III.

REFERENCES

Blaxton, T. A. (in press). The role of the temporal lobes in recognizing nonverbal materials: Remembering versus knowing. *Neuropsychologia.*

Biederman, I., & Cooper, E. E. (1992). Size invariance in visual priming. *Journal of Experimental Psychology: Human Perception and Performance, 18,* 121–133.

Challis, B. H., Chiu, C.-Y., Kerr, S. A., Law, J., Schneider, L., Yonelinas, A., & Tulving, E. (1993). Perceptual and conceptual cueing in implicit and explicit retrieval. *Memory, 1,* 127–151.

Cooper, L. A., Schacter, D. L., Ballesteros, S., & Moore, C. (1992). Priming and recognition of transformed three-dimensional objects: Effects of size and reflection. *Journal of Experimental Psychology: Learning, Memory, and Cognition, 18,* 43–57.

Craik, F.I.M., & Lockhart, R. S. (1972). Levels of processing: A framework for memory research. *Journal of Verbal Learning and Verbal Behavior, 11,* 671–684.

Craik, F.I.M., Morris, L. W., Morris, R. G., & Loewen, E. R. (1990). Relations between source amnesia and frontal lobe functioning in older adults. *Psychology and Aging, 5,* 148–151.

Gardiner, J. M. (1988). Functional aspects of recollective experience. *Memory and Cognition, 16,* 309–313.

Gardiner, J. M. (1991). Memory with and without recollective experience. *Behavioral and Brain Sciences, 14,* 678–679.

Gardiner, J. M., & Java, R. I. (1990). Recollective experience in word and nonword recognition. *Memory and Cognition, 18,* 23–30.

Gardiner, J. M., & Java, R. I. (1993). Recognizing and remembering. In A. Collins, M. A. Conway, S. E. Gathercole, & P. E. Morris (Eds.), *Theories of memory* (pp. 163–188). Hillsdale, NJ: Lawrence Erlbaum Associates.

Gardiner, J. M., & Parkin, A. J. (1990). Attention and recollective experience in recognition memory. *Memory and Cognition, 18,* 579–583.

Graf, P., Mandler, G., & Haden, P. (1982). Simulating amnesic symptoms in normal subjects. *Science, 218,* 1243–1244.

Gregg, V. H., & Gardiner, J. M. (1991). Components of recollective awareness in a long-term modality effect. *British Journal of Psychology, 82,* 153–162.

Gregg, V. H., & Gardiner, J. M. (1994). Recognition memory and awareness: A large effect of study-test modalities on "know" responses following a highly perceptual task. *European Journal of Cognitive Psychology, 6,* 131–147.

Hunt, R. R., & McDaniel, M. A. (1993). The enigma of organization and distinctiveness. *Journal of Memory and Language, 32,* 421–445.

Hunt, R. R., & Toth, J. P. (1990). Perceptual identification, fragment completion, and Free recall: Concepts and data. *Journal of Experimental Psychology: Learning, memory, and Cognition, 16,* 282–290.

Jacoby, L. L. (1978). On interpreting the effects of repetition: Solving a problem versus remembering a solution. *Journal of Verbal Learning and Verbal Behavior, 17,* 649–667.

Jacoby, L. L. (1983a). Perceptual enhancement: Persistent effects of an experience. *Journal of Experimental Psychology: Learning, Memory, and Cognition, 9,* 21–38.

Jacoby, L. L. (1983b). Remembering the data: Analyzing interactive processes in reading. *Journal of Verbal Learning and Verbal Behavior, 22,* 485–508.

Jacoby, L. L. (1991). A process dissociation framework: Separating automatic from intentional uses of memory. *Journal of Memory and Language, 30,* 513–541.

Jacoby, L. L., & Dallas, M. (1981). On the relationship between autobiographical memory and perceptual learning. *Journal of Experimental Psychology: General, 110,* 306–340.

Jacoby, L. L., & Whitehouse, K. (1989). An illusion of memory: False recognition influenced by unconscious perception. *Journal of Experimental Psychology: General, 118*(2), 126–135.

Jacoby, L. L., Woloshyn, V., & Kelley, C. (1989). Becoming famous without being recognized: Unconscious influences of memory produced by dividing attention. *Journal of Experimental Psychology: General, 118,* 115–125.

Java, R. I. (1994). States of awareness following word stem completion. *European Journal of Cognitive Psychology, 6,* 77–92.

Joordens, S., & Merikle, P. M. (1993). Independence or redundancy? Two models of conscious and unconscious influences. *Journal of Experimental Psychology: General, 122,* 462–467.

Kirsner, K., & Smith, M. C. (1974). Modality effects in word identification. *Memory and Cognition, 2,* 637–640.

Light, L. L., & Singh, A. (1987). Implicit and explicit memory in young and older adults. *Journal of Experimental Psychology: Learning, Memory, and Cognition, 13,* 531–541.

Luo, C. R. (1993). Enhanced feeling of recognition: Effects of identifying and manipulating test items on recognition memory. *Journal of Experimental Psychology: Learning, Memory, and Cognition, 19,* 405–413.

Madigan, S. (1983). Picture memory. In J. C. Yuille (Ed.), *Imagery, memory, and cognition: Essays in honour of Allan Paivio* (pp. 65–89). Hillsdale, NJ: Lawrence Erlbaum Associates.

Mandler, G. (1979). Organization and repetition: Organization principles with special reference to rote learning. In L. Nilsson (Ed.), *Perspectives on memory research* (pp. 293–327). Hillsdale, NJ: Lawrence Erlbaum Associates.

Mandler, G. (1980). Recognizing: The judgment of previous occurrence. *Psychological Review, 87,* 252–271.

Mandler, G. (1989). Memory: Conscious and unconscious. In P. R. Solomon, G. R. Goethals, C. M. Kelley, & B. R. Stephens (Eds.), *Memory: Interdisciplinary approaches* (pp. 84–106). New York: Springer-Verlag.

Masson, M.E.J., & MacLeod, C. M. (1992). Reenacting the route to interpretation: Enhanced perceptual identification without prior perception. *Journal of Experimental Psychology: General, 121,* 145–176.

Miller, G. A. (1962). *Psychology: The science of mental life.* New York: Harper & Row.

Nelson, T. O. (1984). A comparison of current measures of the accuracy of feeling-of-knowing predictions. *Psychological Bulletin, 95,* 109–133.

Nelson, T. O. (1992). *Metacognition: Core readings.* Boston: Allyn & Bacon.

Nisbett, R. E., & Wilson, T. (1977). Telling more than we can know. *Psychological Review, 84,* 231–259.

Parkin, A. J., & Russo, R. (1989). Implicit and explicit memory and the automatic/effortful distinction. *European Journal of Cognitive Psychology, 2*(1), 71–80.

Parkin, A. J., & Walter, B. (1992). Recollective experience, normal aging, and frontal dysfunction. *Psychology and Aging, 7,* 290–298.

Rajaram, S. (1993). Remembering and knowing: Two means of access to the personal past. *Memory and Cognition, 21,* 89–102.

Rajaram, S. (1996). Perceptual effects on Remembering: Recollective processes in picture recognition memory. *Journal of Experimental Psychology: Learning, Memory, and Cognition, 22,* 365–377.

Rajaram, S., & Coslett, H. B. (1992, November). *Further dissociations between Remember and Know judgments in recognition memory.* Poster presented at the 33rd Annual Meeting of the Psychonomic Society, St. Louis, MO.

Rajaram, S., & Coslett, H. B. (1993, May). *Effects of size and reflection of objects on Remember/ Know responses in recognition memory.* Paper presented at the 65th Annual Meeting of the Midwestern Psychological Association, Chicago.

Roediger, H. L. (1990). Implicit memory: Retention without remembering. *American Psychologist, 45,* 1043–1056.

Roediger, H. L., & McDermott, K. B. (1993). Implicit memory in normal human subjects. In F. Boller & J. Grafman (Eds.), *Handbook of Neuropsychology* (Vol. 8, pp. 63–131). Amsterdam: Elsevier.

Roediger, H. L., Weldon, M. S., & Challis, B. H. (1989). Explaining dissociations between implicit and explicit measures of retention: A processing account. In H. L. Roediger, & F.I.M. Craik (Eds.), *Varieties of memory and consciousness: Essays in honour of Endel Tulving* (pp. 3–41). Hillsdale, NJ: Lawrence Erlbaum Associates.

Roediger, H. L., Weldon, M. S., Stadler, M. L., & Riegler, G. L. (1992). Direct comparison of two implicit memory tests: Word fragment and word stem completion. *Journal of Experimental Psychology: Learning, Memory, and Cognition, 18,* 1251–1269.

Slamecka N. J., & Graf P. (1978). The generation effect: Delineation of a phenomenon. *Journal of Experimental Psychology: Human Learning and Memory, 4,* 592–604.

Sutherland, S. (1991). *Macmillan dictionary of psychology.* London: Macmillan.

Tulving, E. (1985). Memory and consciousness. *Canadian Psychologist, 26,* 1–12.

von Restorff, H. (1933). Uber die Wirkung von Bereichsbildungen im spurenfeld [On the effects of organization in the trace field]. *Psychologie Forschung, 18,* 299–342.

Wallace, W. P. (1965). Review of the historical, empirical, and theoretical status of the von Restorff phenomenon. *Psychological Bulletin, 63,* 410–424.

Watkins, M. J., & Gibson, J. M. (1988). On the relation between perceptual priming and recognition memory. *Journal of Experimental Psychology: Human Learning and Memory, 14,* 477–483.

Weldon, M. S., & Roediger, H. L. (1987). Altering retrieval demands reverses the picture superiority effect. *Memory and Cognition, 15,* 269–280.

CHAPTER 12

Consciousness and the Limits of Language: You Can't Always Say What You Think or Think What You Say

Jonathan W. Schooler
Stephen M. Fiore
University of Pittsburgh

Language has a rather odd relation with consciousness. On the one hand, many aspects of conscious experience cannot be adequately conveyed in words: the smell of a flower, the appearance of a face, the taste of a fine wine. Despite such inherent limitations, language nevertheless represents the primary tool that we have for demarcating conscious experience. Indeed, verbal *reportability* is the standard criterion for determining whether an event/process was consciously experienced. The verbal report criterion is unquestionably of great value in investigations of consciousness; however, trouble may ensue when verbal reports and conscious awareness are treated as identities. This chapter reviews usage of the reportability criterion and then explores the possible clarifications and implications that follow when subjective awareness and content reportability are distinguished.

The tension between reportability and consciousness is exhibited in this volume's section on implicit learning. For example, Lewicki, Czyzewska, and Hill (chapter 9, this volume) recognize the possible discrepancy between reportability and consciousness, but then dismiss it, observing that "the inability to articulate . . . [implicitly learned] knowledge represents not merely a difficulty with verbalizing (i.e., 'putting into words') something that a person intuitively 'knows' or 'feels,' but rather a fundamental lack of access to the content (meaning) of the relevant rules and principles" (p. 161).

Reber (chapter 8, this volume) takes a slightly more tenuous position with respect to the relation between reportability, consciousness, and the implicit learning task. Reber acknowledges the limitations of the reportability criterion as a measure of consciousness, noting that individuals who are unable to report recently acquired knowledge when simply asked are able to report some aspects of this knowledge under more sophisticated questioning conditions. The slippery relation between verbalizability and consciousness leads Reber to suggest that "the verbalizability criterion is a red herring—a small odoriferous cousin to the sardine that when dragged across one's path, disturbs the scent and diverts one's attention away from the main issues" (p. 141). Although Reber recognizes the limitations of the verbalizability criterion, he nevertheless emphasizes its importance suggesting that the "proper approach is to use an individual's relative inability to provide verbal description of mental content as a kind of 'common sense' marker for increasing the likelihood that implicit processes are present" (p. 146).

Dulany (chapter 10, this volume) also recognizes the potential distinction between reportability and consciousness, however he deals with their potential discrepancy by suggesting that mental episodes may be reportable or nonreportable but, either way, should be viewed as being conscious. Dulany characterizes reportable memories as "deliberative mental episodes," which are made up of propositional contents—"a belief *that* ____" (p. 182)—consisting of more effortful operations such as decision and judgment. These deliberative episodes are contrasted with nonreportable memories that he describes as "evocative mental episodes," which are less effortful and are made up of contents that are "nonpropositional, nonpredicational—a sense *of* ____" (p. 185). Although these nonpropositional episodes may be difficult to articulate, Dulany nevertheless views them as corresponding to conscious experiences, arguing that tasks frequently described as nonconscious (e.g., implicit learning tasks) might be better thought of as relying on "the establishment and use of evocative relations among nonpropositional but fully conscious contents" (p. 189).

In short, the aforementioned authors all grapple with the potential discrepancy between reportability and consciousness, but in the end resolve these differences in rather different ways. Lewicki et al. acknowledge the possible distinction and then dismiss it, arguing that the type of knowledge they discuss is neither conscious nor reportable. Reber concedes that reportability is not a perfect criterion of consciousness but nevertheless argues that it provides a reasonable, albeit imperfect, marker of consciousness. Finally, Dulany uses the distinction between reportability and consciousness in order to discount studies that use reportability as a criterion for implicit knowledge, arguing that such knowledge, though not reportable, is still conscious.

THE AWARE₁/AWARE₂ DISTINCTION

One possible way to reconcile the potential discrepancies between reportability and consciousness is to draw on Dennett's (1969/1986) aware₁/aware₂ distinction. Dennett distinguished two different levels of awareness for a given proposition p at time t. "1) [An individual] A is aware$_1$ that p at time t if and only if p is the content of the input state to A's 'speech center' at time t; 2) A is aware$_2$ that p at time t if and only if p is the content of an internal event in A that is effective in directing current behavior" (Dennett, 1969/1986, pp. 118–119). In short, awareness$_1$ corresponds to internal events that can be reported, whereas awareness$_2$ corresponds to nonreportable internal events that nevertheless influence behavior. To flesh out this distinction, Dennett used the example of a driver traversing a familiar route while holding a conversation. In such a case, the driver would likely have an awareness$_1$ of the conversation as reflected in what was occupying his verbal reports or "speech center." However, he would have an awareness$_2$ of the curves of the road, in the sense that his driving behavior responded to the intricacies of the road and the traffic along the route. Yet, if asked what he could remember about the physical drive, he might respond "Nothing since the route was familiar and I was engrossed in my conversation" (Dennett, 1969/1986, p. 116).

The distinction between awareness$_1$ and awareness$_2$ can be applied to a variety of psychological phenomena that bear on the issue of consciousness. Bower (1990) borrowed Dennett's distinction to illustrate the relation between awareness$_1$ and awareness$_2$ with many different cognitive phenomena. For example, Bower suggested that subjects who are tachistoscopically presented the letter matrices used in the Sperling paradigm (1960) are "aware$_2$ immediately of nearly all of the letters in the display despite their inability to name all of them (in the sense of aware$_1$)" (Bower, 1990, p. 211). Other paradigms Bower saw as illustrating the awareness$_1$/awareness$_2$ distinction include: *subliminal perception* paradigms in which unreportable (no awareness$_1$) words nevertheless influence forced choice recognition behavior (awareness$_2$); *blind sight studies* in which patients with damage to the right occipital lobe report no awareness of the left visual field (awareness$_1$) yet are above chance on forced choice recognition at identifying pictures presented to that field (awareness$_2$); *hypnotically induced blindness studies* in which subjects instructed to behave as if they were functionally blind report no subjective experience of seeing flashed pictures (no awareness$_1$) yet perform below chance on recognition tests (indicating awareness$_2$); and, finally, *split-brain studies* in which commissurotomy patients can verbally identify objects presented to the right visual field indicating awareness$_1$ for the left hemisphere. However, in the latter case when objects are presented to the left visual field, patients are unable to report the object, but can still correctly point to the object, which indicates awareness$_2$ for the right hemisphere.

By explicitly distinguishing between reportable versus nonreportable forms of awareness the awareness$_1$/awareness$_2$ distinction begins to dis-ambiguate the murky relation between reportability and consciousness. However, this distinction still lumps together situations that phenomenologically seem quite different. Specifically, the aware$_1$/aware$_2$ distinction fails to differentiate between behavioral influences that are conscious but not reportable and those that are neither conscious nor reportable. Bower (1990) acknowledged the potential subjective awareness associated with awareness$_2$: "A person may notice and be aware of a face or abstract painting and be able to subsequently select that face or abstract painting from a lineup. Yet he might be unable to describe accurately the face or painting in sufficient detail so that any other person could pick it out" (p. 212). This "minor indeterminacy with Dennett's distinction" notwithstanding, Bower nevertheless goes on to suggest that "by unconscious processes we refer to bodily and psychological events of which we are not aware$_1$" (p. 212).

In our view, the distinction between conscious experiences that cannot be reported versus entirely nonconscious experiences is not a minor glitch that can be acknowledged and then dismissed—rather it gets right to the core of some of the central issues surrounding consciousness. For example, our views of preverbal children seem fundamentally different depending on whether we consider their consciousness to correspond more closely to the awareness that we have for a face that we cannot describe or the awareness that a brain-damaged patient has for his blind field of view. Similarly, anyone who has seen a briefly presented visual array vanish before its contents could be reported knows that this phenomenological experience is quite different from that associated with a subliminally presented word. In short, although the awareness$_1$/awareness$_2$ distinction succeeds in demonstrating the value of using reportability to distinguish between different types of awareness, it fails to consider one critical dimension—individuals' phenomenological experience.

PUTTING CONSCIOUSNESS BACK INTO DEFINITIONS OF CONSCIOUSNESS

Although it might seem self-evident that individuals' perceptions about whether they were aware of an event would be part of our definition of conscious experience, such reports are often treated with skepticism. The major problem associated with including phenomenology in a definition of consciousness is that it assumes a shared agreement about what constitutes a conscious experience—a matter of some issue. Dennett (1991), for example, concluded that "controversy and contradiction bedevil the claims made under these conditions of polite mutual agreement" (p. 67). Ultimately, it must be conceded that the qualia of conscious experience eludes scientific scrutiny: We cannot ascertain the congruence between different individuals' experience of consciousness, anymore than we can tell the degree to which the color blue elicits the same subjective sensation across individuals. The intrinsic ambiguity associated with individuals' subjective reports of their conscious experiences might seem to make such reports entirely uninformative. However, before dismissing subjective reports altogether, two issues should be considered: Do subjects phenomenological reports about their perceived awareness of an experience provide reliable indices that correspond to other measurable cognitive processes? And, if so, can subjects phenomenological reports of awareness be reliably distinguished from their verbal reports about the contents of an experience?

THE VALIDITY OF SUBJECTIVE REPORTS AS A CRITERION FOR CONSCIOUSNESS

Although it may be impossible to know with certainty the true correspondence between different individuals' experience of what they are willing to categorize as "conscious," it is nevertheless possible that such categorization may still be informative. By analogy, even though we cannot know the degree to which two individuals share the same phenomenological experience (qualia) when viewing the color blue, we can still learn a great deal by asking them to report what color they see. And, although there may be some differences in their reports (particularly for nonfocal colors), we nevertheless will find that their reports systematically covary with other variables in a consistent manner (e.g., the responses elicited when pigments labeled blue are mixed with pigments that reflect other wavelengths of light). In a similar manner, we may find that although not privy to individual subjects' qualia of consciousness, we nevertheless observe that subjects' self-reports of consciousness are both reliable and correspond in meaningful ways to other variables. Indeed, although paradigms that rely on subjects' phenomenological reports of awareness are relatively rare, we consider two such paradigms that systematically produce meaningful results consistent with what one would expect if, in fact, subjective reports were measuring conscious experience.

The Know/Remember Distinction

Tulving's (1985) know versus remember paradigm is a cogent example of the effectiveness of using subjective reports as the criterion for judging awareness. In this paradigm, explored by Rajaram and Roediger (chapter 11, this volume; see also Jacoby, Yonelinas, & Jennings, chapter 2, this volume), subjects study a list of words under various conditions, and then are later given a test in which they first decide whether a given word is "old" or "new." If the word is judged "old," they then make a further subjective judgment about whether or not they became "consciously aware again of some aspect or aspects of what happened or what was experienced at the time the word was presented" (Rajaram & Roediger, chapter 11, this volume, p. 216). Under these conditions they are instructed to provide a remember response when they are aware of the context in which the item was learned, and a know response when they "cannot consciously recollect anything about its actual occurrence or what happened or what was experienced at the time of its occurrence."

Although know/remember judgments rely entirely on subjective assessments of conscious states, they nevertheless provide a generally consistent and stable pattern of findings that interact in predictable ways with various cognitive and individual difference variables. For example, Rajaram and Roediger observed that know judgments interact with manipulations typically associated with implicit (unconscious) memory effects. They found that performance on know judgments can be characterized in terms of the influence of perceptual fluency—a construct typically used to account for performance on implicit memory tasks such as word-fragment completion. In contrast, they find that performance on remember judgments can be characterized in terms of the influence of distinctiveness—a construct primarily associated with conscious memory tasks such as recall. Similarly, Jacoby et al. (chapter 2, this volume) observed that know/remember judgments can lead to estimates of conscious and unconscious processes that closely correspond to estimates garnered from more objective procedures, such as comparing fragment completion performance when subjects are given inclusion instructions (use previously seen words) versus exclusion instructions (do not use previously seen words). Thus, although know/remember judgments rely entirely on subjective judgments of awareness, they nevertheless converge with other measures, suggesting that they provide a useful metric for assessing conscious states.

Subjective Perceptual Thresholds

The value of subjective reports of conscious awareness has also been illustrated with research investigating the impact of very briefly presented and then masked visual stimuli—what is commonly, although somewhat controversially, referred to as subliminal perception. For example, Merikle and Cheesman (1986) distinguished between "the subjective threshold, [which is] the level of discriminative responding at which observers claim not to be able to detect perceptual information at better than a chance level of performance, and the objective threshold, [which is] the level of discriminative responding corresponding to chance level performance" (p. 42). Merikle and Cheesman readily conceded that the subjective threshold, in effect, "transfers responsibility of defining awareness to an observer" (p. 42), thereby producing the clear need for providing converging evidence that such a transfer of responsibility is appropriate. And they find such evidence. For example, using a Stroop priming task, Cheesman and Merikle (1986) observed that, when primes are presented below the subjective threshold of awareness, there is evidence of automatic processing of the prime but no evidence of strategic processing. In contrast, when primes are presented above the subjective threshold, subjects performance is qualitatively different, leading to an actual reversal of the standard Stroop effect. This would be expected if subjects are taking advantage of strategies that could be derived from an awareness of the frequency of different types of primes. More recently, Merikle and Joordens (chapter 6, this volume) report that a very similar pattern of results is observed when threshold durations are determined using Jacoby's method of opposition (cf. Jacoby, Yonelinas, & Jennings, chapter 2, this volume) in which subjects performance is compared when they are instructed to complete word fragments, often corresponding to the briefly presented words, under either inclusion ("use the target") or exclusion ("do not use the target") instructions.

In short, both know/remember and subjective perceptual threshold judgments interact with independent variables in a manner that would be expected if they did, in fact, correspond to a criterion for consciousness, and they both provide estimates of unconscious processes that resemble those garnered through other techniques. This converging evidence for the validity of subjective reports of consciousness suggests that the construct of reported awareness is a reasonable one. This is not to say that reported subjective awareness is a flawless measure of consciousness. Indeed, as is discussed later, it is likely that there may be some grey area around precisely where individuals draw the line. It is also likely that there may be alternative ways of assessing subjective awareness (such as Jacoby's method of opposition) that may be more precise than self-reports. Our point here is simply that subjective awareness, as measured by self-reports or superior techniques if available, is a meaningful and measurable construct. It is therefore appropriate to ask whether subjective awareness corresponds to the variable most often used as the metric for consciousness: the ability to report the contents of the cognitive event in question.

CONTENT REPORTABILITY VERSUS SUBJECTIVE AWARENESS

We propose that discussions of consciousness may be facilitated if a distinction is made between two types of reports that can be made about a cognitive event:[1] *Subjective awareness* corresponding to subjects' phenomenological

---

[1] Discussions of consciousness are complicated by the fact that there are various types of information of which an individual might be conscious, including external events, mental processes, mental states, acquired knowledge, and so on. Dennett (1969/1986) and Bower (1990) approached this ambiguity by considering them all subsumed by the notion of "propositions." However, the concept of propositions is laden with implications as well, as demonstrated by Dulany's suggestion (chapter 10, this volume) that evocative remembering is nonpropositional. We there-

judgment of whether or not they believe the cognitive event occurred, and *content reportability,* corresponding to subjects' ability to report the contents of the cognitive event. Our claim is that these two criteria are not simply different levels along a single dimension. Rather, we suggest that subjective awareness and content reportability correspond to fundamentally different psychological dimensions. We base this claim on the observation that these two dimensions are fully dissociable; that is, although there are many cases in which both subjective awareness and content reportability are commensurate, there are also cases in which the two diverge. For example, it is possible to have subjective awareness but lack the ability to report the contents of that awareness. It is even possible to report an absence of subjective awareness and yet, when encouraged, accurately describe the contents of a subjectively nonexistent cognitive event. In other words, the distinction between subjective awareness and content reportability suggests that we should be able to create a 2 × 2 (subjectively aware/not aware × content reportable/content nonreportable) matrix and find instances of situations that apply to each hypothetical cell. Figure 12.1 presents such a grid along with potential candidates for each cell.

Although there is considerable debate regarding precisely what belongs in the two commensurate cells, that is, subjectively aware/reportable versus subjectively unaware/nonreportable (for examples of such debates see the contrasting views offered by Dulany [chapter 10, this volume] vs. Lewicki, Czyzewska, & Hill [chapter 9, this volume] and Reber [chapter 8, this volume]; Ericsson & Simon, 1994, vs. Nisbett & Wilson, 1977; Fowler, 1986, vs. Holender, 1986), the existence of these two cells is relatively noncontroversial, and therefore requires no more discussion here. However, because our distinction relies fundamentally on the existence of situations corresponding to the incommensurate cases—that is, subjectively aware without content reportability and content reportability without subjective awareness—we focus on these cases.

Subjectively Aware/Content Nonreportable

As noted at the outset of our discussion, there are many subjectively conscious experiences that we are nevertheless unable to translate into words. For most of us, it is virtually impossible to convey anything but the bare rudiments of many complex sensory experiences (such as the appearance of a face, the taste of a fine wine, or the sound of a piece of music). Despite our inability to convey the contents of such experiences, we nevertheless are quite likely to

---

fore use the term *cognitive event* to correspond to the class of mental activities of which an individual might in principle be conscious (e.g., the perception of a stimulus, the formation or application of a rule, the recollection of an event, associations, etc.). The contents of cognitions are further discussed in the latter section comparing cognitive processes versus end products.

Subjective Awareness

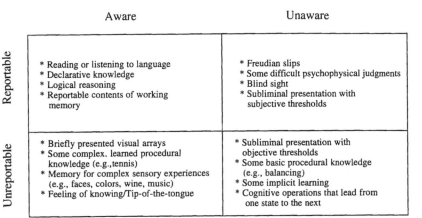

		Aware	Unaware
Content Reportability	**Reportable**	* Reading or listening to language * Declarative knowledge * Logical reasoning * Reportable contents of working memory	* Freudian slips * Some difficult psychophysical judgments * Blind sight * Subliminal presentation with subjective thresholds
	Unreportable	* Briefly presented visual arrays * Some complex, learned procedural knowledge (e.g.,tennis) * Memory for complex sensory experiences (e.g., faces, colors, wine, music) * Feeling of knowing/Tip-of-the-tongue	* Subliminal presentation with objective thresholds * Some basic procedural knowledge (e.g., balancing) * Some implicit learning * Cognitive operations that lead from one state to the next

FIG 12.1.   Matrix distinguishing subjective awareness and content reportability.

report that we were vividly aware of them. In addition to sensory experiences, there are a variety of other cognitive events that would fit into this category. For example, as discussed earlier, there is the Sperling paradigm (1960) in which, even though subjects can only report a fraction of the total items seen in a briefly presented array, they nevertheless reveal a brief awareness of all of the items, as indicated by their near-perfect performance when queried about a randomly selected subportion of the array. Some procedural knowledge, particularly for visual motor skills, may also fit in this cell. In complex learned procedures such as tennis, individuals may be aware of, and use, rather sophisticated knowledge that they are nevertheless unable to articulate—except perhaps by watching themselves engage in the activity. Other examples of situations in which individuals have an awareness of knowledge they cannot fully articulate include "tip-of-the-tongue" experiences (e.g., Brown & McNeill, 1966), intuitive hunches (Bowers, Regehr, Balthazard, & Parker, 1990), affective judgments (e.g., Nisbett & Wilson, 1977; Wilson, chapter 15, this volume), and grammatical judgments (Reber, chapter 8, this volume).

We might also speculate that the cognitive processes of preverbal children and, at least, some animals fit in this category. Such a claim requires a bit of extrapolation because preverbal children and animals cannot report that they are conscious of experiences they are unable to put into words. Indeed, even if they have a phenomenological experience that corresponds to our experience of subjective awareness, they likely lack not just the words for it, but also the self-reflection processes necessary to possess a concept of subjective awareness (cf. Hobson, chapter 19; Johnson & Reeder, chapter 13; Kihlstrom, chapter 24; this volume). Nevertheless, this need not rule out the possibility that they experience subjective awareness so long as we recognize that the ability to report awareness is merely a measure and not the construct itself.

Subjectively Unaware/Content Reportable

Perhaps the most interesting of the various cells implied by our distinction is the one containing situations in which individuals are able to provide verbal reports about the contents of a cognitive event, and yet experience no awareness that their reports were meaningful. A classic example is the Freudian slip. Freud considered such verbal slips as an example of situations in which unconscious sentiments are verbally expressed (cf. Baars, Fehling, LaPolla, & McGovern, chapter 22, this volume). Additional evidence for this cell is implicated by the various demonstrations of situations in which people show behavioral discriminations regarding information for which they report no phenomenological experiences. Merikle and Cheesman (1986) observed that "in difficult detection tasks subjects often claim that they are unaware of the perceptual information, even though their objective detection performance may indicate a considerable ability to respond discriminatively to the stimuli" (p. 42). Many of Bower's (1990) examples of awareness$_2$ without awareness$_1$ (e.g., subliminal priming, blind sight, hypnotic forgetting, and the performance of commissurotomy patients) fit this category. As applied to the awareness$_1$ (verbal reportability) versus awareness$_2$ (behaviorally evidenced) distinction, the fact that these instances correspond to situations in which knowledge is manifested behaviorally, but not reported verbally, is a critical point. However, in our view, the fact that individuals did not report their responses verbally is of little consequence. Indeed, in standard cognitive procedures, behavioral responses such as pressing yes or no on a response key is typically considered a proxy for an actual verbal report. Thus, we view the behavioral responses elicited in subliminal priming and blind sight, as evidence of content reportability without subjective awareness.

Even if one specifically requires verbal reportability as the criterion, evidence can be found to support this category. For example, using a psychophysical judgment paradigm in combination with confidence estimates, Adams (1957) observed that subjects showed better than chance discriminations on judgments rated as guesses. As Fowler (1986) observed, interpreting Adams' finding using the verbalizability criterion for consciousness leads to a rather problematic situation: "Because the psychophysical judgments were made verbally the experimental outcome must be classified as showing discrimination with awareness even though the classification is in conflict with the subject's own (also verbal) assessment" (p. 34).

There is also evidence that individuals who are unaware of seeing a stimulus, can provide more than simple verbal yes/no reports—they can actually identify the stimulus! For example, Sidis (1898) showed subjects printed digits or letters at a distance at which the subjects reported seeing at most a dim black spot. Many even reported that "they might as well shut their eyes and guess" (Sidis, 1898, p. 17). Nevertheless, these subjects were well above chance at verbally guessing the characters.

Individuals' ability to verbally report stimuli that they are unaware of having perceived extends beyond the ability to simply name visual symbols. In a recent demonstration study, we found that individuals are capable of naming briefly flashed words that they were unaware of having seen. Subjects viewed words and nonwords for 32 ms followed by a mask of ampersands. After each presentation, subjects indicated whether or not they believed that a word was presented. Regardless of their previous answer, we then asked them to "Name any word that might have been presented."

Our logic was that a response of "nonword" indicated that the subject did not have a phenomenological experience of seeing a word. Of course, they may have been aware that something was presented, but a judgment of nonword clearly indicates they were subjectively unaware that a word had appeared. In fact, this was probably a conservative measure of phenomenological awareness in that, even when subjects are not subjectively aware, they still would be expected to perform above chance on discriminating words from nonwords (e.g., Cheesman & Merikle, 1986). We expected that subjects who indicated they were not aware that a word was presented would still, at least sometimes, be able to report it accurately. Consequently, our main interest centered on the trials where subjects were presented with a word they believed to be a nonword. Consistent with our predictions, on trials in which subjects indicated that they did not see a word, they nevertheless demonstrated a mean accuracy rate of 18% on their word identification, with individual performance ranging as high as 50%! From our perspective, this finding is just another example of the many previously mentioned situations in which individuals behaviorally reveal knowledge of information of which they are not subjectively aware. The only difference is that, in this case, the behavior is a verbal report. In short, the ability to verbally report the contents of a cognitive event does not mean that event was consciously experienced.

REFINING THE DISTINCTION

The aforementioned discussion suggests that content reportability and subjective awareness, although often co-occurring, do not necessarily go hand in hand. The fact that, across tasks, we can find double dissociations between the two constructs (i.e., situations in which subjective awareness exceeds the ability to report the contents of that awareness, and situations in which content reportability occurs in the absence of subjective awareness) indicates that they are not simply two different points on some single dimension. Rather, content reportability and subjective awareness appear to be distinct constructs, which although frequently overlapping, must be considered separately.

Thus far we have only provided a bare-bones existence proof for the value of the content reportability/subjective awareness distinction, and in so doing have glossed over a number of issues. Given the complexity and marked divergence of opinions on the topic, it seems unlikely that we will be able to alleviate all of the potential concerns that our distinction may raise. Nevertheless, brief consideration of three particularly salient issues, while complicating the picture a bit, helps to further illustrate the distinction's value and indicates some of its other research implications.

Process Versus End Product

Although we have tried to identify some prototypical tasks to illustrate our 2×2 classification scheme, we do not mean to suggest that all psychological tasks neatly fit into one cell or another. Most tasks involve multiple components that may be differentially classified. For example, in characterizing what individuals can report about their task performance, it is often important to distinguish between the end product (i.e., the final conclusion, judgment, or solution) versus the process (i.e., the manner in which that end product was reached). Often an individual may be able to report the end product, and yet be unable to provide much information regarding the steps that led to its formulation (cf. Nisbett & Wilson, 1977). Consider, for example, insight problem solving in which the solution to a problem suddenly pops to mind. Correct insight solutions are readily reportable, yet individuals are often unable to report anything about the steps leading up to successful insights (e.g., Schooler & Melcher, 1995). In contrast, for logical problem solving, individuals are able to report both the specific steps used to solve the problem and the problem solution (Schooler & Melcher, 1995).

The know/remember paradigm corresponds to another situation in which the process versus end product distinction applies. In this paradigm, the end product (whether the word is categorized as "old" or "new") is both conscious and reportable. However, the process (how that conclusion is reached) can vary in both reportability and awareness. For remember responses, individuals are able to report, and are presumably aware of, at least something about the basis for their recognition decisions (i.e., they recall the context, etc.). However, in the case of know judgments, subjects are unable to report any specifics regarding the basis for their decisions, yet they do report an indescribable feeling of familiarity, suggesting that at least a component of the recognition decision process may be conscious, even though it is not reportable.

Implicit learning tasks also reveal this differential reportability of process versus end product, although discussions of implicit learning sometimes overlook this distinction. For example, Shanks and St. John (in press) suggested that subjects' ability in a sequence learning task to predict the next location of an event indicates some awareness of the manner in which the prediction was made. However, the ability to formulate predictions does not require an awareness of the processes that led to those predictions. As Reber (chapter 8, this volume) notes, "Knowing explicitly where a stimulus will occur is certainly not the same thing as explicitly knowing the underlying rule that determines its location" (p. 143).

Continuous Dimensions

A second refinement to our 2 × 2 matrix is that, although we have depicted consciousness and reportability as discrete categories, in fact, both are probably better characterized as continuous dimensions. The notion that consciousness lies on a continuum is well embedded in the phenomenological terms that we use to describe conscious experiences. For example, we describe some events as being in the "center of attention" whereas we may be only "dimly aware" of other events. Similarly, we can refer to some ideas as "foremost in my thoughts," whereas others may linger in "the back of my mind." James (1890) referred to those aspects of consciousness of which we are only dimly aware as the "fringe" and included in this category such experiences as feelings of familiarity, feeling of knowing, the tip-of-the-tongue phenomena, intentions to speak, expectancies, and feelings of being on the right track. (See Farah, O'Reilly, & Vecera, chapter 17, this volume; Galin, 1995; Jackendoff, 1987; Johnson & Reeder, chapter 13, this volume; Kinsbourne, chapter 16, this volume; Mangan, 1991; Reber, chapter 8, this volume, for additional discussions of the continuous nature of consciousness.)

The reportability of experiences also can be characterized as a continuum. The continuous quality of the reportability dimension is particularly apparent when we consider experiences that we have classified in the conscious/nonreportable cell. Strictly speaking, most of these experiences are not entirely nonreportable. For example, we can usually describe some aspects of a memory for a face, or a taste even though we cannot entirely capture it in words. Sometimes we can do a better job of describing such experiences than others, depending on our relative verbal versus perceptual proficiency. As a result, the discrepancy between the reportability and awareness dimensions may vary as a function of individuals' relative expertise. For example, in the domain of face recognition, there is a marked discrepancy between individuals' ability to consciously remember a face (as measured by recognition) versus verbally describe it (as revealed by whether judges can use the description alone to identify the face). Nevertheless, the magnitude of this discrepancy is greater for same race faces than other race faces, presumably because the greater perceptual expertise associated with same race faces is not associated with a commensurately greater verbal expertise (Fallshore & Schooler, 1995). The distinction between the relative degrees of an individuals' ability to remember versus describe a sensory experience highlights the fact that, although both consciousness and reportability are continuous, they nevertheless correspond to different continua.

Reportable Versus Reported: The Effects of Verbalization

The potential discrepancies between the continuums associated with reportability versus awareness lead to yet another clarification of our 2 × 2 table: the distinction between the degree to which a cognitive event is potentially *reportable* versus whether it was actually *reported*. Although the reportability of a cognitive event may lie on a continuum, whether or not it has been reported corresponds to discrete states (reported vs. unreported). Moreover, the act of verbal report can fundamentally influence subjects' awareness of the cognitive event—crystallizing what was articulated and overshadowing the more hazy components that were not (cf. Koestler, 1964).

Evidence for the "crystallizing" effects of reporting comes from a number of studies examining the effects of verbalization on ambiguous stimuli. For example, Brandimonte and Gerbino (1993) observed that verbalizing the appearance of ambiguous forms (e.g., reversible images) interferes with the ability to discover a form's alternative interpretation. Similarly, we have found that insight problem solving, which requires the discovery of a nonobvious alternative approach, is hampered by thinking out loud—in contrast, noninsight or logical problem solving is unaffected by verbalization (Schooler, Ohlsson, & Brooks, 1993).

Evidence that reporting a cognitive event can overshadow its nonreported aspects comes from a number of studies examining the impact of verbalization on tasks involving difficult-to-report knowledge (including many of those included in the subjectively aware/content nonreportable cell of Figure 12.1). For example, describing a previously seen face can interfere with an individuals' ability to distinguish that face from verbally similar distractors (Schooler & Engstler-Schooler, 1990). Comparable verbal overshadowing effects have now been observed in quite a few domains that rely on nonreportable knowledge or processes (for a recent review, see Schooler, Ryan, Fallshore, & Melcher, in press).

The reportability/reported distinction is also nicely illustrated when we consider it in the context of the continuous nature of the awareness and reportability dimensions. Specifically, the extent to which an individual's performance is disrupted by verbal report depends on the magnitude of the discrepancies between the awareness versus reportability dimensions. For example, as mentioned earlier, memory for same race faces is associated with a greater discrepancy between awareness and reportability than other race faces. And, consistent with the present considerations, same race face recognition is more impaired by verbalization than is other race recognition (Fallshore & Schooler, 1995). Similarly, Melcher and Schooler (1996) found that verbalization impaired the wine recognition of nonprofessional drinkers (who had drinking experience but relatively little verbal wine knowledge). However, verbalization had little effect on either nondrinkers (individuals with relatively little perceptual or verbal expertise) or wine professionals (individuals with marked perceptual and verbal expertise). These findings suggest that when the discrepancy between awareness and reportability is great, as when individuals have perceptual expertise without commensurate verbal expertise, then the disruptive effects of verbalization can be substantial. However, when the discrepancy between these two dimensions is less (i.e., when either verbal expertise is high or perceptual expertise is low), there is little consequence of committing an experience to words.

CONCLUSIONS

Language has a powerful hold on consciousness. There seems little doubt of the profound debt that our sense of consciousness has to language. Many of the defining qualities of our conscious experience (e.g., reflection—Johnson & Reeder, chapter 13, this volume; self-concept—Kihlstrom, chapter 24, this volume; symbolic reasoning—Dulany, chapter 10, this volume) may be closely wed to language. Given languages' centrality to consciousness, it is understandable that individuals may mistakenly confuse what they are able to report with what they are conscious of, thereby leading them to ignore that which was not adequately described. It is also understandable, although perhaps a bit less so, that researchers have considered what subjects are able to report about an experience as the primary index for assessing what is conscious. However, as researchers, we must be ever vigilant not to confuse the construct with the measure. Language may provide a critical window to consciousness, nevertheless, what one reports and what one is conscious of are not the same thing.

Distinguishing consciousness from reportability might seem to make scientific investigations of consciousness all the more elusive, as the measure with the seemingly greatest face validity is shown to be inadequate. However, as we have tried to illustrate, self-reported subjective awareness can help to fill the void created by abandoning content reportability as the sole measure of what is conscious. Moreover, distinguishing subjective awareness and content reportability enables us to refine our understanding of conscious experience, differentiating between situations that might otherwise be inappropriately lumped together. Most importantly, this distinction leads us to ask questions that we might have otherwise overlooked. Is it possible to report a cognitive event and yet not be conscious of it? Are there costs to reporting cognitive events that are not entirely reportable? The answers to questions such as these elucidate the tenuous relation between language and thought. Appreciating the limits of language may help us avoid the confusion (both in measurement and in cognition) that can ensue when language is used as a proxy for consciousness.

ACKNOWLEDGMENTS

The writing of this chapter was supported by an NIMH grant to the first author.

REFERENCES

Adams, J. K. (1957). Laboratory studies of behavior without awareness. *Psychological Bulletin, 54*, 383–405.

Bower, G. H. (1990). Awareness, the unconscious, and repression: An experimental psychologist's perspective. In J. L. Singer (Ed.), *Repression and dissociation* (pp. 209–231). Chicago: University of Chicago Press.

Bowers, K. S., Regehr, G., Balthazard, C., & Parker, K. (1990). Intuition in the context of discovery. *Cognitive Psychology, 22*, 72–110.

Brandimonte, M. A., & Gerbino, W. (1993). Mental image reversal and verbal recoding: When ducks become rabbits. *Memory and Cognition, 21*, 23–33.

Brown, R., & McNeill, D. (1966). The "tip of the tongue" phenomenon. *Journal of Verbal Learning and Verbal Behavior, 5*, 325–337.

Cheesman, J., & Merikle, P. M. (1986). Distinguishing conscious from unconscious perceptual processes. *Canadian Journal of Psychology, 40*, 343–367.

Dennett, D. (1986). *Content and consciousness*. London: Routledge & Kegan Paul. (Original work published 1969)

Dennett, D. (1991). *Consciousness explained*. Boston: Little, Brown.

Ericsson, K. A., & Simon. H. A. (1994). *Protocol analysis: Verbal reports as data*. Cambridge, MA: MIT Press.

Fallshore, M., & Schooler, J. W. (1995). Verbal vulnerability of perceptual expertise. *Journal of Experimental Psychology: Learning, Memory and Cognition. 21*, 1608–1623.

Fowler, C. A. (1986). An operational definition of conscious awareness must be responsible to subjective experience. *Behavioral and Brain Sciences, 9*, 33–35.

Galin, D. (1995). *The structure of awareness: Contemporary applications of William James' forgotten concept of "the fringe."* Unpublished manuscript, University of California.

Holender, D. (1986). Semantic activation without conscious identification in dichotic listening, peripheral vision, and visual masking: A survey and appraisal. *Behavioral and Brain Sciences, 9*, 1–66.

Jackendoff, R. (1987). *Consciousness and the computational mind*. Cambridge, MA: MIT Press.

James, W. (1890). *The principles of psychology*. New York: Dover.

Koestler, A. (1964). *Act of creation*. London: Hutchinson.

Mangan, B. (1991). *Meaning and the structure of consciousness: An essay in psycho-aesthetics*. Unpublished doctoral dissertation, University of California.

Melcher, J. M., & Schooler, J. W. (1996). The misremembrance of wines past: Verbal and perceptual expertise differentially mediate verbal overshadowing of taste memory. *The Journal of Memory and Language, 35*, 231–245.

Merikle, P. M., & Cheesman, J. (1986). Consciousness is a "subjective" state. *Behavioral and Brain Sciences, 9,* 42–43.

Nisbett, R. E., & Wilson, T. D. (1977). Telling more than we can know: Verbal reports on mental processes. *Psychological Review, 84,* 231–259.

Schooler, J. W., & Engstler-Schooler, T. Y. (1990). Verbal overshadowing of visual memories: Some things are better left unsaid. *Cognitive Psychology, 22,* 36–71.

Schooler, J. W., & Melcher, J. M. (1995). The ineffability of insight. In S. Smith, T. B. Ward, and R. A. Finke (Eds.), *The creative cognition approach* (pp. 97–133). Cambridge, MA: MIT Press.

Schooler, J. W., Ohlsson, S., & Brooks, K. (1993). Thoughts beyond words: When language overshadows insight. *Journal of Experimental Psychology: General, 122*(2), 166–183.

Schooler, J. W., Ryan, R. D., Fallshore, M., & Melcher, J. M. (in press). Knowing more than you can tell: The relationship between language and expertise. In R. E. Nisbett & J. Caverni (Eds.), *The psychology of expertise.* Amsterdam: Elsevier.

Shanks, D. R., & St. John, M. F. (in press). Characteristics of dissociable human learning systems. *Behavioral and Brain Sciences.*

Sidis, B. (1898). *The psychology of suggestion.* New York: Appleton.

Sperling, G. (1960). The information available in brief visual presentations. *Psychological Monographs, 74,* 1–29.

Tulving, E. (1985). Memory and consciousness. *Canadian Psychologist, 26,* 1–12.

Metacognitive Processes

V

CHAPTER 13

Consciousness
as Meta-Processing

Marcia K. Johnson
John A. Reeder
Princeton University

The invitation to participate in the 25th Carnegie Symposium on Cognition asked participants to consider fundamental issues of consciousness and encouraged us to "let our hair down." We signed up for the task because the company was good and, like many cognitive labs, our lab has always worked on consciousness, whether or not we called it that. That is, we have explored questions about the properties and conditions of certain subjective, mental experiences. For example, what distinguishes simply processing syntax and word meanings from the dramatically different phenomenal experience of comprehending prose (Bransford & Johnson, 1973)? What phenomenal qualities lead people to believe some mental experiences derive from actual, autobiographical events and not from imagination or inference (Johnson & Raye, 1981)? The phenomenal experiences of understanding and remembering certainly are part of what is meant by consciousness.

How might such subjective topics be approached? The most developed example from our lab is our work on the phenomenology and processes involved in source monitoring. This concerns how people attribute particular experiences to sources—for example, how they tell memories for actual events from memories for imagined events, or how they identify Margaret rather than Mieke as the origin of a statement. Source monitoring is central to understanding not only the conscious experience of remembering, but also the conscious experience of believing. Our strategy over the years involves what we have called an *experimental phenomenological approach* (Johnson, 1988b; see also Brewer, 1986). As Flanagan (1992) advocated, we have had faith that understanding subjective experience will accrue from converging evidence from phenomenological (e.g., Johnson, Foley, Suengas, & Raye, 1988), experimental cognitive (e.g., Johnson, Foley, & Leach, 1988; Lindsay & Johnson, 1989), neuropsychological (e.g., Johnson, 1991a), and clinical (Johnson, 1988b) approaches (see Johnson, Hashtroudi, & Lindsay, 1993, for a review). This same strategy can, of course, be applied to many domains (e.g., object perception, neglect, sleep and dreaming), each yielding a view of consciousness within the context of a particular set of problems.

To add to the developing picture, we considered contributing a discussion of our empirical and theoretical work on source monitoring. However, a recent review is available (Johnson et al., 1993) and, in any event, that would hardly have constituted "letting our hair down." More important, it is not clear that a characterization of consciousness in any particular domain will satisfy the impulse to understand Consciousness, with a big *C,* which energizes so much of the thought on this topic. After all, a great deal is known about vision from much elegant research, but it is not generally agreed that vision researchers have solved the problem of Consciousness. What many cognitive psychologists and philosophers seem to want is a way to think about the problem of Consciousness in general terms.

In looking for some unifying principles or concepts, one approach would be to single out a specific *process* or *module* of a cognitive model to identify with consciousness. Along these lines, Klatzky (1984) suggested that information-processing theories have offered three types of candidates for consciousness: an *executive* that controls the flow of information, a *short-term memory* stage of processing, or the state of information being in *focal attention.* A recent variant of the stage or state view (depending on definition) adds consciousness as a module in a subsystems model (e.g., Schacter, 1989). An alternative to identifying consciousness with a particular process or structure is to equate it with a particular type of *content.* According to Kihlstrom (1987), for example, consciousness occurs whenever the concept of "self" is an active element of working memory.

These solutions all seem less than compelling, however. There certainly are states of heightened awareness in which you seem to lose consciousness of yourself; that is part of the attraction of filling your consciousness with intellectual puzzles, novels, scenery, or a beloved. Thus, as important as the "I" of Consciousness is, "I" cannot be the whole story. In addition, it is difficult to see how a single process, structural module, or type of content could give rise to the varieties of consciousness—the differences between, for example, being aware and deciding, or being aware and being aware that one is aware.

A third approach, used here, is to consider *systemwide functional relations* that could give rise to the varieties of consciousness. Our point of departure is the Multiple-Entry, Modular memory system (MEM; Johnson, 1991a,

1991b; Johnson & Hirst, 1993). MEM is a systems-level architecture,[1] designed in a top-down fashion to accommodate a wide range of cognitive phenomena and to incorporate many recurring ideas about processes from the cognitive literature (Johnson, 1983, 1991a, 1991b; Johnson & Hirst, 1991, 1993; Johnson & Multhaup, 1992). This chapter first considers some broad themes that run through many discussions of consciousness and that express what it is that theorists seem to be looking for in treating consciousness as a topic independent of its many domain-specific manifestations. We take these themes to represent what needs to be explained. Then we consider whether or how MEM could accommodate these central themes. We conclude that consciousness arises not from any single component or type of content, but from processing relations among components within a general cognitive architecture that permit certain interactions.

ASPECTS OF CONSCIOUSNESS

Consciousness is notoriously hard to define, but as Churchland (1988) and others noted, theoretical and empirical progress on complicated problems do not depend on starting with firm definitions. Rather than look for a single definition of consciousness, here we have abstracted three general aspects of cognition that commonly appear in discussion of consciousness and that capture much of what psychologists and philosophers seem to have in mind when they consider the concept of consciousness. For short, we have labeled these *awareness, control,* and *representational complexity.*

Awareness

Awareness (as in "to be aware of X") is perhaps the most commonly identified aspect of consciousness. For example, Searle (1992) called awareness a near synonym for consciousness (p. 84) and Reber (1985) identified the state of

---

[1]One way to build a model of a cognitive system is to start with a proposal about data representations and processes, develop simulations for particular tasks, and expand or complicate the model as new features or functions are considered (e.g., the addition of a procedural memory system and representations for temporal strings and spatial images as Anderson's ideas evolved into ACT*; Anderson, 1983). Another way, followed here, is to use available empirical data and theoretical ideas to outline a systems-level architecture or framework that specifies a comprehensive range of cognitive functions, including constraints on them, before investing in any particular simulation strategy (e.g., Moscovitch & Umiltà, 1991; Schacter, 1989; Squire, 1992; Tulving, 1985). Presumably, computational instantiations of such top-down designs eventually would be able to incorporate ideas from computational approaches (e.g., Anderson, 1983; Hintzman, 1986; Holyoak & Thagard, 1989; Metcalfe Eich, 1985; Newell, Rosenbloom, & Laird, 1989; Rumelhart & McClelland, 1986; Simon, 1989; Sejnowski & Rosenberg, 1987).

awareness as the most common usage of "consciousness" in his dictionary of psychological terms (p. 148). Typically, this aspect of consciousness is associated with various kinds of information contents, such as percepts, memories, emotions, and dreams (e.g., Marcel, 1988, p. 123). As we are using it here, then, awareness includes perceptual experiences such as tasting wine, the experience of activities such as swimming, the sense of knowing facts (canaries are yellow), or remembering experiences (the train ride to Bressanone).

Control

Control is that aspect of consciousness that gives us a sense that we can start and stop our own activity. Control is often thought of as a strategic organization of cognitive processes (Umiltà, 1988). And it often entails the notion of monitoring, that is, the idea that outcomes of cognitive operations are compared to criteria, expectations, or goals and that these comparisons influence subsequent processing. Engaging in controlled action and thought gives rise to a sense of efficacy, or intentional, purposeful activity. Umilta (1988, p. 334) suggested that control is the aspect of consciousness "most likely to have a causal role." The notion of control is central to Shallice's (1978) conceptualization of "modular supervisory processes," which organize and augment the activity of other processes, to Johnson-Laird's (1983) idea of consciousness as the "operating system," and to Baddeley's (1992a, 1992b) "central executive."

Representational Complexity

There are several other characteristics of consciousness that we think can be subsumed by the notion that consciousness can range in complexity: that is, it can focus on single mental elements or encompass multiple representations. For example, in addition to awareness and control, an important aspect of consciousness is analytic capability, such as when individuals bring information to bear on making decisions (Johnson & Hirst, 1993). This aspect of consciousness involves seeing alternative points of view and attempts to resolve conflict. Comparing, deciding, evaluating, and seeing another's point of view depend on being able to maintain and negotiate among two or more complex representations (e.g., DeLoache & Burns, 1994). Deliberation is a similar idea that was of particular interest to Popper (Popper & Eccles, 1977), who suggested that consciousness is "where an aim or purpose ... can be achieved by alternative means, and when two or more means are tried out, after deliberation" (p. 125). Calvin (1989), too, suggested this capacity of organisms to select among multiple actions is the most promising characteristic of conscious experience to examine scientifically.

A special case of complexity arising from multiple representations is the idea that consciousness can be reflective (Jackendoff, 1987), or "self-reflective" (e.g., Stuss, 1991). That is, a concept of self can function as a complex representation in relation to other complex concepts in one's awareness, controlled action, or complex deliberations. Thus, consciousness has been described as occurring when a mental representation of the self is linked with the mental representation of an event (Kihlstrom, 1987) or when an active mental model includes a model of the self (Johnson-Laird, 1988).

We do not, however, want to equate the idea of self only with this latter, complexity, aspect of consciousness. The notion of self might arise from all aspects of consciousness. That is, it is in such activities as tasting wine, engaging in purposeful behavior, and deciding among alternatives that one experiences an "I" or a "self," not only when the self is taken as an explicit object of consciousness.

A Semantic Organization of Mental Concepts

Do these aspects of consciousness identified by psychologists and philosophers reflect the way people understand their own mental experience? This question can be explored by looking at the relations among English words used to refer to mental experience to see if there is converging evidence supporting these three aspects. Of course, studying words is not necessarily equivalent to studying their referents. Yet, one function of words is that they represent any particular culture's efforts to refer to phenomenally salient aspects of experience and, therefore, they are a rich source of evidence and ideas.

A set of 47 words expressing concepts related to *mind* was selected and subjects were asked to make similarity ratings for pairs of words. That is, for each pair of words, subjects were asked to rate the pair for similarity in meaning on a scale that ranged from 0 to 9 where the endpoints were "extremely dissimilar" and "extremely similar." Each subject received a subset of the items, and across 738 Princeton undergraduates, all possible pairs were rated between 32 and 47 times. A multiple dimensional scaling analysis on these similarity ratings was performed, asking for a solution with three dimensions, to see if they converged on the same or different aspects of consciousness as those already discussed. The outcome seems to fit quite well with the critical aspects of consciousness abstracted from the literature.

Figure 13.1 represents two of the three dimensions, with each concept located in two-dimensional space. This plot seems to reflect a dimension of control on the horizontal axis and something that looks like representational complexity on the vertical axis. That is, concepts such as *emotion, impression,* and *dream* are at the low end of control, whereas concepts such as *manipulation, reason,* and *logic* are at the high end. On the vertical axis, concepts such as *sense, attention,* and *awareness* are low in complexity, whereas concepts such as *deliberation, reflection,* and *synthesis,* are at the high end of complexity. That is, mental experiences that can be based on single repre-

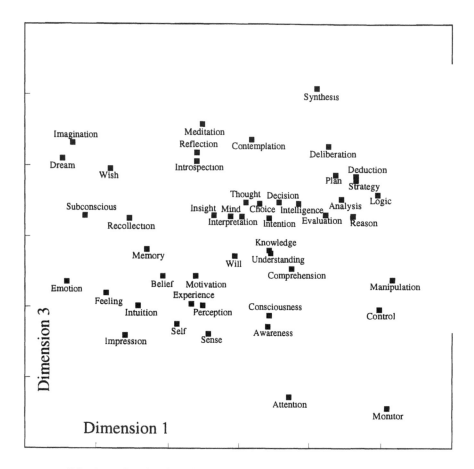

FIG. 13.1.   Results of a Multiple Dimensional Scaling analysis of similarity ratings for English words related to *mind*, showing aspects of "control" (horizontal axis) and "representational complexity" (vertical axis).

sentations (a smell, an object, etc.) are lower in complexity than mental experiences that involve multiple representations (comparing, contrasting).
Thus, an *emotion* or *impression* is low both in control and complexity, whereas
logic, analysis, deliberation, and *plan* are high in control and complexity.
Notice that the self-reflective activity of *introspection* is high on the complexity scale as well. Overall, this plot nicely points out that control and complexity
are not always correlated: Mental activities can be relatively high in control
but low in complexity (e.g., simple monitoring or attention) or high in complexity but low in control (e.g., dreams, wishes, some forms of imagination).
    Figure 13.2 again has the complexity dimension on the vertical axis, and
now a dimension on the horizontal axis that appears to match what we are call-

ing the aspect of awareness. As discussed earlier, the aspect of consciousness here called *awareness* is strongly associated with conscious *contents* (e.g., percepts and memories). Baars (1988) further delineated the concept of conscious contents by distinguishing it from the concept of goal contexts. The horizontal dimension in Fig. 13.2 is consistent with this distinction. Words to the left (e.g., *motivation, wish, intention, plan*) are associated with motivation and potential action; words to the right (e.g., *recollection, knowledge, perception, deduction*) refer to conscious contents or to cognitive operations that primarily produce contents.

Although other dimension names or alternative approaches to analyzing the relations among mental terms are certainly possible (see, e.g., D'Andrade,

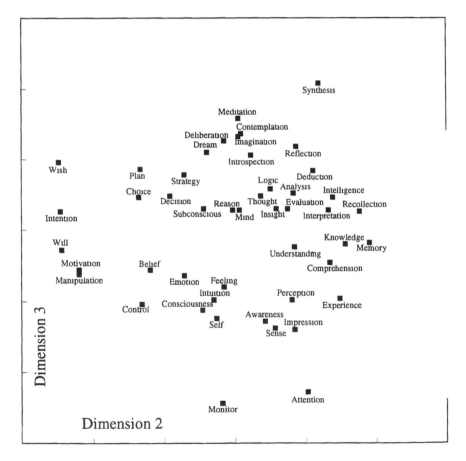

FIG. 13.2.   Results of a Multiple Dimensional Scaling analysis of similarity ratings for English words related to *mind*, showing aspects of "awareness" (horizontal axis) and "representational complexity" (vertical axis).

1987; Rips & Conrad, 1989), on the whole, the Multiple Dimensional Scaling analysis depicted in Figs. 13.1 and 13.2 is consistent with the central ideas derived from the literature. Major aspects of consciousness include awareness, control, and a complexity factor that reflects the extent to which multiple representations contribute to phenomenal experience. It should not be surprising that everyday language and scholarly inquiry would converge on common themes when the topic is phenomenal experience.

A GENERAL COGNITIVE ARCHITECTURE: MEM

If awareness, control, and representational complexity are accepted as distinct aspects of consciousness, then what are the various cognitive processes and interactions that produce these phenomenal qualities of consciousness? This chapter proposes that these aspects of consciousness can be viewed as a derived product of the right system-level cognitive architecture. There is some set of cognitive architectures that could do this. This chapter suggests that these various aspects of consciousness can be characterized within a particular cognitive architecture—the MEM system proposed by Johnson and colleagues (Johnson, 1983, 1990, 1992; Johnson & Chalfonte, 1994; Johnson & Hirst, 1991, 1993; Johnson & Multhaup, 1992).

The MEM Framework

The MEM framework is a process-oriented approach to describing memory in particular and cognition in general. The ideas embodied in MEM are derived from a wide range of empirical findings, theoretical proposals, and recurring insights about cognition—for example, that there are processes for locating and identifying objects (e.g., Biederman, 1987; Mishkin, 1982; Rock & Gutman, 1981); organizing, elaborating, and finding relations (e.g., Gentner, 1988; Mandler, 1967; Tulving, 1962); maintaining activation (e.g., Atkinson & Shiffrin, 1968; Baddeley, 1986); retrieving information (e.g., Reiser, 1986); and guiding thought and action (e.g., Bartlett, 1932; Miller, Galanter, & Pribram, 1960; Shallice, 1988). The particular structural arrangement of processes in MEM is consistent with major phenomena of current interest, such as dissociations among memory measures (e.g., Jacoby & Dallas, 1981), patterns of breakdown in cognition from brain damage (e.g., Cohen & Squire, 1980; Stuss & Benson, 1986), and differences in the controllability of processes (e.g., Hasher & Zacks, 1979; Norman & Shallice, 1986; Posner & Snyder, 1975); and with central ideas about the relation between cognition and emotion (e.g., Lazarus, 1982; Mandler, 1984; Roseman, 1984; Smith & Ellsworth, 1985), and about the development (e.g., Flavell & Wellman, 1977; Kail, 1984;

Schacter & Moscovitch, 1984) and evolution (e.g., Ekman, 1984; Reber, 1989) of cognition.

We begin with an overview of MEM and related evidence (see also Johnson, 1983, 1990, 1991a, 1991b, 1992; Johnson & Chalfonte, 1994; Johnson & Hirst, 1991, 1993; Johnson & Multhaup, 1992). Next we show how MEM could give rise to the three major aspects of consciousness described previously, and then we consider how disruptions in consciousness might come about within the MEM framework.

The lower level elements of the MEM framework are component processes. In MEM, a component process constitutes a particular manipulation of information. The current version of MEM includes 16 such component processes that operate on information in qualitatively different ways (Fig. 13.3). The performance of any mental task involves some combination of these component processes, and memory is viewed as a record of the operation of the component processes. Furthermore, MEM includes an organizational framework that groups component processes according to broad functional classes or subsystems. In the nomenclature of MEM, the P-1 and P-2 subsystems include relatively simple and more complex *perceptual* processes, respectively, whereas the R-1 and R-2 subsystems include simple and complex *reflective* processes. The organizational class, or subsystem, to which a component process belongs reflects our speculations about the type of information it preferentially acts on, other component processes with which it frequently operates, its degree of computational sophistication or elaborateness, the evolutionary stage at which it might first have arisen, and the brain mechanisms by which it might be mediated. Furthermore, the concept of separate but interacting subsystems is critical for addressing consciousness and the "homunculus."

As shown in Fig. 13.3c, the component processes of P-1 are *locating* stimuli, *resolving* stimulus configurations, *tracking* stimuli, and *extracting* invariants from perceptual arrays. P-1 processes are involved in learning and memory, but we usually are not directly aware of the perceptual information that is operating as cues. P-1 processes underlie such skills as adjusting to a person's foreign accent or anticipating the trajectory of a baseball.

The component processes of P-2 are *placing* objects in spatial relation to each other, *identifying* objects, *examining* or perceptually investigating stimuli, and *structuring* or abstracting a pattern of organization across temporally extended stimuli. P-2 processes are responsible for much of our phenomenal experience of the external world. They provide us with a subjective world of objects and events, organized meaningfully based on prior experience. Relative to P-1, the processes of P-2 can operate on more complex data structures. For example, *locating* (a P-1 process) can be applied to an undifferentiated external stimulus, whereas *placing* (a P-2 process) computes relative positions of two differentiated and usually identified objects.

(a)

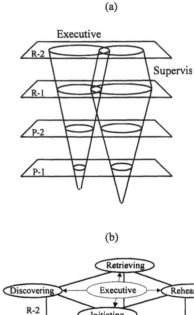

(b)

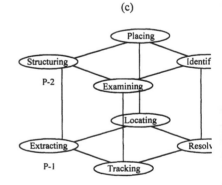

FIG. 13.3.   The Multiple-Entry Modular (MEM) memory system: (a) The reflective and perceptual subsystems (R-1 and R-2, P-1 and P-2, respectively) interact with each other through the executive and supervisory functions of agendas; (b) component processes of the reflective subsystems, R-1 and R-2; (c) component processes of the perceptual subsystems, P-1 and P-2.

(c)

The component processes of the two reflective subsystems, R-1 and R-2, allow us to go beyond the perceptual world (Fig. 13.3b). They permit the manipulation of self-generated and externally derived information and memories and allow us to anticipate future events, imagine possible alternatives, and so forth. Both R-1 and R-2 involve component processes that allow people to sustain, organize, and revive information. R-1 and R-2 differ in the complexity of tasks they can handle. For example, R-1 processes would be sufficient for coming up with the idea of having a party, but R-2 processes would be necessary for planning one. In R-1 the component processes are *refreshing* information so that it remains active and one can easily shift back to it, *shifting* attention to something potentially more useful, *noting* relations, and *reactivating* information that has dropped out of consciousness. Component processes in R-2 include *rehearsing, initiating, discovering,* and *retrieving*. Relative to R-1, R-2 processes typically deal with more complex data structures. For example, *noting* (R-1) can compute overlapping relations from associations activated by two items (e.g., *dog* and *cat* both activate *animal*). In contrast, *discovering* (R-2) finds relations that are less direct, for example, relations that depend on other relations as in computing analogies (e.g., Gentner, 1988).

The component processes derive great functional power from the fact that they can be marshaled and executed by agendas. Agendas recruit processes in the service of goals—a combination of goals and component processes constitutes what we call an *agenda*. An agenda can be thought of as a script or recipe. That is, a recipe is somewhat more flexible than a program; its instantiation allows for some opportunistic flexibility and improvisation. Most agendas are learned through experience.

Agendas coordinate *control* and *monitoring*. Control is, in effect, the execution of the operations as specified in the agenda. Monitoring is accomplished by matching consequences of executed processes against expectations embodied in or activated by the agenda. In Fig. 13.3, the *Supervisor* and *Executive* are ways of referring, collectively, to potential R-1 and R-2 agendas, respectively. Thus, there is no unitary executive mechanism in either subsystem; the operation of an agenda creates a "virtual" executive (cf. Dennett's, 1991, virtual "captain").

One way the subsystems of MEM interact with one another is through activated agendas (see Fig. 13.3a). For example, an agenda initiated in R-2 such as *look for a restaurant,* might activate relevant perceptual schemas from perceptual memory (e.g., look for a building with ground-level window, tables visible, menu in window).[2] The agenda might also activate reflective plans adapted to the current situation (e.g., check the restaurant guide for this part of town). Typically, agendas have greater access to reflective memory records

---

[2]Like the reflective agendas of R-1 and R-2, perceptual schemas are ways of representing learned coordination among component processes within P-1 and P-2.

than to perceptual memory records, and greater access to P-2 than to P-1 records (as suggested by the sizes of the ellipses at the intersects of cones and planes in Fig. 13.3a). This greater access could reflect a higher level of change in activation, a faster speed of access, access to a larger domain of representations, or some combination of these.

An especially important aspect of reflection is that the supervisor and executive agendas in R-1 and R-2 can recruit (control and monitor) each other, as depicted by their overlap. Interaction between R-1 processes and those in R-2 greatly increases the complexity of tasks that can be handled (e.g., interaction between R-1 and R-2 provides one mechanism for sequencing subgoals). More central to the issues addressed by this symposium, interaction between R-1 and R-2 is the basis for phenomenal experiences such as reflecting on reflection (i.e., introspection) and self-control, which are among the most complex forms of consciousness that we experience. More generally, consciousness is associated with interactions across subsystems, as discussed later.

MEM proposes a midlevel vocabulary for the minimum number of component processes—and their schematic, architectural arrangement—necessary to characterize a fully functioning adult cognitive system. This, of course, is only one "grain-size" for analysis and understanding. Each component process in MEM could be broken down further—for example, abstracting "geons" (Biederman, 1987) might be a subcomponent of *identifying,* or the component process of *noting* could be decomposed into types of relations (e.g., parts, category membership) or activities (comparing, contrasting). At the same time, MEM's component processes (e.g., *refreshing, shifting, noting, reactivating*) could be used to decompose more general-level concepts in the literature, such as "organizing" or "elaborating" (e.g., Johnson & Hirst, 1993).

At the intermediate level of description, we intend for MEM to be exhaustive—that is, our goal is to represent a full set of processes required for a cognitive architecture described at this level of analysis. For example, locating in vision and locating in audition are clearly not the same, but *locating* is a fundamental operation in both modalities. Or, to take another example, we do not include "inferencing" as a specific component process in MEM but rather assume inferencing takes place in all subsystems. *Extracting* invariants or *structuring* patterns can be viewed as types of perceptual inference processes. More complex inferencing derives from the joint activity of component reflective processes. For example, *noting* the relation of an activated semantic associate to presented text might lead to the conclusion that an unmentioned hammer was probably used to pound a nail; in playing the board game "Clue," *initiating* a systematic move through the rooms in combination with *retrieving* the outcomes of previous plays might lead to *discovering* that the murderer was Miss Scarlet in the kitchen with the knife. If attempts to use MEM to characterize new domains lead to the discovery that the current component processes are not sufficient, there would be a rationale for replacing those that have been postulated with more general processes or for adding new ones. Although MEM is a working hypothesis and not a final proposal, it entails the idea that there is a relatively limited set of processes that constitute a useful conceptual analysis of memory at this intermediate level.

Alternatively, it may seem that there are not too few components in MEM, but rather there are too many. That is, perhaps MEM's higher order processes could be decomposed into MEM's lower order processes. MEM's architecture is not, however, nested in this fashion. For example, we would not expect to find that an R-1 component process could be fully decomposed into P-2 component processes or that a P-2 component process could be decomposed into P-1 component processes. Rather, decomposing MEM's component processes requires a more specific level of analysis that is not represented in MEM. For example, *shifting* may subsume a "disengage" process (Posner, Petersen, Fox, & Raichle, 1988) that is not represented at MEM's level of analysis.

Although not nested, there are relations between component processes across MEM's subsystems. We have speculated that higher level processes along the vertical columns (see Figs. 13.3b and 13.3c) in MEM might have evolved out of earlier component processes in the same column (Johnson & Hirst, 1993). Nevertheless, the critical idea here is "evolved." Higher order processes might draw on the outcomes of lower ones without being decomposable into these same lower processes. Evolution implies that something has been added. For example, *placing* involves not simply a string of *locating* computations, but rather the capacity to simultaneously represent two or more items within a common frame.

Empirical Evidence

As with any framework, the level of analysis was determined in part by the types of empirical findings we were attempting to account for. In the initial generation of MEM (Johnson, 1983), a distinction between perceptual and reflective records was offered as a solution to, among other problems, the apparent paradox that memory seems both accurate and inaccurate. The MEM model proposed that veridical memory is obtained when a test taps into representations created by perceptual processes designed to detect highly probable recurrences and invariants (Brunswik, 1956; Gibson, 1966) and to record perceptual features, accounting for such findings as modality-specific effects (Craik & Kirsner, 1974; Hintzman & Summers, 1973) and contributing to verbatim memory (Kintsch & Bates, 1977). In contrast, memory distortions arise when tests tap into representations created by reflective processes engaged when people anticipate, draw inferences, reminisce, plan, and otherwise manipulate ideas. Because reflective activity leaves a record, people sometimes mistake their embellishments of information for fact (e.g., Johnson, Bransford, & Solomon, 1973). In addition, evidence for a distinction between perceptually derived and reflectively generated representations includes subjects' frequent success in discriminating between these types of representations and differential effects of variables on memory for perceptually derived and reflectively generated events (Johnson & Raye, 1981).

The distinction between P-1 and P-2[3] was postulated to account for phenomena such as the effects of prior exposure, especially modality-specific effects and subthreshold effects, which do not necessarily produce phenomenal experience (e.g., Fowler, Wolford, Slade, & Tassinary, 1981; Jacoby & Dallas, 1981; Kunst-Wilson & Zajonc, 1980; D. Scarborough, Cortese, & H. Scarborough, 1977), the cumulation of information over trials each alone too brief to permit stimulus identification (Haber & Hershenson, 1965), memory for some aspects of unattended information (Rock & Gutman, 1981), and more exotic findings such as long-term retention of random dot patterns that had not yet been synthesized into phenomenal objects (Stromeyer & Psotka, 1970) and blindsight (Weiskrantz, Warrington, & Sanders, 1974). Such findings seemed to point to a division between perceptual processing that leaves a record that can be used in subsequent perceptual processing and that plays a critical role in perceptual motor tasks and skill learning (P-1), and perceptual processing that results in phenomenally recoverable memories of objects and events (P-2). In contrast, the reflection system was assumed to provide the major mechanisms underlying the creative combination and voluntary recovery of information such as illustrated by organizational activities in free recall (e.g., Mandler, 1967; Tulving, 1968).

In short, although most tasks are unlikely to tap into one subsystem only, dissociations among memory measures are nevertheless likely because different measures draw disproportionately on different memory subsystems; for example, recall draws relatively more on reflective records, recognition on P-2 records, and priming or perceptual identification tasks on P-1 records. This basic framework, although primarily generated by considering findings from normal adult cognition, was also applied to findings reported from studies of amnesia drawing on reviews by Hirst (1982), Moscovitch (1982), and Baddeley (1982), as well as a number of other theoretical proposals (e.g., Cermak, 1977; Huppert & Piercy, 1982; Warrington & Weiskrantz, 1970; Wickelgren, 1979). Using the MEM framework, it was proposed that the pattern of amnesic deficits—profound deficits in recall along with relatively preserved perceptual and motor skill performance (e.g., rotary pursuit, reading mirror images of words, identifying the objects in degraded pictures)—could be accounted for by postulating the anterograde amnesia results from a deficit in

---

[3]P-1 and P-2 processes were initially called "sensory" and "perceptual," respectively (Johnson, 1983).

the reflective subsystem with the perceptual subsystems relatively intact. Thus, it would be expected that amnesics show preserved performance in tasks largely served by the P-1 and P-2 subsystems and increasing deficits as tasks require greater amounts of reflection at encoding or at test.

The distinction between R-1 and R-2 was introduced later to capture something of the processing complexity and nuance ignored by more global ideas such as reflection, effort, attention, capacity, and control (Johnson, 1990, 1991a, 1991b; Johnson & Hirst, 1991). First, global categories such as "reflection" or single dimensions such as "control" do not explicate foundational concepts such as chunking (Miller, 1956; see Johnson, 1990). Furthermore, executing two layers of organization (e.g., categorizing and alphabetizing within categories) requires the ability to negotiate two agendas. Once two interactive reflective subsystems are postulated that each have mechanisms for maintaining, reviving, and organizing information, it becomes relatively easy to incorporate the fact that disruption of reflective processes, like disruption of perceptual processes, can be quite selective. For example, memory disorders come in a number of varieties and degrees of severity (e.g., Johnson & Hirst, 1991). Some patients appear to show selective disruption of *refreshing* and/or *rehearsing* ("short-term memory" deficits, Vallar & Baddeley, 1984; Warrington, 1982). Others appear to show selective disruption of *reactivating*, a reflective activity that is critical for binding features and consolidating them into complex memories (anterograde amnesia, Johnson & Chalfonte, 1994; Squire, 1987). Still others appear to show selective disruption of those aspects of remembering that depend on processes such as *shifting, noting,* and *retrieving* that are critical for organizing and self-cuing ("frontal" or "executive" deficits, Baddeley, 1982). A more complex view of the relation between reflection and memory should help integrate findings implicating many brain areas in memory disorders: for example, temporal and diencephalic lesions may have a relatively larger impact on R-1 processes and frontal lesions a relatively larger impact on R-2 processes (Johnson, 1990).

The distinction between R-1 and R-2 also accommodates proposed differences in judgment processes that have been useful for analyzing a variety of tasks, including old/new recognition and source monitoring. Johnson and Raye (1981), for example, made a distinction between reality monitoring judgments based, on the one hand, on relatively quick assessment of activated trace characteristics (e.g., amount of perceptual detail) and, on the other hand, judgments based on additional retrieval and more extended reasoning (e.g., is this plausible given other things I know?). Within the MEM framework, the former result from R-1 processes and the latter from R-2 processes. This distinction is similar to Chaiken, Lieberman, and Eagly's (1989) distinction between heuristic and systematic processing, and encompasses the idea that heuristic judgments might be based on familiarity whereas systematic judgments involve more active recall processes (e.g., Gardiner & Java, 1990; Jacoby & Dallas, 1981; Mandler, 1980).

In addition, patterns of confabulation in brain-damaged patients are consistent with the idea that R-1 and R-2 can be selectively disrupted (Johnson, 1991a). For example, the confabulations of some patients are realistic and may include confusions among elements of actual autobiographical events, whereas the confabulations of other patients are bizarre or fantastic. Relatively realistic confabulation could result from deficits in R-1 processes, for example, overly lax criteria for source monitoring in which ideas that come to mind are treated as reflecting true episodic memories based on familiarity or minimal amounts of perceptual and contextual detail (see also Moscovitch, 1989). Lax criteria could result if such processes as *shifting* to more appropriate criteria or *noting* other trace characteristics are impaired (as well as, of course, from reduced motivation to engage in reality monitoring). In contrast, more bizarre confabulations appear to be produced by disruptions of R-2 processes, for example, deficits in *retrieving* additional semantic and episodic information that could be used for *discovering* inconsistencies or implausible aspects of what is remembered.

Generally speaking, we have been encouraged by the continuing accumulation of evidence that is consistent with the MEM framework (although many studies were motivated, of course, by other conceptions of cognition and memory). Dissociations among memory measures are now among the central, to-be-explained facts of the field (e.g., Richardson-Klavehn & Bjork, 1988). There are numerous demonstrations that indirect measures of memory that do not depend on phenomenally available event records (e.g., perceptual identification, pattern drawing) show savings and such savings can be independent of performance on direct measures such as recognition or recall (e.g., Musen & Treisman, 1990). Even "unattended" information has persisting effects, supporting the idea that P-1 processing alone is sufficient to create memory records (Eich, 1984; DeSchepper & Treisman, 1996), but not sufficient to produce the P-2 records required for, say, frequency judgments about complex events defined by the conjunctions of features (Johnson, Peterson, Chua-Yap, & Rose, 1989).

The proposition that amnesics have relatively intact P-1 and P-2 subsystems and thus should have intact performance on perceptually based tasks also has received considerable support. Amnesics show savings in processing random dot stereograms (Benzing & Squire, 1989) and priming in a variety of other perceptually based tasks (e.g., Musen & Squire, 1992; Schacter, Cooper, Tharan, & Rubens, 1991). Amnesics show preserved effects of frequency of exposure on recognition (Johnson, Kim, & Risse, 1985; Johnson & Multhaup, 1992), preserved recognition when reflective processing is minimized by task demands (Weinstein, 1987; also cited in Johnson, 1990), or when the reflective component of recognition is subtracted out (Verfaellie &

Treadwell, 1993; see Jacoby, 1991), and relatively preserved recognition in relation to recall (Hirst, Johnson, Kim, Phelps, Risse, & Volpe, 1986; Hirst, Johnson, Phelps, & Volpe, 1988; but see Haist, Shimamura, & Squire, 1992). Amnesics also show deficits in acquisition of affect (Johnson et al., 1985) and in skill learning (Phelps, 1989) that appear to be related to the degree of reflection required by the task.

A by-product of the intense interest generated by memory researchers' rediscovery of indirect or savings measures of memory is that exploration of reflective processes has lagged somewhat. However, the idea that a full account of memory must include a specification of the role of complex reflective processing involved in encoding and retrieval is beginning to have its own revival, as evidenced by the increasing interest in controlled (i.e., reflective) processes in skill acquisition (Seger, 1994), and in the role of frontal functions in memory (e.g., Craik, L. W. Morris, R. G. Morris, & Loewen, 1990; Kopelman, 1989; Moscovitch & Umiltà, 1991; Shimamura, Janowsky, & Squire, 1990; Tulving, Kapur, Craik, Moscovitch, & Houle, 1994; see also Johnson et al., 1993).

Along with accumulating empirical evidence, there is also an encouraging convergence of theoretical ideas consistent with some of MEM's fundamental assumptions, for example, that memory is a record of processing (Kolers & Roediger, 1984); that characterizing this processing in terms of component processes is a central goal for cognitive-level theories (e.g., Moscovitch, 1992); that processing and subsystems accounts are not necessarily incompatible (e.g., Schacter, 1990); that there are subsystems that represent perceptual information (Musen & Treisman, 1990; Tulving & Schacter, 1990); that some memory records can mediate responses in the absence of awareness (Kihlstrom, Barnhardt, & Tataryn, 1992); and that memory subsystems reflect an evolutionary history in which veridical representations of perceptual stimuli appeared earlier than embellished, reflectively generated representations (Reber, 1989). Finally, there appears to be increasing recognition that interactions among subsystems are as important as dissociations. For example, perceptual subsystems are especially important in mediating certain types of perceptual-motor action but perceptual and reflective processes interact in skill learning (Seger, 1994).

Characterizing the relations among empirical findings and common theoretical themes remains a central goal of our efforts. We believe that a midlevel conceptual schema such as MEM is desirable for a sense of the "whole" of cognition. With fewer analytic units, the full, flexible range of cognition is difficult to depict and to see, and with too many more, the cohesive, integrated nature of cognition is lost in detail. As is discussed in the next section, such a midlevel conceptual schema is especially appropriate for explicating Consciousness with a big "C."

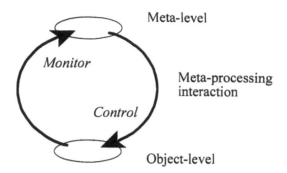

FIG. 13.4.   Meta-processing: the monitoring and controlling of an object level by a meta-level.

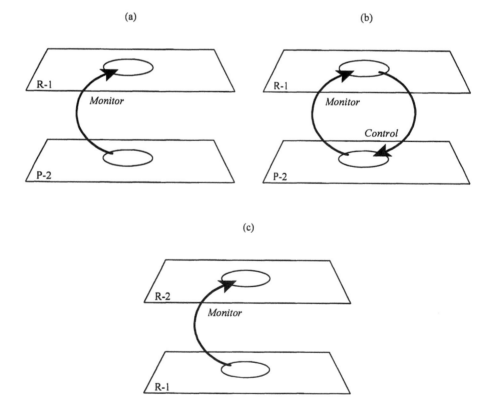

FIG. 13.5.   Meta-processing interactions between two MEM subsystems: (a) awareness or "looking at" characterized as R-1 monitoring P-2; (b) control or "looking for" characterized as R-1 monitoring and controlling P-2; (c) awareness of thinking characterized as R-2 monitoring R-1.

CONSCIOUSNESS AND MEM

Consider the proposition that in MEM consciousness occurs when an active agenda recruits component processes across subsystems. In order to help characterize these across-subsystem transactions that we propose underlie consciousness, a useful distinction is one made by Nelson and Narens (1990) between meta-level and object-level processing (Fig. 13.4). According to Nelson and Narens, two processing relations or transactions can hold between these levels: monitoring and control (cf. Kihlstrom, 1984, p. 150). In *monitoring,* the meta-level's model of the situation is changed as it observes the object level. Whereas the object level is not affected by being monitored, through *control,* the meta-level causes processes in the object level to be activated, continued, or terminated. (Although it seems unlikely that any process or representation can be "examined" without affecting it in some way, this point is not critical for the present discussion.) In MEM, any subsystem can be an "object level"; here the focus is on the meta-level roles of R-1 and R-2. As different subsystems play the meta- and the object-level role in different situations, different types of conscious experience result.

Examples of Meta-processing Across Subsystems

As an example of this "meta-processing," take the case where P-2 is the object level for an R-1 agenda; if R-1 is only monitoring P-2, you would have a type of conscious experience we call *awareness,* say, the experience of watching or "looking at" fish in an afternoon of relaxed snorkeling (Fig. 13.5a).[4] If R-1 is also controlling P-2, you might have the sense of "looking for" a particular object (e.g., a sea urchin; Fig. 13.5b). In this case, a perceptual schema in P-2 for what urchins look like may be activated as a result of an R-1 agenda that is executed for checking the parts of a reef where you know it is more likely that you will find sea urchins. Consider another example where R-1 is the object level for R-2 (Fig. 13.5c). If only monitoring is taking place, you would have an awareness of observing your thoughts (e.g., a sense of being aware that you are thinking that there may be sea urchins at the reef). Or, if R-2 is monitoring while R-1 is monitoring and controlling P-2 (Fig. 13.6a), you would have a sense of observing yourself trying to do something (e.g., being aware of yourself trying to find a sea urchin). If R-2 were also controlling R-1, you might have the experience of planning (e.g., you might tell yourself to look first at

---

[4]These hypothetical situations illustrate varieties of phenomenal experience such as the difference between looking for something and being aware that you are looking for something. The section "Disruptions of Consciousness," goes beyond this appeal to everyday experience and suggests that a MEM-based account of consciousness can be applied to the types of empirical evidence yielded by studies of brain-damaged patients.

the near reef and, if no urchins are found, venture out to the far reef) (Fig. 13.6b).

More complex conscious experiences arise when a subsystem monitors and controls two or more object-level representations simultaneously. These object-level representations might be from different subsystems or from a single subsystem. For example, R-1 might monitor representations from P-2 and R-2; here you might be aware that you are both watching a sea urchin and try-

(a)

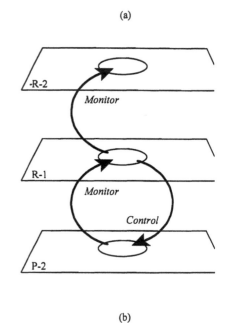

(b)

FIG. 13.6.   Recursive meta-processing interactions among three MEM subsystems: (a) self-observation of purposeful activity characterized as R-2 monitoring R-1 as R-1 monitors and controls P-2; (b) self-regulation of purposeful activity (e.g., planning) characterized as R-2 monitoring and controlling R-1 as R-1 monitors and controls P-2.

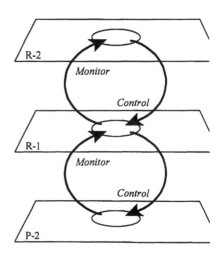

ing to retrieve whatever information you have been told about them (e.g., retrieve names of people who might have an interest in snorkeling in order to remember what they might have told you) (Fig. 13.7a). If R-2 were monitoring and controlling which R-1 agenda is followed (look for urchins or look for flounder), you might have the experience of deciding (Fig. 13.7b). If you were deliberating between R-1 agendas, such as finding sea urchins and going back

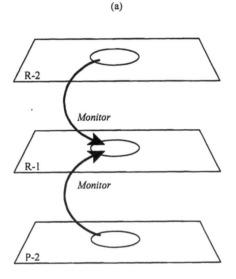

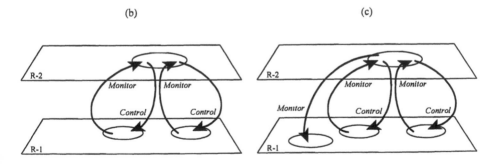

FIG. 13.7. Representationally complex meta-processing interactions between MEM subsystems: (a) simultaneous perceptual (e.g., looking at) and reflective (e.g., retrieving related information) awareness characterized as R-1 monitoring both P-2 and R-2; (b) deciding characterized as R-2 monitoring and controlling two R-1 representations (e.g., agendas); (c) self-control characterized recursively as R-1 monitoring R-2 as R-2 monitors and controls two R-1 representations (e.g., agendas).

to your hotel room to write a paper, you might experience self-control. (Notice that the difference between deciding and self-control is often in the nature of the representations and values attached to them.) Complex consciousness is sometimes recursive: for example, if R-1 were monitoring your decision (executed by R-2 processes) to continue looking for urchins or write a paper, you might be aware you are exercising self-control (Fig. 13.7c).

Awareness, Control, and Representational Complexity in MEM

As illustrated by the previous examples, interactions between subsystems can take various forms. At the most general level, these forms correspond to the three fundamental aspects of consciousness outlined in an earlier section: awareness, control, and representational complexity. In MEM, awareness derives from meta-processing that consists largely of monitoring the outcomes of component processes. Control derives from meta-processing in which agendas recruit component processes and engage in corresponding error correction processes as outcomes are matched to agendas. Representational complexity derives from monitoring and control of processes applied to two or more representations, as in noting conflicts among representations or discovering compromise representations.[5] Thus, within MEM, such differences in conscious experiences as *looking at* (monitoring), *looking for* (control), and *comparing* (representational complexity) are relatively straightforward.

It is also possible to characterize differences within a particular aspect of consciousness. For example, within the domain of awareness, one can be *looking at* (aware of) either external events (e.g., R-1 monitoring P-2) or internal events (e.g., R-2 monitoring R-1). That is, one's awareness can be directed outward toward the perceptual environment or inward toward one's thoughts. Within the domain of control, one can be *looking for* an external object (e.g., R-1 controlling P-2) or for an idea or memory (e.g., R-1 controlling R-2). Within the domain of representational complexity, *comparing* can involve representational complexity when two or more perceptually derived representations are compared (e.g., similarity judgments between pictures), when two or more reflectively generated representations are compared (e.g., similarity judgments between abstract concepts such as awareness and control), or when perceptually derived and reflectively generated representations are compared (e.g., similarity judgments between an actress and the idea of the character one has imagined based on reading the play). In short, with relatively few as-

---

[5]Although the focus here is on complexity introduced by multiple representations, even for individual representations, complexity may increase from P-1 to R-2 in that reflective representations and level 2 (P-2 and R-2) representations tend to encompass and integrate more elements than perceptual and level 1 (P-1 and R-1) representations, respectively.

sumptions, the MEM architecture can yield the diverse subjective experiences implicit in what we mean by consciousness.

Consciousness and Attention

In MEM consciousness and attention are distinguished. As discussed elsewhere (Johnson & Hirst, 1993), "attention" is engaging component processes such as *locating, noting,* and so forth. There is no separate, single attentional mechanism (see also Allport, 1993). Component processes engaged by well-learned agendas within a subsystem would not necessarily give rise to a conscious experience (e.g., to take a familiar example, sometimes in driving to work, driving involves "attention" but is not necessarily conscious). Thus, consciousness can be distinguished from attention in that consciousness includes the idea that processes are engaged and monitored by another subsystem, whereas attention does not necessarily involved transactions across subsystems.

Evolution and Development of Consciousness

The evolution and development of consciousness can be represented in MEM as well. Assume that the cognitive system evolved in the order P-1, P-2, R-1, and R-2 and that development from infancy to adulthood proceeds in the same order. This means the conscious experiences of an organism functioning with, say, P-1 and P-2 only would be quite different from the conscious experiences of an organism functioning with R-1, which in turn would be different from the experiences of one functioning with R-2 as well. Note that in MEM, consciousness does not depend on a single mechanism—such as a single, centralized executive system or language. Rather, consciousness depends on the capacity for transactions between subsystems that have a certain architectural arrangement. Therefore, prelinguistic humans and other animals are presumed to have conscious experiences. Insofar as language helps us represent and manipulate multiple complex concepts, it increases the range of possible conscious experiences.

The Function of Consciousness

As we are using it here, *consciousness* is a summary term for the various phenomenal experiences arising from cognitive activities. Nevertheless, this does not mean that consciousness is "epiphenomenal." Because consciousness has phenomenal properties, it can serve the causal, *motivational* functions of phenomenal experiences—we act in order to bring into being, or to maintain, or to escape certain conscious experiences. For example, someone who is cooking engages in a rich mixture of thoughts and actions in creating a crème brulée in order to bring about a particular state of conscious experience later. In fact, much of what we do (drinking wine, going to the movies, snorkeling, concocting theories) is done to manipulate the aspects of consciousness we call awareness, control, and cognitive complexity.

Disruptions of Consciousness

The fact that different subsystems mediate different aspects of perception and reflection underlies the fascinating variety of phenomena that fall in the category of dissociations and disruptions of consciousness. In addition to the previously mentioned dissociations among memory measures in neurologically intact and amnesic patients and the characteristics of confabulation in brain-damaged patients (e.g., see Dalla Barba, 1993), there are such phenomena as blindsight (above chance ability of patients to guess the locations, presence or absence, or direction of motion of stimuli they do not feel that they see: Marcel, 1983; Weiskrantz, 1980), prosopagnosia (implicit recognition of faces that subjects do not appear to identify: Tranel & Damasio, 1985; Young & De Haan, 1992), agnosias (failing to recognize what an object is: Farah, 1990), dissociations of action and perception (e.g., normal motor adjustments of hand position in reaching when perception of the object is disrupted: Goodale, Milner, Jakobson, & Cary, 1991), neglect (e.g., failure to notice stimuli occurring on one side of space: Bisiach, 1988), and anosognosia (unawareness of deficit: Prigatano & Schacter, 1991). In the MEM framework, such phenomena occur as a consequence of selective breakdowns within subsystems or breakdowns in transactions between subsystems. Furthermore, disruptions within or between subsystems could come about in several ways and each should produce a somewhat different pattern of deficit, accounting for the wide range of symptoms observed.

For example, if P-2 processes were only partially disrupted (say, *placing* were disrupted), and assuming that R-1 or R-2 agendas could still engage in monitoring P-2, then the patient might have the experience of disordered perception (e.g., an agnosia) in which perception seems disorganized and piece-meal because parts of objects cannot be placed in relation to each other. Of course, the more component processes of P-2 that are disrupted, the more disorganized or incomplete the phenomenal experience. Blindsight may reflect a more general deficit within the P-2 subsystem, which disturbs the patient's ability to compute phenomenal representations of objects. Intact P-1 processes, however, would permit *locating, tracking* (motion detection), and some *resolving* of figure-ground information.

On the other hand, suppose that all the P-2 computations necessary to yield a representation (i.e., component processes of *identifying, placing, examining,* and *structuring*) are intact, but it is no longer possible for R-1 to monitor or control the relevant representations formed by P-2. For example, there may be a deficit in the *shifting* component process of R-1. Such a failed R-1/P-2 transaction would constitute a form of neglect or unawareness of a potential phenomenal object. Similarly, suppose that R-1 was unable to refresh representations in an area of P-2. *Refreshing* is a component process that prolongs activation—information is kept active for brief periods so that it remains highly available to be integrated with new incoming information. If components of a complex stimulus are not refreshed, they would fade quickly and not be available to be organized into stable, conscious percepts. Or, suppose reactivation of P-2 were disrupted. *Reactivation* is a component process by which information that has become inactive is revived to an active state; as suggested previously, anterograde amnesia could result from a disruption in the component process of reactivation (Johnson, 1992; Johnson & Chalfonte, 1994; Johnson & Hirst, 1991, 1993).

In cases of disrupted R-1 processes (e.g., *shifting, refreshing, reactivating*), if R-1 agendas were intact, then the patients would realize there was something wrong with their perception or with their memory. That is, they would have insight into their deficit because expectations generated by R-1 agendas would be experienced as unfulfilled; the *shift, refresh,* or *reactivation* signal would have been sent, but no return result received. Suppose instead, or in addition, the fault were in the agenda itself—it might be incapable of recruiting the *refreshing* or *reactivating* process. That is, no signal would be sent in the first place. Then the subject would not experience any unfulfilled intention (i.e., would not experience the absence of a return signal or an "error message"). In this case, the subject should show a type of unawareness of deficit or anosognosia. In short, how and which particular transactions were disrupted would determine the nature of both the deficit and whether the patient would or would not show insight about the deficit. Unawareness of deficit is a central issue for theories of consciousness, and we are only beginning to appreciate the implications of how specific anosognosia can be in particular cases (see Prigatano & Schacter, 1991).

Disrupted interactions between subsystems produce not only disorders of perception and memory, but also disorders of action and thought. For example, disruptions in interactions between reflection and perceptual subsystems could produce deficits in agenda-initiated action or conscious control. One such deficit might be the inability to voluntarily (i.e., through reflective agendas) activate motor and/or perceptual schemas, although they could still be activated through perceptual cues (Heilman & Rothi, 1985; McCarthy & Warrington, 1990). For example, a patient might not be able to mime how to use a hammer but might be able to show how to use it if given a hammer. Furthermore, given the critical role that interactions between R-1 and R-2 normally play in complex behaviors requiring planning, decision making, and problem solving, disruptions in either or both of these subsystems or in their interaction could have profound consequences. The growing body of evidence often subsumed under the idea of disrupted frontal lobe functions or "dys-executive syndrome" (e.g., Baddeley, 1986; Stuss & Benson, 1986; Shallice, 1988) are the type of problems that should result from disrupted R-1/R-2 interactions.

Finally, consider the implications of the idea that R-1 and R-2 monitoring and control of each other are in some fundamental way optional. This suggests that we might be able to engage in quite complex cognition and behavior with-out necessarily observing ourselves doing so. Dissociative states like hypnosis and multiple-personality disorder (Kihlstrom, 1985; Schacter & Kihlstrom, 1989) may reflect a capacity that most individuals have not developed for sus-pending the normally constant and recursive interaction between R-1 and R-2 processes, or a capacity for restricting the representational domains across which monitoring and control are done.

All disruptions of cognition are potentially informative about the nature of consciousness. However, the disruptions of cognition that seem most like dis-orders of consciousness, as opposed to disorders of perception or thought, are those that involve deficits in transactions between subsystems.

CONCLUSIONS

This chapter has considered issues of consciousness from the perspective of the MEM framework. From this point of view, awareness is the consequence of monitoring between subsystems and would include, as Marcel (1988) sug-gested, internal as well as external events as objects of consciousness. Simi-larly, Bisiach's (1988) "C-2" consciousness referred to the idea that some parts or processes of a system have access to other parts or processes, per-mitting the "monitoring of internal representations" (p. 12). As noted earlier, this general idea is embodied in most conceptions of a central executive.

A single, central executive system, however, cannot alone account for the variety in types of consciousness one experiences or the complex patterns of breakdown in consciousness that are possible. Recognizing this limitation, Stuss (1991) posited three monitoring systems at different levels of mental processes. Similarly, within MEM, there is not a single executive, rather there are sets of agendas organized in such a fashion that monitoring and control are possible across subsystems. Specific qualitative characteristics of the re-sulting interactions will determine the nature of the conscious experience. For example, the difference between control and self-control depends in part on what is the object-level target of the control. Complex phenomenal experi-ences such as deliberation and choice depend on being able to represent and keep active two or more alternatives. Furthermore, the recursive quality that consciousness can have—for example, thinking about ourselves, thinking about ourselves thinking about ourselves, and so on—is possible because we have two functional reflective subsystems capable of communicating (i.e., of mutual meta-processing) by taking turns as the object and meta-level. Thus, we do not need an infinite number of "homunculi," but only two cooperating subsystems to create the subjective experience of "layered" self-reflection.

The specific qualitative characteristics of our conscious experiences will depend on the nature of the information that is serving as the object level and on which subsystem is serving as the meta-level. All cognitive activity results in some sort of experience. But the subset of those experiences that have been most salient to philosophers and psychologists and are identified as Conscious experiences arise from the very structure of our cognitive architecture.

ACKNOWLEDGMENTS

Preparation of this chapter was supported by NIA Grants AG09253 and AG09744 and NIH Grant MH50131. We would like to thank Phil Johnson-Laird, Tom Nelson, and Carol Raye for comments on an earlier draft; Frank Durso and Mark Edwards for helping us consider alternative ways to analyze the "mind" words; and Tina Loose, Christine Voegeli, and Rebecca Ratner for collecting the similarity ratings.

REFERENCES

Allport, A. (1993). Attention and control: Have we been asking the wrong questions? A critical review of twenty-five years. In D. E. Meyer & S. Kornblum (Eds.), *Attention and performance XIV, Synergies in experimental psychology, artificial intelligence, and cognitive neuroscience* (pp. 183–218). Cambridge, MA: MIT Press.

Anderson, J. R. (1983). *The architecture of cognition.* Cambridge, MA: Harvard University Press.

Atkinson, R. C., & Shiffrin, R. M. (1968). Human memory: A proposed system and its control processes. In K. W. Spence & J. T. Spence (Eds.), *The psychology of learning and motivation: Advances in research and theory* (Vol. 2, pp. 89–195). New York: Academic Press.

Baars, B. J. (1988). *A cognitive theory of consciousness.* Cambridge, England: Cambridge University Press.

Baddeley, A. (1982). Amnesia: A minimal model and an interpretation. In L. S. Cermak (Ed.), *Human memory and amnesia* (pp. 305–336). Hillsdale, NJ: Lawrence Erlbaum Associates.

Baddeley, A. (1986). *Working memory.* (Oxford Psychology Series No. 11). London: Oxford University Press.

Baddeley, A. D. (1992a). Consciousness and working memory. *Consciousness and Cognition, 1,* 3–6.

Baddeley, A. D. (1992b). What is autobiographical memory? In M. A. Conway, D. C. Rubin, H. Spinnler, & W. Wagenaar (Eds.), *Theoretical perspectives on autobiographical memory* (pp. 13–29). The Netherlands: Kluwer.

Bartlett, F. C. (1932). *Remembering: A study in experimental and social psychology.* London: Cambridge University Press.

Benzing, W., & Squire, L. R. (1989). Preserved learning and memory in amnesia: Intact adaptation-level effects and learning of stereoscopic depth. *Behavioral Neuroscience, 103,* 548–560.

Biederman, I. (1987). Recognition-by-components: A theory of human image understanding. *Psychological Review, 94,* 115–147.

Bisiach, E. (1988). The (haunted) brain and consciousness. In A. J. Marcel & E. Bisiach (Eds.), *Consciousness in contemporary science* (pp. 101–120). Oxford, England: Clarendon.

Bransford, J. D., & Johnson, M. K. (1973). Considerations of some problems of comprehension. In W. Chase (Ed.), *Visual information processing* (pp. 383–438). New York: Academic Press.

Brewer, W. F. (1986). What is autobiographical memory? In D. C. Rubin (Ed.), *Autobiographical memory* (pp. 25–49). Cambridge, England: Cambridge University Press.

Brunswik, E. (1956). *Perception and the representative design of psychological experiments* (2nd ed.). Berkeley: University of California Press.

Calvin, W. H. (1989). *The cerebral symphony.* New York: Bantam.

Cermak, L. S. (1977). The contribution of a "processing" deficit to alcoholic Korsakoff patients' memory disorder. In I. M. Birnbaum & E. S. Parker (Eds.), *Alcohol and human memory* (pp. 195–208). Hillsdale, NJ: Lawrence Erlbaum Associates.

Chaiken, S., Lieberman, A., & Eagly, A. H. (1989). Heuristic and systematic information processing within and beyond the persuasion context. In J. S. Uleman & J. A. Bargh (Eds.), *Unintended thought* (pp. 212–252). New York: Guilford.

Churchland, P. S. (1988). Reduction and the neurobiological basis of consciousness. In A. J. Marcel & E. Bisiach (Eds.), *Consciousness in contemporary science* (pp. 273–304). Oxford, England: Clarendon.

Cohen, N. J., & Squire, L. R. (1980). Preserved learning and retention of pattern analyzing skill in amnesia: Dissociation of knowing how and knowing that. *Science, 210,* 207–210.

Craik, F. I. M., & Kirsner, K. (1974). The effects of speaker's voice on word recognition. *Quarterly Journal of Experimental Psychology, 26,* 274–284.

Craik, F. I. M., Morris, L. W., Morris, R. G., & Loewen, E. R. (1990). Relations between source amnesia and frontal lobe functioning in older adults. *Psychology and Aging, 5,* 148–151.

Dalla Barba, G. (1993). Confabulation: Knowledge and recollective experience. *Cognitive Neuropsychology, 10,* 1–20.

D'Andrade, R. (1987). A folk model of the mind. In D. Holland & N. Quinn (Eds.), *Cultural models in language and thought* (pp. 112–148). Cambridge, England: Cambridge University Press.

DeLoache, J. S., & Burns, N. M. (1994). Symbolic functioning in preschool children. *Journal of Applied Developmental Psychology, 15,* 513–527.

Dennett, D. C. (1991). *Consciousness explained.* Boston: Little, Brown.

DeSchepper, B., & Treisman, A. (1996). Visual memory for novel shapes. *Learning, Memory, and Cognition, 22,* 1–21.

Eich, E. (1984). Memory for unattended events: Remembering with and without awareness. *Memory and Cognition, 12,* 105–111.

Ekman, P. (1984). Expression and the nature of emotion. In K. Scherer & P. Ekman (Eds.), *Approaches to emotion* (pp. 319–343). Hillsdale, NJ: Lawrence Erlbaum Associates.

Farah, M. J. (1990). *Visual agnosia.* Cambridge, MA: MIT Press.

Flanagan, O. (1992). *Consciousness reconsidered.* Cambridge, MA: MIT Press.

Flavell, J. H., & Wellman, H. M. (1977). Metamemory. In R. V. Kail, Jr. & J. W. Hagen (Eds.), *Perspectives on the development of memory and cognition* (pp. 3–33). Hillsdale, NJ: Lawrence Erlbaum Associates.

Fowler, C. A., Wolford, G., Slade, R., & Tassinary, L. (1981). Lexical access with and without awareness. *Journal of Experimental Psychology: General, 110,* 341–362.

Gardiner, J. M., & Java, R. I. (1990). Recollective experience in word and nonword recognition. *Memory and Cognition, 18,* 23–30.

Gentner, D. (1988). Analogical inference and analogical access. In A. Prieditis (Ed.), *Analogica* (pp. 63–88). Los Altos, CA: Morgan Kaufmann.

Gibson, J. J. (1966). *The senses considered as perceptual systems.* Boston: Houghton.

Goodale, M. A., Milner, A. D., Jakobson, L. S., & Cary, D. P. (1991). A neurological dissociation between perceiving objects and grasping them. *Nature, 349,* 154–156.

Haber, R. N., & Hershenson, M. (1965). Effects of repeated brief exposures on the growth of a percept. *Journal of Experimental Psychology, 4,* 318–330.

Haist, F., Shimamura, A. P., & Squire, L. R. (1992). On the relationship between recall and recognition memory. *Journal of Experimental Psychology: Learning, Memory and Cognition, 18,* 691–702.

Hasher, L., & Zacks, R. T. (1979). Automatic and effortful processes in memory. *Journal of Experimental Psychology, 108,* 356–388.

Heilman, K. M., & Rothi, L. J. G. (1985). Apraxia. In K. M. Heilman & E. Valenstein (Eds.), *Clinical neuropsychology* (pp. 131–150). New York: Oxford University Press.

Hintzman, D. L. (1986). "Schema abstraction" in a multiple-trace memory model. *Psychological Review, 93,* 411–428.

Hintzman, D. L., & Summers, J. J. (1973). Long-term visual tracers of visually-presented words. *Bulletin of the Psychonomic Society, 1,* 325–327.

Hirst, W. (1982). The amnesic syndrome: Descriptions and explanations. *Psychological Bulletin, 91,* 435–460.

Hirst, W., Johnson, M. K., Kim, J. K., Phelps, E. A., Risse, G., & Volpe, B. (1986). Recognition and recall in amnesics. *Journal of Experimental Psychology: Learning, Memory and Cognition, 12,* 445–451.

Hirst, W., Johnson, M. K., Phelps, E. A., & Volpe, B. T. (1988). More on recognition and recall in amnesia. *Journal of Experimental Psychology: Learning, Memory and Cognition, 14,* 758–762.

Holyoak, K., & Thagard, P. (1989). Analogical mapping by constraint satisfaction. *Cognitive Science, 13,* 295–355.

Huppert, F. A., & Piercy, M. (1982). In search of the functional locus of amnesic syndromes. In L. S. Cermak (Ed.), *Human memory and amnesia.* Hillsdale, NJ: Lawrence Erlbaum Associates.

Jackendoff, R. (1987). *Consciousness and the computational mind.* Cambridge, MA: MIT Press.

Jacoby, L. L. (1991). A process dissociation framework: Separating automatic from intentional uses of memory. *Journal of Memory and Language, 30,* 513–541.

Jacoby, L. L., & Dallas, M. (1981). On the relationship between autobiographical memory and perceptual learning. *Journal of Experimental Psychology: General, 110,* 306–340.

Johnson, M. K. (1983). A multiple-entry, modular memory system. In G. H. Bower (Ed.), *The psychology of learning and motivation* (Vol. 17, pp. 81–123). New York: Academic Press.

Johnson, M. K. (1988a). Discriminating the origin of information. In T. F. Oltmanns & B. A. Maher (Eds.), *Delusional beliefs* (pp. 34–65). New York: Wiley.

Johnson, M. K. (1988b). Reality monitoring: An experimental phenomenological approach. *Journal of Experimental Psychology: General, 117,* 390–394.

Johnson, M. K. (1990). Functional forms of human memory. In J. L. McGaugh, N. M. Weinberger, & G. Lynch (Eds.), *Brain organization and memory: Cells, systems and circuits* (pp. 106–134). New York: Oxford University Press.

Johnson, M. K. (1991a). Reality monitoring: Evidence from confabulation in organic brain disease patients. In G. P. Prigatano & D. L. Schacter (Eds.), *Awareness of deficit after brain injury: Clinical and theoretical issues* (pp. 176–197). New York: Oxford University Press.

Johnson, M. K. (1991b). Reflection, reality monitoring and the self. In R. Kunzendorf (Ed.), *Mental imagery: Proceedings of the 11th Annual Conference of the American Association for the Study of Mental Imagery* (pp. 3–16). New York: Plenum.

Johnson, M. K. (1992). MEM: Mechanisms of recollection. *Journal of Cognitive Neuroscience, 4,* 268–280.

Johnson, M. K., Bransford, J. D., & Solomon, S. K. (1973). Memory for tacit implications in sentences. *Journal of Experimental Psychology, 98,* 203–205.

Johnson, M. K., & Chalfonte, B. L. (1994). Binding complex memories: The role of reactivation and the hippocampus. In D. L. Schacter & E. Tulving (Eds.), *Memory systems 1994* (pp. 311–350). Cambridge, MA: MIT Press.

Johnson, M. K., Foley, M. A., & Leach, K. (1988). The consequences for memory of imagining in another person's voice. *Memory and Cognition, 16,* 337–342.

Johnson, M. K., Foley, M. A., Suengas, A. G., & Raye, C. L. (1988). Phenomenal characteristics of memories for perceived and imagined autobiographical events. *Journal of Experimental Psychology: General, 117,* 371–376.

Johnson, M. K., Hashtroudi, S., & Lindsay, D. S. (1993). Source monitoring. *Psychological Bulletin, 114*(1), 3–28.

Johnson, M. K., & Hirst, W. (1991). Processing subsystems of memory. In R. G. Lister & H. J. Weingartner (Eds.), *Perspectives in cognitive neuroscience* (pp. 197–217). New York: Oxford University Press.

Johnson, M. K., & Hirst, W. (1993). MEM: Memory subsystems as processes. In A. Collins, S. Gathercole, M. Conway, & P. Morris (Eds.), *Theories of memory* (pp. 241–286). Hillsdale, NJ: Lawrence Erlbaum Associates.

Johnson, M. K., Kim, J. K., & Risse, G. (1985). Do alcoholic Korsakoff's syndrome patients acquire affective reactions? *Journal of Experimental Psychology: Learning, Memory and Cognition, 11,* 22–36.

Johnson, M. K., & Multhaup, K. S. (1992). Emotion and MEM. In S. A. Christianson (Ed.), *The handbook of emotion and memory: Current research and theory* (pp. 33–66). Hillsdale, NJ: Lawrence Erlbaum Associates.

Johnson, M. K., Peterson, M. A., Chua-Yap, E., & Rose, P. (1989). Frequency judgments and the problem of defining a perceptual event. *Journal of Experimental Psychology: Learning, Memory and Cognition, 15,* 126–136.

Johnson, M. K., & Raye, C. L. (1981). Reality monitoring. *Psychological Review, 88,* 67–85.

Johnson-Laird, P. (1983). *Mental models.* Cambridge, MA: Harvard University Press.

Johnson-Laird, P. (1988). *The computer and the mind.* Cambridge, MA: Harvard University Press.

Kail, R. (1984). *The development of memory in children* (2nd ed.). New York: Freeman.

Kihlstrom, J. F. (1984). Conscious, subconscious, unconscious: A cognitive perspective. In K. S. Bowers & D. Meichenbaum (Eds.), *The unconscious reconsidered* (pp. 149–211). New York: Wiley-Interscience.

Kihlstrom, J. F. (1985). Hypnosis. *Annual Review of Psychology, 36,* 385–418.

Kihlstrom, J. F. (1987). The cognitive unconscious. *Science, 237,* 1445–1452.

Kihlstrom, J. F., Barnhardt, T. M., Tataryn, D. J. (1992). The psychological unconscious: Found, lost, and regained. *American Psychologist, 47,* 788–791.

Kintsch, W., & Bates, E. (1977). Recognition memory for statements from a classroom lecture. *Journal of Experimental Psychology: Human Learning and Memory, 3,* 150–168.

Klatzky, R. L. (1984). *Memory and awareness.* New York: Freeman.

Kolers, P. A., & Roediger, H. L., III. (1984). Procedures of mind. *Journal of Verbal Learning and Verbal Behavior, 23,* 425–449.

Kopelman, M. D. (1989). Remote and autobiographical memory, temporal context memory and frontal atrophy in Korsakoff and Alzheimer patients. *Neuropsychologia, 27,* 437–460.

Kunst-Wilson, W. R., & Zajonc, R. B. (1980). Affective discrimination of stimuli that cannot be recognized. *Science,* 557–558.

Lazarus, R. S. (1982). Thoughts on the relations between emotion and cognition. *American Psychologist, 37,* 1019–1024.

Lindsay, D. S., & Johnson, M. K. (1989). The eyewitness suggestibility effect and memory for source. *Memory and Cognition, 17,* 349–358.

Mandler, G. (1967). Organization and memory. In K. W. Spence & J. T. Spence (Eds.), *The psychology of learning and motivation* (Vol. 1, pp. 327–372). New York: Academic Press.

Mandler, G. (1980). Recognizing: The judgment of previous occurrence. *Psychological Review, 87,* 252–271.

Mandler, G. (1984). *Mind and body: Psychology of emotion and stress.* New York: Norton.

Marcel, A. J. (1983). Conscious and unconscious perception: An approach to the relations between phenomenal experience and perceptual processes. *Cognitive Psychology, 15,* 238–300.

Marcel, A. J. (1988). Phenomenal experience and functionalism. In A. J. Marcel & E. Bisiach (Eds.), *Consciousness in contemporary science* (pp. 121–158). Oxford, England: Clarendon.

McCarthy, R. A., & Warrington, E. K. (1990). *Cognitive neuropsychology: A clinical introduction.* New York: Academic Press.

Metcalfe Eich, J. (1985). Levels of processing, encoding specificity, elaboration, and CHARM. *Psychological Review, 92,* 1–38.

Miller, G. A. (1956). The magical number seven plus or minus two: Some limits on our capacity for processing information. *Psychological Review, 63,* 81–97.

Miller, G. A., Galanter, E., & Pribram, K. H. (1960). *Plans and the structure of behavior.* New York: Holt, Rinehart & Winston.

Mishkin, M. A. (1982). A memory system in the monkey. *Philosophical Transaction of the Royal Society of London, 298B,* 85–95.

Moscovitch, M. (1982). Multiple dissociation of function in amnesia. In L. S. Cermak (Ed.), *Human memory and amnesia* (pp. 337–370). Hillsdale, NJ: Lawrence Erlbaum Associates.

Moscovitch, M. (1989). Confabulation and the frontal systems: Strategic versus associative retrieval in neuropsychological theories of memory. In H. L. Roediger III & F. I. M. Craik (Eds.), *Varieties of memory and consciousness: Essays in honour of Endel Tulving* (pp. 133–160). Hillsdale, NJ: Lawrence Erlbaum Associates.

Moscovitch, M. (1992). Memory and working with memory: A component process model based on modules and central systems. *Journal of Cognitive Neuroscience, 4,* 257–267.

Moscovitch, M., & Umiltà, C. (1991). Conscious and nonconscious aspects of memory: A neuropsychological framework of modules and central systems. In R. G. Lister & H. J. Weingartner (Eds.), *Perspectives on cognitive neuroscience* (229–266). New York: Oxford University Press.

Musen, G., & Treisman, A. (1990). Implicit and explicit memory for visual patterns. *Journal of Experimental Psychology: Learning, Memory, and Cognition, 16,* 127–137.

Musen, G., & Squire, L. R. (1992). Nonverbal priming in amnesia. *Memory and Cognition, 17,* 1095–1104.

Nelson, T. O., & Narens, L. (1990). Metamemory: A theoretical framework and some new findings. In G. H. Bower (Ed.), *The psychology of learning and motivation* (Vol. 26, pp. 125–173). New York: Academic Press.

Newell, A., Rosenbloom, P. S., & Laird, J. E. (1989). Symbolic architectures for cognition. In M. I. Posner (Ed.), *Foundations of cognitive science* (pp. 93–131). Cambridge, MA: MIT Press.

Norman, D. A., & Shallice, T. (1986). Attention to action: Willed and automatic control of behavior. In R. J. Davidson, G. E. Schwartz, & D. Shapiro (Eds.), *Consciousness and self-regulation* (Vol. 4, pp. 1–18). New York: Plenum.

Phelps, E. A. (1989). *Cognition skill learning in amnesics.* Unpublished doctoral dissertation, Princeton University, Princeton, NJ.

Popper, K. R., & Eccles, J. C. (1977). *The self and its brain.* Boston: Routledge & Kegan Paul.

Posner, M. I., Petersen, S. E., Fox, P. T., & Raichle, M. E. (1988). Localization of cognitive operations in the human brain. *Science, 240,* 1627–1631.

Posner, M. I., & Snyder, C. R. R. (1975). Attention and cognitive control. In R. L. Solso (Ed.), *Information processing and cognition: The Loyola Symposium* (pp. 55–85). Hillsdale, NJ: Lawrence Erlbaum Associates.

Prigatano, G. Schacter, D. L. (1991). *Awareness of deficit after brain injury* (pp. 176–197). New York: Oxford University Press.

Reber, A. S. (1985). *The Penguin dictionary of psychology.* New York: Viking Press.

Reber, A. S. (1989). Implicit learning and tacit knowledge. *Journal of Experimental Psychology: General, 118,* 219–235.

Reiser, B. J. (1986). The encoding and retrieval of memories of real-world experiences. In J. A. Galambos, R. P. Abelson, & J. B. Black (Eds.), *Knowledge structures* (pp. 71–99). Hillsdale, NJ: Lawrence Erlbaum Associates.

Richardson-Klavehn, A., & Bjork, R. A. (1988). Measures of memory. *Annual Review of Psychology, 39,* 475–543.

Rips, L. J., & Conrad, F. G. (1989). Folk psychology of mental activities. *Psychological Review, 96,* 187–207.

Rock, I., & Gutman, D. (1981). The effect of inattention on form perception. *Journal of Experimental Psychology: Human Perception and Performance, 7,* 275–285.

Roseman, I. J. (1984). Cognitive determinants of emotion: A structural theory. In P. Shaver (Ed.), *Review of personality and social psychology* (pp. 11–36). Beverly Hills, CA: Sage.

Rumelhart, D. E., & McClelland, J. L. (1986). *Parallel distributed processing: Vol. 1. Exploration in the microstructure of cognition.* Cambridge, MA: MIT Press.

Scarborough, D., Cortese, C., & Scarborough, H. (1977). Frequency and repetition effects in lexical memory. *Journal of Experimental Psychology: Human Perception and Performance, 3,* 1–17.

Schacter, D. L. (1989). On the relation between memory and consciousness: Dissociable interactions and conscious experience. In H. L. Roediger III & F. I. M. Craik (Eds.), *Varieties of memory and consciousness: Essays in honour of Endel Tulving* (pp. 355–389). Hillsdale, NJ: Lawrence Erlbaum Associates.

Schacter, D. L. (1990). Perceptual representation systems and implicit memory. *Annals of the New York Academy of Sciences, 608,* 543–571.

Schacter, D. L., Cooper, L. A., Tharan, M., & Rubens, A. B. (1991). Preserved priming of novel objects in patients with memory disorders. *Journal of Cognitive Neuroscience, 3,* 118–131.

Schacter, D. L., & Kihlstrom, J. F. (1989). Functional amnesia. In F. Boller & J. Grafman (Eds.), *Handbook of neuropsychology* (Vol. 3, pp. 209–231). New York: Elsevier.

Schacter, D. L., & Moscovitch, M. (1984). Infants, amnesics, and dissociable memory systems. In M. Moscovitch (Ed.), *Infant memory* (pp. 173–216). New York: Plenum.

Searle, J. R. (1992). *The rediscovery of the mind.* Cambridge, MA: MIT Press.

Seger, C. A. (1994). Implicit learning. *Psychological Bulletin, 115,* 163–196.

Sejnowski, T., & Rosenberg, C. (1987). Parallel networks that learn to pronounce English text. *Complex Systems, 1,* 145–168.

Shallice, T. (1978). The dominant action system: An information-processing approach to consciousness. In K. S. Pope & J. L. Singer (Eds.), *The stream of consciousness: Scientific investigations into the flow of human experience* (pp. 117–157). New York: Plenum.

Shallice, T. (1988). *From neuropsychology to mental structure.* New York: Cambridge University Press.

Shimamura, A. P., Janowsky, J. S., & Squire, L. R. (1990). Memory for the temporal order of events in patients with frontal lobe lesions and amnesic patients. *Neuropsychologia, 28,* 803–813.

Simon, H. A. (1989). *Models of thought* (Vol. 2). New Haven, CT: Yale University Press.

Smith, C. A., & Ellsworth, P. C. (1985). Patterns of cognitive appraisal in emotion. *Journal of Personality and Social Psychology, 48,* 813–838.

Squire, L. R. (1987). *Memory and brain.* New York: Oxford University Press.

Squire, L. R. (1992). Memory and the hippocampus: A synthesis from findings with rats, monkeys, and humans. *Psychological Review, 99*(2), 195–231.

Stromeyer, C. F., III, & Psotka, J. (1970). The detailed texture of eidetic images. *Nature* (London), *225,* 346–349.

Stuss, D. T. (1991). Self, awareness and the frontal lobes: A neuropsychological perspective. In J. Strauss & G. R. Goethals (Eds.), *The self: Interdisciplinary approaches* (pp. 255–278). New York: Springer-Verlag.

Stuss, D. T., & Benson, D. F. (1986). *The frontal lobes.* New York: Raven.

Tranel, D., & Damasio, A. (1985). Knowledge without awareness: An autonomic index of facial recognition by prosopagnosics. *Science, 228,* 1453–1454.

Tulving, E. (1962). Subjective organization in free recall of "unrelated" words. *Psychological Review, 69,* 344–354.

Tulving, E. (1968). Theoretical issues in free recall. In T. R. Dixon & D. L. Horton (Eds.), *Verbal behavior and general behavior theory* (pp. 2–36). New York: Prentice-Hall.

Tulving, E. (1985). Memory and consciousness. *Canadian Psychology, 26,* 1–12.

Tulving, E., Kapur, S., Craik, F. I. M., Moscovitch, M., & Houle, S. (1994). Hemispheric encoding/retrieval asymmetry in episodic memory: Position emission tomygraphy findings. *Proceedings of the National Academy of Sciences, 91,* 2016–2020.

Tulving, E., & Schacter, D. L. (1990). Priming and human memory systems. *Science, 247,* 301–396.

Umiltà, C. (1988). The control operations of consciousness. In A. J. Marcel & E. Bisiach (Eds.), *Consciousness in contemporary science* (pp. 334–356). Oxford, England: Clarendon.

Vallar, G., & Baddeley, A. D. (1984). Fractionation of working memory: Neuropsychological evidence for a phonological short-term store. *Journal of Verbal Learning and Verbal Behavior, 23* (2), 151–161.

Verfaellie, M., & Treadwell, J. R. (1993). Status of recognition memory in amnesia. *Neuropsychology, 7,* 5–13.

Warrington, E. K. (1982). The double dissociation of short- and long-term memory deficits. In L. S. Cermak (Ed.), *Human memory and amnesia* (pp. 61–76). Hillsdale, NJ: Lawrence Erlbaum Associates.

Warrington, E. K., & Weiskrantz, L. (1970). Amnesic syndrome: Consolidation or retrieval? *Nature (London), 228,* 628–630.

Weinstein, A. (1987). *Preserved recognition memory in amnesia.* Unpublished doctoral dissertation, State University of New York at Stony Brook.

Weiskrantz, L. (1980). Varieties of residual experience. *Quarterly Journal of Experimental Psychology, 210,* 207–209.

Weiskrantz, L., Warrington, E. K., & Sanders, M. D. (1974). Visual capacity in the hemianopic field following a restricted occipital ablation. *Brain, 97,* 709–728.

Wickelgren, W. A. (1979). Chunking and consolidation: A theoretical synthesis of semantic networks, configuring in conditioning, S–R versus cognitive learning, normal forgetting, the amnesic syndrome, and the hippocampal arousal system. *Psychological Review, 86,* 44–60.

Young, A. W., & De Haan, E. H. F. (1992). Face recognition and awareness after brain injury. In A. D. Milner & M. D. Rugg (Eds.), *The neuropsychology of consciousness* (pp. 69–90). London: Academic Press.

CHAPTER 14

Why the Mind Wanders

Daniel M. Wegner
University of Virginia

The essential achievement of the will . . . is to attend to a difficult object and hold it fast before the mind.
—William James, *Principles of Psychology* (1890, p. 266)

The mind wanders. Ideally, it does not wander so far that it forgets it is reading this chapter. But consciousness does have an inevitable drift, changing its contents moment by moment. The focus seems to move relentlessly, shimmering and fidgeting no matter how hard we may try to concentrate on a thought, preserve an image, or otherwise freeze the instant. Not only does it seem quite impossible to hold a particular thought or percept fully in mind for an indefinite period, it also seems futile to attempt to keep consciousness away from a chosen target by fixing our minds on something else. Consciousness simply cannot hold itself still.

The persistent flow of consciousness prompted James (1890) to use the descriptive metaphor of a "stream," and this quality has been recognized by contemporary commentators as one of the key phenomena of consciousness that must be explained (e.g., Baars, 1988; Johnson-Laird, 1988). Why, after all, must it be like this? Why can't we just push the psychological equivalent of a "still frame" button on a videotape recorder and stop all this wriggling and hopping about? The purpose of this chapter is to suggest a theoretical framework within which the necessity of a wandering consciousness is made clear, and through which predictions can be made about where the mind will wander. In this framework, the constant metamorphosis of consciousness turns

295

out to be a natural product of the mechanism that allows consciousness to control itself.

First, the chapter considers several explanations of the wandering mind, with a view toward clarifying what it is that must be explained and what prior explanations have contributed. Then, mechanisms of the self-control of consciousness are explored from the perspective of the theory of *ironic processes of mental control* (Wegner, 1994). The implications of this theory are traced for the way that wandering happens in two forms of the intentional fixation of consciousness—intentional concentration and intentional thought suppression—and evidence is offered from laboratory studies of these cases. This discussion leads to a consideration of the peculiar fact that wandering typically moves against our attempts to control consciousness, not just in random directions. The chapter concludes with a consideration of brain functions in these processes. Mindwandering, as becomes evident, is a conscious manifestation of contrary unconscious processes created when we attempt to control the direction of consciousness.

PERSPECTIVES ON MENTAL WANDERING

There is no shortage of observations by psychologists that the mind does wander (see, e.g., Bills, 1931; Giambra, 1991; Vallacher & Nowak, 1994). James (1890, Vol. 2, p. 421) remarked forcefully that "*no one can possibly attend continuously to an object that does not change,*" and offered an account of this that was already traditional and widely accepted in his era (cf. Carpenter, 1875). He held that attention has both voluntary and involuntary manifestations, and that any voluntary or willful direction of the mind could be overcome by the powerful and involuntary attraction of attention by other objects. This approach suggests that each movement of attention can be chalked up alternatively to voluntary mental control or to involuntary environmental guidance.

The environment does guide the mind quite effectively, as one can easily attest in viewing a stirring film or listening to an absorbing piece of music. These instances were appreciated by James as objects that do change, and it is no doubt true that many cases of mental movement occur as the result of the guidance of mind by changing stimulation. The mind can wander even from these attractions, however, suggesting that the distinction between voluntary and involuntary attraction of attention can even be applied to changing stimulation. The Jamesian account would suggest that voluntary control of attention is simply incomplete, and so can be overridden by involuntary forces whether attention is being focused voluntarily on some stationary object or has been voluntarily attached to some moving environmental event. This perspective leaves open the question of what it is that creates the regular involuntary movement of attention, short of saying that certain objects are just more naturally attractive than anything to which we might voluntarily attend.

The Jamesian approach thus accounts for what draws attention, but fails to account for why attention must be drawn every few moments. It does not explain why voluntary attention is so perpetually weak in the face of involuntary attention. It seems plausible, for example, that in a sufficiently boring environment the frequency of involuntary attractions could be reduced to any arbitrarily low level, and as a result, voluntary consciousness could be made to fix steadily for lengthy periods of time. This is not, however, what happens. In conditions of slightly reduced sensory input, the mind continues to wander, often with greater vigor than during normal input (e.g., Klinger, 1978; Pope, 1978). Even when sensory stimulation is almost fully obstructed during sleep, wandering continues in the form of dreaming. Indeed, dreaming is such raucous wandering that it begins to seem that the environment is more of a help in the prevention of wandering than the culprit behind it (Hobson, 1988). Wandering is not just the result of weakness of will in the face of absorbing environmental stimulation, but rather is compelled somehow, perhaps even required, by the architecture of the mind.

If the processes of mind dictate wandering, then useful purposes might be served. It is possible, for example, that wandering might be built into the functioning of consciousness as a means of preventing debilitating habituation. In suggesting this possibility, Baars (1988) noted that the wandering of consciousness could serve the same sort of purpose that physiological nystagmus or eye tremor serves in keeping the sensors of the eye fresh and sensitive to experience. Just as redundant stimulation leads to habituation of sensory structures, redundancy in conscious experience might lead to the habituation of consciousness and so to insensitivity. Effects such as semantic satiation (Amster, 1964; the tendency of a word frequently repeated to seem different or somehow meaningless) and repetition blindness (Kanwisher & Potter, 1990; the tendency not to detect or recall repetitions of words in rapid serial presentation) suggest that habituation could be a danger not only at neural or sensory levels but also at semantic or conceptual levels of conscious experience. In other words, perhaps the mind wanders to keep it from getting weary with monotony.

This sort of functional, teleological explanation of wandering is not fully satisfying, however, because it offers no suggestion of a mechanism whereby wandering is achieved. Like postulating that the mind wanders because of a need for variety or simply because it is alive, the habituation–prevention theory does not allow prediction of the course of wandering. Baars (1988, p. 205) made an effort in this direction by proposing a model that traces conscious wandering to certain nonconscious processors that stop being interested in the conscious contents. But this is an explanation of conscious wandering in terms of the wandering of something else, and the wandering of these lower level units then begs for explanation. Although understanding wandering as "refreshment" of the system may still be useful (see Vallacher & Nowak, 1994), this approach has not yet engendered a causal theory.

One step toward a causal theory of mindwandering is suggested in the work of Jonides (1981). He presented subjects with arrows in various parts of the visual field as a means of directing their attention and found that peripheral arrows are processed more automatically than central ones. That is, when given instructions to ignore an attention-directing cue, subjects were able to comply when the cue appeared in the center of the display, but they were less able to do so when the cue appeared in the periphery. This suggests that the mind may wander because things that are not in its focus are inevitably more compelling than those in its focus. The evidence provided by Jonides pertains only to visual attention, of course, and it does not suggest why items on the periphery might be more likely to draw attention automatically than those in the center of the mind. But this finding provides a good hint, which is carried further in the theory of ironic processes of mental control (Wegner, 1994).

WANDERING AS IRONY

The mind wanders when we want to control it. The peculiar next step that the ironic process theory suggests is that the mind wanders as the result of our attempts to control it. Although this assertion may sound suspiciously like an Eastern religious insight into the achievement of mental peace (see Taylor, 1978), it arises from a decidedly Western scientific analysis of the self-control of mental states.

The theory begins with the supposition that consciousness can control itself. It is the nature of this control that conscious preferences for mental states that appear in mind at one time (e.g., preferences to concentrate on something, to avoid thinking about a painful sensation, to get into a better mood, etc.) can function to create those preferred states at a subsequent time. This much has been surmised in a number of theories of the self-control of consciousness (e.g., Carver & Scheier, 1982; Logan, 1985; Uleman, 1989; Wegner, 1989), and the ability of consciousness to control itself has been mentioned as one of its defining features (Lefebvre-Pinard, 1982; Oatley, 1988; Umiltà, 1988). Indeed, there is a growing research literature suggesting that mental control is a useful construct for understanding many domains of psychology (Wegner & Pennebaker, 1993).

The ironic process theory offers the idea that each instance of mental control is implemented through the production of a control system that consists of two processes. These include an *intentional operating process* that searches for mental contents yielding the desired state, and an *ironic monitoring process* that searches for mental contents signaling the failure to achieve the desired state. The control of anything involves changing it to a certain criterion, after all, and processes are thus needed to provide both the change and the assessment of progress in reaching the criterion. The two processes suggested here thus resemble the "operate" and "test" units traditionally included as components of control systems (Miller, Galanter, & Pribram, 1960; Powers, 1973) or production systems (Newell & Simon, 1972).

The intentional operating process is what we sense as our conscious activity when we exert mental control. Imagine, for example, deciding to attend to the period at the end of this sentence. The intentional operating process searches for the period. If we are not looking at the period, then the operating process is what finds the period; if we are looking at the period, then the operating process is the effortful attempt to continue looking at it and thinking about it. Such an operating process takes effort and remains in awareness during its operation. Thus, it has some of the properties normally associated with conscious or "controlled" mental processes (Bargh, 1984, 1989; Hasher & Zacks, 1979; Logan, 1988; Posner & Snyder, 1975; Shiffrin & Schneider, 1977). Because this operating process absorbs cognitive capacity, it is susceptible to interference from distraction and can easily be sidetracked or terminated. Fortunately, there is a monitoring process to keep track of this.

The ironic monitoring process is not normally sensed as part of the activity of mental control, as its functioning is unconscious and relatively less demanding of mental effort. In this sense, it resembles an automatic cognitive process (cf. Wegner, 1992). Unlike the intentional operating process, the monitor does not come and go over time with variations in the allocation of mental effort, and instead stands continually watchful of lapses in the intended control as long as the intention to engage in control is in effect. In the case of the intention to concentrate on the period at the end of this sentence, for instance, the monitor would search for any item that was not the period (e.g., noises in the next room, thoughts of lunch, etc.). The monitor searches for failures of control by examining preconscious mental contents arising from memory and/or sensation, and when items indicating failed control are found it restarts the operating process. In this way, the cyclic interplay of the operating and monitoring processes implements the intended control and we concentrate our attention on the dot.

The watchfulness of the monitor is also the source of ironic effects, however, and it is in this sense that the monitor is an *ironic* process. Because the monitor searches for potential mental contents that signal failure of mental control, it increases the accessibility of these contents to consciousness (cf. Higgins, 1989). Just like an externally encountered prime, the ironic monitor increases the likelihood that the primed content will enter the conscious mind and become available for report. In the usual functioning of the operating and monitoring processes, of course, the ironic monitor is relatively less effective than the conscious operator in introducing items to consciousness. The conscious operating process prevails by and large, and the ironic monitor primarily serves its watchdog function. So, we watch the dot for the most part, only occasionally to glance off or think of other things.

The ironic process theory suggests that the mind does not just *wander*, then, but rather that it is alternating between intentional and ironic contents. The intentional operating process and the ironic monitoring process both act to increase the accessibility of their associated search targets to consciousness, as both processes bring items from preconscious sources in memory and sensation into consciousness as part of their usual functioning. Each of them acts as a conduit of sorts between what could be conscious and what is conscious. When there is plentiful mental capacity, the intentional operating process can be very effective, and so will largely dominate consciousness with its output and balance any sensitivity produced by the monitor. The monitor runs continually once the intention to control the mind has been implemented, however, and for this reason it can create wanderings even when the operating process is performing well under conditions of full mental capacity. More commonly, however, it is when the operating process is undermined by other processes that also consume cognitive resources that the ironic monitoring process is uncovered to yield significant episodes of the ironic wandering of mind.

IRONIC WANDERING OF CONCENTRATION

If the mind wanders because of ironic processes, then it should be possible to enhance the wandering with the imposition of a very direct manipulation. When a distracter or cognitive load is imposed during concentration, the intentional operating process should be undermined and the ironic monitoring process should have relatively greater influence. With load, then, a person who is trying to concentrate should experience excessive wandering. Such wandering would not merely take the form of a reduction of processing of the target of concentration—although that should certainly happen as a result of interference with the intentional operating process. Wandering should also take the form of a relative release of the ironic monitor. Irrelevant items that are not the intended target of concentration, and are instead the focus of the ironic monitor, should become more accessible to consciousness as the result of the mental load.

In a way, the theory suggests that mental loads should produce a paradoxical state of mind in which unattended items are especially accessible. With a load, we should all be in the position of Alice in *Through the Looking Glass:* "The shop seemed to be full of all manner of curious things—but the oddest part of it all was that, whenever she looked hard at any shelf, to make out exactly what it had on it, that particular shelf was always quite empty, though the others round it were crowded as full as they could hold" (Carroll, 1982/1872, p. 175).

One test of whether such a state of mind might be induced experimentally was made by Wegner and Erber (1991), who invited subjects to study a map containing the names of 40 unfamiliar African cities. Subjects were asked to concentrate their attention on half the cities (e.g., those highlighted in yellow on the map) and not on the other half (highlighted in blue) because a later test would ostensibly cover only the yellow ones. During the study period, cognitive load was varied in that some subjects were given a 9-digit number to hold in mind and recall at the end of their studying, whereas others were given no number. After studying and then spending some time on a filler task, all subjects completed a recognition test for the entire map in which they were to indicate on a list whether each of the cities (as well as 40 other ones) had appeared on the map.

The obvious expectation for this study was that cognitive load would reduce memory for the cities that were the target of concentration, and this was indeed found. Subjects under load who were concentrating on the highlighted cities recognized them less well (by a recognition index of hits minus false alarms) than did those with no load. The less obvious but more specifically ironic effect was also observed: Subjects under load who were concentrating on the highlighted cities later recognized more of the other set of cities than did those who were not under load (see Fig. 14.1). This could be explained by suggesting that the load manipulation undermined the operating process, and

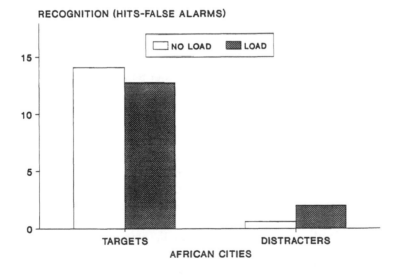

FIG. 14.1.   Recognition of target and nontarget items following study with or without mental load. Based on data from Wegner and Erber (1991).

thus allowed the monitoring process to come forward and produce the ironic effect. Subjects trying to concentrate under mental load, in other words, ended up memorizing the distracters.

Zukier and Hagen (1978) reported a parallel result in their study of the effects of distraction on learning in children. In their research, distracting conditions were found to enhance recall of incidental information while reducing recall of task-relevant information. These studies lend some credence to the insight offered by one of my students on the irony of concentration in cramming for an exam. He noted that when he arrived at an exam with "just a few last things to look over" in the stressful moments before the test, he ended up not only failing to concentrate on the items, but unfortunately, hearing with near superhuman acuity all the conversations going on at each side. It may well be that the intention to concentrate creates conditions under which mental load enhances the monitoring of irrelevancies.

Taken together, however, these studies do not substantiate this contention very effectively. Specifically, they fail to rule out the possibility that subjects under load or distraction simply forget the task instructions and so attend more often to irrelevant items. It could be, after all, that distractions muddy the distinction between to-be-attended and to-be-ignored items, and greater processing of the to-be-ignored items enhances memory for them. It is still remarkable in some sense that adding a memory load can increase subjects' memory for anything, especially incidental items, but it is not clear that these findings necessitate postulating an ironic monitoring process.

Stronger evidence for the ironic monitoring view of concentration comes from research on the automatic accessibility of concentration targets in the Stroop (1935) interference paradigm (Wegner, Erber, & Zanakos, 1993, Exp. 2). In this study, subjects were asked to imagine a personal episode that resulted in a success or a failure, and to write 5 to 6 sentences about it in a 5-min period. Then, they were asked to spend another 5-min period either thinking about that episode or trying not to think about it. As they continued to follow this instruction, all subjects then performed a Stroop task at a computer monitor, responding with different keypresses to signal whether words appearing on the screen were in red or blue. In addition, as a manipulation of cognitive load, prior to each word presentation either a 5-digit or a 2-digit number appeared on the screen for the subject to remember during the trial and report aloud afterward. The words appearing on the screen included 8 occurrences of the target ("success" or "failure") embedded with 64 nontarget words unrelated to success or failure.

Now as a rule, the latency to name colors in this situation is interpreted as a sign of cognitive accessibility of the meaning of the word. Just as one might hesitate ever so slightly in color naming if one's own name appeared as the word on the screen, one hesitates in naming the color of other words that are highly accessible. The ironic process prediction for the "think" condition in

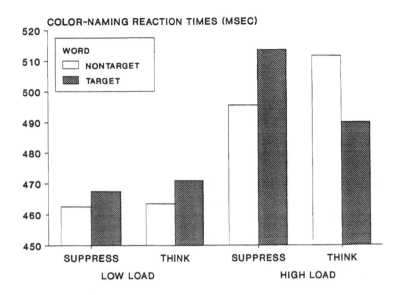

FIG. 14.2.   Color-naming reaction times (msec) for target and nontarget words during instructions to suppress or to think about the target under high or load mental load. Based on data from Wegner, Erber, and Zanakos (1993). ©1993 by American Psychological Association.

this experiment, then, is that subjects trying to think of their target who are given a high cognitive load should show greater accessibility of nontarget words than target words. As shown in Figure 14.2, this is exactly what was found. This difference was not observed under conditions of low cognitive load, or under conditions of thought suppression. It appears, then, that the attempt to concentrate on a target increases the person's sensitivity to anything that is not the target. This finding is not susceptible to the argument noted for the earlier studies—that load simply makes people forget the task, as in this case it was found that interference for nontarget words was significantly greater than for target words. A task-forgetting interpretation would only predict parity for these conditions.

The results for thought suppression in this study are also remarkable. This research indicates that trying not to think about something can increase the accessibility of that target to consciousness under conditions of cognitive load. This is, of course, another prediction of the ironic process theory. When one tries to suppress a thought, the intentional operating process is turned to the task of searching for distracters. The ironic monitor, in contrast, is aimed to search for the target because it is the target's appearance that indicates failed mental control. With load, then, the suppression target should become highly accessible. The prediction of the theory is that the suppression of a thought under conditions of mental load should increase the accessibility of the thought, even beyond the accessibility of that thought given concentrated attention. This odd result of thought suppression was examined in the studies discussed next.

IRONIC WANDERING OF SUPPRESSION

The mind wanders, not just away from where we aim it, but also toward what we forbid it to explore. In fact, it is in the case of suppression that the ironic failure of mental control is especially evident to the would-be controller. The failure to suppress the thought of a white bear, for example, is announced by the ironic monitor whenever a single search target is encountered—the white bear. This target is thus made relatively more accessible by the ironic monitor in suppression than are any of the wide array of nontargets that are each only slightly highlighted by the ironic monitor in concentration. The ironic monitor in suppression is applied to a relatively smaller range of search targets, making the search more effective (cf. Newman, Wolff, & Hearst, 1980; Sternberg, 1966). And, the ironic monitor in suppression is aimed at a cue that serves as an obvious reminder of the needed operation—the unwanted thought; this should make the monitor more effective as well.

The upshot of this reasoning is that suppression should produce strong ironic effects, measurable as the increased accessibility of the suppressed thoughts to consciousness. This has been observed in a variety of formats, beginning with the finding that suppressed thoughts recur frequently in stream-of-consciousness reports (Wegner, Schneider, Carter, & White, 1987). Similar observations have been made without a reporting requirement. When subjects are asked to suppress thoughts that are exciting (say, of sex), they show skin conductance level (SCL) reactivity rivaling the strength of reactions that occur when they are asked explicitly to entertain those thoughts (Wegner, Shortt, Blake, & Page, 1990). Evidence from a range of studies in which subjects are asked to suppress many different kinds of thoughts indicate that this manipulation dramatically increases the ease with which these thoughts are subsequently brought to mind (see Wegner, 1989, 1992).

Wegner and Erber (1992) termed this the *hyperaccessibility* of suppressed thoughts. In their first experiment, mental load was manipulated by imposing time pressure on subjects' word association responses. The subjects were asked to think or not to think about a target word (e.g., *house*), and over several trials their tendency to respond associatively with that target word to related prompts (e.g., *home*) and unrelated prompts (e.g., *adult*) was observed. Suppressing subjects who were under time pressure to report associates responded often with the target word to target-relevant prompts—blurting out the very word they had been trying not to think about. They did this more often than did suppressing subjects who were not under time pressure to give their associations. This high level of access was termed *hyperaccessibility* because suppression with time pressure even boosted responses of the target word to target-relevant prompts over the level of subjects under time pressure who were actively trying to think about the target.

This observation of hyperaccessibility can be attributed to the operation of ironic processes in both the suppression and concentration conditions. For subjects performing suppression, time pressure undermines the effortful operating process that looks for distracters, releasing the relatively less effortful ironic monitor to sensitize the person to the unwanted thought. For subjects performing concentration, time pressure undermines the effortful operating process that looks for the target thought, releasing the ironic monitor to sensitize the person to distracters. Thoughts suppressed under load thus enhance accessibility beyond that of thoughts concentrated on under load, so to increase the frequency of target associates in the suppression condition.

Wegner and Erber's (1992) second experiment tested the ironic process prediction in the Stroop interference paradigm, much as in the aforementioned study by Wegner et al. (1993). As in that study, subjects who were suppressing a target word (e.g., *house*) under high cognitive load showed interference with color naming when the target word appeared on the screen, as compared to nontarget words and as compared to target-related words (e.g., *home*). Subjects who were suppressing the target with low load, or who were concentrating on the target in either load condition, did not show evidence of differential interference. This study indicated, then, that suppressing a word during cognitive load promoted relatively effortless cognitive access to the target word. It seems that when the range of the ironic monitor is sharply focused by the intention to suppress, it is easy for a mental load to undo the intended operation and reveal the monitor's activity. This experiment did not show enhanced accessibility of nontargets during concentration that was found by Wegner et al. (1993). The arguments noted earlier do suggest that ironic effects of suppression should be stronger than ironic effects of concentration because the range of ironic search targets is smaller (see also Wegner, 1994). However, the theory predicts an ironic concentration effect here (at least a minor one), and its absence suggests that further inquiry is needed.

The overall conclusion suggested by this and other suppression research (Wegner, 1989, 1992) is that the suppression of thoughts is difficult. The mind wanders back to the suppressed thought repeatedly, apparently as a result of an ironic monitoring process that promotes the hyperaccessibility of the suppressed thought. These experiments suggest that it is only with enough mental capacity that suppression may be at least modestly effective. In other words, plentiful time and distraction may allow people to work themselves into the position of experiencing intrusions of their unwanted thoughts only very rarely. With the occurrence of mental loads or stresses, however, the mind does not merely wander toward suppressed thoughts, it seems to lurch back to them with a vengeance.

WANDERING INTO TRAFFIC

Ironic processes appear to make the mind go precisely where it does not want to go. This may be why we often find that the very thing we do not want to say, feel, think, or do comes forward to assert itself most obstinately when we are distracted or distressed. The phenomena of Freudian slips that are precisely the least appropriate thing to say in a given situation might also be explained in this way (Baars, 1985). Cognitive busyness or time pressure could interfere with many processes of self- presentation, deception, self-regulation, or self-control that depend on mental control for their success, and so promote social blunders, unintentional disclosures of deceit, or self-control lapses that are not entirely random. Rather, because the most unwelcome mental states are typically chosen as targets for suppression, and the most desired states are chosen as the focus of concentration, ironic effects will expose us to the caprice of our least desired states of mind.

There are a distressing number of such unwanted states, but a specific research example may suffice to communicate the point here. This is the case of trying not to be sexist (Wegner, Erber, & Bowman, 1994). Now, there is a growing body of research on the idea that certain untoward expressions (such as sexist, racist, or otherwise prejudiced remarks) may be subject to the opposing forces of automatic and controlled cognitive processes (see, e.g., Devine, 1989; Fiske, 1989; Perdue & Gurtman, 1990). Although prejudices may be brought to mind as an automatic result of knowledge of a pejorative stereotype, for example, it is believed that controlled cognitive processes typically will also come forward to counteract or undermine such expressions. By this logic, everyone may be automatically prejudiced, but some fight it through controlled processes and so express unprejudiced attitudes and behaviors. Bargh (1990) summarized this view by suggesting that "stereotype and trait construct activation . . . can be *prevented* from influencing responses, given sufficient motivation and effort" (p. 95).

The ironic process framework would suggest, however, that the motivation to be unprejudiced could well backfire if effort cannot be expended. If ironic processes are engaged in the pursuit of the mental control of prejudice, it might be that expressions of prejudice could occur merely because of the monitoring process. The theory would predict, for example, that subjects given the task of trying not to be sexist might even be especially inclined toward sexist responses under conditions of cognitive load.

Wegner, Erber, and Bowman (1994, Exp. 2) encouraged one group of subjects to try not to be sexist as the subjects completed a series of sentence stems. Some of the stems prompted completions relevant to sexism, as they were derived from items on the *Attitudes Toward Women Scale* (*ATWS*, Spence & Helmreich, 1972). So, for instance, subjects heard someone say "Women who go out with lots of men are . . . ," and were asked to complete the sentence. An egalitarian completion might be something like "popular," whereas a sexist completion might be something like "cheap." Other subjects for comparison were given no special instruction on how to respond. For some sentence completions, mental load was imposed by asking for immediate responses; for others, mental load was reduced by allowing subjects up to 10 sec to respond. The frequency of responses rated as sexist by coders was examined in each condition.

As would be expected, the rate of sexist sentence completions under low load was indeed substantially reduced when subjects were admonished not to be sexist as compared to no instruction. But the rate of sexist completions was significantly increased by the instruction not to be sexist under conditions of high load. This result was observed for both male and female respondents, and it also did not differ between subjects who were high in sexist attitudes as measured by the ATWS and those who were low in such sexist attitudes. In short, the attempt not to be sexist under time pressure increased the likelihood that sexist comments would be made, regardless of the person's sex or attitudes toward women. It makes sense, then, that ironic processes might be responsible for some fair proportion of the daily errors we least intend, from sexist remarks to faux pas of every kind.

There is reason to believe that certain psychopathologies might be traceable to ironic processes working under similar conditions. An individual who dearly desires to gain some form of mental control over an undesired symptom, and who attempts to exert this control under conditions of cognitive load, is likely to create ironic effects that could be quite unexpected—and that could prompt further attempts at control that serve only to compound the problem. This analysis might be useful in understanding the conditions that produce obsessive thinking, anxiety disorders such as phobias or generalized anxiety, insomnia, depression, overeating, and posttraumatic stress disorders. In each of these instances, people are confronted with symptoms of some kind—recurrent thoughts, unpleasant emotions, inability to perform some desired behavior or avoid an undesired behavior—and they may choose, not unwisely it would seem, to try to control the occurrence of the symptom.

If such attempts to control symptoms occur under conditions of stress, fatigue, or other forms of mental load, then ironic processes could be unleashed. It is known that stresses can exacerbate many of these conditions (e.g., Jacobs & Nadel, 1985; Polivy, 1990). It makes sense that people might produce some fairly deviant unwanted states and actions if they thought they were only trying to help themselves, and so continued only to dig themselves deeper with continued control in the face of failure. This perspective may provide a useful way of understanding apparently paradoxical effects such as panic attacks induced among panic-prone individuals by the instruction to relax (Adler, Craske, & Barlow, 1987) or sleep induced in insomniacs by the instruction to stay awake (Turner & Ascher, 1979). Perhaps people with problems like these are already trying so hard to control themselves that additional inducements unleash ironic reversals of the control.

Ironic effects could eventually become the sources of new stress and mental load themselves, as when a person becomes highly anxious and preoccupied with a symptom such as insomnia, panic, overweight, or the like. When the ironic effect produces its own load, no amount of further effort aimed at gaining mental control would derail the system from its self-perpetuating ironic outcome. In essence, ironic processes could create a mind that wanders permanently toward some desperately unwanted state.

THE WAY THE MIND STOOD STILL

This suggestion of "permanent" wandering begins to sound suspiciously like stability. A mind that always wanders toward some particular state or constellation of thoughts is at least attracted to a stationary point, if not always resident at that point. The examples of extreme states such as obsession and phobia suggest that there may be some forms of mental fixedness that are afforded by the ironic process model.

Indeed, there are a number of circumstances in which the mind does not seem to wander. Although the focus of the chapter to this point has been on the seeming inevitability of wandering, the alert reader will probably have generated several potential counterexamples. What about cases of meditation or trance states in which people assert that their minds are empty or otherwise effectively stopped? What about the occurrence of mental "blanking" in which the mind seems to have no contents for a significant period of time? What about cases of "fixed ideas" or other obsessional states in which people do not seem to be able to avoid thinking about one thing for excessive intervals?

These cases appear to fall into two groups. First, there are instances when wandering may stop because mental control is not exerted. Second, there are instances when wandering ceases because ironic processes of mental control are in extreme effect due to the crippling of their complementary operating processes. Exceptions to the rule of wandering, in other words, occur without mental control or in opposition to mental control.

Consider the first of these options: The relaxation of mental control should diminish the pace of mindwandering, perhaps even to a standstill. Without any intention to control consciousness, there is no operating process and no monitoring process either, so the constant war between them that normally jiggles the focus of attention should not occur. The relaxation of wandering should result when people have either relaxed control voluntarily or have become so fatigued or distracted that intentions to control the mind are not even formulated or implemented. It makes sense, then, that phenomenal descriptions of "going blank" are found among people using meditation techniques that involve specifically rescinding mental control (Taylor, 1978), as well as among depressed individuals who do not have the energy to exert mental control (Watts, MacLeod, & Morris, 1988). The peace of mind that comes from no more wandering may result from no more control.

The relaxation or repeal of mental control should not be confused with the exercise of control in the pursuit of a blank state of mind. Pennebaker (1993) asked a group of subjects to clear their minds completely for a period of 30 sec, for example, with the instruction that they make note each time they experienced any thought during this period. They reported a mean of 5.29 thoughts in this interval, and most indicated they never achieved anything akin to a state of empty-mindedness. A very few did report successful blanking, however, suggesting that they may have indeed suspended mental control. It would be interesting to learn what sort of mental translations or interpretations of the instructions were made by subjects who experienced these different outcomes, as this might allow a step toward the study of the intentional suspension of mental control and its associated ironic processes.

It is worthwhile to note in this regard that mindwandering appears to lessen with age. Giambra (1989) reported laboratory studies showing that older people are less inclined to experience "task- unrelated thoughts," and suggested by way of explanation that unconscious cognitive activity may decline with aging and so create fewer intrusions. There is also the possibility, though, that mental control is gradually suspended or reduced in vigor with age. With lessened energy devoted to concentration and/or suppression, older people may experience a release of sorts from the ironic processes that normally compel wandering. A reduction in the desire to concentrate may actually improve concentration.

The second way in which wandering stops, as already noted, involves individuals who have not relaxed control but instead have exerted control in such a way that their minds are invariably drawn to the same point. Although it is still fair to say that their minds are wandering, in the sense that they have no desire to keep returning to that attractive point, it does seem that they have achieved a certain sameness of consciousness, a fixed outlook that does not appear all that wanderful. The state of obsessive preoccupation or fixation on an idea, then, is the other escape from "free" wandering that can be produced by ironic processes. When people try suppressive forms of mental control, that target the avoidance of a thought, feeling, or action, they may find themselves returning so frequently to that target that they achieve a seemingly stable mental focus. The various psychopathologies mentioned earlier fit this model, as they all represent states of mind that are deeply unwanted by their hosts, and are thus the target of constant suppression. To the degree that such recurrent suppression occurs in the presence of constant load, or creates its own mental load conditions, it produces a state of overcontrolled consciousness—an obsessive return to exactly that which the control is attempting to eliminate.

It is interesting to reflect, in this light, on the relation between mental control and the wandering of mind. It appears that mindwandering is a symptom, of sorts, indicating the ongoing operation of everyday mental control. We know the mind is being controlled with some modicum of success when it wanders from time to time. When it does not seem to wander, in contrast, this is a signal that mental control has either lapsed entirely, or that it has entered a hopeless and self-defeating feedback loop that leaves it spinning wildly only to undermine its own efforts. The mind that does not wander is the mind that does not control itself.

THE BRAIN WANDERS, TOO

As a final exercise for this chapter, it is worth examining physical evidence of the postulated processes. Do we know anything about the brain that would allow us to evaluate the ironic process model? Although brain research specifically aimed at testing this theory has not yet been conducted, there is psychophysiological and neuropsychological evidence pertinent to the theory that suggests its plausibility.

Perhaps the most sweeping proposition of the ironic process theory is that each attempt at mental control produces not only an active operating process but also an ironic monitor that searches for errors. It is thus noteworthy that recent studies of brain psychophysiology support the existence of a general error-monitoring system like this one. In particular, the analysis of human event-related brain potentials (ERPs) indicates a regular ERP associated with errors in reaction-time tasks (Gehring, Coles, Meyer, & Donchin, 1990). This error-related brain activity is observed shortly after the onset of electromyographic (EMG) activity in the muscles of the limb that is about to make the error, and it peaks about 100 msec following its onset. The error-related ERP takes the form of a sharp, negative-going deflection of up to 10 mV in amplitude and is largest at electrodes placed over the front and middle of the scalp. The response is enhanced when subjects strive for accurate performance, and is also related to attempts to compensate for the erroneous behavior. Such an ERP would make sense as an indicator of the proposed ironic monitoring process.

Neuropsychological findings also appear supportive of an ironic error monitor that can be disabled given certain patterns of damage to the brain. Luria

(1966) identified such a dysfunction among patients with massive lesions of the frontal lobes. The "frontal syndrome" he described amounted to a breakdown of voluntary activity accompanied by an inability to discern when actions are in error. He noted that a preponderance of cases of frontal lobe damage resulted in an inability to respond even to direct commands. A patient asked to squeeze a bulb repeatedly, for example, might squeeze a few times, after which the pressure of the squeeze gradually diminished. The patient might repeat verbally "yes, squeeze" on each trial without making any movement. In other patients, the movements transformed over trials into a series of related, uncontrolled movements, or the bulb is squeezed without stopping to the point that the patient must be instructed to let go. Luria noted that characteristically, a patient asked to "squeeze 3 times," for example, would later respond to queries on the instructions by saying "yes, I squeezed 3 times," even though there were actually 6 squeezes, or perhaps none at all.

More contemporary neuropsychological theorizing suggests that such a syndrome is part of a lapse in "frontal control" (Stuss & Benson, 1987) that may permeate several cognitive and memorial systems in frontal lobe pathology. A key feature of such failed control is the patient's unawareness of errors of action, a seeming obliviousness to even the most conspicuous mistakes. Beyond the "local" unawareness of a specific error, frontal syndrome patients may also exhibit a more "global" unawareness of the implications of their overall handicap (Zaidel, 1987). In one case, a patient "would sit idly for long periods of time. The only activities the patient initiated on his own were simple card games and backgammon. . . . [Yet] the patient had very little insight into his condition. He had been told about his professional activities and how successful he had been. . . . [Still, he] considered himself fully recovered and was unperturbed by the obvious incongruity between his premorbid status and his present situation" (Goldberg, Bilder, Hughes, Antin, & Mattis, 1989, p. 689).

The inability to appreciate the errors of intentional action is very much like what one would expect of a person deprived of an ironic monitoring system. The decision of whether this is an apt portrayal of the individuals who have frontal lobe damage must await further research. For now, it is worth noting that the lives of people deprived of error monitoring come to an abrupt standstill. Apart from the occasional movement produced as an automatic or habitual response to irrelevant stimuli, frontal syndrome patients live without doing anything. Certainly, there seems to be no wandering of the mind. These observations serve as a reminder that ironic monitoring has a fundamental role in consciousness and cognition. This chapter has cast ironic processes as villians in an erstwhile critique of the wandering mind, but such processes can also be understood as keys to the capacity of consciousness to move anywhere at all.

CONCLUSIONS

A reader who successfully finishes a chapter on the wandering of mind is to be congratulated. Just in case you glazed over at some point, here is a summary of what you missed: The wandering of the mind has classically been ascribed to the interest value of the wide variety of stimuli that impinge on it. With this approach, James and many others noted the involuntary character of much of mental life and have explained mindwandering as though our intentional mental control had little to do with it. A contrasting approach is offered by the ironic process model outlined in this chapter. This account includes among the involuntary forces that produce a wandering mind a special class of forces: the involuntary attractions of mind that occur as the result of the mind's monitoring of its own acts of control. This theory indicates that consciousness may be prone to wander when we try to hold it in place while we check to make sure it has not moved. In this checking, and particularly when our conscious mental energies are taxed, we inadvertently draw our minds toward precisely where they least intend to go. Our voluntary mental control becomes the culprit in our involuntary mindwandering.

ACKNOWLEDGMENTS

Research reported in this chapter was supported by Grant BNS 90-96263 from the National Science Foundation and Grant MH49127 from the National Institute of Mental Health. Thanks are due to Robert Kleck, Darren Newtson, and Toni Wegner for their comments on these ideas.

REFERENCES

Adler, C. M., Craske, M. G., & Barlow, D. H. (1987). Relaxation-induced panic (RIP): When resting isn't peaceful. *Integrative Psychiatry, 26*, 94–112.

Amster, H. (1964). Semantic satiation and generation: Learning? Adaptation? *Psychological Bulletin, 62*, 273–286.

Baars, B. J. (1985). Can involuntary slips reveal one's state of mind?—With an addendum on the problem of conscious control of action. In T. M. Shlechter & M. P. Toglia (Eds.), *New directions in cognitive science* (pp. 242–261). Norwood, NJ: Ablex.

Baars, B. J. (1988). *A cognitive theory of consciousness.* New York: Cambridge University Press.

Bargh, J. A. (1984). Automatic and conscious processing of social information. In R. S. Wyer, Jr., & T. K. Srull (Eds.), *Handbook of social cognition* (Vol. 3, pp. 1–43). Hillsdale, NJ: Lawrence Erlbaum Associates.

Bargh, J. A. (1989). Conditional automaticity: Varieties of automatic influence in social perception and cognition. In J. S. Uleman & J. A. Bargh (Eds.), *Unintended thought* (pp. 3–51). New York: Guilford.

Bargh, J. A. (1990). Auto-motives: Preconscious determinants of social interaction. In E. T. Higgins & R. M. Sorrentino (Eds.), *Handbook of motivation and cognition* (Vol. 2, pp. 93–130). New York: Guilford.

Bills, A. G. (1931). Blocking: A new principle of mental fatigue. *American Journal of Psychology, 43,* 230–245.

Carpenter, W. G. (1875). *Principles of mental physiology.* New York: Appleton.

Carroll, L. (1982). *Complete illustrated works of Lewis Carroll.* London: Chancellor. (Original work published 1872)

Carver, C. S., & Scheier, M. F. (1981). *Attention and self-regulation: A control-theory approach to human behavior.* New York: Springer-Verlag.

Devine, P. (1989). Stereotypes and prejudice: Their automatic and controlled components. *Journal of Personality and Social Psychology, 56,* 5–18.

Fiske, S. T. (1989). Examining the role of intent, toward understanding its role in stereotyping and prejudice. In J. S. Uleman & J. A. Bargh (Eds.), *Unintended thought* (pp. 253–283). New York: Guilford.

Gehring, W. J., Coles, M. G. H., Meyer, D. E., & Donchin, E. (1990). The error-related negativity: An event-related brain potential accompanying errors. *Psychophysiology, 27,* S34. (Abstract)

Giambra, L. M. (1989). Task-unrelated thought frequency as a function of age: A laboratory study. *Psychology and Aging, 4,* 136–143.

Goldberg, E., Bilder, R. M., Hughes, J. E. O., Antin, S. P., & Mattis, S. (1989). A reticulo- frontal disconnection syndrome. *Cortex, 25,* 687–695.

Hasher, L., & Zacks, R. T. (1979). Automatic and effortful processes in memory. *Journal of Experimental Psychology: General, 108,* 356–388.

Higgins, E. T. (1989). Knowledge accessibility and activation: Subjectivity and suffering from unconscious sources. In J. S. Uleman & J. A. Bargh (Eds.), *Unintended thought* (pp. 75–123). New York: Guilford.

Hobson, J. A. (1988). *The dreaming brain.* New York: Basic Books.

Jacobs, W. J., & Nadel, L. (1985). Stress-induced recovery of fears and phobias. *Psychological Review, 92,* 512–531.

James, W. (1890). *Principles of psychology.* New York: Holt.

Johnson-Laird, P. N. (1983). A computational analysis of consciousness. *Cognition and Brain Theory, 6,* 499–508.

Jonides, J. (1981). Voluntary versus automatic control over the mind's eye's movement. In J. Long & A. Baddeley (Eds.), *Attention and performance* (Vol. 9, pp. 187–203). Hillsdale, NJ: Lawrence Erlbaum Associates.

Kanwisher, N. G., & Potter, M. C. (1990). Repetition blindness: Levels of processing. *Journal of Experimental Psychology: Human Perception and Performance, 16,* 30–47.

Klinger, E. (1978). Modes of normal conscious flow. In K. S. Pope & J. L. Singer (Eds.), *The stream of consciousness* (pp. 225–258). New York: Plenum.

Lefebvre-Pinard, M. (1983). Understanding and the auto-control of cognitive functions: Implications for the relationship between cognition and behavior. *International Journal of Behavioral Development, 6,* 15–35.

Logan, G. D. (1985). Executive control of thought and action. *Acta Psychologica, 60,* 193–210.

Logan, G. D. (1988). Toward an instance theory of automatization. *Psychological Review, 95,* 492–527.

Luria, A. R. (1966). *Higher cortical functions in man.* New York: Basic Books.

Miller, G. A., Galanter, E., & Pribram, K. H. (1960). *Plans and the structure of behavior.* New York: Holt.

Newell, A., & Simon, H. A. (1972). *Human problem solving.* Englewood Cliffs, NJ: Prentice-Hall.

Newman, J. P., Wolff, W. T., & Hearst, E. (1980). The feature-positive effect in adult human subjects. *Journal of Experimental Psychology: Human Learning and Memory, 6,* 630–650.

Oatley, K. (1988). On changing one's mind: A possible function of consciousness. In A. J. Marcel

& E. Bisiach (Eds.), *Consciousness in contemporary science* (pp. 369–389). Oxford, England: Clarendon Press.

Pennebaker, J. W. (1993). [The effect of instructions not to think.] Unpublished research data.

Perdue, C. W., & Gurtman, M. B. (1990). Evidence for the automaticity of ageism. *Journal of Experimental Social Psychology, 26,* 199–216.

Polivy, J. (1990). Inhibition of internally cued behavior. In E. T. Higgins & R. M. Sorrentino (Eds.), *Handbook of motivation and cognition* (Vol. 2, pp. 131–147). New York: Guilford.

Pope, K. S. (1978). How gender, solitude, and posture influence the stream of consciousness. In K. S. Pope & J. L. Singer (Eds.), *The stream of consciousness* (pp. 259–299). New York: Plenum.

Posner, M. I., & Snyder, C. R. R. (1975). Attention and cognitive control. In R. L. Solso (Ed.), *Information processing and cognition* (pp. 55–85). Hillsdale, NJ: Lawrence Erlbaum Associates.

Powers, W. T. (1973). *Behavior: The control of perception.* Chicago: Aldine.

Shiffrin, R. M., & Schneider, W. (1977). Controlled and automatic human information processing: II. Perceptual learning, automatic attending, and a general theory. *Psychological Review, 84,* 127–190.

Spence, J. T., & Helmreich, R. (1972). Who likes competent women? Competence, sex-role congruence of interests, and subjects' attitude toward women as determinants of interpersonal attraction. *Psychology of Women Quarterly, 5,* 147–163.

Sternberg, S. (1966). High speed scanning in human memory. *Science, 153,* 652–654.

Stroop, J. R. (1935). Studies of interference in serial verbal reactions. *Journal of Experimental Psychology, 18,* 643–662.

Stuss, D. T., & Benson, D. F. (1987). The frontal lobes and the control of cognition and memory. In E. Perecman (Ed.), *The frontal lobes revisited* (pp. 141–158). New York: IRBN Press.

Taylor, E. (1978). Asian interpretations: Transcending the stream of consciousness. In K. S. Pope & J. L. Singer (Eds.), *The stream of consciousness* (pp. 31–54). New York: Plenum.

Turner, R. M., & Ascher, L. M. (1979). Paradoxical intention and insomnia: An experimental investigation. *Behavioral Research and Therapy, 17,* 408–411.

Uleman, J. S. (1989). A framework for thinking intentionally about unintended thoughts. In J. S. Uleman & J. A. Bargh (Eds.), *Unintended thought* (pp. 425–449). New York: Guilford.

Umiltà, C. (1988). The control operations of consciousness. In A. J. Marcel & E. Bisiach (Eds.), *Consciousness in contemporary science* (pp. 334–356). Oxford, England: Clarendon Press.

Vallacher, R. R., & Nowak, A. (1994). The stream of social judgment. In R. R. Vallacher & A. Nowak (Eds.), *Dynamical systems in social psychology* (pp. 251–277). San Diego, CA: Academic Press.

Watts, F. N., MacLeod, A. K., & Morris, L. (1988). Associations between phenomenal and objective aspects of concentration problems in depressed patients. *British Journal of Psychology, 79,* 241–250.

Wegner, D. M. (1989). *White bears and other unwanted thoughts.* New York: Viking.

Wegner, D. M. (1992). You can't always think what you want: Problems in the suppression of unwanted thoughts. In M. Zanna (Ed.), *Advances in experimental social psychology* (Vol. 25, pp. 193–225). San Diego, CA: Academic Press.

Wegner, D. M. (1994). Ironic processes of mental control. *Psychological Review, 101,* 34–52.

Wegner, D. M., & Erber, R. (1991). *Hyperaccessibility of implicitly suppressed thoughts.* Unpublished manuscript.

Wegner, D. M., & Erber, R. (1992). The hyperaccessibility of suppressed thoughts. *Journal of Personality and Social Psychology, 63,* 903–912.

Wegner, D. M., Erber, R., & Bowman, R. E. (1994). *On trying not to be sexist.* Manuscript in preparation.

Wegner, D. M., Erber, R., & Zanakos, S. (1993). Ironic processes in the mental control of mood and mood-related thought. *Journal of Personality and Social Psychology, 65,* 1093–1104.

Wegner, D. M., & Pennebaker, J. W. (Eds.). (1993). *Handbook of mental control.* Englewood Cliffs, NJ: Prentice-Hall.

Wegner, D. M., Schneider, D. J., Carter, S., III, & White, L. (1987). Paradoxical effects of thought suppression. *Journal of Personality and Social Psychology, 58,* 409–418.

Wegner, D. M., Shortt, J. W., Blake, A. W., & Page, M. S. (1990). The suppression of exciting thoughts. *Journal of Personality and Social Psychology, 58,* 409–418.

Zaidel, E. (1987). Hemispheric monitoring. In D. Ottoson (Ed.), *Duality and the unity of the brain* (pp. 247–281). London: Macmillan.

Zukier, H., & Hagen, J. W. (1978). The development of selective attention under distracting conditions. *Child Development, 49,* 870–873.

CHAPTER 15

The Psychology of
Meta-Psychology

Timothy D. Wilson
University of Virginia

The job of a discussant is similar to the child's game in which you have to guess why a group of objects go together. The discussant's job, however, is often more difficult than determining that a suit of armor, stone tool, and quill pen are relics of antiquity, or that a hula hoop, yo-yo, and frisbee are all faddish toys. The common themes of chapters in one volume are not always so evident. Clearly the research programs described by Johnson and Reeder (chap. 13, this volume) and Wegner (chap. 14, this volume) are neither relics nor passing fads, but vibrant, enduring approaches to classic problems. On a first reading, however, their common themes may not be readily apparent in that they seem to be addressing rather different psychological problems.

This state of affairs is typical of the literature on consciousness, in that researchers are investigating quite different phenomena from distinct theoretical assumptions, using different methodologies. The chapters in this volume, for example, concern such diverse topics as dreams, connectionist approaches to mental representation, and the study of the hippocampus as a possible neural substrate for consciousness. One reason for this diversity is the catch-all nature of the term *consciousness*. As noted by Johnson and Reeder, this term has taken on an amazingly wide range of meanings.

Even chapters grouped under the same heading, such as Johnson and Reeder's and Wegner's on metacognition, represent rather different research traditions. I will begin my comments by noting some of the differences in the general approaches taken by Johnson and Reeder and Wegner, and then point to some of the similarities. I will then broaden the discussion by noting that there are several recent research programs on metacognition that appear diverse at first glance, but have some intriguing similarities in terms of their assumptions about the relation between conscious and nonconscious processing. These different research programs might profit from recognizing these common assumptions.

JOHNSON AND REEDER'S MEM MODEL
AND WEGNER'S IRONIC PROCESSING MODEL

Johnson and Reeder's and Wegner's chapters both reflect extremely rich and productive research programs that, although relatively new, have already proven to be influential. They attack the problem of consciousness, however, from rather different angles. There are two general ways in which to approach the study of consciousness. One, as noted by Johnson and Reeder, is to approach Consciousness with a capital C, developing a model that encompasses all the many meanings of the term, including awareness, control and executive functions, and representational complexity. Johnson and Reeder's Multiple-Entry Modular (MEM) model is an excellent example of this approach in that it is an ambitious attempt to sketch the entire cognitive architecture out of which consciousness arises. The power and explanatory breadth of this model are very impressive, and it has a great deal of intuitive appeal. Especially intriguing is the ability of the model to represent the reflective and recursive nature of consciousness, such as the examples depicted in their Fig 13.7. Of the many recent attempts to develop comprehensive models of consciousness, Johnson and Reeder's approach shows great promise.

Though the MEM model leads to some fascinating predictions about consciousness as an emergent property of interactions between subsystems, it is not entirely clear, at this stage of its development, how to test these predictions empirically. This is the major stumbling block of almost all general models of consciousness: We *know* consciousness exists, we can *describe* what it is and what it does, and, with the aid of Johnson and Reeder's MEM model, we can map out very specifically how it arises from particular mental modules. But when we put on our hats as experimental psychologists, how can we translate these models and predictions into testable hypotheses?

An alternative approach is to begin with the detailed study of a specific phenomenon that can be studied empirically. From careful observations of this phenomenon, models of conscious and unconscious experience can be built from the bottom up. Such an approach is in one sense less satisfying, because it does not attempt to be a grand theory, explaining all of the myriad properties of consciousness. It has the advantage, however, of discovering new, interesting phenomena that we can sink our teeth into and know how to study.

This is the approach taken by Wegner with his ironic processing model. He focused on mental control, one of the functions of consciousness mentioned by Johnson. He began with a very interesting empirical phenomenon, namely the ironic consequences that occur when people attempt to suppress a thought (Wegner, Schneider, Carter, & White, 1987). From this a general model of thought suppression resulted (Wegner, 1992), which in turn grew into a more general model of mental control and ironic processing (Wegner, 1994). In this volume, Wegner explores the implications of this model to the phenomenon of mind wandering—yet another extension of the model. Thus, what began as the investigation of a very specific phenomenon (people's ability to suppress thoughts of a white bear) has blossomed into a very broad theory of mental control, a theory that makes specific predictions about the conditions under which people can and cannot control a wide variety of phenomena, including their motor behavior, sleep onset, sexist comments, and attraction toward others (see Wegner, 1994).

Despite the differences in their models, Johnson and Reeder and Wegner are weaving some common threads. Both models involve the interaction of mental processes that are relatively conscious with those that are relatively nonconscious. Further, both models stress the importance of metacognitions, or cognitions about one's own cognitions. In fact, they share this emphasis with several other recent research programs, many of which are not typically thought to have much in common. A number of researchers, including Johnson and Reeder and Wegner, are investigating people's conscious beliefs about their own minds, and the relation of these metacognitions to nonconscious processing.

Figure 15.1 portrays a simple model of conscious and nonconscious processing, which has been adapted from Nelson and Narens (1990). The top box represents people's conscious beliefs about their minds, labeled *lay meta-psychology*. These beliefs include metacognitions, or cognitions about one's own cognitions. I prefer the term *meta-psychology* to metacognition here (Flavell, 1979) because it is important to consider not only people's beliefs about their beliefs, but their beliefs about all aspects of their minds, including their memories, affective responses, and the causes of their behavior.

Some schools of psychology have assumed that this box is all there is, or at least all that is needed to study to understand "outputs" of interest. The early structuralists such as Wundt and Titchener assumed that the mind could be understood by studying what was accessible via introspection (albeit not everyday introspection, but reports by trained introspectionists; see Boring, 1953; Lieberman, 1979). It became apparent that this method was inadequate to study the mental processes of interest to the structuralists—namely, the bases of sensation and perception—and their version of the introspectionist method quickly grew out of favor. Many areas of modern psychology, however, still rely on a simpler form of introspection, whereby people are asked to report their attitudes, emotions, beliefs, and memories. And, it is

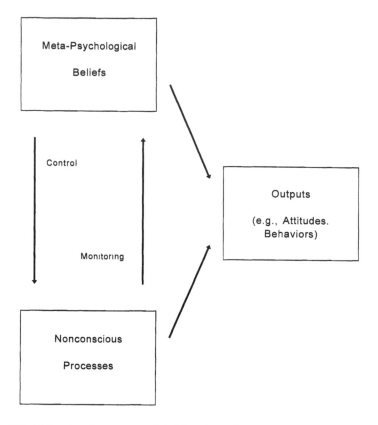

FIG. 15.1. A rudimentary model of the relation between meta-psychological beliefs and nonconscious mental processes.

clear that many of these introspective reports are reliable, valid, and predictive of behavior (Lieberman, 1979).

In the past 20 years, however, it has become increasingly apparent that a great deal of mental processing is inaccessible to introspection. These mental processes are represented by the lower box in Fig. 15.1, labeled nonconscious processes. This is not the place to review the evidence for nonconscious processing; suffice it to say that the idea that people are unaware of a substantial amount of their mental processing has a firmer toehold in psychology than ever before (Jacoby, Lindsay, & Toth, 1992; Kihlstrom, 1987; Lewicki, 1986; Nisbett & Wilson, 1977; Posner & Rothbart, 1989). For present purposes, the important issue is the relation between nonconscious processes and more conscious parts of the system.

Several recent approaches have addressed this question by considering meta-psychological beliefs, or people's conscious beliefs about how their minds operate, and how these beliefs relate to nonconscious processing. As seen in Fig. 15.1, there are several interesting relations between meta-psychological beliefs and nonconscious processing, including ways in which both can lead to "outputs" such as attitudes and behavior, and ways in which conscious beliefs might monitor the outputs of nonconscious processing and exert control over these nonconscious processes. The remainder of the chapter describes research programs that can be fit into this rubric, including Johnson and Reeder's and Wegner's. It is instructive to examine the similarities and differences between these diverse areas of research.

CURRENT RESEARCH ON LAY META-PSYCHOLOGY AND NONCONSCIOUS PROCESSING

Nonconscious Processing and Causal Theories

Nisbett and Wilson (1977) emphasized the separation between conscious and nonconscious processing, illustrating that a good deal of mental processing was inaccessible to conscious awareness. At the time this was a somewhat radical statement, because few theorists in cognitive and social psychology acknowledged (at least explicitly) that there was a "nonconscious" processing box. An equally important goal, however, was to document the importance of people's causal theories—namely, their beliefs about the determinants of their judgments, emotions, beliefs, and behaviors (i.e., the top, meta-psychology box; see Fig. 15.2). Given that people often have limited access to the mental processes mediating their responses, what do they do when asked to report these processes? Interestingly, they do not throw up their hands and profess ignorance, but rather generate reasons quite freely. A major source of these reasons, argued Nisbett and Wilson (1977), was shared causal theories (see also Wilson & Stone, 1985). For example, when asked about the causes of their daily moods people often report that their moods improve on Fridays, presumably because they have learned from their culture that such a boost in mood is common (e.g., "TGIF," or "thank God it's Friday"). People endorse such a theory despite little evidence that mood does, in fact, improve on Fridays (Wilson, Laser, & Stone, 1982).

As seen in Fig 15.2, Nisbett and Wilson (1977) had little to say about possible relations between meta-psychological beliefs (causal theories) and nonconscious processes (that is, there are no lines drawn between the "causal theory" and "nonconscious processing" boxes). In our zeal to illustrate the separation between the systems, we portrayed causal theories as epiphenomenal—explanations generated by the conscious system that, while interesting, did not have much of an effect on anything of importance. This is illustrated in Fig. 15.2 by dividing the output box into two: one that is a function of causal

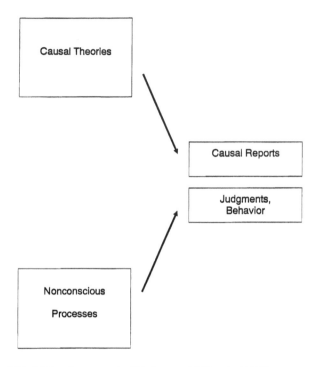

FIG. 15.2.   A portrayal of Nisbett and Wilson's (1977) arguments about the lack of awareness of mental processes.

theories, the other a function of nonconscious processes. It seems unlikely, however, that our brains would consist of two totally independent systems, one conscious and one nonconscious, with no lines of communication or control between them. As research on nonconscious processing has flourished, models of meta-psychology have become more sophisticated, with interesting hypotheses about communications between conscious and nonconscious systems.

Metamemory

Nelson and Narens (1990) presented a model of metamemory that conforms nicely to the scheme depicted in Fig. 15.1, which is not surprising given that Fig. 15.1 is adapted from their article. The lay meta-psychology with which they were concerned was a person's beliefs about how his or her memory operates, including theories about how information is acquired, retained, and retrieved. Nonconscious processing, in this case, refers to the actual mental processes governing human memory. Like Nisbett and Wilson (1977), Nelson and Narens (1990) noted that lay beliefs are not entirely correct: Lay beliefs about how memory operates do not always correspond to the actual workings of memory. Unlike Nisbett and Wilson (1977), they emphasized the control and monitoring functions of metamemory, pointing out that metamemory is not purely an epiphenomenon without impact. The meta-system monitors the outputs of the nonconscious system, by, for example, judging how much one knows about a topic and how easily new material can be learned. It can also exert control over the nonconscious system, by, for example, determining how much time is spent studying new material.

Thus, the output box—in this case, people's performance on memory tasks—is a joint function of the conscious and unconscious systems. Performance depends not only on the capabilities and limits of the memory system to encode, retain, and retrieve information, but also on the success of the metamemory system at monitoring and controlling the memory system. Interestingly, the metamemory system is sometimes not very adept at performing these executive functions. Several studies have shown that people's judgments of how well they have learned new material, right after studying it, are not very accurate. Consequently, people's decisions about which items need further study are often nonoptimal (e.g., Nelson & Dunlosky, 1991; Nelson & Leonesio, 1988; Vesonder & Voss, 1985). For our purposes, the important point is that researchers in the tradition of metamemory and metacognition (e.g., Flavell, 1979; Nelson & Narens, 1990) were among the first to draw attention to the necessity of understanding the way in which meta-beliefs monitor and control nonconscious processing.

Johnson and Reeder's MEM Model

The MEM model fits nicely into the framework of Fig. 15.1, as shown in Fig. 13.4 of Johnson and Reeder's article (this volume). Like Nelson and Narens (1990), Johnson and Reeder focus on the control and monitoring functions of meta-level processes. They extend the analysis of metamemory and actual memory processes to mental functioning more generally, presenting a broad, general model of mental life. A distinctive feature of the model is that it postulates several different subsystems of reflection and perception, most of which can play the role of the meta-system versus the "object" (nonconscious processing) system. There is a great deal of richness and specificity to the model that are very intriguing, and which certainly warrant further inquiry. One important departure of the MEM model from the simplistic diagram in Fig. 15.1 is that consciousness is not said to reside in one "box," but arises from meta-processing *between* systems.

Wegner's Ironic Processing Model

Wegner's research requires a larger shoehorn to squeeze into the model depicted in Fig. 15.1, in that the boundaries between meta-level processing and nonconscious processing are not drawn as sharply. As shown in Fig. 15.3, the top figure could be considered the conscious system that initiates attempts to concentrate on or suppress a given thought. This system attempts to control thought via the intentional operating process, which attempts to find thoughts that meet the desired state of affairs. In the case of mind wandering, the intentional operating process attempts to focus attention on one object, such as the words on the page of a book. The ironic monitor searches for failures to achieve the desired state of affairs, such as evidence that the mind has wandered from the words on the page to thoughts about chocolate cake or sex or the clothes that need to be picked up from the dry cleaners.

One way in which Wegner's model does not entirely fit the template shown in Fig. 15.1 is that the mental states in the bottom box are not always unconscious. In the example of mind wandering, people would be aware that they are reading the words on a page; that is, the target of the intentional operator is conscious. Further, failures of mental control also become conscious; we realize that we've just spent the last 5 minutes thinking about cake. The ironic monitor, however, operates largely out of awareness. When the system is running smoothly, the ironic monitor finds unwanted thoughts that have not yet

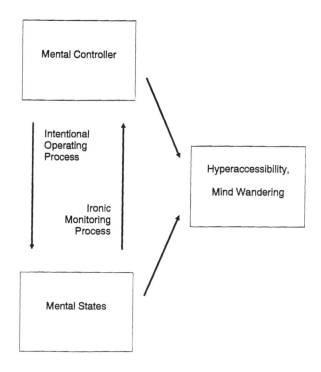

FIG. 15.3.   Wegner's (1994) model of ironic processing.

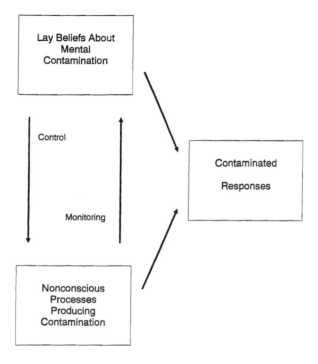

FIG. 15.4.   Wilson and Brekke's (1994) model of mental contamination.

reached consciousness, and manages to alert the intentional operating system automatically. Thus, though the lines of what is conscious and what is not are drawn somewhat differently from the models already discussed, there is still a good deal of commonality between Wegner's model and the others: People's conscious beliefs are an important part of the system (in this case, people's conscious goals), and these beliefs operate jointly with nonconscious processes to produce interesting outcomes (e.g., thought suppression).

Mental Contamination

Wilson and Brekke (1994) proposed a model of mental contamination that can also be fit into the current framework, as shown in Fig. 15.4. This model is concerned with the conditions under which people end up with an unwanted judgment, emotion, or behavior, due to nonconscious or uncontrollable mental processes that contaminate these responses. To understand how and when contamination occurs, it is critical to consider both the nonconscious processes producing the response and lay beliefs about contamination. For example, suppose we wanted to predict whether the grade a college professor assigns to a student's paper is biased by how much the professor likes the student. To make this prediction we obviously need to consider the mental processes that produce halo effects. There is evidence that people's liking can distort their judgments of a person, and these processes are largely nonconscious (e.g., Wetzel, Wilson, & Kort, 1981).

We also need to consider, however, people's meta-psychological beliefs about halo effects. Professors who have the theory that they are susceptible to halo effects are more likely to take steps to avoid these effects by monitoring and controlling their mental processes. Thus, in this model the control and monitoring functions of lay beliefs represent attempts to watch for evidence of biased processing, and attempts to control one's mental processes in such a way to avoid contamination. Wilson and Brekke (1994) noted that control and monitoring are often unsuccessful in this area, because it is difficult to know when a response has been contaminated and difficult to control one's thoughts sufficiently to avoid bias. Nonetheless, there are circumstances under which such monitoring and control are successful.

Effects of Verbalization on Memory and Problem Solving

Two related lines of research focus on what happens when the conscious system is stimulated, by asking people to verbalize their mental processes or to introspect about the reasons for their attitudes. The first area is Schooler's work on the negative effects of verbalization (Schooler, this volume; Schooler & Engstler-Schooler, 1990; Schooler, Ohlsson, & Brooks, 1993). Schooler and his colleagues demonstrated that attempting to verbalize a process that is normally not performed verbally is disruptive. For example, people who verbalized their memories for faces were less accurate at identifying the faces in a recognition task than people who did not verbalize their memories (Schooler & Engstler-Schooler, 1990). In terms of the present framework, this research is distinctive because it looks at what happens when the conscious, verbal system is stimulated, overshadowing processing that is normally done outside of awareness (or at least is not verbalizable).

As seen in Fig. 15.5, there is not much emphasis in this model on the monitoring or control functions of the verbal system. Instead, the verbal, meta-psychological system is hypothesized to operate in parallel with the nonverbal system. The intriguing issue is which system produces the response of interest, such as recognition memory. Normally, people seem to know to let the nonverbal system do the task; in that they do not spontaneously verbalize their memories for faces. In this case, the arrow between nonverbal processing and the output boxes is activated. When the verbal system is stimulated, however, it in some sense takes over, overshadowing the nonverbal system. In this case, the arrow between the lay belief and output boxes is activated. As seen shortly, however, the nonverbal system appears still to operate, and under some conditions it can still be tapped.

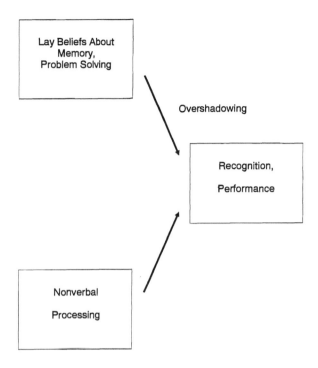

FIG. 15.5.   Schooler and Engstler-Schooler's (1990) model of
verbal overshadowing.

Effects of Analyzing Reasons on Attitudes

My colleagues and I have found that jump starting the conscious system in a
different way also changes people's responses. We ask people to introspect
about the reasons for their attitudes and find that after people introspect their
attitudes have changed (e.g., Wilson, Bybee, Dunn, Hyman, & Rotondo, 1984;
Wilson & Dunn, 1986; Wilson, Kraft, & Dunn, 1989; Wilson & Kraft, 1993;
Wilson et al., 1993; Wilson & Schooler, 1991). For example, Wilson and Kraft
(1993) found that asking students to think about the reasons why their dating relationships were going the way they were changed their attitudes about
these relationships.

The explanation of this effect is related to the framework presented in Fig.
15.1, as shown in Fig. 15.6. The bottom box represents the actual determinants of people's attitudes and behavior, which are not completely available to
the conscious explanatory system. When people do not reflect about the reasons for their attitudes, these are the processes that determine how they feel
and what they do (i.e., the arrow between the bottom box and the output box
is activated). When people reflect about reasons, they focus on lay beliefs

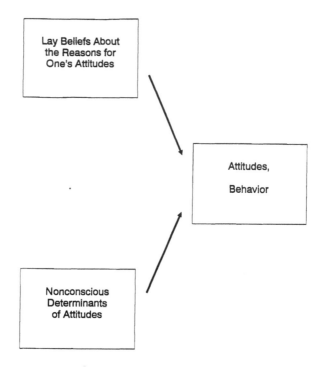

FIG. 15.6.   Wilson et al.'s (1989) model of reasons-generated attitude change.

about the causes of their attitudes (the top box), and they infer that their attitudes are consistent with these reasons. That is, the arrow between the top box and their attitudes and behaviors is now activated. We have obtained evidence for these proposed processes in several studies (e.g., Wilson, Hodges, & LaFleur, 1995). Interestingly, the attitudes people report right after analyzing reasons is often different from their later behavior, presumably because the attitudes are a function of lay beliefs about reasons, whereas behavior, measured later, is more a function of mental processes that are less accessible (see Wilson, Dunn, Kraft, & Lisle, 1989).

POINTS IN COMMON, POINTS OF DIVERGENCE

I have attempted to show that several diverse areas of research can be fit (squeezed?) into a common framework, in which conscious, lay beliefs are contrasted to nonconscious mental processes. It is interesting to consider how research programs on different topics, not typically thought to be related, can be brought under the same umbrella. Such an exercise is particularly fruitful if it leads to cross-fertilization between research areas. I now sketch a few ways these different areas can learn from one another, both theoretically and methodologically.

A quick glance at Figs. 15.1–15.6 illustrates interesting differences between the research programs. In some, the arrows between the conscious and nonconscious boxes, representing the control and monitoring functions, are given central importance. This is the case with Nelson and Naren's (1990) work on metamemory, Johnson and Reeder's MEM model, and Wegner's ironic processing model. In others, these control and monitoring functions are either downplayed or are missing from the model, such as Nisbett and Wilson's work on causal theories, Schooler's work on verbalization, and Wilson's work on introspecting about reasons. The latter research programs have gotten some mileage by demonstrating that there are circumstances under which conscious processing operates in parallel with nonconscious processes, with little interaction between the systems. As noted earlier, however, it seems unlikely that two independent systems would have evolved with no lines of communication between them. The extent to which these lines are activated or used when people verbalize memories or reasons is an interesting question that deserves further attention.

Another way in which the models differ concerns a methodological procedure with theoretical implications, namely, the extent to which conscious processes are allowed to operate in experimental settings. In some research programs, attempts are made to prevent conscious, controlled processing with the use of cognitive load manipulations. This serves the methodological purpose of allowing a clearer look at automatic, nonconscious processing, unobscured by conscious processing that would normally occur in parallel. It also serves the theoretical purpose of simulating real-life instances in which cognitive capacity is limited, thereby demonstrating how the system works when people are preoccupied or tired. A clear example of this research strategy is Wegner's work on mental control. When people are placed under cognitive load the intentional operating process is suppressed, with very interesting consequences. The ironic monitor continues to look for and find unwanted thoughts, but now there is no process in place to get rid of these thoughts. Wegner (1992, 1994) argued that this process accounts for several real-life responses of considerable interest, such as obsessions and prejudice (see also Gilbert, 1991, 1993).

Some of the other areas discussed here might profit from using similar cognitive load manipulations. For example, it would be interesting to see what would happen to recall and recognition if metamemory processing were reduced with a load manipulation. Under some circumstances, this should hinder memory, because it would prevent the adaptive and functional control and monitoring functions of metamemory (e.g., Nelson & Dunlosky, 1991). Under other circumstances, metamemory seems to get things wrong, leading to incorrect allocations of such important responses as study time (e.g., Vesonder & Voss, 1985). It is possible that under these circumstances a load manipulation would improve performance by limiting maladaptive control and monitoring processing.

Other areas of research take the opposite tack: Instead of trying to shut down conscious processing, an attempt is made to enhance it. This is the approach taken in Wilson's work on introspecting about reasons and in Schooler's work on verbalization. In both cases, making the conscious system play a greater role in processing has been found to have some intriguing (and often detrimental) consequences. It might be interesting to apply this sort of manipulation to other areas, to see if similar effects are found when conscious processes are stimulated.

It appears that there is a continuum, across research paradigms, representing the extent to which meta-psychological processing is allowed to operate. In some studies, conscious processing is blocked with the use of cognitive load manipulations, preventing meta-psychological beliefs from having much of an effect. In others, it is allowed to operate as it naturally does, with no attempt to block or stimulate it. In still others, meta-psychological processing is stimulated by instructing people to introspect or verbalize. Interestingly, these different strategies are rarely used within a single paradigm. It might be fruitful to examine one phenomenon across the full range of this continuum, observing what happens when conscious processing is blocked versus stimulated. For example, Schooler and Engstler-Schooler (1990) found that stimulating the system by asking people to verbalize their memories increased recognition errors, but that these errors could by avoided by "turning off" the verbal system at the test phase (with the use of a load manipulation).

The three-box framework presented in Fig. 15.1, and applied to several research programs, is clearly simplistic. This is illustrated nicely by Johnson and Reeder's MEM model, which adds intriguing layers of complexity to my elementary portrayal. Further, there are undoubtedly other research programs that have not been discussed here, but that could be placed under this rubric. Nonetheless, the research programs discussed have important commonalities. Though these commonalities might not be as obvious as those shared by relics of antiquity or faddish toys, attempts to squeeze them into a common framework will lead to a cross-fertilization of ideas, enriching our understanding of the importance of lay meta-psychology.

ACKNOWLEDGMENTS

The writing of this article was facilitated by National Institute of Mental Health Grant MH41841. I would like to thank Tom Nelson for helpful comments on a draft of this article.

REFERENCES

Boring, E. G. (1953). A history of introspection. *Psychological Bulletin, 50*, 169–187.

Flavell, J. H. (1979). Metacognition and cognitive monitoring: A new area of cognitive-developmental inquiry. *American Psychologist, 34*, 906–911.

Gilbert, D. T. (1991). How mental systems believe. *American Psychologist, 46*, 107–119.

Gilbert, D. T. (1993). The assent of man: Mental representation and the control of belief. In D. M. Wegner & J. W. Pennebaker (Eds.), *The handbook of mental control* (pp. 57–87). Englewood Cliffs, NJ: Prentice-Hall.

Jacoby, L. L., Lindsay, D. S., & Toth, J. P. (1992). Unconscious influences revealed: Attention, awareness, and control. *American Psychologist, 47*, 802–809.

Kihlstrom, J. F. (1987). The cognitive unconscious. *Science, 237*, 1445–1452.

Lewicki, P. (1986). *Nonconscious social information processing.* Orlando, FL: Academic Press.

Lieberman, D. A. (1979). Behaviorism and the mind. *American Psychologist, 34*, 319–333.

Nelson, T. O., & Dunlosky, J. (1991). When people's judgments of learning (JOLs) are extremely accurate at predicting subsequent recall: The "delayed-JOL effect." *Psychological Science, 2*, 267–270.

Nelson, T. O., & Leonesio, R. J. (1988). Allocation of self-paced study time and the "labor-in-vain effect." *Journal of Experimental Psychology: Learning, Memory, and Cognition, 14*, 676–686.

Nelson, T. O., & Narens, L. (1990). Metamemory: A theoretical framework and new findings. In G. Bower (Ed.), *The psychology of learning and motivation: Advances in research and theory* (Vol. 26, pp. 125–173). New York: Academic Press.

Nisbett, R. E., & Wilson, T. D. (1977). Telling more than we can know: Verbal reports on mental processes. *Psychological Review, 84*, 231–259.

Nelson, T. O., & Leonesio, R. J. (1988). Allocation of self-paced study time and the "labor-in-vain effect." *Journal of Experimental Psychology: Learning, Memory, and Cognition, 14*, 676–686.

Posner, M. I., & Rothbart, M. K. (1989). Intentional chapters on unintended thoughts. In J. S. Uleman & J. A. Bargh (Eds.), *Unintended thought* (pp. 450–469). New York: Guilford.

Schooler, J. W., & Engstler-Schooler, T Y. (1990). Verbal overshadowing of visual memories: Some things are better left unsaid. *Cognitive Psychology, 22*, 36–71.

Schooler, J. W., Ohlsson, S., & Brooks, K. (1993). Thoughts beyond words: When language overshadows insight. *Journal of Experimental Psychology: General, 122*, 166–183.

Vesonder, G. T., & Voss, J. F. (1985). On the ability to predict one's own responses while learning. *Journal of Memory and Language, 24*, 363–376.

Wegner, D. M. (1992). You can't always think what you want: Problems in the suppression of unwanted thoughts. In M. P. Zanna (Ed.), *Advances in experimental social psychology* (Vol. 25, pp. 193–225). San Diego: Academic Press.

Wegner, D. M. (1994). Ironic processes of mental control. *Psychological Review, 101*, 34–52.

Wegner, D. M., Schneider, D. J., Carter, S., II, & White, T. (1987). Paradoxical effects of thought suppression. *Journal of Personality and Social Psychology, 53*, 5–13.

Wetzel, C. G., Wilson, T. D., & Kort, J. (1981). The halo effect revisited: Forewarned is not forearmed. *Journal of Experimental Social Psychology, 17*, 427–439.

Wilson, T. D., & Brekke, N. C. (1994). Mental contamination and mental correction: Unwanted influences on judgments and evaluations. *Psychological Bulletin, 116*, 117–142.

Wilson, T. D., Bybee, J. A., Dunn, D. S., Hyman, D. B., & Rotondo, J. A. (1984). Effects of analyzing reasons on attitude-behavior consistency. *Journal of Personality and Social Psychology, 47*, 5–16.

Wilson, T. D., & Dunn, D. S. (1986). Effects of introspection on attitude-behavior consistency: Analyzing reasons versus focusing on feelings. *Journal of Experimental Social Psychology, 22,* 249–263.

Wilson, T. D., Dunn, D. S., Kraft, D., & Lisle, D. J. (1989). Introspection, attitude change, and attitude-behavior consistency: The disruptive effects of explaining why we feel the way we do. In L. Berkowitz (Ed.), *Advances in experimental social psychology* (Vol. 19, pp. 123–205). Orlando, FL: Academic Press.

Wilson, T. D., Hodges, S. D., & LaFleur, S. J. (1995). Effects of introspecting about reasons: Inferring attitudes from accessible thoughts. *Journal of Personality and Social Psychology, 69,* 16–28.

Wilson, T. D., & Kraft, D. (1993). Why do I love thee?: Effects of repeated introspections about a dating relationship on attitudes toward the relationship. *Personality and Social Psychology Bulletin, 19,* 409–418.

Wilson, T. D., Kraft, D., & Dunn, D. S. (1989). The disruptive effects of explaining attitudes: the moderating effect of knowledge about the attitude object. *Journal of Experimental Social Psychology, 25,* 379–400.

Wilson, T. D., Laser, P. S., & Stone, J. I. (1982). Judging the predictors of one's own mood: Accuracy and the used of shared theories. *Journal of Experimental Social Psychology, 18,* 537–556.

Wilson, T. D., Lisle, D., Schooler, J., Hodges, S. D., Klaaren, K. J., & LaFleur, S. J. (1993). Introspecting about reasons can reduce post-choice satisfaction. *Personality and Social Psychology Bulletin, 19,* 331–339.

Wilson, T. D., & Schooler, J. (1991). Thinking too much: Introspection can reduce the quality of preferences and decisions. *Journal of Personality and Social Psychology, 60,* 181–192.

Wilson, T. D., & Stone, J. I. (1985). Limitations of self-knowledge: More on telling more than we can know. In P. Shaver (Ed.), *Review of personality and social psychology* (Vol. 6, pp. 167–183). Beverly Hills, CA: Sage.

VI

Neuropsychological and Neurobiological Approaches

CHAPTER 16

What Qualifies a Representation for a Role in Consciousness?

Marcel Kinsbourne
New School for Social Research, New York

Theoretical interest in defining and explaining consciousness has surged dramatically in recent years (CIBA Foundation, 1993; Dennett, 1991; Marcel & Bisiach, 1988; Searle, 1990). Indirectly, empirical studies in cognitive psychology have also addressed consciousness by the intensive analysis of nonconscious information processing, both by normal people (Schacter, 1992, and several chapters in this volume) and by patients with neuropsychological syndromes that selectively constrain awareness, such as amnesia, blindsight, and unilateral neglect (McGlynn & Schacter, 1989; Moscovitch & Umiltà, 1994). By a subtractive logic, nonconscious processing supposedly casts light on consciousness. These debates all revolve around issues best addressed by neuropsychological research.

NEUROPSYCHOLOGY AS THE SCIENCE OF CONSCIOUSNESS

Conceptually, neuropsychology occupies the cognitive neuroscience middle ground, between the software simulations of artificial intelligence (AI), and the hardware analyses of neurophysiology. Neither AI nor neurophysiology as directly investigates consciousness. Neuropsychology is the essential discipline for determining the relation between human experience and brain organization. Neuropsychologists can adapt concepts and methods from cognitive and brain science toward the goal of clarifying the relevant neuronal organization. The recent emphasis on parallel distributed processing in AI and on heterarchical rather than hierarchical organization in neurophysiology encourages efforts toward a viable neuropsychological theory of consciousness.

THE CENTERED BRAIN

The assumption that information flow into and through the brain is convergent and integrative and therefore hierarchical seemed too obvious to classical neuropsychologists to merit debate. Correspondingly, scientists and philosophers have long taken it for granted that consciousness is housed in a single and unique area of exquisite specialization, deeply recessed from the input to and output from the brain (Sherrington, 1933). Descartes nominated the pineal gland as interface between body and soul on account of its unique midline location. (Though his contemporary, the surgeon, Willis, rejected this idea because many animals, unquestionably lacking souls, also have pineals.) The centered point of view extrapolates into the substance of the brain convergent peripheral sensory channels. Similarly, it back projects divergent motor channels that transmit decisions for action. At the level of object recognition the hypothetical "grandmother cell" incorporates the set of features unique to a particular recognizable thing, assembling them from convergent inputs. The highest (most deliberate) mental functions were referred to a literal "highest" level in the cerebrum. The hypothesized confluence of input did seem to peter out in ill-defined "association areas," whence intentions would somehow consequently arise, and actions would follow.

Every theorist routinely professes disbelief in an inner conscious homunculus, to which information, suitably preprocessed, is displayed. Such a construct only kicks the problem up nonexistent stairs. Nonetheless, Sherrington (1933, p. 25) complained that "when trying to allocate nerve-action with mental activity, we face something which not only transmits signals but reads them. Otherwise the signaling, whatever its complexity, remains just one red lamp showing itself to another red lamp." In more contemporary language, "phenomenal experience is a specific construction to which previously activated schemata have contributed" (Mandler, 1985, p. 35). To contribute, they have to be entered into a specialized system. Even contemporary neuropsychological theory envisages specialized "supervisory" structures that "enter" information into consciousness (e.g., Edelman, 1989; J. A. Gray, in press; Johnson-Laird, 1983; Posner & Rothbart, 1992; Shallice, 1988) so that some central entity can "read" it. The ghost still clings tenaciously to the innards of the machine. Consciousness is its analogue to ectoplasm.

A SPECIALIZATION FOR SUBJECTIVITY?

If consciousness has causal efficacy, as some theorists have held (Eccles, 1970; Sherrington, 1946; Sperry, 1965) then one must explain how it interacts with neural systems. Sherrington (1933, p. 28) asked, "May it be that in those parts of the brain which may be called mental, nerve actions exist still unknown to us, and that these may correlate with mind?" Such dedicated nerve actions still remain undiscovered. The idea that there is a specialized conscious awareness system, module or "workspace" that functions as a central information exchange (Baars, Fehling, LaPolla, & McGovern, chapter 22, this volume; De Haan, Bauer, & Grove, 1992) arises naturally from the assumption that it takes highly specialized processing to engender awareness from the activity of neurons. The elitist notion that consciousness requires its own dedicated neural paraphernalia reflects the intuition that elaborate, even impenetrably mysterious processes (Penrose, 1989) are involved. This consciousness chauvinism requires that only a very specialized neural mechanism can participate in such a supremely sophisticated activity (as generating meaning, Searle, 1992; Searle's claim that only conscious states can be meaningful is thoroughly refuted by the chapters on unconscious processing in this volume).

The implication that consciousness is functionally the pinnacle of an ascending gradient of understanding and the origin of intentions also tends to place it center stage in the brain. But, as Jackendoff (1987) pointed out, conscious representations are typically "intermediary." Input is experienced at a perceptual level, with semantic implications not explicit in the phenomenology. Action is initiated preconsciously, before the individual is aware of his intention to act (e.g., Libet, 1985). If there were an all-knowing homunculus, it would not be the locus of consciousness.

The reifications of consciousness can be subjected to an empirical test. If experience is a product (be it independently instrumental or epiphenomenal) that is generated by neuronal activity, then the brain activity and the experience should be dissociable. A pathological state should exist in which the neurons continue to function, but experience fails to happen. This is why philosophers are so interested in processing without awareness, as in "blindsight" (Weiskrantz, 1986). The hope is to be able to argue that the occipital lesion in blindsight peeled away consciousness, leaving the processing of visual stimuli intact, and thereby to demonstrate that consciousness has a separate or separable existence. In fact, the processing that remains possible without awareness in blindsight is grossly deficient as compared to conscious processing, notably in that it is not integrated with the state of the organism. It is not related to context. Therefore, it is not voluntarily initiated, but has to await the behest of the experimenter. Other instances of unconscious processing are similarly fragmented. No true dissociation between fully functional neuronal activity in control of behavior and lost consciousness (or the reverse) is on record. In the total absence of supportive evidence, it is fruitless to pursue the type of speculation that treats consciousness as *sui generis*, be it a force field (Libet, 1994), a quantum state (Hameroff, 1994), or "just one more physically based, biological information bearing medium" (Mangan, 1993, p. 13).

SUBJECTIVITY IS NEURAL ACTIVITY

A more parsimonious alternative interpretation is that subjectivity is not a product of neuronal activity. It is the activity of the circuitry itself. It is an individual perspective on personal neuronal activity; it is what it is like for neurons to be in a particular state. The human brain does not generate or produce consciousness. It is conscious (or, at least some of it, in some of its possible functional states, is conscious). Nothing additional is produced and there is nothing beyond the neural activity that can be studied by the methods of science, which operate in the public domain. In fact, it is a category error to suppose that any physical state beyond the state of the neurons could further explain subjective phenomena. Science finds general reasons for specific facts. How could one ever explain why a particular publicly observable neural state (or its collapse, Nunn, Clarke, & Blott, 1994) results in particular private subjective impression? How could one possibly predict from the levels of physical (observable) variables the state of (unobservable) subjective variables? No individual can truly know what it is like to be another individual (of the same species, not to mention of another species as remote in brain organization as a bat; Nagel, 1974), without himself assuming the exact same brain state, which would be tantamount to being that individual, and no longer himself. Indeed, subjectivity is so closely tied to the state of the moment (the specious present) that one cannot even recapture, in any detail, what it was like to have been oneself on a previous occasion. Recollecting is really recreating, and this is inextricably tied into the present within which it is accomplished. Therefore, it is in principle impossible to verify that any predicted subjective state based on physical observations has in fact occurred, that is, to render the physical-subjective relation "transparent" (as worded by Nagel, 1993). But then it is not the cognitive neuroscientist's responsibility to acquaint one organism with how things seem to another or to determine why things seem the way they do. The neuropsychologist need not pursue consciousness as though it were a separate entity. The neuropsychologist's task is to determine the set of brain states of which subjective states are concomitants. By determining the detailed nature of each neural state that corresponds to each subjective state we explain, not only in practice but also in principle, consciousness. Such an ambitious effort is currently well beyond us. But, to make a start, we can ask: What are some properties of circuitry that is conscious? What do states of consciousness have in common, how do they systematically differ from nonconscious states, which (if any) neural systems are uniquely engaged during conscious activity, and which (if any) never are?

CONSCIOUS PROCESSES
OR CONSCIOUS REPRESENTATIONS?

Lashley (1956) drew attention to the introspectively obvious fact that when we identify a stimulus or solve a problem, we are not aware of the process of identifying or solving that preceded the percept or the solution that "pops into mind" (see also Prinz, 1992). The mental machinery works outside consciousness (e.g., Velmans, 1991). The brain does not sense its own electrical activity, but represents an extrinsic reality. The ingredients of our experience are largely "finished products," not products in the making. They are the representations that result when the microgenesis (Brown, 1988) is done. This is not to concede that processing and representation are distinct, and utilize different neural systems. Alternatively, representations perpetually transform in a self-organizing neural network (Globus & Arpaia, 1994) to generate the ongoing succession of state changes that characterizes a living self-organizing network. The ongoing transformations (or settlings of the network) are not conscious, only the transformed brain state that results. Representation does not imply that symbolic representations exist in the brain; it refers to brain states that code for action appropriate to the existing environment, thus representing those stimulus features of current adaptive significance (see Bickhard, 1993; Brooks, 1991). A model of conscious brain states must therefore in this sense invoke representations, and must explain why only some of the functionally active representations in the brain normally code for contents of which we become aware. Is it because only some cell assemblies are specialized to do so (Hobson, chapter 19, this volume), or are there circumstances under which any cell assembly might participate? This chapter attempts to determine the boundaries of consciousness by establishing what the brain can accomplish unconsciously or preconsciously.

DOES NONCONSCIOUS PROCESSING CAST LIGHT ON CONSCIOUSNESS?

Studies of nonconscious (implicit) information processing generally do not explicitly target the phenomenon of consciousness. Their more limited objective is to determine the extent to which tasks can be accomplished without benefit of awareness. This literature would cast light on consciousness if nonconscious processing regularly occurs first, and consciousness is the result of an additional processing stage. But, as Jacoby (Jacoby, Yonelinas, & Jennings, chapter 2, this volume) points out in the context of memory studies, implicit and explicit processing may not be sequential but independent. If so, studying unconscious processes may cast little light on consciousness. A recent instance is reported by Neuman, Niepel, Tappe, and Koch (in press). When auditory and visual stimuli were paired, the auditory stimulus resulted in a speedier motor response. But the visual stimulus was given precedence in temporal order judgment. Neuman et al. (in press) suggested that the auditory information is initially transported more quickly and gains earlier access to the motor response facility. But the visual activation "overtakes" it in the race to the second, more distant "finish line," consciousness. An alternative view is that there is no single race along a single track, and no single finish line (Dennett & Kinsbourne, 1992). It so happens that the auditory activation more rapidly entrains with the motor system, and the visual activation more readily entrains in the "dominant focus" of consciousness (discussed later). According to this account (which is hypothetical, but testable), prior access to motor reactions, which can be programmed before the subject is conscious of the stimulus (discussed by Velmans, 1991), has no bearing on awareness. Farah (Farah, O'Reilly, & Vecera, chapter 17, this volume) suggested that any representation the content of which would normally be conscious, but on account of brain damage is not, must be defective in itself. This must often be the case. But Schacter (1992) pointed out that in amnesia, nonconscious priming of memory not only occurs but is no less efficient than in normals. This suggests that a representation may in itself be sufficiently intact to influence subsequent behavior, but nonetheless be precluded from contributing to consciousness by the pathology. The crucial variable may be duration. Possibly a representation needs to be longer activated for its contents to contribute to consciousness, than solely to exert nonconscious effects, even in the intact brain. If so, the literature on nonconscious information processing is relevant only in so far as it clarifies neural states that are substrates for the emergence, or microgenesis, of conscious states. But it is now feasible, both conceptually and methodologically, to sidestep priming effects in favor of discovering how and where the neural substrate of consciousness is distributed and how its activity is organized in the brain.

ARE REPRESENTATIONAL CONTENTS CONSCIOUS BY VIRTUE OF THEIR LOCALIZATION?

What properties enable a representation to donate its contents to the field of awareness? Two overriding concepts compete. One has already been mentioned: The represented information has to be "entered" into a specialized part of the brain (a privileged location), which is dedicated to transducing mere neuronal activity into subjective experience (e.g., Schacter, McAndrews, & Moscovitch, 1988). And, there are monitoring systems that do the "entering" (Bisiach, 1988; Gray, in press; Johnson & Reeder, chapter 13, this volume; Reber, chapter 8, this volume; Stuss, 1991). The alternative to this concept of consciousness embedded in a "centered brain" is the idea that the contents of representations become conscious not by virtue of where the corresponding cell assemblies are localized in the brain, but when the cell assemblies are in specific states (regardless of where in the brain the circuitry that embodies them is to be found). This "uncentered" view has the corollary that it downplays the uniqueness and unity of the human brain–awareness relation by attributing to widespread and diverse neuronal circuitry the capability of generating conscious experiences (as long as it meets strict functional conditions). This opens the flood gates to the possibility that there are multiple consciousness in one mind, and that different kinds of minds are conscious (consciousness as a graded characteristic; Hobson, chapter 19, this volume).

MULTIPLE SITES FOR CONSCIOUSNESS

It should have been obvious to neuropsychologists from when they first addressed the issue that no one place in the brain is dedicated to the generation of awareness. There is no known site at which a lesion eliminates awareness in general while conserving the full human repertoire of mental operations. Nor does any focal cerebral lesion eliminate all perception and decision. Notably even loss of both frontal lobes—the most promising sites for "supervisory" structures—does not abolish consciousness (Eslinger & Damasio, 1985). That can only be accomplished by the destruction of subcortical activating systems (see Hobson, chapter 19, this volume; and Bogen, 1995, who persuasively advocate the intralaminar nuclei of the thalamus as necessary, though not sufficient, for consciousness; see also Kinsbourne, 1995). Thus the awareness inherent in the outputs of mental operations seems to arise from the operations themselves rather than after their proceeds are transported or transduced "into consciousness."

As neurophysiological knowledge accumulates, the potential neural substrate of consciousness retreats from most of the brain to ever smaller residual parts thereof, as newly discovered sensory representations crowd out what was, largely by default, called association cortex. At a recent count, 32 visual maps had been identified, as well as almost 300 reciprocal interconnections (Felleman & Van Essen, 1991). In contrast, there is a dearth of multimodal cortex and of convergence zones. This arrangement is more consistent with parallel (heterarchical) than hierarchical interaction. The maps extract different perceptual attributes, and feed back on to each other to represent the appearance of the moment, with reciprocal feedback through the limbic system (Eichenbaum, Otto, & Cohen, 1994). The temporarily stable state of the network is experienced as the percept. It is certainly modulated by prefrontal and related subcortical structures, but it does not need to be brought to the attention of some extraneous consciousness module downstream. The assumption that motor cortex is hierarchically organized must also be abandoned (Strick, 1988). Not only is there no Central Experiencer, there is no Central Intender either. However, as Horrobin (1990) noted, beliefs generally persist within the biomedical community long after evidence is available to destroy them, and most theorists still imply a centered location for consciousness at the same time as they disavow this belief.

Alternatively, because the notion of "distributed" circuitry in the brain is current, can a special status for "conscious representations" be salvaged by proposing a distributed (though still dedicated) basis for consciousness? On this assumption the conspicuous absence of a single consciousness-ablating lesion is no longer a problem. But any such retreat, however circumspect from the centered brain, faces a new difficulty. If different aspects of reality are represented in distinct areas, how are they "bound" together to form the unified percept? This is the "binding problem" (e.g., Hardcastle, 1994). A more economical move is to dispense altogether with the assumption that there is cerebral circuitry dedicated to holding "bound" information (representational contents) in consciousness. The mental operations that generate any representations may activate the cell assemblies in such a way as to render their contents potentially conscious.

THE UNCENTERED BRAIN

If there is no consciousness module, what determines which of the many sets of representations contribute to consciousness? Accepting that the cerebrum is a massively parallel and recursive system, some have assumed that binding, by one or other means, suffices for the bound material to constitute objects, or even contents in consciousness (Crick & Koch, 1990; Damasio, 1989). But even if binding of properties within or across modalities is indeed a necessary step toward awareness of an object, it cannot be sufficient. This is because objects can prime subsequent behavior while they are themselves outside awareness, and presumably their attributes must have been bound unconsciously to enable them to do so. Being bound does make them more effective candidates for consciousness, however, as follows: The uncentered "integrated cortical field" model (Kinsbourne, 1988, 1993a), proposes that any representation may, if it meets certain specifiable conditions, contribute its contents to experience without its coded information having to be transported to some privileged place in the brain. The attractor state of the neural network embodies the represented contents. It is, at any moment, the outcome of rivalry between competing behaviors. What determines whether the contents of a representation become conscious is not the intrinsic nature of the contents, nor its location in the brain, but how amenable it is to being integrated into the dominant neuronal activity. The state of the network is all there is. No additional binding is called for; there is nothing for which the represented information need be packaged. The "dominant focus" (Kinsbourne, 1988) is not localized; its components are its constituent representing cell assemblies, however dispersed in the brain. These change from moment to moment. The limited capacity of consciousness is not the consequence of limitation in any resource other than the scope of the neural network itself (Kinsbourne,

1981). The network is either engaged intensively by a limited set of representations in focal attention, or more superficially invested in more of a hodge podge of contents in diffuse attention. Peripheral or "fringe" information is no less well specified than information that is in focal attention. It differs in the nature of its contents, not in "how conscious it is" (Kinsbourne, 1982). In my version of the uncentered brain, each content is represented once only at the appropriately specialized point in the brain. There is no evidence that any cell assembly "knows" all that another knows, and more beside. Representations contribute their contents to awareness if they are sufficiently activated for a sufficient period of time and are integrated into the *dominant focus* of neural activity—that is, the predominant pattern of neuronal firing constituting the brain state (network attractor state), which is in current control of response processes, including verbal and imaged thought and expression.

QUALIFYING FOR CONSCIOUSNESS

Consider three properties that might enable a representation to contribute to awareness: *duration, activation,* and *congruence.* Damage to its neural substrate might disqualify a representation from being activated and enduring enough (though it might still exert unconscious, e.g., priming, effects; cf. Farah et al., chapter 17, this volume). Implicit effects (Jacoby et al., chapter 2, this volume; Merikle & Joordans, chapter 6, this volume; Reber, chapter 8, this volume) would be mediated by representations that fall short of eligibility for consciousness in one or more of the three respects discussed later.

Assuming that no specific neuronal circuits are dedicated to consciousness, under what conditions might any circuit, the functional state of which represents something of which people can become aware, be included in the neural basis of ongoing experience?

Sufficient Duration

Binding may occur when cell assemblies fire in temporal synchrony (e.g., Crick & Koch, 1990; C. M. Gray & Singer, 1989). If so, such a synchrony may be initiated either virtually instantly, when a novel, massive, or biologically prepotent sudden input (e.g., pain) commands neural space, or over time, as asynchronously firing neuron assemblies settle into synchronous oscillation. In either case, the causative events capture awareness.

The gradual settling process is subject to multiple revision as pertinent information continues to enter the system (Multiple Drafts—Dennett & Kinsbourne, 1992). Earlier "drafts" may be conscious and remembered: I see a "cat" on the highway, which as I come closer clarifies into a sliver of whitewall tire. I remember both cat and tire. Or the earlier drafts may not be remembered (it being indeterminate whether or not they were briefly conscious), being rapidly superseded by a more ecologically valid and enduring interpretation. Thus, the growth of information about a percept, its "clearing up" allegedly in terms of a primal sketch yielding to 2½-D (viewer-centered) vision (Marr, 1982), is not reflected by a conscious succession of approximations to the final gestalt. The preliminary stages are too brief. They are too brief to become entrained or, though entrained, they are too evanescent to be remembered moments later when the observer reports his experience (which, it is in principle impossible to know; see Dennett & Kinsbourne, 1992). Other nonconscious representations are also staging posts on the way to forming the representations that enter consciousness, as when an automatized action sequence yields an experience only as its final product. The stage at which perceptual representation organizes action is normally likely to be more enduring, and if it is the case that conscious representations are characterized by common coding of perception and action (Prinz, 1992), their enduring nature may be why they are conscious—although no percept can be assumed to be canonical. Draft revision continues indefinitely in memory (Dennett & Kinsbourne, in press).

But if the formation of the percept is aborted on account of pathology somewhere along the information-processing sequence (the "settling" process), then an intermediary representation, such as a primal sketch in vision, perhaps does linger into awareness. This might afford the patient with recovering cortical blindness (Poppelreuter, 1923), a visual experience denied to visually intact people. The idea that pathology unveils evanescent preattentive events is a mainstay of Brown's (1988) explanation of much neuropsychological symptomatology as due to curtailed microgenesis of perception and action. Similar considerations apply to speech perception (Jackendoff, 1987). We perceive a segmented transformation, by top-down influence, of the objectively continuous flow of auditory signals generated by the speaker's vocalizations. Maybe an initial representation that codes the acoustic signals is so rapidly superseded by the corresponding phonological representation that the former is not experienced (or remembered as an experience). The auditory agnosic may, however, experience the speech sounds in an unelaborated manner that more closely reflects their untransformed physical character (Saffran, Marin, & Yeni-Komshian, 1976). The experience is perhaps like that of listening to a speaker in a foreign language except that the sense of familiarity may be preserved. The same may apply to syntactic analysis. This organizes the phonologically coded input but is immediately superseded by representations of meaning, which are indissolubly incorporated into the experience of listening to speech. When a Wernicke's aphasic listens to speech, does the patient's sluggish semantic system expose his auditory awareness to phonology and syntax unorganized by meaning? More generally, abstract representations, where they exist, may remain unexperienced because they are so rapidly instantiated in a sensory brain code.

The representations that in succession result in speech output (Levelt, 1989) are also not ordinarily experienced, perhaps for the same reason. The semantic preverbal representation may so rapidly be superseded by the relevant syntax and phonology that its unique contribution to consciousness is undermined. An ill-defined feeling of intention lingers in awareness. Again, certain language impairments could be sought, in which speech realization is so sluggish that the antecedent semantic representations enter awareness in some as-yet-unknown form. Expressive aphasics certainly act as though they know what they want to say. A normal analogue may be the occasionally experienced "feeling of knowing" while groping for a word in the tip-of-the-tongue phenomenon. The phonological representation is so rapidly replaced by the phonetic representation (both in inner and overt speech; Levelt, 1989) that it vanishes too quickly to become integrated into the dominant focus of awareness.

In fluent activities of the body, the intermediary representations generated by the component processes are too soon superceded to become conscious. A possible exception is in phantom limb, following amputation. The conventional explanation that it results from noisy discharge from the centrally labeled nerve endings embedded in the stump is inadequate, because it cannot explain why the limb, rather than being experienced as an inert appendage, is felt to move in coordination with the rest of the body. For instance, if the patient stumbles, he or she extends a nonexistent leg so as to retain balance (cited in Kinsbourne, 1995a). The intention to move, a mental image of the projected movement, is not normally experienced before and separate from the movement (which itself may go unnoticed). But in phantom limb, the movement is not realizable, and therefore the intentional representation lingers into consciousness as an illusory movement of the missing limb.

An act of the will, reinforced by practice, can even bring into awareness (and under conscious control) what are normally purely local, unconscious reactions. Examples are blushing, blanching, swallowing at the esophageal stage, and changing the rate of the heart beat. These are all instances of the potential eligibility for awareness and intention even of the most unpromising of candidate representations.

Sufficient Activation

Damaged Representations as Underactivated.   Neuropsychological deficits are often state dependent. Anomic patients at times can and at times cannot name a given object, amnesics sometimes do and sometimes do not recover a given memory, apraxics intermittently succeed in a particular action sequence. The necessary representations, therefore, cannot have been eliminated. Presumably they are underactivated. If so, appropriate cueing may strengthen them sufficiently to enable them to entrain with the dominant focus, and so should any maneuver that increases their activation. A vivid illustration derives from the study of unilateral neglect of space and person.

Left Neglect.   Large acute right posterior cerebral lesions not infrequently result in a dramatic withdrawal of attention from the left side of surrounding space and from left body parts, notably the hand. There is a corresponding exaggeration of attention directed rightward. The patient is aware of the right end of things and the rightmost of scattered objects, but not the left. Has the lesion disconnected input from the left side of space from a conscious awareness center? Once one realizes that it is not the left half of space but the left end of things that is implicated (Kinsbourne, 1970a, 1993b), this becomes an awkward concept: Is it plausible that there are separate representations and pathways for the right and left ends of things? A formulation in terms of the strength of activation is more explanatory. The right hemisphere contributes activation to those neurons that by their activity represent the left end of any laterally extended stimulus or display, and the left does the reverse (enabling shifts of attention along the object by shifts of the balance of activation between hemispheres, regardless of the hemispace in which the object is on view).

The attentional bias can be minimized by cues that stimulate the right brain to shift attention leftward. But more relevantly, it is modifiable by maneuvers quite unrelated to the visual task that enhance the right hemisphere's activation level relative to that of the left. Both sensory and cognitive manipulations have been found to be effective.

Correcting Neglect by Counteracting Hemisphere Imbalance.   Best documented is unilateral vestibular stimulation. Eliciting a unilateral caloric vestibular response by irrigating the ear contralateral to the lesion with lukewarm water, or the ipsilateral one with warm water, has the remarkable effect of temporarily eliminating neglect of the left side of the ambient environment (Rubens, 1985; Silberpfennig, 1949) and of the patient's own body (Vallar, Sterzi, Bottini, Cappa, & Rusconi, 1990). This maneuver activates the damaged hemisphere, and evidently this suffices to restore the equilibrium between opponent processors that enables one to orient to any point on one's own body surface or the surrounding world. Clearly, the cerebral lesion had not eliminated left-sided representations, because they so readily reappear. Nor has it disconnected input from a conscious awareness center. We can infer this from the little discussed incidental finding that the patient has no memory, in one state, of what it was like to be in the other. If input reaches awareness under the influence of ear irrigation, then why can it not subsequently be recollected? If the alternate states of consciousness are nothing more than alternate states of the same neural network, then this dissociation is predictable. In the neglect state, patients cannot recall (reimage) what it was like to see both sides of things; in the temporarily normalized state, they cannot represent what it was like not to see both sides of things.

Another way to manipulate the activation balance between the hemispheres is to attenuate visual stimulation of the left hemisphere by occluding the right eye. This obstructs the input to the right nasal retina, which is the major source of visual stimulation to the left hemisphere, Deuel and Collins (1983) pioneered this method of accelerating recovery in a monkey model of neglect, and Butter and Kirsh (1992) reported that right-eye patching can reduce the severity of left neglect in human patients. Reducing left hemisphere activation can enhance right hemisphere functioning by releasing the right hemisphere from excessive inhibition. This would increase the activation of representations in the right brain.

Kinsbourne (1970b) suggested a cognitive manipulation to modify right–left hemisphere imbalance. Previous work had indicated that imposing a light verbal load on a normal subject primes the left hemisphere and swings attention rightward (Kinsbourne, 1970a; see also Kinsbourne, 1972, 1973). Conversing with a neglect patient might therefore aggravate the left neglect by inducing a verbal mental set and thus further activating his left hemisphere and suppressing the right. Neglect should be mitigated if a nonverbal (right hemispheric) task is imposed. Heilman and Watson (1978) tested this prediction and confirmed it experimentally. Again, the idea that left neglect arises from underactivated representing of the left of things is consistent with this outcome. (The opposite should of course apply to right neglect.)

When a representation is demonstrably intact, even though rendered inaccessible by disease, then the lesion cannot have eliminated the neural substrate of the representation. Either the representation was depleted but not destroyed (Farah et al., chapter 17, this volume), or it had been multiplexed in neighboring cortex, where it remains accessible, though less readily (John, 1980). Neuropsychological syndromes may be classifiable into those in which the lesion impairs circuitry that activates specific representations at a distance, and those in which the substrate of the representation is itself destroyed. In the latter case, it should be impossible to restore the percept or action in question by any means. If the cell assemblies that represent specific information are widely diffused, then they must still be most concentrated at those points in the brain where focal lesions cause specific deficits. If the focal points are destroyed, then it should require a more potent activating influence to elicit the behaviors in question but it should not be completely impossible to do so (see Bach-y-Rita, 1993; Lashley, 1958, for comparable points of view).

Congruence of Representations. Represented contents can escape awareness, even though they have the necessary duration and level of activation, if attention is diverted elsewhere. For example, a sudden sharp pain renders the individual oblivious of whatever else is happening, eliminating from consciousness the contents of representations that were adequate a moment earlier. Another instance is counterirritation, in which local pain is relieved by a somewhat more painful stimulation elsewhere on the body. A shift in attention disengages the previously controlling complex of representations from awareness. Thus, the determinant is not the absolute level of cell assembly activity, but the relative activation of competing "drafts." What they compete for is participation in the dominant focus. Each newly incorporated content shapes or modifies the existing field of awareness, as the potter's fingers continually modify the artefact. Conversely, if contextual effects are powerful, contents that are not at all energized from sources external to the body can nonetheless be experienced if they correspond to the brain's "inference" about what must be the case (e.g., apparent movement, "filling in"). The effect of context in the form of expectancy on the percept is strongly implicated in a range of "microtemporal" phenomena in touch and vision, as discussed by Dennett and Kinsbourne (1992). They cited effects, such as apparent movement and the "cutaneous rabbit," in which static stimuli staggered in time give rise to the convincing illusion of a moving stimulus and stimulus sequence respectively. This is the brain's retrospective interpretation of what occurred, in the light of what would be expected, that is, of what makes ecological sense. Every time the mind (brain) goes "beyond the information given" (Bruner, 1983), a representation has been integrated into awareness in a manner that is congruent with a contextual effect. The brain basis of such "filling-in" is discussed by Kinsbourne (1994).

CAN REPRESENTATIONS OUTSIDE CONSCIOUSNESS INFLUENCE REMEMBERING?

According to the dominant focus model, the dominant organized pattern of brain activation (its dominant attractor state), which contributes its contents to consciousness, is in control of response processes. Other concurrently active but unentrained representations are relatively fragmentary and unorganized. Although these representations cannot control ongoing behavior, they can and often do modify response predispositions.

In episodic remembering, the contents of a prior experience is reexperienced by reactivation, in part, of the prior dominant focus. Subjects respond on the basis of this conscious recollection. However, their response may also be influenced by prior experiences that are not, at the time of remembering, consciously recollected. These are components of a prior dominant focus that falls short of reassuming control of awareness, but nonetheless bias ("prime") responding. The individual has "source amnesia" for the basis of such responding. Much of cognitive development is based on primed responding accompanied by source amnesia.

If the aforementioned account is correct, then none of the three models considered by Jacoby et al. (chapter 2, this volume) is viable. Upgrading of a representation into awareness could hardly decrease its potential for implicitly biasing responding (exclusivity model). The claim that contents, which is conscious, would have equally effectively exerted control on behavior were it unconscious (redundancy model), ignores the advantage for memory of consciously remembered context. For explicit and implicit remembering to be "independent" (i.e., "doubly dissociated") implies parallel but nonoverlapping conscious and unconscious organized sources of recovered information, whereas according to the dominant focus hypothesis, representations are not marked for conscious versus nonconscious efficacy but can play both roles. The view espoused here favors a fourth alternative, more akin to Farah's (chapter 17, this volume) approach to percepts; a previously experienced contents can bias subsequent behavior even though it is not included in a reactivated dominant focus that recapitulates (or reinvents) the state of awareness at that time. But it will influence behavior in stronger and more diverse ways if it is consciously recollected, not least of which is that it brings its prior context into awareness with it. The contextual embeddedness of the conscious memory explains why it is influenced by many more variables than is implicit memory.

DYSUNITY OF CONSCIOUSNESS

As there are numerous different response modes, the possibility exists that different organized patterns of representation gain control of distinct response systems at the same time. Whereas it appears that at any one time a single dominant focus incorporates and thus informs the verbal response system (giving rise to an illusory continuity of patchily remembered experiences as a constructed and extended narrative about the self; Dennett, 1991), separate complexes of cell assembly activity may control other response systems. These may be unconscious only from the vantage point of the verbal system. Such divided control over responses has been demonstrated in discriminative reaction time measurement using multiple effectors, voice, blink and button press, which often conflicted with each other (Marcel, 1993). It also occurs in dichotic listening, in which the subject reports a neutral word, whereas a synchronized emotional word, though unreported, may nonetheless control facial expression, via right hemisphere activation (Wexler, Warrenburg, Schwartz, & Jammer, 1992). James (1950) remarked that "organized systems of paths can be thrown out of gear with others, so that the processes in one system give rise to one consciousness, and those of another system to another simultaneously existing consciousness" (p. 399), providing a possible brain basis for multiple personality disorder (Humphrey & Dennett, 1989).

Focal brain lesions can skew or deplete the contents of consciousness, but they do not strip consciousness as such from the operations of the brain. It follows that conscious experience attends the operation of wide areas of neuronal network. Indeed, dual consciousness illustrates the flexibility of neural networks. Split brain research reveals distinct and even at times conflicting awarenesses within cerebral hemispheres, and regardless of whether one of these is more refined than the other, consciousness is revealed as investing more than a single focal point. That the usual claim is for only two, left and right, brain consciousnesses (Bogen, 1969) may be more a consequence of where the surgeon dare cut than of what might conceivably result from interruption of continuity within the cortical neuronal network (Kinsbourne, 1982). Multiple transections might generate multiple (albeit depleted) consciousnesses down to some critical lower limit on the necessary complexity of the network. Any disconnection syndrome (Geschwind, 1965) implies that one part of the brain can be aware of information that is inaccessible to another. How then can awareness be unitary (or even countable in whole numbers—see Nagel, 1971)?

Referring to what he saw as "the crude explanation of 'two selves' by 'two hemispheres,'" James (1950) remarked that "the selves may be more than two, and the brain systems severally used for each must be conceived as interpenetrating each other in very minute ways" (p. 400).

MINIMAL COMPLEXITY
THAT SUSTAINS CONSCIOUSNESS

Sherrington (1933), writing of the "great new nerve net," the activity of which coincides with mental experience, remarked that "no microscopical, no physical or chemical means detect there anything radically other than in nerve nets elsewhere. All is as elsewhere, except greater complexity" (p. 22). It seems reasonable to suppose that some minimal level of complexity of interaction between neurons is a necessary condition for consciousness on the part of the animal in question. Of course, no one knows what that threshold level is, and therefore no one knows how far back in phylogeny (or in human ontogeny) there existed not even a glimmer of awareness. Based on the dominant focus idea, it can be speculated that only those brains are conscious in which multiple cognitive and emotional processing systems are simultaneously active and provide context for one another. But for reasons already noted, it may be a category error to suppose that the subjectivity of primitive neural systems could ever be proven, or disproven.

CONCLUSIONS

This chapter has argued that any representation could, in principle, contribute its contents to experience without any other (supervisory) system having to project its contents on any metaphorical blackboard or screen for viewing in the brain's Cartesian theater. Consciousness does not arise within a distinct information-bearing medium, but characterizes the mundane functioning of wide areas of the brain. When the brain is intact and fully functional, however, many representations are transformed so quickly after they are formed that they do not have enough time to establish themselves firmly in consciousness, that is, in a stable resonant state of the neural network (dominant focus). When pathology obstructs a particular processing stage, then the antecedent representation may persist long enough to be experienced (by being incorporated in the dominant focus), affording the patient a subjective state denied the intact individual. Other representations are typically not sufficiently activated. Yet others depend on just the right contextual conditions to overcome and replace the current dominant focus; hence, alternate states of consciousness reach their extreme in multiple personality disorder.

The explicit study of consciousness by empirical means calls for the testing of hypotheses about what qualifies a representation to contribute to the individual's subjective experience. A logical point of departure in neuropsychology would be to scrutinize the effects of brain lesions for tell-tale aspects of the resulting pattern of misperception and misguided action (with concurrent monitoring by electrophysiological and metabolic measurement). From such observations one might be able to construct a description of the type of mental operation that generates conscious representations and of the subset of brain states that incorporates them. Ideally, one would also determine what it takes to restore to awareness the contents of representations that have been precluded from it on account of pathology, and what it takes to remove from awareness novel representations that are there on account of pathology. One would thereby further elucidate the qualifications required of a candidate representation for consciousness.

REFERENCES

Bach-y-Rita, P. (1993). Recovery from brain damage. *Journal of Neurological Rehabilitation, 6,* 191–199.

Bickhard, M. H. (1993). Representational content in humans and machines. *Journal of Experimental and Theoretical Artificial Intelligence, 5,* 285–333.

Bisiach, E. (1988). The (haunted) brain and consciousness. In A. J. Marcel & E. Bisiach (Eds.), *Consciousness in contemporary science* (pp. 101–120). Oxford, England: Clarendon.

Bogen, J. E. (1969). The other side of the brain: II. An appositional mind. *Bulletin of the Los Angeles Neurological Society, 34,* 135–162.

Bogen, J. E. (1995). On the neurophysiology of consciousness: I. An overview. *Consciousness and Cognition. 4*, 52–62.

Brooks, R. A. (1991). Intelligence without representation. *Artificial Intelligence, 47*, 139–160.

Brown, J. W. (1988). *Life of the mind: Selected papers.* Hillsdale, NJ: Lawrence Erlbaum Associates.

Bruner, J. (1983). *In search of mind.* New York: Harper & Row.

Butter, C. M., & Kirsh, N. (1992). Combined and separate effects of eye patching and visual stimulation on unilateral neglect following stroke. *Archives of Physical and Medical Rehabilitation, 73*, 1135–1139.

CIBA Foundation Symposium 174. (1993). *Experimental and theoretical studies of consciousness.* Chichester, England: Wiley.

Crick, F., & Koch, C. (1990). Towards a neurobiological theory of consciousness. *Seminars in Neuroscience, 2*, 263–275.

Damasio, A. R. (1989). Time-locked multiregional retroactivation: A systems-level proposal for the neural substrates of recall and recognition. *Cognition, 33*, 25–62.

Dennett, D. C. (1991). *Consciousness explained.* Boston: Little, Brown.

Dennett, D. C., & Kinsbourne, M. (1992). Time and the observer: The where and when of consciousness in the brain. *Behavioral and Brain Sciences, 15*, 183–247.

Dennett, D. C., & Kinsbourne, M. (1995). Multiple drafts: An eternal golden braid. *Behavioral and Brain Sciences, 15*, 810–811.

De Haan, E. H. F., Bauer. R. M., & Grove, K. W. (1992). Behavioral and physiological evidence for covert recognition in a prosopagnosic patient. *Cortex, 28*, 77–95.

Deuel, R. K., & Collins, R. C. (1983). Recovery from unilateral neglect. *Experimental Neurology, 81*, 733–748.

Eccles, J. C. (1970). Facing reality. Heidelberg: Springer.

Edelman, G. M. (1989). *The remembered present: A biological theory of consciousness.* New York: Basic Books.

Eichenbaum, H., Otto, T., & Cohen, N. J. (1994). Two functional components of the hippocampal memory system. *Behavioral and Brain Sciences, 17*, 449–518.

Eslinger, P., & Damasio, A. (1985). Severe disturbance of higher cognition after bilateral frontal ablation. *Neurology, 35*, 1731–1741.

Felleman, D. J., & Van Essen, D. C. (1991). Disinhibited hierarchical processing in the primate cerebral cortex. *Cerebral Cortex, 1*, 1–47.

Geschwind, N. (1965). Disconnexion syndromes in animal and man. *Brain, 88*, 237–294.

Globus, G. G., & Arpaia, J. P. (1994). Psychiatry and the new dynamics. *Biological Psychiatry, 35*, 352–364.

Gray, J. A. (in press). The contents of consciousness: A neuropsychological conjecture. *Behavioral and Brain Sciences.*

Gray, C. M., & Singer, W. (1989). Stimulus-specific neuronal oscillations in orientation columns of cat visual cortex. *Proceedings of the National Academy of Sciences, 86*, 1698–1702.

Hameroff, S. R. (1994). Quantum coherent in microtubules: A neural basis for emergent consciousness? *Journal of Consciousness Studies, 1*, 91–118.

Hardcastle, V. G. (1994). Psychology's binding problem and possible neurobiological solutions. *Journal of Consciousness Studies, 1*, 66–90.

Heilman, K. M., & Watson, R. T. (1978). Changes in the symptoms of neglect induced by changing task strategy. *Archives of Neurology, 35*, 47–49.

Horrobin, D. (1990). Discouraging hypotheses slows progress. *The Scientist*, 13–14.

Humphrey, N., & Dennett, D. C. (1989). Speaking for our selves. *Raritan, 9*, 68–98.

Jackendoff, R. (1987). *Consciousness and the computational mind.* Cambridge, MA: MIT Press.

James, W. (1950). *Principles of psychology.* New York: Dover.

John, E. R. (1980). Multipotentiality: A statistical theory of brain function—evidence and implications. In J. M. Davidson & R. J. Davidson (Eds.), *Psychobiology of consciousness* (pp. 129–146). New York: Plenum.

Johnson-Laird, P. (1983). *Mental models: Towards a cognitive science of language, inference, and consciousness.* Cambridge, MA: Cambridge University Press.

Kinsbourne, M. (1970a). The cerebral basis of lateral asymmetries in perception. *Acta Psychologica, 33,* 193–201.

Kinsbourne, M. (1970b). A model for the mechanism of unilateral neglect of space. *Transactions of the American Neurological Association, 95,* 143–145.

Kinsbourne, M. (1972). Eye and head turning indicate cerebral lateralization. *Science, 176,* 539–541.

Kinsbourne, M. (1973). The control of attention by interaction between the cerebral hemispheres. In S. Kornblum (Ed.), *Attention and performance* (Vol. 4, pp. 239–256). New York: Academic Press.

Kinsbourne, M. (1981). Single channel theory. In D. H. Holding (Ed.), *Human skills* (pp. 65–90). Chichester, England: Wiley.

Kinsbourne, M. (1982). Hemispheric specialization and the growth of human understanding. *American Psychologist, 37,* 411–420.

Kinsbourne, M. (1988). Integrated field theory of consciousness. In A. J. Marcel & E. Bisiach (Eds.), *The concept of consciousness in contemporary science* (pp. 239–256). New York: Oxford University Press.

Kinsbourne, M. (1993a). Integrated cortical field model of consciousness. In *Experimental and theoretical studies of consciousness.* Ciba Foundation Symposium 174 (pp. 43–51). New York: Wiley.

Kinsbourne, M. (1993b). Orientational bias model of unilateral neglect: Evidence from attentional gradients within hemispace. In I. H. Robertson & J. C. Marshall (Eds.), *Unilateral neglect, clinical and experimental studies* (pp. 63–86). New York: Lawrence Erlbaum Associates.

Kinsbourne, M. (1994). Models of consciousness: serial or parallel in the brain? In M. S. Gazzaniga (Ed.), *The cognitive neurosciences* (pp. 1321–1329). Cambridge, MA: MIT Press.

Kinsbourne, M. (1995a). Awareness of one's own body: A neuropsychological hypothesis. In J. Bermudez, A. J. Marcel, & N. Eilan (Eds.), *The body and the self* (pp. 205–224). Cambridge, MA: MIT Press.

Kinsbourne, M. (1995b). The intralaminar thalamic nuclei: Subjectivity pumps or attention-action coordinators? *Consciousness and Cognition, 4,* 167–171.

Lashley, K. S. (1956). Integrative functions of the cerebral cortex. *Physiological Review, 13,* 1–42.

Lashley, K. (1958). Cerebral organization and behavior. In H. C. Solomon, S. Cobb, & W. Penfield (Eds.), *The brain and human behavior* (pp. 1–18). Research Publications, Association of Research on Nervous and Mental Disorders, No. 36. Baltimore, MD: Williams & Wilkins.

Levelt, W. J. M. (1989). *Speaking.* Cambridge, MA: MIT Press.

Libet, B. (1985). Unconscious cerebral initiative and the role of conscious will in voluntary action. *Behavioral and Brain Sciences, 8,* 529–566.

Libet, B. (1994). A testable field theory of mind–brain interaction. *Journal of Consciousness Studies, 1,* 119–126.

Mandler, G. (1985). *Cognitive psychology: An essay in cognitive science.* Hillsdale, NJ: Lawrence Erlbaum Associates.

Mangan, B. (1993). Dennett, consciousness and the sorrows of functionalism. *Consciousness and Cognition, 2,* 1–17.

Marcel, A. J. (1993). Slippage in the unity of consciousness. In *Experimental and theoretical studies of consciousness.* Ciba Foundation Symposium 174 (pp. 168–180. New York: Wiley.

Marcel, A. J., & Bisiach, E. (Eds.). (1988). *The concept of consciousness in contemporary science*. New York: Oxford University Press.

Marr, D. (1982). *Vision*. New York: Freeman.

McGlynn, S. M., & Schacter, D. L. (1989). Unawareness of deficits in neurological syndromes. *Journal of Clinical and Experimental Neuropsychology, 11,* 143–205.

Moscovitch, M., & Umiltà, C. (Eds.). (1994). Conscious and nonconscious information processing. *Attention and performance* (Vol. 15). Cambridge, MA: MIT Press.

Nagel, T. (1971). Brain bisection and the unity of consciousness. *Synthese, 22,* 396–413.

Nagel, T. (1974). What is it like to be a bat? *Philosophical Review, 38,* 435–450.

Nagel, T. (1993). What is the mind–body problem? In *Experimental and theoretical studies of consciousness*. Ciba Foundation Symposium 174 (pp. 1–13). New York: Wiley.

Neuman, O., Niepel, M., Tappe, Th. & Koch, R. (in press). Temporal order judgment and reaction time to visual and auditory stimuli of different intensities: Further evidence for dissociation. *Perception and Psychophysics*.

Nunn, C. M. H., Clarke, C. J. S., & Blott, B. H. (1994). Collapse of a quantum field may affect brain function. *Journal of Consciousness Studies, 1,* 127–139.

Penrose, R. (1989). *The emperor's new mind: Concerning computers, minds and the laws of physics*. Oxford, England: Oxford University Press.

Poppelreuter, W. (1923). Zur Psychologie und Pathologie der optischen Wahrnehmung [On the psychology and pathology of visual perception]. *Zeitschrift für die gesamte Neurologie und Psychiatrie, 83,* 26–152.

Posner, M. I., & Rothbart, M. K. (1992). Attention mechanisms and conscious experience. In A. D. Milner & M. D. Rugy (Eds.), *Neuropsychology of consciousness* (pp. 91–111). New York: Academic Press.

Prinz, W. (1992). Why don't we perceive our brain states? *European Journal of Cognitive Psychology, 4,* 1–20.

Rubens, A. B. (1985). Caloric stimulation and unilateral visual neglect. *Neurology, 35,* 1019–1024.

Saffran, E. M., Marin, O.S.M., & Yeni-Komshian, G. (1976). An analysis of speech perception and word deafness. *Brain and Language, 3,* 209–228.

Schacter, D. L. (1992). Implicit knowledge: New perspectives on unconscious processes. *Proceedings of the National Academy of Sciences, USA, 89,* 11113–11117.

Schacter, D. L., McAndrews, M. P. & Moscovitch, M. (1988). Access to consciousness: Dissociations between implicit and explicit knowledge in neuropsychological syndromes. In L. Weiskrantz (Ed.). *Thought without language* (pp. 242–278). Oxford, England: Clarendon.

Searle, J. R. (1990). Consciousness, explanatory inversion, and cognitive science. *Behavioural and Brain Sciences, 13,* 585–642.

Searle, J. R. (1992). *The rediscovery of the mind*. Cambridge, MA: MIT Press.

Shallice, T. (1988). Information-processing models of consciousness: Possibilities and problems. In A. J. Marcel & E. Bisiach (Eds.), *Concept of consciousness in contemporary science* (pp. 305–333). London: Oxford University Press.

Sherrington, C. S. (1933). *The brain and its mechanism*. Cambridge, England: Cambridge University Press.

Sherrington, C. S. (1946). *Man on his nature*. Cambridge, England: Cambridge University Press.

Silberpfennig, J. (1949). Contributions to the problem of eye movements: III. Disturbance of ocular movements with pseudo hemianopsia in frontal tumors. *Confinia Neurologica, 4,* 1–13.

Sperry, R. W. (1965). Mind, brain and humanist values. In J. R. Platt (Ed.), *New views of the nature of man* (pp. 71–92). Chicago: University of Chicago Press.

Strick, P. L. (1988). Anatomical organization of multiple motor areas in the frontal lobe. In S. G. Waxan (Ed.). *Functional recovery in neurological disease: Vol. 47. Advances in neurology* (pp. 293–312). New York: Raven.

Stuss, D. T. (1991). Self, awareness and the frontal lobes: A neuropsychological perspective. In J. Strauss & G. R. Goethals (Eds.), *The self: Interdisciplinary approaches* (pp. 255–278). New York: Springer.

Vallar, G., Sterzi, R., Bottini, G., Cappa, S., & Rusconi, M. L. (1990). Temporary remission of left hemianesthesia after vestibular stimulation: A sensory neglect phenomenon. *Cortex, 26,* 123–131.

Velmans, M. (1991). Is human information processing conscious? *Behavioral and Brain Sciences, 14,* 651–726.

Weiskrantz, L. (1986). *Blindsight: A case study and implications.* Oxford, England: Oxford University Press.

Wexler, B. E., Warrenburg, S., Schwartz, G. E., & Jammer, L. D. (1992). EEG and EMG responses to emotion-evoking stimuli processed without conscious awareness. *Neuropsychologia, 30,* 1065–1079.

CHAPTER 17

The Neural Correlates of Perceptual Awareness: Evidence From Covert Recognition in Prosopagnosia

Martha J. Farah
University of Pennsylvania

Randall C. O'Reilly
Shaun P. Vecera
Carnegie Mellon University

Dissociations between perception and awareness of perception have been reported in a variety of neuropsychological disorders (see Farah, 1994, for a review). These dissociations are of great interest for what they can tell us about the neural correlates of perceptual awareness: What does the brain need to enable conscious, aware perception that it does not need to enable unconscious perception? One of the best-studied and most dramatic perception/awareness dissociations is found in *prosopagnosia*. This chapter reviews the evidence on unconscious, or "covert," recognition in prosopagnosia; proposes a new explanation for the phenomenon, which has been implemented in a connectionist computer simulation; and attempts to discriminate among the different explanations. It concludes with a discussion of the implications of each of these explanations of covert recognition for the neural correlates of conscious awareness.

Prosopagnosia is an impairment of face recognition that can occur relatively independently of impairments in object recognition, and that is not caused by impairments in lower level vision, or memory. Prosopagnosic subjects are impaired in tests of face recognition (including naming faces or classifying them according to semantic information, such as occupation) and are also impaired in everyday life situations that call for face recognition. Furthermore, by their own introspective reports, prosopagnosics do not feel as though they recognize faces. However, when tested using certain indirect techniques, some prosopagnosic subjects show evidence of face recognition. These subjects apparently recognize faces without being aware that they are

recognizing them. The evidence for so-called covert face recognition in prosopagnosia involves both psychophysiological and behavioral measures.

EVIDENCE FOR COVERT FACE RECOGNITION
IN PROSOPAGNOSIA

An early and impressive example of a psychophysiological evidence for covert
face recognition comes from Bauer (1984), who measured a prosopagnosic's
skin conductance while the subject viewed a series of photographs of familiar
faces. While viewing each photograph, the subject heard a list of names read
aloud, one of which was the name of the person in the photograph. For normal subjects, the skin conductance response (SCR) is greatest to the name
belonging to the pictured person. Bauer found that, although the prosopagnosic patient's SCRs to names were not as strongly correlated with the names
as a normal subject's, they were nevertheless significantly correlated. In contrast, the patient performed at chance levels when asked to select the correct
name for each face. In the absence of theories relating psychophysiological indices to mechanistic accounts of cognition or neural information processing,
it is difficult to use the psychophysiological findings to constrain a mechanistic model of covert recognition. Therefore, this chapter focuses primarily on
the behavioral data implicating covert recognition of faces in prosopagnosia.

The most widely documented form of covert face recognition occurs when
prosopagnosics are taught to associate names with photographs of faces. For
faces and names that were familiar to the subjects prior to their prosopagnosia, correct pairings are learned faster than incorrect pairings (e.g., De
Haan, Young, & Newcombe, 1987b). That is, prosopagnosics require fewer
learning trials when the pairing is correct than when it is incorrect, which suggests the subject does possess at least some knowledge of the people's facial
appearance. Furthermore, this has been shown to be true even when the stimulus faces were selected from among those the patient had been unable to
identify in a preexperiment stimulus screening test.

Evidence of covert recognition has also come from reaction time tasks in
which the familiarity or identity of faces are found to influence processing
time. In a visual identity match task, pairs of faces are presented and the subject must decide as quickly as possible whether they are the same face; normal subjects are faster to match familiar faces than unfamiliar. De Haan,
Young, and Newcombe (1987a) found that a prosopagnosic subject was also
faster at matching pairs of previously familiar faces. In contrast, he was unable
to name any of the previously familiar faces. De Haan et al. then went on to
show another similarity between the performance of the patient in this task
and that of normal subjects. If the task is administered to normal subjects with
either the external features (e.g., hair and jaw line) or the internal features
(e.g., eyes, nose, and mouth) blocked off, with instructions to match on the

visible parts of the face, normal subjects show an effect of familiarity only for the matching of internal features. The same result was obtained with the prosopagnosic subject.

In another reaction time (RT) study, De Haan, Young, and Newcombe (1987a, 1987b) found evidence that photographs of faces could evoke covert semantic knowledge of the depicted person, despite the inability of the prosopagnosic patient to report such information about the person when tested overtly. Their task was to categorize a printed name as belonging to an actor or a politician as quickly as possible. On some trials an irrelevant (i.e., to be ignored) photograph of an actor's or politician's face was simultaneously presented. Normal subjects are slower to categorize the names when the faces come from a different occupation category relative to a no-photograph baseline. Even though their prosopagnosic patient was severely impaired at categorizing the faces overtly as belonging to actors or politicians, he showed the same pattern of interference from different-category faces. This suggests that he was unconsciously recognizing the faces sufficiently to derive occupation information from them. A related finding was reported by Young, Hellawell, and De Haan (1988) in a task involving the categorization of names as famous or nonfamous. Both normal subjects and a prosopagnosic patient showed faster RTs to the famous names when the name was preceded by a picture of a semantically related face (e.g., the name "Diana Spencer" preceded by a picture of Prince Charles) than by an unfamiliar or an unrelated face. The patient was able to name only 2 of the 20 face prime stimuli used.

INTERPRETATIONS OF COVERT RECOGNITION, AND THEIR IMPLICATIONS FOR THE NEURAL CORRELATES OF PERCEPTUAL AWARENESS

Bauer (1984) offered the first interpretation of covert face recognition. He suggested that there may be two neural systems capable of face recognition (as shown in Fig. 17.1), only one of which is associated with conscious awareness. According to Bauer, the ventral cortical visual areas, which are damaged in prosopagnosic patients, are the location of normal conscious face recognition. The dorsal visual areas are hypothesized to be capable of face recognition as well, although they do not mediate conscious recognition but instead affective responses to faces. Covert recognition is explained as the isolated functioning of the dorsal face system. What does this interpretation imply about the neural correlates of awareness? It suggests that awareness is a function of some perceptual systems and not others. This is analogous to explanations of blindsight in which the preserved visual abilities are attributed to the subcortical visual system, and this visual system is hypothesized to operate normally without awareness (see Weiskrantz, 1986). The difficulty of this interpretation is that it suggests that covert recognition should be confined to recognizing the affective status of a face (e.g., whether or not the person is liked or disliked). In contrast, the findings reviewed earlier show that names and occupations of people can also be extracted from their faces during covert recognition by prosopagnosics.

Another interpretation of covert recognition is that the perceptual processing underlying face recognition per se is normal, but has been disconnected from other brain systems necessary for conscious awareness. For ex-

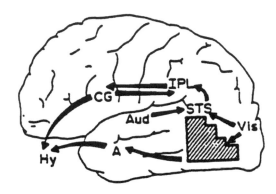

FIG. 17.1. Two routes to face recognition hypothesized by Bauer (1984).

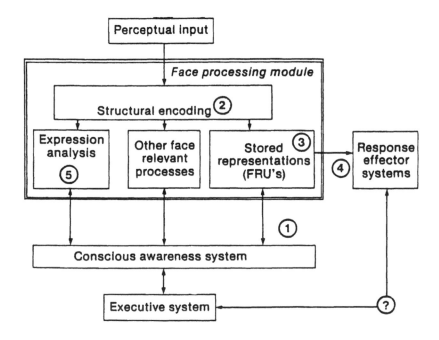

FIG. 17.2. A model of face perception and awareness, proposed by De Haan, Bauer, and Greve (1992), within which covert face recognition is explained by a disconnection at location 1.

ample, De Haan, Bauer, and Greve (1992) interpreted covert recognition in terms of the components shown in Fig. 17.2, in which separate systems subserve face recognition and conscious awareness thereof. According to their model, the face-specific visual and mnemonic processing of a face (carried out within the "face processing module") proceeds normally in covert recognition, but the results of this process cannot access the "conscious awareness system" because of a lesion at location number 1. Like the previous account, this one attributes the presence or absence of awareness to the involvement of a particular brain system. Whereas in the previous account the brain system needed for awareness of face recognition was also involved in face recognition per se, in this account there is a dedicated conscious awareness system. Thus, according to this interpretation, the neural correlate of perceptual awareness is the activity of a brain system that is specifically dedicated to enabling awareness.

A related hypothesis is that an intact face recognition system has been totally or partially disconnected from the rest of the cognitive system, and is therefore unable to contribute its outputs to the global, integrated state of the system. This has been proposed independently and in somewhat different terms by Tranel and Damasio (1988), and by Burton, Young, Bruce, Johnston, and Ellis (1991). Figure 17.3 illustrates the computer simulation of Bur-

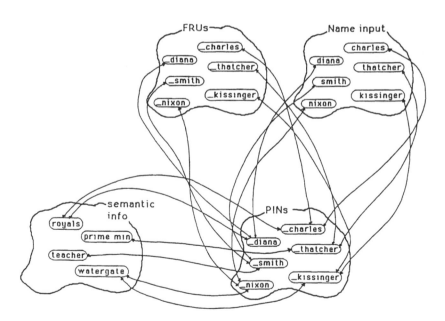

FIG. 17.3.   A schematic diagram of the connectionist model of Burton, Young, Johnston, and Ellis (1991), which explains covert face recognition in terms of an attenuation of the connections between FRUs and PINS.

ton et al. According to this interpretation of covert recognition, normal processing within modality-specific visual areas is not sufficient for awareness; this information must reach other brain systems before we experience awareness of it.

Both of the last two accounts feature a disconnection between intact face recognition and other brain systems. The primary motivation in each case is the assumption that face recognition is truly spared in covert recognition. Given that recognition is spared—yet the prosopagnosics have no awareness of recognition and fail face recognition tasks that require overt, conscious responses—a disconnection seems the obvious cause. However, it is unknown whether recognition is normal according to the covert measures (see Farah, 1994). Furthermore, prosopagnosics generally have impaired perception of faces (Farah, 1990); this includes those subjects who show covert recognition. For example, although the prosopagnosic subject showed faster matching of familiar faces than unfamiliar in the experiment described earlier, his response times and error rates were considerably higher than normal subjects' at the simple task of matching the faces.

These observations lead to a consideration of an alternative way of interpreting covert face recognition, with different implications for the neural correlates of perceptual awareness.

COVERT FACE RECOGNITION AS THE FUNCTIONING OF A DAMAGED, BUT NOT OBLITERATED, FACE RECOGNITION SYSTEM

Farah, O'Reilly, and Vecera (1993) tested the hypothesis that when face recognition mechanisms themselves are damaged, they would manifest their residual knowledge on just the types of tasks that have been used to measure covert recognition. Thus, this account differs from Bauer's (1984) hypothesis in requiring just one face recognition system. It also differs from those in which face recognition is disconnected from other brain systems by placing the damage within face recognition per se, not downstream. Perhaps more to the point of this chapter, it differs from all of the other accounts in suggesting that perceptual awareness, and the ability to perform overt perceptual tasks, are functions of the quality of the information processing within the perceptual system.

Why would a network retain covert recognition of faces at levels of damage that lead to poor or even chance levels of overt recognition? The general answer lies in the nature of knowledge representation in neural networks. Network representations consist of a pattern of activation over a set of highly interconnected neurons, or neuronlike model units. The extent to which the activation of one unit causes an increase or decrease in the activation of a neighboring unit depends on the "weight" of the connection between them. For the network to learn that a certain face representation goes with a certain name representation, the weights among units in the network are adjusted so that presentation of either the face pattern in the face units or the name pattern in the name units causes the corresponding other pattern to become activated. Upon presentation of the input pattern, all of the units connected with the input units will begin to change their activation in accordance with the activation value of the units to which they are connected and the weights on the connections. As activation propagates through the network, a stable pattern of activation eventually results, determined jointly by the input activation and the pattern of weights among the units of the network.

The account of covert face recognition described here is based on the following key idea: The set of the weights in a network that cannot correctly associate patterns because it has never been trained (or has been trained on a different set of patterns) is different in an important way from the set of weights in a network that cannot correctly associate patterns because it has been trained on those patterns and then damaged. The first set of weights is random with respect to the associations in question, whereas the second is a subset of the necessary weights. Even if it is an inadequate subset for performing the association, it is not random; it has some degree of knowledge of the associations "embedded" in it. Hinton and colleagues (Hinton & Plaut, 1987; Hinton & Sejnowski, 1986) showed that such embedded knowledge can be demonstrated when the network relearns, suggesting the findings of savings in relearning face–name associations may be explained in this way. In general, consideration of the kinds of tests used to measure covert recognition suggests that the covert measures would be sensitive to this embedded knowledge. The most obvious example is that a damaged network would be expected to relearn associations it originally knew faster than novel associations because of the nonrandom starting weights. Less obvious, but confirmed by simulations, the network would settle faster when given previously learned inputs than novel inputs, even though the pattern into which it settles is not correct, because the residual weights come from a set designed to create a stable pattern from that input. Finally, to the extent that the weights continue to activate partial and subthreshold patterns over the nondamaged units in association with the input, then these resultant patterns could prime (i.e., contribute activation toward) the activation of patterns by intact routes.

SIMULATION OF COVERT FACE RECOGNITION

Overt and covert face recognition were modeled using the five-layer recurrent network shown in Fig. 17.4. The "face input" units subserve the initial visual representation of faces, the "semantics" units subserve representation of the

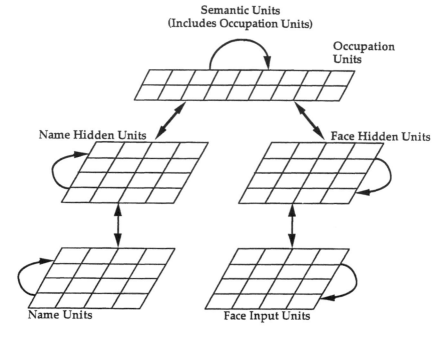

FIG. 17.4. A schematic diagram of the connectionist model of Farah, O'Reilly, and Vecera (1993), which explains covert face recognition in terms of damage to face recognition itself (face inputs units and/or face hidden units).

semantic knowledge of people that can be evoked by either the person's face or name, and the "name" units subserve the representation of names. Hidden units were used to help the network learn the associations among patterns of activity in each of these three layers. These are located between the "face" and "semantic" units (called the "face hidden" units) and between the "name" and the "semantic" units (the "name hidden" units). Thus, there are two pools of units that together comprise the visual face recognition system in our model, in that they represent visual information about faces: the face input units and the face hidden units.

The connectivity among the different pools of units was based on the assumption that in order to name a face, or to visualize a named person, one must access semantic knowledge of that person. Thus, face and name units are not directly connected, but send activation to one another through hidden and semantic units. All connections shown in Fig. 17.4 are bidirectional.

Faces and names are represented by random patterns of 5 active units out of the total of 16 in each pool. Semantic knowledge is represented by 6 active units out of the total of 18 in the semantic pool. The only units assigned an interpretation are the "occupation units" within the semantic pool. One repre-

(A) Hidden Unit Lesions

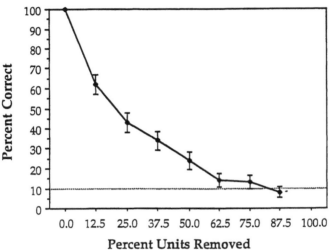

(B) Input Unit Lesions

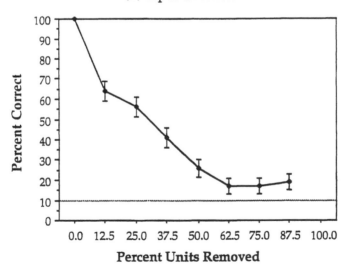

FIG. 17.5. Performance of the model before and after differing degrees of damage on an overt face recognition task, forced-choice naming.

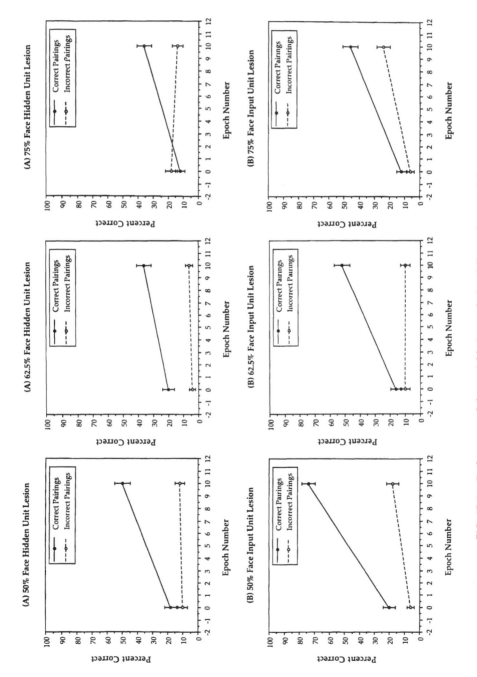

FIG. 17.6. Performance of the model before and after differing degrees of damage on re-learning correct and incorrect face–name associations.

sents the semantic feature "actor" and the other represents the semantic feature "politician." The network was trained to be able to associate an individual's face, semantics, and name whenever one of these was presented, using the Contrastive Hebbian Learning algorithm (e.g., Movellan, 1990). After training, the network was damaged by removing units.

Figure 17.5 (p. 365) shows the performance of the model in a 10-alternative forced-choice naming task for face patterns, after different degrees of damage to the "face input" and "face hidden" units. At levels of damage corresponding to removal of 62.5% and 75% of the "face" units in a given layer, the model performs at or near chance on this overt recognition task. This is consistent with the performance of prosopagnosic subjects who manifest covert recognition. Such individuals perform poorly, but not invariably at chance, on overt tests of face recognition.

In contrast, the damaged network showed faster learning of correct face–name associations. When retrained after damage, the network consistently showed more learning for correct pairings than incorrect in the first 10 training epochs, as shown in Fig. 17.6. The damaged network also completed visual analysis of familiar faces faster than unfamiliar faces. When presented with face patterns after damage, the face units completed their analysis of the input (i.e., the face units settled) faster for familiar than unfamiliar faces, as shown in Fig. 17.7. And finally, the damaged network showed semantic interference from faces in a name classification task. Fig. 17.8 shows that when the network was presented with name patterns, and the time to classify them according to occupation (i.e., the number of processing cycles for the occupation units to reach threshold) was measured, classification time was slowed when a face from the incorrect category was shown, relative to faces from the correct category and, in some cases, to a no-face baseline.

CONCLUSIONS

This chapter has reviewed a number of interpretations of covert recognition in the literature on prosopagnosia, and found two different views of the neural correlates of perceptual awareness implicit in these previous interpretations. One view is that awareness is dependent on particular brain systems. According to this view, the neural correlate of perceptual awareness should be described in structural terms: Awareness is the function of a particular system. In the case of Bauer's (1984) hypothesis, the ventral visual system serves the role of both recognizing faces and enabling conscious awareness of face recognition. In the case of De Haan et al.'s (1992) hypothesis, there is a system needed for conscious awareness that is exclusively dedicated to that purpose. A different view is that perceptual awareness is possible only when visual areas and other, nonvisual areas are in a state of communication and integra-

(A) Hidden Unit Lesions

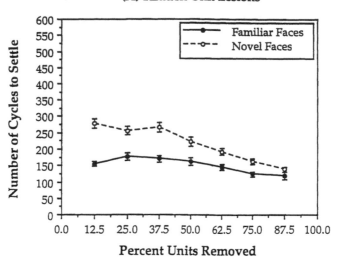

(B) Input Unit Lesions

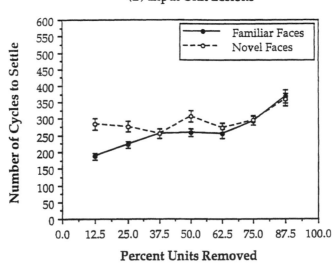

FIG. 17.7.   Performance of the model before and after differing degrees of damage on visual analysis of familiar and unfamiliar faces.

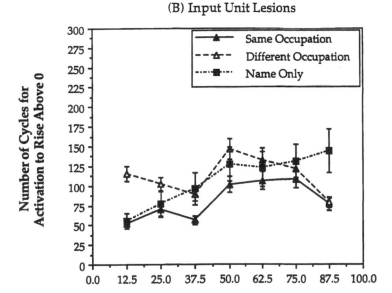

(A) Hidden Unit Lesions

(B) Input Unit Lesions

FIG. 17.8.   Performance of the model before and after differing degrees of damage on name classification when primed with faces.

tion. According to this view, the neural correlate of perceptual awareness should be described in state terms: It is a relatively global state of a set of systems, in which information within each individual system is integrated or associated with corresponding information in other systems. This view can be found in the computer simulation model of Burton et al. (1991) and in the speculations of Tranel and Damasio (1988).

This chapter has proposed a third view, which like the second, emphasizes state rather than structural correlates of awareness. According to this view, perceptual awareness is correlated with the quality of representation within the perceptual system itself. The low-quality representations of a damaged perceptual system are capable of supporting recognition in some tasks. However, they are incapable of supporting performance in many other tasks, and also incapable of supporting the experience of conscious awareness.

How should we decide among the alternative views? Although any conclusions about the neural correlates of awareness must be tentative at present, we (predictably) find the third account is most satisfactory. In the absence of evidence for truly normal levels of covert recognition performance, there is no reason to hypothesize separate brain systems needed for awareness above and beyond those needed for perception, and parsimony favors eliminating nonessential components of any theory. Deciding between the second and third views is more difficult, and the preference for the third derives primarily from the ability of the simulation to capture a reasonably wide range of phenomena by simply degrading visual representations. An additional consideration, discussed earlier, is that even when prosopagnosics show covert recognition, their perception of faces seems degraded rather than normal.

Kinsbourne (1988, chapter 16, this volume) articulated an integration hypothesis for explaining unconscious perception in visual neglect that is quite compatible with the hypothesis on the role of representational quality in enabling conscious awareness of perceived stimuli. It differs from the integration accounts described earlier in emphasizing the role of representational quality in achieving integration among brain systems. According to Kinsbourne, degraded representations will be less able to influence the global information-processing state of the brain. This is compatible with our model of covert face recognition in that the failure of degraded representations to support overt recognition is a consequence of their inability to activate the corresponding, associated patterns in distant parts of the network. One could therefore view the hypothesis described here and Kinsbourne's as "notational variants."

In closing, note that awareness, or lack thereof, has not been simulated in this model. All that has been simulated is the dissociation in performance on tests of overt and covert recognition. Other mechanistic accounts of covert recognition share this limitation, and it is difficult to imagine how to implement the equivalent of subjective awareness in a computer simulation. However, the relation between representational quality and the subjective component of overt recognition is able to find more direct support in studies of subliminal perception in normal subjects. The experimental manipulations that bring about unconscious perception in normal subjects invariably involve degrading the percept by brief stimulus presentations, masking, noise, or divided attention.

ACKNOWLEDGMENTS

This research was supported by ONR Grant N00014-93-I0621, NIMH Grant R01 MH48274, NINDS Grant R01 NS34030, Alzheimer's Disease Association/ Hearst Corporation Pilot Grant PRG 93-153, and the University of Pennsylvania Research Foundation.

REFERENCES

Bauer, R. M. (1984). Autonomic recognition of names and faces in prosopagnosia: A neuropsychological application of the guilty knowledge test. *Neuropsychologia, 22*, 457–469.

Burton, A. M., Young, A. W., Bruce, V., Johnston, R. A., & Ellis, A. W. (1991). Understanding covert recognition. *Cognition, 39*, 129–166.

De Haan, E.H.F., Bauer, R. M., & Greve, K. W. (1992). Behavioral and physiological evidence for covert recognition in a prosopagnosic patient. *Cortex, 28*, 77–95.

De Haan, E.H.F., Young, A. W., & Newcombe, F. (1987a). Face recognition without awareness. *Cognitive Neuropsychology, 4*, 385–415.

De Haan, E.H.F., Young, A., & Newcombe, F. (1987b). Faces interfere with name classification in a prosopagnosic patient. *Cortex, 23*, 309–316.

Farah, M. J. (1990). *Visual agnosia: Disorders of object recognition and what they tell us about normal vision.* Cambridge, MA: MIT Press.

Farah, M. J. (1994). Visual perception and visual awareness after brain damage: A tutorial review. In M. Moscovitch & C. Umiltà (Eds.), *Conscious and unconscious information processing: Attention and performance XV.* Cambridge, MA: MIT Press.

Farah, M. J., O'Reilly, R. C., & Vecera, S. P. (1993). Dissociated overt and covert recognition as on emergent property of lesioned neural networks. *Psychological Review, 100*, 571–588.

Hinton, G. E., & Plaut, D. C. (1987). Using fast weights to deblur old memories. *Proceedings of the 9th Annual Conference of the Cognitive Science Society*, Seattle.

Hinton, G. E., & Sejnowski, T. J. (1986). Learning and relearning in Boltzmann machines. In D. E. Rumelhart & J. L. McClelland (Eds.). *Parallel distributed processing: Explorations in the microstructure of cognition.* Cambridge, MA: MIT Press.

Kinsbourne, M. (1988). Integrated field theory of consciousness. In A. J. Marcel & E. Bisiach (Eds.), *Consciousness in contemporary science.* Oxford, England: Clarendon.

Movellan, J. R. (1990). Contrastive Hebbian learning in the continuous Hopfield model. In D. S. Touretzky, G. E. Hinton, & T. J. Sejnowski (Eds.), *Proceedings of the 1989 connectionist models summer school.* San Mateo, CA: Morgan Kaufman.

Tranel, D., & Damasio, A. R. (1988). Nonconscious face recognition in patients with prosopagnosia. *Behavioral Brain Research, 30*, 235–249.

Weiskrantz, L. (1986). *Blindsight: A case study and implications.* Oxford, England: Oxford University Press.

Young, A. W., Hellawell, D., & De Haan, E.H.F. (1988). Cross-domain semantic priming in normal subjects and a prosopagnosic patient. *Quarterly Journal of Experimental Psychology, 38A*, 297–318.

CHAPTER 18

Déjà Vu
All Over Again?

Clark Glymour
Carnegie Mellon University

One January a few years ago, shortly after the governor of Arizona had been impeached and the Exxon Valdez had spilled its cargo around Port Arthur, I had one of those uncanny experiences reserved for the people who read old news. Paging through the San Jose *Mercury* for January 1917, I came upon an article describing the impeachment of the governor of Arizona and a report of a large oil spill at Valdez, Alaska. Nietzche, it seems, was on to something, the Eternal Return, the no news under the sun, the history repeats itself sort of thing. I have had similarly uncanny experiences over the last few years reading bits of the literature of cognitive science as it has emerged in our time, and reading in the same years the literature of physiology, psychology, and psychiatry in the closing years of the 19th century. The parallelism is nowhere more striking than with the views of Kinsbourne and Farah (this volume).

One hundred years ago a vital enterprise aimed to use experiment and observation directed on children and adults to uncover the fundamental mechanisms—the very computational mechanisms—through which the brain makes thought, emotion, and action. That enterprise involved scientists of diverse tastes, perspectives, and aptitudes; those trained in physics and those not; those who approached the mind through the biology of the brain and those who thought of the mind in functional terms, and those who did both; those who experimented with tissues; those who experimented with people; and those who did no experiments but did observe tissues or people. The late 19th-century cognitive science community was a rich and diverse miniature of ours. Between 1870 and 1900 in Vienna alone theories of aspects of cognition

373

were offered by Meynert, Wernicke, Bruer, Exner, and Freud. Wernicke, Exner, and Freud wrote about the nature of consciousness. Theoretical views close to those advanced by Kinsbourne and Farah were represented in that community, and substantiated by the same styles of argument and the same sorts of evidence. Because Freud is the most famous and the most translated, he is the best example.

Kinsbourne's and Farah's chapters bear comparison with two of Freud's essays. One, a short book on aphasia, was published in 1891; the other, an unpublished (by Freud) manuscript entitled "Project for a Scientific Psychology," was drafted around 1895. The project assumes that particular arrays of cell bodies "bind" (as we might now say) parts or aspects of representations, that beliefs and thought processes are physiological processes in which some physical activation (Freud called it "Q" for quantity) are distributed and redistributed about the neural system as a whole. Intensity of a thought is determined by the amount of this activation it involves. Learning is by facilitation, or what would nowadays be called "Hebbian" synapses. Memory consists in a comparatively stable distribution of Q and the adjustments of synaptic junctions (Freud called them "contact barriers") that have come about through learning. Consciousness is distributed: Adjacent to every cell that does processing or is involved in memory, there is a "consciousness cell" responsive to the activation of the cell it adjoints and also responsive to special wavelike properties of Q. The patterns of activation of these special cells constitute conscious thought. This is not Kinsbourne's theory—Freud, for example, had nothing to say about the role of duration of action in producing awareness— but the two theories are close, both in content and in generality. Many of the questions one might reasonably have put to Freud about details, about the consistency of the story, and about alternative theories, might also be put to Kinsbourne.

The relation between Farah's chapter and Freud's views may seem preposterous because Farah's chapter centers on a computer simulation and Freud had no computer and no idea of simulation. Farah's chapter is one of several simulation studies following her book on visual agnosia. Farah (1990) argued that visual agnosias have been misclassified and classification that pays close attention to the empirical phenomena argues strongly that these deficits are produced by damage to a connectionist system. Her later papers illustrated, extended, and made concrete that hypothesis and related claims. For an extraordinary experience, take a weekend and on one day read Farah's (1990) *Visual Agnosia* and on the other read Freud's (1891) *On Aphasia*. I am confident Farah had never read a word Freud wrote before the turn of the century, and yet the parallelisms between the two book, written almost exactly a century apart, make one think Freud and Farah were collaborators, transcentury e-mail colleagues. Freud argued that aphasias have been misclassified and that appropriate classification that pays close attention to the empirical phenomena argued strongly that these deficits are produced by damage to a connectionist system. He argued in much the same way as does Farah, and at much the same level of detail about connectionist theory. Farah has richer empirical detail, and she has the computer to make her explanatory proposals concrete, but those matters aside her theoretical position and the kind of evidence she can muster for it are very much those of Freud in 1891.

Optimistic morals could, and perhaps should, be drawn from this parallelism between Freud in the 1890s and Kinsbourne and Farah in the 1990s. For example, the theoretical ideas are so appropriate to the phenomena that the theory is bound to dominate as soon as the basic anatomy and physiology of the nervous system is understood and a supply of clinical cases is available. That view is supported by the fact that Freud's connectionist approach was widely shared by many of his contemporaries, for example, by Freud's colleague Exner, by James, and by Cajal.

I want to temper the optimism with a more pessimistic challenge. After the turn of the century, cognitive neuropsychology collapsed under criticism from physiologists and behaviorists. Much of the criticism was careless and unsubstantiated, and it may have succeeded only because of the positivist mood of the scientific time. Sound or not, the critics' coherent theme was exactly that the theoretical goals of 19th-century neuropsychology were radically underdetermined by the kind of evidence available to neuropsychologists, and perhaps by any evidence that might be available to any science. I despise the contemporary philosophical impulse to make scientific uncertainty into inevitable mystery, but it does seem to me past time to supplement the advocacy of various theoretical perspectives in cognitive psychology with some careful and explicit consideration of what questions can be resolved, under what assumptions, by what sorts of evidence.

Nineteenth-century physicists and engineers speculated about how flight is possible; their contemporaries in psychology and physiology gave some thought to the nature of consciousness. We now know a great deal about how something can fly, most if it learned within the last 100 years. Is our contemporary understanding of consciousness comparably improved with answers to some of our fundamental questions: Just how does being conscious of something work, at every level of detail? What makes someone at some moment have one conscious content rather than another? How, exactly, do we have mental states with contents at all? Are we conscious because of what we are made of or because of what we do? Can only things made as we are do whatever it is that makes us conscious? These questions are rhetorical, and of course our understanding and control of consciousness does not really answer them. But, why not?

We can observe things take off in flight—projectiles, birds, insects—and we can manipulate and vary features of natural and artificial objects to understand conditions that make things able to fly. We find physical principles that contribute to the capability of flight: conservation of momentum, buoyancy, pressure differences for fluid flow over airfoils. We find a variety of mechanisms for applying these principles to produce flight, and a variety of materials from which the mechanisms can be made. We find physical constraints on flight peculiar to each category, to each mechanism, and to each material constitution. What, if anything, makes investigating consciousness any different?

One difference is that we have a way to tell whether a thing, anything at all, is flying, but we have only a very special way of telling whether a very special sort of thing is conscious. Each of us knows about consciousness principally for two reasons: Others tell us they are conscious, and each of us, anatomically and physiologically very much like other people, is conscious. (Maybe that is not always true. Newell once told me that many psychologists hated Watson because his sort of behaviorism was a lie, and Watson knew it was a lie. The more charitable hypothesis is that Watson was not lying about himself.) Some people believe various other creatures are conscious, and they may be right, but it is only from humans that we can get generally uncontested indications of specific and varied contents of consciousness. This fact alone is bound to make inquiry into consciousness very different from inquiry into flight. Any ancient Egyptian kid with a piece of papyrus could experiment on conditions relevant to flight; he could fold it, bend it, twist it, sail it, drop it, weight it, and so on. There is nothing we can do with any artifact to test whether it is conscious of anything. Imagine, if you can, the obstacles to a science of flight if no artifact could ever be tested, if our evidence about flight were confined to the behavior of a few species of birds. Now imagine that we are not allowed deliberately to damage any birds—no feather plucking, no wing clipping. That is now things are with the science of consciousness. We are confined to identifying the physical and behavioral correlates of conscious states so far as we can without damaging people, we are limited to examining correlations between kinds of accidental brain damage and limitations of consciousness, and to whatever analogical evidence computer simulations and animal experiments may provide. (Experiments with animals have the oddity that the more relevant to consciousness you think they are, the more moral qualms you should have about doing them.)

This is not a recipe for dramatic scientific progress. The evidence is limited, the system under study so extraordinarily complex, and its manifold features so closely correlated, that we should not expect that general physical or physiological principles will answer our questions the way general dynamical principles answer our questions about the motions of stars, nor should we expect that features will sufficiently separate themselves that we will uncover the necessary components of conditions sufficient for conscious contents. We should expect a growth in understanding of physical and physiological correlates of aspects of consciousness, perhaps the elimination of some hypotheses about the physical basis of consciousness, a growing wealth of information about how normal consciousness may come apart under experimental duress or brain damage, more and more articulation of computational analogues, but no solution to the fundamental questions about consciousness that have always troubled people.

Cognitive psychology could use more modest studies of the power and limits of inquiry into particular kinds of issues. For example, there should be more investigation into the class of possible computational models for phenomena associated with visual cognitive deficits, and into what patterns of actual or possible data would (or do) decide among them. We should know under what assumptions it is true, and under what not, that with lesioning connectionist models can generate any imaginable pattern of normal behavior and deficits. We should know whether normal behavior and patterns of deficits can tell us anything—and, if so, how much and under what assumptions—about processing architecture involving network modules of the kind Farah postulates. One response to these demands is that whiners should be workers. Fair enough: An answer to the last question is given in the *British Journal for Philosophy of Science* (Glymour, 1994), and a good start toward an answer to the second, encouraged by Kinsbourne, has been obtained in recent work (Glymour, in press; Glymour, Richardson, & Spirtes, 1995). But there are lots more questions.

REFERENCES

Farah, M. J. (1990). *Visual agnosia: Disorders of object recognition and what they tell us about normal vision.* Cambridge, MA: MIT Press.

Freud, S. (1891). *Zur Auffassung der Aphasien: Eine kritische Studie.* Leipzig: F. Deuticke.

Freud, S. (1895). *Project for a scientific psychology.* Unpublished manuscript.

Glymour, C. (1994). On the methods of cognitive neuropsychology. *British Journal for the Philosophy of Science, 45,* 815–835.

Glymour, C. (in press). Group data in cognitive neuropsychology. *British Journal for the Philosophy of Science.*

Glymour, C., Richardson, T., Spirtes, P. (1995). *Neural nets, damaged brains and explanatory limits.* Manuscript submitted for publication.

CHAPTER 19

Consciousness as a
State-Dependent Phenomenon

J. Allan Hobson
Harvard Medical School

After a long period of taboo, it is once again permissible, and even fashionable, to discuss *consciousness*. Having the long-suppressed *C*-word come out of the closet is both good news and bad. It is good news because our intellectual freedom must benefit from our ability to talk about a human attribute that we all hold dear, especially one like consciousness that we privately never doubted even at the height of our public intimidation by the positivists and the behaviorists. It is bad news in that we may not have made sufficient scientific progress during the censorship era to avoid the many egregious and embarrassing faux pas that have so consistently tripped up the free thinkers of earlier eras (Descartes, 1978; Eccles, 1964; Sherrington, 1955; and, for a review, see Dennett, 1991).

If we are not careful, we are likely to make fools of ourselves once again and, once again, lose our freedom of *C*-word speech. Under the circumstances, it would be wise to use our new right of *C*-word discourse both cautiously and modestly. Like a teenager who has just gotten a driver's license, we would do well to drive around the block several times before heading out on the freeway. Before attempting a global theory, we might consider more limited and modest approaches to the question of consciousness. And even such limited and modest approaches would best be based on new developments in science that are of undoubted relevance to consciousness even if they do not entirely explain it.

To introduce a second analogy, if we would each look at the development of our own field for the conceptual and empirical equipment needed for a successful assault on the consciousness problem, we could thereby resist the fatal temptation to rush to the peak before the base camp was constructed. We could also regard our goal as not only partial and modest (rather than complete and overly ambitious), but also as cooperative and mutually enhancing (rather than competitive and mutually exclusive).

Assuming that whatever definition of consciousness is used, our states of consciousness are dependent on the states of our brain, it will be the aim of this chapter to review what has been learned in the last half century about the mechanisms of brain state control. Whatever consciousness is, we know for sure that it is dependent on the state of the brain. By learning more about how changes in the brain state determine changes in our state of consciousness, we may be able to understand what brain processes are most significant for consciousness (Hobson & Stickgold, 1995).

To be effective and economical the review is necessarily strategically selective. Instead of trying to summarize the detailed data available in comprehensive monographs (Hobson, 1988, 1989; Hobson & Steriade, 1986; Steriade & McCarley, 1990), this chapter focuses on some general themes that emerge from the behavioral neurobiology of waking, sleeping, and dreaming. It then discusses those themes within the framework of a relatively restricted and simple model that accounts for the differences in conscious state between waking and dreaming in a preliminary way.

In brief and anticipatory summary, the chapter suggests that consciousness, in its state-dependent aspect, is particularly sensitive to three kinds of changes in brain function: activation level, input–output gating, and chemical modulation. Using the now well-confirmed findings of basic sleep research, it attempts to explain the robust changes that all of us experience when our conscious state shifts from that of waking to that of dreaming. True to the principle of modesty, the conscious state theory presented is nothing more than a rough first draft, subject to change in accord with any new philosophical, psychological, or physiological data or critique that can show it to be in error or, better yet, improve on it.

WHAT IS CONSCIOUSNESS?

Aware that any attempt to define consciousness is likely to be met with vociferous objection, this chapter nonetheless offers up a simple idea: Consciousness, as it is fully experienced in awake humans, is the brain's representation of the outside world, the body, and the self. I mean that an individual's brain, in its present state (awake, thinking, and writing), is able to represent its informational content to itself and to any other brain that is similarly awake, thinking, and reading instead of writing. Using your consciousness, you can now reflect on the definition that was the product of my consciousness.

This simple idea is appealing because it enables us to proceed quite directly from the already uncontestable idea that the brain represents the external world in its neuronal activation states. I see, you see, and our pet cats see when light patterns excite retinal ganglion cells; these cells then relay their on or off center signals to the lateral geniculate body, which in turn sends codes representing edges to the visual cortex (Hubel, 1959). This picture, though still quite incomplete, is already crystal clear: The visual world is represented in the activation state of the visual cortex. This fact is clearly relevant to what we call perception, an important component of consciousness.

By a simple logical step we need only assume that a brain with enough neurons to assign to the task can make a second and simultaneous representation of what it sees (or hears, or smells, or feels) and thereby activate a vision of its vision. It can even activate an awareness of its being able to perceive that vision. Of course, it is crucial to have an understanding of how this second step occurs. The point is that the paradigm for working out the details is already clear enough to imagine the general nature of its ultimate formulation. Consciousness involves the brain's representation of its representations of the world, our bodies, and our selves.

The assumption that only adult humans have enough extra neurons to create secondary ideational representations of their primary sensory representations could be called the large secondary network tenet of my conscious state hypothesis. The assumption that its order is secondary (or higher) could be called the hierarchical organization tenet of the conscious state theory.

Before proceeding to build on what is admittedly only a very partial and incomplete account, let me add that the idea of a representation of a representation is not any more complicated than that of representation itself. In other words, if I am conceptually comfortable (now) with the idea that my vision is nothing more (and nothing less) than a neuronal activation pattern, then why should I balk at a higher order vision of my vision that is also nothing more (nor less) than a neuronal activation pattern?

This view of full human consciousness as depending on large secondary networks does not imply either that primary networks are not involved in conscious experience, or that animals without large secondary networks have no consciousness. These two related issues are discussed in more detail; suffice it to say that the view in this chapter is thus compatible with Kinsbourne's (chapter 16, this volume) insistence on the essential participation of primary networks in conscious experience and with Johnson's (Johnson & Reeder, chapter 13, this volume) emphasis on "metarepresentation," which would occur in large secondary networks.

We humans see and we know that we see. We can control what we see. We can even close our eyes and summon an image of what we have seen. Indeed, we can imagine images that have never been seen at all. An account of that process that utilizes individuals' ability to describe their vision can be given using words that represent some of the content of their consciousness. Later, in building a model to represent the state-dependency of consciousness, numbers are used to show that the ideas presented here, represented mathematically, are not only internally consistent, but conform to directly observable data regarding easily quantified conscious states.

Because dreaming is the altered state of consciousness used as a device for a better understanding of how waking consciousness is achieved, the remarkable visual perceptions of our dreams are our most convincing evidence of the entirely representational nature of our seeing. In the dark, with our eyes closed, we see persons, places, and actions with impressive clarity. This fact alone guarantees the identity of our vision with our brain activation states. And, because there is no formed external stimulus for the dream images, it further shows that our brain activation patterns are alone sufficient to support visual perception. We later return to consider some other differences between dream vision and wake vision, which are important clues to the mechanisms by which waking consciousness is achieved.

SOME STRUCTURAL AND FUNCTIONAL ASSUMPTIONS

Before taking up the three subjects of brain activation, input–output gating, and chemical modulation, we must confront the fact that not all brains that have state control have fully developed conscious states. To account for this dissociation we need to make a second, simple assertion: Of all living things, only adult humans have enough extra brain cells connected and activated in such a way as to achieve the luxury of representing our representations. Thus, only adult humans are capable of communicating the contents of consciousness to ourselves and to one another via words and numbers. This fact seems obvious, but it often runs afoul of assumptions—and some interesting evidence—that our own infants and many other animals have some significant aspects of conscious experience (such as visual perception in the example of a pet cat, or person recognition in the case of a baby boy, or problem solving and even sign recognition in the case of some primates). In other words, consciousness has many parts (or components) and many degrees of expression in both phytogemy and ontogemy.

That consciousness is componential, graded, and developmentally cumulative—both within and across species—should not in itself inspire debate among biological scientists. Only biblical theories would ever really hold otherwise. Thus we can be reasonably comfortable with the commonsense argument that because only adult humans evince propositional and abstract language, only adult humans have conscious experience that is both propositional and abstract. Although painstaking experiments may well reveal surprising approximations of these functions in immature humans and in other animals, such findings would only serve to reinforce the large secondary network activation principle that the ontogenetic and phylogenetic data appear to support.

When we examine the effects of changing activation level, input–output gating, and modulation on our consciousness, it becomes evident that even in the case of human beings, having a large secondary network is not enough to guarantee consciousness. A large network is necessary, but is not sufficient. Likewise, activation is necessary—but not sufficient, possibly because hierarchical organization breaks down when input–output gating changes and/or when the modulatory chemical microclimate of the large secondary network changes, as it does most dramatically when we fall asleep and dream.

In addition to the ontogenetic and phylogenetic evidence that consciousness depends on the size and organization of large secondary neuronal networks, we have the striking witness of our own daily lives. Our own consciousness changes, and it changes radically, over the course of each 24-hour day. Some of the brain mechanisms of these changes are now known at a first draft level that tells us much about the conditions that must be met for large secondary networks are to operate synchronously and hierarchically. Synchronous operations is necessary to the unity of consciousness, its integrative capacity. Hierarchical operation is necessary to the deliberate direction of consciousness, to its guided focal capacity.

A COMPONENTIAL ANALYSIS OF CONSCIOUSNESS

Consciousness integrates at least the 10 functional components defined in Table 19.1. Sensations arising outside and inside the body are represented as sensory neuronal activation patterns that impinge on the thalamocortical system and compete for attention: Attention gates the representation of percepts in conscious awareness. Once admitted to consciousness, perceptual data

TABLE 19.1
Definition of Components of Consciousness

Attention	Selection of input data
Perception	Representation of input data
Memory	Retrieval of stored representations
Orientation	Representation of time, place, and person
Thought	Reflection upon representations
Narrative	Linguistic symbolization of representations
Emotion	Feelings about representations
Instinct	Innate propensities to act
Intention	Representations of goals
Volition	Decisions to act

need to be referenced with respect to the individual's immediate and local context (via the orienting response) and to the individual's intermediate range context (via time, place, and person orientation). Of course, a sufficiently strong stimulus can come immediately to dominate consciousness without apparent filtering. In such cases, attention is "sized" or "captured" by external data.

Short- and long-term memory are necessarily involved in this contextual frame-fitting function called *orienting* and *orientation*. Exogenous and endogenous perceptual data are also processed in parallel by the emotional system of the limbic area, particularly the amygdala. This module contributes specific feeling tones with their positive or negative valence to our unified conscious experience.

At this early stage of conscious experience we have already recruited and integrated five functions: sensation, perception, attention, orientation, and memory. And this far, we believe, most of our fellow mammals can easily go. So can our own babies and very young children.

But now a break occurs. If we are to reflect on the impressions we have received and processed, then we must be able to keep our impressions represented in awareness and to conduct a variety of analytic operations on them. These include making qualitative statements to our selves (implying both language and a self who speaks), quantifying our thoughts and feelings (implying their enumeration if not their mathematization), and weighing these evaluations in the balance for volitional decision making. This function is generally referred to as *working memory*.

Before we deliberately change the course of our ongoing behavior (which in itself implies a structured schema or plan), we must compare the new information and its higher order representations to long-range goals. Many such volitional schemata are narrative and may be detected by us as internal language. Now we have added persistent memory, internal attention, analytic thought, linguistic representation, narrative discourse, and volition to the already rich mix of primary conscious components. We are inclined to attribute these secondary, higher order consciousness component functions to adult humans (and to waking adult humans at that).

In addition to cognitive judgments, emotion-related instinctual representations may also enter consciousness. Should we approach or avoid, fight or flee, feed or fast, drink or thirst, court or shun? And, if one or another new motor program is called for, can we enact it? Now that we are once again on the output side, with instinctual priority as a possible determinant of action, we are back with our fellow animals and perhaps only slightly the safer, or wiser, for our passage through the vast neuronal networks in which our primary representations are represented at second and higher order. But over the long term, there can be no doubt that abstract consciousness functions to give its possessor an enormous strategic advantage whether our goal is to control and compete or to reflect and contemplate.

PHENOMENOLOGICAL DIFFERENCES BETWEEN WAKING, SLEEPING, AND DREAMING

When we first go to sleep, the level of activation of almost all brain systems falls. This initial deactivation phase of sleep is called non-rapid eye movement sleep (NREM). It becomes evident that the inactivation is a simple function of decreases in brain stem neuronal activity. As a consequence, virtually every component of consciousness is weakened, as shown in Fig. 19.1 and Table 19.2. As sensory system activation declines, thresholds to stimulation rise and external inputs have difficulty gaining access to central processes. Attention fails altogether and perception of the external word becomes weak to nonexistent. Central processes are likewise retarded: Thought, if there is any, is nonprogressive and/or perserverative. Upon awakening from NREM sleep, orientation, memory, and even our ability to give a narrative account of our impoverished state of consciousness is impaired by sleep inertia. Because the brain is immediately activated (in an electrical sense) while consciousness is lagged suggests that other factors besides activation contribute to consciousness. This chapter later suggests that two such factors are input–output gates and chemical modulation. And it further hypothesizes that chemical modulation, being quasihormonal, is most likely to have the longest time constant of these three processes.

TABLE 19.2
Changes in Consciousness During Sleep

Function	NREM Sleep	REM Sleep
Input level	Low	Blocked
Attention	Poor	Poor
Perception	Weak	Strong (internal)
Memory	Poor	Poor
Orientation	Poor	Unstable and imprecise
Thought	Perseverative	Illogical
Insight	Poor	Delusional
Narrative	Poor	Confabulatory
Emotion	Weak	Episodically strong
Instinct	Weak	Strong
Intentions	Weak	Confused
Volition	Poor	Episodically strong
Output	Low	Blocked

Emotions (feelings) and instincts (primordial and innate behaviors) are generally absent from NREM sleep, but they sometimes show surprising intensification during the process of arousal. These intensifications are experienced as the night terrors of children and the incubi (or pure emotion nightmares) of adults. Intention (planned acts), volition (voluntary motor commands), and motor output itself is likewise generally in complete abeyance in NREM. But, all these functions can be surprisingly and automatically activated—as in sleep walking, sleep talking, and bruxism (tooth grinding). These motoric sleep disorders are relatively common in the second and third decade of life when the tendency of the cortex to become disorganized by sleep is greatest. More elaborate fuguelike states, sometimes with violent outcomes, may occur in the somnambulistic states of incomplete arousal that characterize all of the abnormal NREM phenomena.

To understand these paradoxes, it is important to realize the complex dynamics of awakening from deep sleep and to appreciate the dissociative power of sleep inertia. In brief, it seems that some parts of the brain (the subcortex, e.g.) can be activated while others (the cortex, e.g.) remain inactivated. Under the circumstances, each of us is capable of automatic behaviors that can occasionally assume Frankensteinian proportions. In this early phase of sleep, our consciousness is not entirely lost but it is certainly partial and decidedly degenerate.

The degradation of consciousness in NREM sleep itself, and the aberrations of consciousness that occur on incomplete and disoriented arousals from NREM sleep illustrate the principal of herarchical organization that the large secondary nets must possess if consciousness is to be fully realized. To clarify this point, an individual may see perfectly well upon arousal from sleep—or even during a sleep-walking episode—but that person is certainly not fully conscious under these circumstances despite the evident activation of many primary networks. On this view, consciousness is possible only if the upper brain networks are activated so that they can exert top-down control upon the rest of the brain and body. It is now well established that the cortical activation process is characterized by strong excitatory depolarization of thalamocortical neuronal ensembles with associated high-frequency (40-Hz) low-voltage brain waves (Gray & Singer, 1989; Singer, 1989; Steriade & Llinás, 1988). This rhythm is a candidate for the global synchronization of the large secondary network with primary ones and may solve the binding problem (Damasio, 1989) that the unity of our conscious experience poses.

So far most of what we have described can be attributed to deactivation pure and simple. But after about 70 min of NREM sleep, the brain begins to reactivate and consciousness is gradually resuscitated but now assumes the peculiar and distinctive cognitive form of dreaming (see again Fig. 19.1). This process is called REM sleep. Because activation has returned to most brain systems, it is instructive to examine the cognitive scorecard of Table 19.2

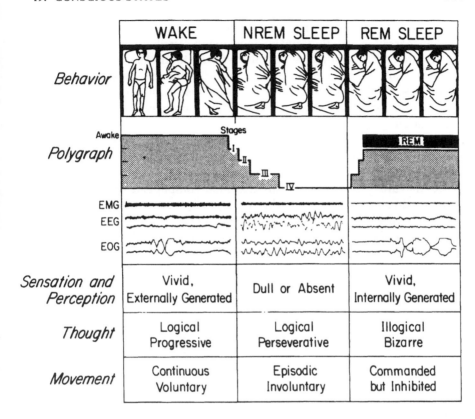

FIG. 19.1.   Behavioral states in humans. States of waking, NREM sleep, and REM sleep have behavioral, polygraphic, and psychological manifestations. In behavior channel, posture shifts (detectable by time-lapse photography or video) can occur during waking and in concert with phase changes of sleep cycle. Two different mechanisms account for sleep immobility: disfacilitation (during Stages I–IV of NREM sleep) and inhibition (during REM sleep). In dreams we imagine that we move but we do not. Sequence of these stages represented in polygraph channel. Sample tracings of three variables used to distinguish state are also shown: electromyogram (EMG), which is highest in waking, intermediate in NREM sleep, and lowest in REM sleep; and electroencephalogram (EEG) and electrooculogram (EOG), which are both activated in waking and REM sleep and inactivated in NREM sleep. Each sample record is 20 sec. Three lower channels describe other subjective and objective state variables. From Hobson & Steriade (1986). Copyright 1986 by American Physiological Society. Reprinted with permission.

more carefully. Those functions that are strong in dreaming (like internal perception, narration, emotion, and instinct) may be attributed to the reactivated brain systems. Those functions that are deficient in REM sleep (like attention, memory, thinking, orientation, and intention) will need some other explanation. These weakened conscious state components fail because of a global change in the chemical microclimate of the brain here called *chemical demodulation*.

Still another specific explanation is needed to account for the fact that sensory input and motor output are not simply deactivated in REM sleep but are actively blocked. These functions are blocked by REM-specific inhibitory processes that are defined and quantified as input–output gate changes.

In our dreams we have a remarkable range of perception, especially vision. We also have a sense of continuous movement despite the fact that our sensory inlets and motor outlets are closed. And instead of being able to inspect, look at, or pay attention to our dream images, they flood our senses with recurrent surges of new data that keep us constantly disoriented and befuddled! And rather than being able to think constructively or deliberately about what we are dreaming, we make weak, post-hoc, and logically flawed inferences about the reasons for our bizarre dream images. Most remarkable of all is our failure to recognize our dreaming state for what it is; we are consistently duped into believing that we are awake. In sum, our dream consciousness is as delusional as it is hallucinatory.

In keeping with the pell-mell sequence of images is a marked instability of our orientation function: persons, times, and places are fused, plastic, incongruous and discontinuous. To integrate this orientational chaos, our narrative capacity rises to the magnitude of its monumental synthetic task by being enhanced to confabulatory levels. Everything fits, or so we believe, into a story-like frame. In dreams, we see the narrative capacity of our consciousness working at its integrative peak. The intensification of dream emotion, especially fear, seems to help us explain and integrate even the most improbable and impossible events. Via dreams, our feelings thus show themselves to be powerful cognitive organizers and even—paradoxically—rationalizers. In dreams it would seem that feelings so clearly direct thought as to be considered reasons unto themselves. So too are instinctual programs powerful organizers of dream cognition. Consider sex and aggression: Once these fictive behavioral forces are activated in our brains, dream plots become focused and truly single-minded, at least for a time. We are made lustful—and terrified—by them.

None of these dream automata are under planned or even temporary voluntary control. Indeed, if we try to gain control of this flow of perceptions, emotions, and fictive behavior, then we almost always notice that whatever glimmers of insight may occasionaly sparkle, quickly fade as the dream resumes its relentless course. To escape the villains who contrive to pursue us, we may try to run faster only to find our legs inert or rubbery. And our lust may propel us to orgiastic consummation no matter how inappropriate this fictive coupling would seem to wake-state morality; or, we may fail our dream conjunctions despite our most desparate fictive clasping.

THE PHYSIOLOGICAL BASIS OF DIFFERENCES BETWEEN WAKING AND DREAMING CONSCIOUSNESS

This chapter argues for our being satisfied, for now, with a partial theory of consciousness. Now consider also accepting the limiting notion of a gradual continuum of consciousness levels: from lower to higher animals, from babies to adult humans, and from adult humans waking to adult humans dreaming.

The gradualism principle buys more than accomodation to an obvious developmental aspect of consciousness. It buys the comparative approach. Instead of fruitless debate about whether or not animals are conscious, we can take scientific advantage of the fact that whatever kind of consciousness they have it is state dependent. We know that their brains are endowed with state control mechanisms that we can investigate. And by using the data that we obtain, we can develop realistic models about the state-dependent aspects of our own consciousness. A summary of the most important findings is presented in Fig. 19.2.

To better appreciate the brief overview of neurophysiology that follows, consider an intellectual strategy of subtraction that is formally akin to that used by positron emission tomography (PET) and other brain image engineers. First, we look at our cognitive scorecard to identify those cognitive functions that are enhanced in dreaming and those functions that are impaired (compared to waking); then we look at the brain physiology in REM (again, compared to waking) for functions that are enhanced and for functions that are diminished. Finally, we hypothesize a causal relation between whatever positive–positive and negative–negative correlations we observe. And those causal hypotheses will drive future experiments. In other words, we view the diurnal changed in our state of consciousness as an experiment of nature that is trying to tell us something about how consciousness is engineered.

As Table 19.3 indicates, two input–output gating functions are actively enhanced in REM sleep: Primary sensory afferent nerve endings in the spinal cord and brain stem undergo dramatic presynaptic inhibition; data from the external world are thus excluded from our dream consciousness. Dream consciousness entirely depends on a reactivation of stored sensory (and motor) representations. The aside "(and motor)" is included because motor output is also blocked in REM this time by 10 mV of hyperpolarization of brain stem and spatial motorneurons. Our awareness of dream movement is thus entirely

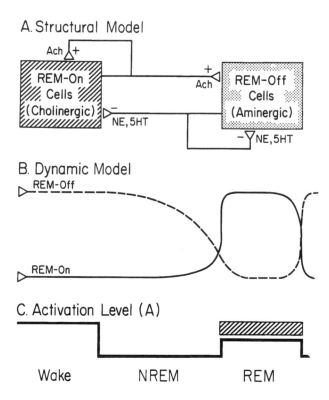

FIG. 19.2.   Physiological mechanisms determining alterations in activation
level (A). (A) Structural Model of Reciprocal Interaction. REM-on cells of the
pontine reticular formation are cholinoceptively excited and/or cholinergically
excitatory (ACH+) at their synaptic endings (open boxes). Pontine REM-off
cells are noradrenergically (NE) or serotonergically (5HT) inhibitory (–) at
their synapses (filled boxes). (B) Dynamic Model. During waking the pontine
aminergic system is tonically activated and inhibits the pontine cholinergic system. During NREM sleep aminergic inhibition gradually wanes and cholinergic
excitation reciprocally waxes. At REM sleep onset aminergic inhibition is shut
off and cholinergic excitation reaches its high point. (C) Activation Level (A).
As a consequence of the interplay of the neuronal systems shown in A and B,
the net activation level of the brain (A) is at equally high levels in waking and
REM sleep and at about half this peak level in NREM sleep. From Hobson
(1992). Copyright 1992 by Lawrence Erlbaum Associates. Reprinted with permission.

fictive and emanates from motor pattern command signals rather than from feedback data related to their enactment.

Only rarely, and under exceptional circumstances, are the blockades of sensory input and motor output overcome. This fact allows us to conclude that a very impressive level of consciousness can be achieved by the brain with no help from the external world. In REM, sleep consciousness is entirely offline, doing its own thing as it were.

When we look at the constellation of functions that are enhanced in REM

TABLE 19.3
Physiological Basis of Differences Between Waking and Dreaming

Function	Nature of Difference	Causal Hypothesis
Sensory input	Blocked	Pre-synaptic inhibition
Perception (external)	Diminished	Blockade of sensory input
Perception (internal)	Enhanced	Disinhibition of networks storing sensory representations
Attention	Lost	Decreased aminergic modulation causes a (decrease in) signal to noise ratio
Memory (recent)	Diminished	Because of aminergic demodulation activated representations are not restored in memory
Memory (remote)	Enhanced	Disinhibition of networks storing mnemonic representations increases access to consciousness
Orientation	Unstable	Internally inconsistent orienting signals are generated by cholinergic system
Thought	Reasoning ad hoc Logical rigor weak Processing hyper-associative	Loss of attention memory and volition leads to failure of sequencing and rule inconstancy Analogy replaces analysis
Insight	Self-reflection lost (failure to recognize state as dreaming)	Failure of attention, logic, and memory weaken second (and third) order representations
Language (internal)	Confabulatory	Aminergic demodulation frees narrative synthesis from logical restraints
Emotion	Episodically strong	Cholinergic hyperstimulation of amygdala and related temporal lobe structures triggers emotional storms, which are unmodulated by aminergic restraint
Instinct	Episodically strong	Cholinergic hyperstimulation of hypothalamus and limbic forebrain triggers fixed action motor programs, which are experienced fictively but not enacted
Volition	Weak	Top down motor control and frontal executive power cannot compete with disinhibited sub-cortical network activation
Output	Blocked	Post-synaptic inhibition

sleep we can appreciate how far brain activation alone can carry consciousness. We can have formed perceptions, enhanced remote memory, rapid orientation shifts, confabulatory narration, and we can experience powerful emotion and fictive instinctual acts.

But we also see what activation alone cannot provide. In dream consciousness we do not have focused attention; recent memory; stable orientation; self-reflective awareness; or logical, analytic thought. And, we are utterly incapable of either planned or voluntarily executed action!

In this account of what activation alone can and cannot provide, the dramatic difference in the consciousness balance sheet suggests either that there are two quite different kinds of brain activation (one in waking and one in REM) and/or that in REM some systems are overactivated while others are underactivated. In fact, both assumptions are correct. And they boil down to essentially the same underlying cause; a shift in the ratio of neuromodulatory chemicals (Fig. 19.2).

In waking, the aminergic systems of the brain stem are spontaneously, continuously, and responsively active; in REM, they are shut off by an active inhibitory process that is probably gaba-ergic. As a function of this shut-down of aminergic systems in REM, the cholinergic systems of the brain stem become disinhibited and excite the brain with strong tonic and phasic activation signals. The net result is that, in REM sleep, the brain is aminergically demodulated and cholinergically hypermodulated.

The electrical activation, indicating the turning on of what we imagine to be the huge secondary cortical networks responsible for the representation of representations is shared by both the waking and REM sleeping brains. Yet our consciousness in the two states is quite different. Even allowing that some part of that difference is due to the changes in input–output gating, we are forced to conclude that an electrical activation process alone is not enough to account for waking consciousness. Cognitive theories that use the activation concept as if it were equivalent to brain activation must take note of this important qualification. In other words, here is a crucial difference between activation (in the electrical sense of turning the brain on) and modulation (in the sense of determining the kind of information processing that will be performed in any particular activated state).

Aminergic modulation of the networks in waking appears to provide both restraint (via inhibition of the membrane) and direction (via chemical signals to the nucleus) of each neuron in the large secondary networks whose activation constitutes our conscious states (Hobson & Steriade, 1986; Steriade & McCarley, 1990). One way in which such modulation could enhance the hierarchical ordering of the activated secondary networks is by providing a more even balance of excitation and inhibition throughout the network: between competing sensory channels, between competing patterns of motor output, and even between competing cortical sectors. In addition to this finer tuning of parallel pathways, it could also shift the vertical balance of power in the brain to favor top-down (volitional, reflective, logical) operations over bottom-up (reflexive, impulsive, emotional) operations.

Such, indeed, is the difference between waking and dreaming. Aminergically demodulated in REM, the secondary forebrain cortical networks are subject to invasion and takeover by the cholinergic fiends from brain stem hell. These signals appear to be quite capable of revving up the secondary networks and thereby simulating both sensory perceptions and fictive motor actions so convincingly that we have the delusion that we are awake. They also turn on emotions and instincts and they organize these wild and untamed elements into the impressively synthesized scenarios of dream consciousness utilizing only what is at hand in the way of representations of representations. We dream away, almost never guessing that we are not awake.

A THREE-DIMENSIONAL MODEL OF CONSCIOUS STATE: AIM

According to the basic assumptions, there are three independent functions that determine the conscious state of the brain. They are *activation level* (A), *input–output* gating (I), and *neuromodulatory concentration ratio* (M).

> If A, the electrical activation level is high the brain will be in waking or REM sleep.
>
> If I, input–output gating is high the brain will be in REM sleep. If I is low, the brain will be in waking.
>
> If M, the aminergic to cholinergic modulatory ratio is high, the brain will be in waking. If M is low the brain will be in REM.

If all these three functions are at intermediate levels, the brain will be in NREM sleep.

Hobson (1990, 1992, 1993) described in detail how the numerical value of A, I, and M can be derived from the physiological data. Here it must suffice to say that:

> A can be estimated from either the degree cortical EEG desynchronization or the firing level of mesencephelic reticular neurons.
>
> I can be directly estimated from the degree of suppression of either electromyographic activity or spinal motorneuron excitability. It also measures the relative intensity of internal to external stimulus strength within the brain.
>
> M can be most accurately measured at the cellular level as the firing level

of aminergic neurons (in the DR and LC). It can also be quantified as the ratio of aminergic to cholinergic modulation strength.

This model is unique in that it lends itself to easy visualization (Fig. 19.3). If each of the three functions is taken to be one of the three axes of a cube, then a three-dimensional state space is defined in which the state of the brain at any instant of time is a point.

The succession of points representing AIM varies continuously in time as the three functions fluctuate; this gives the model a realistic dynamism. Time

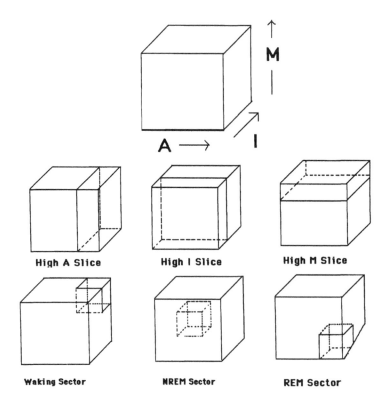

FIG. 19.3. (Visualizing AIM) Utilizing numerical values for activation (A) input-output gating (I) and modulation (M) a three-dimensional state space model can be constructed (top cube). By dividing each of the three axes into thirds, the cube can be divided into 27 slices, 3 of which are shown in the central array of AIM cubes. By convention, the high A (activation) slice is the right-hand third of the space, the high I (input–output gating) is the front third, and the high M (modulation) is the top third. On this account, each of the cardinal three conscious state domains (waking, NREM, and REM) occupy discrete sectors of the state space.

becomes the fourth dimension of the state space when the succession of points representing AIM is displayed sequentially; this sequence of points, in turn, describes trajectories through the state space. As time passes, AIM migrates from the wake, though the NREM, to the REM sector of the state space then homes back to wake, and begin the cycle over again.

One heuristic value of the AIM model is that it is theoretically extensible to all possible states of the brain-mind, including abnormal and extreme states like coma and psychosis and exceptional states like hypnosis and meditation. These features of the model have been discussed elsewhere, but it must be admitted that they have as yet been very inadequately explored.

To help drive these points home, consider how the AIM model behaves when conscious state changes from wake to dream. In both states, the activation level is high and therefore AIM will in both cases occupy the high A slice of the space (a slice being that one third of the space that contains all high values of A). See the middle row of Fig. 19.3 to visualize this slice concept.

We can now further subdivide the high A slice by cutting it into three rows one of which represent high values of I and then cut the slice the other way into three columns, one of which represents high values of M. Only one ninth of the high A space will now contain all of the AIM values of waking (high I, high M) and only one other ninth will contain all of the AIM values of REM sleep (low I, low M). Notice that the W and R segments are at diametrically opposite corners of the high A slice with waking at the back upper right and REM at the front lower right corner of the state space.

From this short exercise in mental imagery manipulation it should become clear that the model is highly sensitive to the robust differences between our states of consciousness when we are awake and when we are dreaming. Moreover, the model offers the possibility of an ultimate identity between our conscious states as we perceive them and the unobservable states of our brain in which they arise.

SUMMARY AND CONCLUSIONS

1. Consciousness is a graded, global integration of multiple cognitive functions yielding a unified representation of the world, our bodies, and our selves.

2. The kind (or state) of consciousness that we experience varies most dramatically as our brain changes in state over the sleep–waking cycle:
   a. The level of consciousness changes as a function of activation.
   b. The focus of consciousness changes as a function of input–output gating.
   c. The form of consciousness changes as a function of modulatory neurotransmitter ratios.

3. The Three Functions, Activation (A), Input–Output Gating (I), and Modulation (M) can be used to construct a three-dimensional state-space model (called AIM) that predicts conscious state from the physiological parameters.

4. Consciousness is the forebrain's representation of the world, our bodies, and our selves. It is always a construction whose level, focus, and form depends on the brain stem.

REFERENCES

Damasio, A. R. (1989). The brain binds entities and events by multiregional activation from convergence zones. *Neural Computation, 1*, 123–132.

Dennett, D. C. (1991). Consciousness explained (Paul Weiner, Illus.). Boston: Little, Brown.

Descartes, R. (1978). *His moral philosophy and psychology* (John J. Bloom, Trans.). New York: New York University Press.

Eccles, J. C. (1964). *Brain and conscious experience.* New York: Springer-Verlag.

Gray, C. M., & Singer, W. (1989). Stimulus-specific neuronal oscillations in orientation columns of cat visual cortex. *Proceedings of the National Academy of Science, USA, 86,* 1698–1702.

Hobson, J. A. (1988). *The dreaming brain.* New York: Basic Books.

Hobson, J. A. (1989). *Sleep.* New York: Scientific American Library.

Hobson, J. A. (1990). Activation, input source, and modulation: A neurocognitive model of the state of the brain-mind. *Sleep and Cognition, 2,* 25–40.

Hobson, J. A. (1992). A new model of brain-mind state: Activation level, input source, and mode of processing (AIM). In J. S. Antrobus & M. Bertini (Eds.), *The neuropsychology of sleep and dreaming* (Vol. 12, pp. 227–245). Hillsdale, NJ: Lawrence Erlbaum Associates.

Hobson, J. A., & Steriade, M. (1986). The neuronal basis of behavioral state control. In F. E. Bloom (Ed.), *Handbook of physiology—The nervous system. Vol. IV: Intrinsic regulatory systems of the brain* (pp. 701–823). Bethesda, MD: American Physiological Society.

Hobson, J. A., & Stickgold, R. W. (1995). The conscious state paradigm: A neurocognitive approach to waking, sleeping and dreaming. In M. Gazzaniga (Ed.), *The cognitive neurosciences* (pp. 1373–1389). Cambridge, MA: MIT Press.

Hubel, D. (1959). Single unit activity in striate cortex of unrestrained cats. *Journal of Physiology, 147,* 226–240.

Sherrington, C. (1955). *Man on his nature.* New York: Doubleday.

Singer, W. (1989). Search for coherence: A basic principle of cortical self-organization. *Concepts in Neuroscience, 1*(1), 1–26.

Steriade, M., & Llinás, R. R. (1988). The functional states of the thalamus and associated neuronal interplay. *Physiology Review, 68,* 649–742.

Steriade, M., & McCarley, R. W. (1990). *Brainstem control of wakefulness and sleep.* New York: Plenum.

C H A P T E R

Dimensions of Consciousness:
A Commentary on
Kinsbourne and Hobson

Jonathan D. Cohen
Carnegie Mellon University

One of the central debates within the recent history of cognitive science has concerned the nature of cognitive representations: Can they be identified as discrete and invariant elements within the cognitive system, or are they inextricably intertwined with one another, and with the processes in which they are used? Kinsbourne and Hobson offered views about consciousness that suggest a similar debate. Put simply: Can consciousness be usefully identified with a distinguishable subcomponent of the system in which it arises, or does it necessarily involve interactions among all (or an arbitrarily varying subset of) components? Of course, this debate raises a number of important associated questions (e.g., what do we mean by "distinguishable subcomponent," and where do we draw the line regarding the involvement of others). Some of these questions are vexing, and neither Hobson nor Kinsbourne presume to answer all of them in detail. Nevertheless, each touches on issues that are central to any understanding of consciousness and, at least on the surface, their points of view appear to diverge. Kinsbourne argues for what he called the *heterarchical,* or highly distributed nature of consciousness, whereas Hobson focuses on the importance of structured hierarchies, at both the neurobiological and cognitive levels. This chapter suggests that, in fact, their views are quite compatible, and even complement one another. First however, it is useful to review each perspective.

KINSBOURNE'S HETERARCHICAL VIEW
OF CONSCIOUSNESS

According to the central thesis of Kinsbourne's chapter, consciousness cannot be identified with any delimitable component of the cognitive processing system. Rather, it reflects functional attributes of processing that can involve any combination of subcomponents. Kinsbourne is quite clear about the extremity of his view: "No one place in the brain is dedicated to the control of awareness." He offers two basic arguments in favor of this view. First, he points out that it has not been possible to demonstrate a dissociation between consciousness on the one hand, and a full preservation of all other mental faculties on the other. Localized lesions are capable of impairing consciousness for certain types of information, but only at the expense of impaired (though perhaps not obliterated) performance (see Farah, this volume, for a careful exploration of this topic). Thus, consciousness is intimately bound to the processing work done by the brain. Second, he points out that consciousness is associated even with very low levels of processing. This view corresponds well with contemporary theories about phenomena such as mental imagery. These are clearly conscious phenomena that have been shown to directly engage the same primary systems involved in sensory perception (e.g., Farah, Hammond, Levine, & Calvanio, 1988) As an embellishment on this point, he refers to an intriguing class of phenomena: the "unmasking" of conscious experiences of primary sensory and motor processes, when—due to damage of higher level systems—these are released from top-down influences that ordinarily prevent them from expressing themselves in consciousness. He offers this as an explanation for the phantom limb phenomenon, suggesting that the illusion is actually the rising to consciousness of the intention to move the limb, in the absence of the usual, and overriding, experience of the action itself. Thus, according to Kinsbourne, consciousness can arise anywhere within the system, so long as the representations involved meet certain functional criteria.

Kinsbourne can take heart in the impact that connectionist approaches have had on cognitive theory, as they share some fundamental similarities. Both emphasize that representations and processing are best understood as distributed phenomena involving large numbers of processing components, and characterized by the interactions that occur between these, above and beyond the identity or "location" of any component by itself. For example, connectionist theories have asserted that there are no unitary, identifiable representations of cognitive "rules," but rather these arise indissolubly from interactions among multiple processing units. Similarly, Kinsbourne argues that there is no identifiable component of the system that alone is responsible for consciousness, but rather that it arises indissolubly from events and interactions that can occur anywhere within the system.

HOBSON'S HIERARCHICAL VIEW
OF CONSCIOUSNESS

Hobson takes a very different approach to the topic. He focuses much of his consideration on two important, and inescapable observations: First, nervous systems are characterized by a decidedly hierarchical organization, which is expressed increasingly as one progresses up both the phylogenetic and ontogenetic ladders. And, second, as adult humans, we have a unique capacity for self-representation, expressed in our capacity for self-awareness and self-reflection. Hobson suggests that these observations are closely related to one another. He proposes that "consciousness, as it is fully experienced in awake humans, is the brain's representation to itself . . . of the outside world, the body and the self." Hobson asserts that this capability arises from the hierarchical complexity present in adult human brains, and cashes this idea out specifically in structural terms. He hypothesizes the existence of a large, secondary network, responsible for the metarepresentational functions associated with consciousness in adult humans. Thus, he offers a decidedly hierarchical view of the cognitive architecture, and of its relation to consciousness. In this respect, his view coincides closely with classical theories of cognition, that postulate specific components associated with higher cognitive processing (e.g., a central executive), and with which consciousness has often been identified.

HETERARCHICAL VERSUS HIERARCHICAL
VIEWS OF CONSCIOUSNESS

On the surface, it would appear that Kinsbourne and Hobson offer contradicting views of how and where consciousness emerges within a processing system. Kinsbourne's view is radically distributed or, in his words, heterarchical. In contrast, Hobson emphasizes the hierarchical organization of our nervous system, and an intimate relation between consciousness and the capacity for self-representation that relies on a large secondary network sitting at the top of this hierarchy. Thus, Hobson seems to disagree with Kinsbourne by pointing to both the hierarchical and structurally localized aspects of consciousness.

Despite these apparent differences, I believe that both of these authors actually share similar views, and have simply chosen to emphasize different aspects of the topic. Kinsbourne, as his title suggests, focuses on the general properties that determine whether representations are conscious. Hobson, also true to his title, focuses on the dimensions that make for different types of consciousness. The critical point here is that we take Hobson's emphasis on hierarchies not as a basis for deciding whether a system has or does not have consciousness, but rather as a way of characterizing the kind of consciousness it can have. Thus, if we allow that several different levels, or types of consciousness are possible, Kinsbourne is interested in what they have in common, whereas Hobson is interested in how they differ. It is useful to consider further each of their views from this perspective, as each has its strengths and weaknesses.

KINSBOURNE: DIMENSIONS THAT DEFINE THE DEGREE OF CONSCIOUSNESS

Kinsbourne focuses on the set of properties that conscious states share in common, and the types of processing mechanisms that support such states, and is less interested in (though does not ignore) the qualitative differences that may exist among such states. Thus, he identifies a set of functional parameters, or dimensions, that specify when a set of representations will give rise to consciousness: the intensity, duration, and coherence of activity patterns, regardless of their location within the system. His ideas can be applied to systems with and without metarepresenational capacity, and to the various states that such systems may assume. This is a potentially powerful and integrative approach that suggests specific dimensions along which degrees of consciousness might be quantified.

At the same time, as I am sure Kinsbourne would be quick to admit, it makes it clear how far we have to go. Most of the details remain to be specified. For example, can extremes in one of the parameters (e.g., intensity or coherence) make up for failures in another (duration)? That is, can different combinations of values satisfy the criterion for consciousness? This is suggested by the fact that in some places Kinsbourne argues that conscious representations must be sufficiently enduring as to be stable, whereas in others he allows that they may be fleeting if they are sufficiently coherent and intense. Given this tradeoff, exactly what equation or combination of values defines the boundary conditions of conscious representations? Similar questions arise with regard to the indexing of coherence. How can this be formalized and measured: degree of mutual excitation, temporal synchrony, and/or the co-activation of "third-party" representations? And at what scale must coherence arise? Furthermore, are all of these continuous variables, or are there intrinsic discontinuities, or at least dramatic nonlinearities, that allow us to delimit qualitative boundaries between conscious and nonconscious states? And, finally, how can we account, in terms of *specific mechanisms*, for the seemingly directed flow of wakeful consciousness ("focus of activation"), and the emergence of purposeful, goal-directed behavior from an inherently distributed system? These are questions that must be sorted out in more detailed theoretical work leading ultimately to formal analyses or computer simulation of candidate systems, and empirical studies of behavior and neurophysiological function that confirm the validity of these theories.

HOBSON: DIMENSIONS THAT DEFINE THE TYPE OF CONSCIOUSNESS

Hobson's view complements that of Kinsbourne. Allowing for a variety among conscious states (as both he and Kinsbourne do), Hobson points to a number of the critical dimensions that characterize the differences among these states. One of these dimensions is hierarchical complexity within the system. Thus, simple systems may be capable of, but at the same time also restricted to simple, apperceptive consciousness. At the other extreme, possessing a large secondary network with metarepresentational capacity, adult humans are capable of self-awareness, self-reflection, and symbolic communication. These capabilities distinguish the nature of consciousness from that of animals or infants. Of course, this is not the only dimension along which consciousness can vary. For example, it can also vary by the nature of its content (e.g., visual vs. olfactory, sensory vs. emotional).

Perhaps more interestingly, and more importantly, Hobson points to the distinction between consciousness in the waking state (characterized by focused attention and volitional control) as compared to dream states (in which "the secondary forebrain cortical networks are subject to invasion and takeover by the cholinergic fiends from brain stem hell"). Here, he is quite specific about the variables that define the relevant dimensions, which he formalizes in his AIM model (activation, input–output gating, and neuromodulation). This model provides a fascinating framework for characterizing different states of consciousness, and ways in which they might be disturbed in psychopathological disorders.

All of these dimensions are clearly important ones, which Hobson points out, map cleanly onto fundamental features of our brains such as their hierarchical organization and their neuromodulatory systems. Nevertheless, as with Kinsbourne's contribution, they also point out how much detail remains to be discovered. Hobson offers some intriguing directions for quantifying the variables of his model, but doing so promises to be hard work, at best. Other components of the theory remain to be specified as well. For example, what form does the large secondary network take: Is it structurally distinct (e.g., in a particular lobe of the brain and/or set of subcortical structures), or tightly interwoven with primary processing systems distributed throughout the brain. And, of course, this view faces all of the vexing questions that attend hierarchical theories of cognition: How, if strictly hierarchical, does it escape the regress of a homunculus (that is, another, higher level system that controls its functioning)?

SUMMARY AND CONCLUSIONS

Though Kinsbourne and Hobson emphasize different, and superficially opposing, features of consciousness, in fact their views complement one another in important ways; both point out a fundamental feature of consciousness: It lies along a continuum. The presentations are interesting because they each explore different facets of this continuum: degree and kind. Kinsbourne suggests how the rich diversity of experiences that come under the category of consciousness might merit status as a scientifically meaningful construct, by defining a set of dimensions that identify the features common to all conscious representations. Representations can vary continuously along these dimensions, defining a continuum in the degree of consciousness. Hobson's view gives order to the rich variety of forms (e.g., primary vs. metarepresentational, wakeful vs. dreaming) that such representations may assume. Again, representations can vary continuously along these dimensions, here defining a continuum in the kind of consciousness. Indeed, one can imagine a rich set of interactions that might occur when representations vary in both degree and kind.

Both views also reveal the tremendous amount of work that needs to be done. Here I would like to echo the call for more refined analyses and empirical studies made by Glymour in his commentary on Farah's and Kinsbourne's chapters. I concur fully with his concern that contemporary theories not lead us down the same path that classical theories of the mind have led us in the past. By remaining overly general, or targeting phenomena that could not yield to analysis or empirical investigation, they were untestable. To paraphrase an old saw: "Consciousness is in the details."

However, this call should not be mistaken for pessimism. Although we are still far from our destination, the direction in which Kinsbourne and Hobson are leading us is a safe and productive one. They have identified theoretical and neurobiological variables that can, and in some cases already have, come under careful scrutiny. New theoretical tools are available (both mathematical and computational) as are empirical tools (such as multielectrode recording techniques in animals and functional neuroimaging techniques in humans) that are more sophisticated than our predecessors ever had available. This alone does not guarantee success. However, science has consistently shown itself to be a cyclical process, in which good ideas stubbornly re-emerge, until they find a receptive environment (both theoretically and empirically). No one can say with certainty that such times have arrived with respect to a scientific understanding of consciousness. However, it is incumbent upon us, with the new tools we have in hand, to try again. To torture the old saw: Who knows, the details may soon be in consciousness!

REFERENCES

Farah, M. J., Hammond, K. M., Levine, D. L., & Calvanio, R. (1988). Visual and spatial imagery: Dissociable systems of representation. *Cognitive Psychology, 20,* 439–462.

Theoretical Issues and Approaches

VII

CHAPTER 21

Prospects for a
Unified Theory of Consciousness
or, What Dreams Are Made of

Owen Flanagan
Duke University

What are the prospects for a unified theory of consciousness? This is a question that needs to be asked but about which there is a certain ambivalence. This ambivalence comes from the fact that it seems like a bad idea to speculate too much about the future history of science, and because the question of whether there can be a unified theory of consciousness invites such speculation. On the other hand, this is a period of renewal in the scientific study of consciousness and some vision of the shape of theory is needed at such times.

The prospects for a theory of consciousness are not bad and the prospects for a unified theory are not bad either—but they depend a great deal on the sort of unity one hopes for or demands. So is the glass half empty or half full? This chapter recommends the half-full attitude. Be an optimist about the prospects for a theory of consciousness, not because the prospects for such a theory are so certain, but because optimism will make trying to solve this problem seem worth the effort.

The aim of this chapter is first, to exemplify and defend the method proposed in *Consciousness Reconsidered* (1992) for tackling the problem of consciousness, what was called "the natural method." And second, it responds to three different objections to the very idea that there could be a theory, especially a unified theory of consciousness: "the hodge-podge objection," "the heterogeneity objection," and "the superficiality objection." The example of dreams is used throughout to illustrate distinctive scientific and philosophical concerns and as a good example of one of the multifarious types of conscious experience.

NO HODGE-PODGES ALLOWED

First consider the following three claims:

1. *Consciousness exists.* This claim is meant to be as innocent as possible, so it is probably better to put it this way: There exist conscious mental states, events, and processes, or, if you like, there are states, events, and processes that have the property of being conscious.

2. *Consciousness has depth, hidden structure, hidden and possibly multiple functions, and hidden natural and cultural history.* Consciousness has a first-personal phenomenal surface structure. But from a naturalistic point of view, the subjective aspects of consciousness do not exhaust the properties of consciousness. Part of the hidden structure of conscious mental states involves their neural realization. Conscious mental states supervene on brain states. These brain states are essential aspects or constituents of the conscious states, as are the phenomenal aspects of these states. But, of course, nothing about neural realization is revealed at the phenomenal surface, not even that there is such realization. The phenomenal surface often hints at or self-intimates the causal role of conscious states. Indeed, it often seems as if we place certain conscious intentions onto the motivational circuits in order to live as we wish. But the phenomenology leaves us clueless as to how these conscious intentions actually get the system doing what it does; and, of course, experience intimates nothing about the causal origins and evolutionary function, if there are any, for the different kinds of consciousness.

3. *Conscious mental states, processes, events—possibly conscious supervisory faculties, if there are any—are heterogeneous in phenomenal kind.* Again this claim is meant to be to be innocent and uncontroversial. *Consciousness* is a superordinate term meant to cover all mental states, events, and processes that are *experienced*—that is, all the states that there is something it is like for the subject of them to be in. So, for example, sensory experience is a kind in the domain of consciousness that branches out into the types of experience familiar in the five sensory modalities. Each of these branches further, so that, for example, visual experience branches off from sensory experience and color experiences branch off from it. My experience of red, like yours, is a token color experience, that is, a token of the type red sensation. So sensory experience divides into five kinds and then branches and branches again and again as our taxonomic needs require. But consciousness covers far more than sensory experience. There is, in the human case, linguistic consciousness, and there is conscious propositional attitude thought that seems heavily linguistic but may or may not be; there are sensory memories, episodic memories, various types of self-consciousness, moods, emotions, dreams, and much else besides.

The first two claims, that consciousness exists and that it has depth and hidden structure, suggest the need for a theory. Theories are needed for what exists, for what is real, and this is especially so when what interests us is ubiquitous and has here-to-fore hidden structure, hidden function, and a mysterious natural (and cultural) history.

But the third claim, the heterogeneity claim, has led many to express skepticism about the prospects for a theory, certainly for a unified theory of consciousness. But heterogeneity in itself is not a good basis for skepticism.

Consider elementary particle physics. Here heterogeneity of particle types reigns to a degree that surpasses normal capacities to count types. But elementary particle physics flourishes; heterogeneity of particle types does not thwart it.[1] Or, to take a different example, the periodic table just is a table of the heterogeneous element types that constitute chemistry. So elementary particle physics and chemistry exist and provide theoretical cohesion to heterogeneous kinds in their domain of inquiry. These examples are sufficient to show that heterogeneity of the phenomena in the domain-to-be-explained has in itself no bearing on whether theory can be developed for that domain. The inference from the premise that there are zillions of different kinds of elementary particles, heterogeneous in structure and function, to the conclusion that there can be no elementary particle physics yields a false conclusion because the implicit premise about heterogeneity thwarting theory development is false. The point is that the heterogeneity of forms of consciousness in itself should not worry us as we proceed to try to develop a theory of consciousness.

Some naturalists are skeptical about the prospects for a theory of consciousness not simply because of heterogeneity but because of the kind of heterogeneous array that the superordinate term *consciousness* names—it names a heterogeneous hodge-podge (Wilkes, 1988a, 1988b). A hodge-podge, according to the dictionary, is a "heterogeneous mixture of incongruous and ill-suited elements." Churchland (1983) compared "consciousness" to the kind "dirt" or the kind "things-that-go-bump-in-the night." Wilkes (1988b, p. 33) thought that "conscious phenomena" is more like the arbitrary set consisting of "all the words with 'g' as the fourth letter" than it is like a superordinate category such as "metal," "mammal," "fish" or like a subcategory of such a superordinate category: "gold," "whale," "flounder."

The problem is not with heterogeneity as such but with the hodge-podginess of the heterogeneous array. Perhaps one can develop a scientific theory for a heterogeneous set of phenomena, especially if the phenomena constitute a natural kind, but one cannot develop a scientific theory for a hodge-podge.

Wilkes' argument turns on the conviction that in order for a superordinate category to be nonarbitrary it must display a certain coherence, but that con-

---

[1] Of course, many elementary particle physicists worry about this lack of unity at the highest level.

scious phenomena fail to display the required coherence. "What sort of coherence? Well, even though most of the interesting laws may concern only the subclasses, there might be some laws that are interestingly—non trivially—true of all the subclasses; or even if this is not so, the laws that concern the subclasses may have significant structural analogy or isomorphism" (Wilkes, 1988b, pp. 33–34).

Wilkes' concern can be explained in terms of the two examples used so far. The heterogeneous types of elementary particle physics and chemistry have shown themselves to have a certain theoretical unity as evidenced by certain cross-kind similarities. For example, the kinds on the periodic table are typed according to atomic number, atomic weight, specific gravity, melting point, and boiling point. In some cases, these are not yet known; and in the case of elementary particle physics, typing is based on characteristic mass and quantum properties such as charge and spin. Furthermore, the phenomena-in-question adhere to a set of coherent laws, that is, the laws of quantum physics in one case and the periodic law in the other case.[2] Chemistry and elementary particle physics are in reasonably good shape because they unify heterogeneous phenomena that are interestingly embedded in connected layers of laws of nature—this making them nonhodge-podgy. We do not think celestial mechanics is suspect because of the heterogeneity of the composition, size, and gravitational force of bodies in our solar system. Nor do we think that the astronomical variety of subatomic particles, within the three main classes, forecloses the possibility of quantum theory. A theory of consciousness will in the end be part of a unified theory of the mind. This is compatible with the theory making generalizations suited to whatever deep local idiosyncrasies exist. Physics tells us that bodies at extremely far distances from each other traveling close to the speed of light are subject to regularities very different from objects moving around our little spherelike home. It would be no more surprising and no more damaging to the success of the science of the mind if it tells us that visual consciousness obeys very different laws and is subserved by different neural mechanisms than is conscious reflection on one's love life. In fact, this is exactly what we expect.

One thing worth noticing is that if this correctly characterizes the criterion of being nonhodge-podgy, then there is no requirement that a theory must be unified at the most general or highest level, because neither chemistry nor physics are at present, and possibly not even in principle, unified at that level. The important question is whether the category and the subcategories that constitute the kind consciousness show signs, taken together or individually, of being a hodge-podge.

---

[2]The periodic law says that the physical and chemical properties of the elements depend on atomic structure and thus are, for the most part, periodic functions of their atomic numbers.

Wilkes (1988a) wrote that "scientific research, it would seem, can manage best if it ignores the notion of consciousness. . . . 'Consciousness' . . . is a term that groups a thoroughly heterogeneous bunch of psychological phenomena [and] is unsuited *per se* for scientific or theoretical purposes" (pp. 192–196).

But Wilkes can not mean that heterogeneity dashes hopes for a theory of consciousness. She must be thinking that consciousness is a hodge-podge. According to Wilkes, some set of phenomena is a hodge podge if it lacks a certain kind of coherence. In particular, the demarcation criterion distinguishing phenomena-suited-for-respectable-theory and those not so suited rests on this idea that "even though most of the interesting laws may concern only the subclasses, there might be some laws at least that are interestingly—nontrivially —true of all the subclasses; or even if this is not so, the laws that concern the subclasses may have significant structural analogy or isomorphism" (Wilkes, 1988b, pp. 33–34).

Being unsure exactly what the standard here is, especially the last part, then what exactly counts as "significant structural analogy or isomorphism"? Imagine, contrary to the truth, that all learning was governed by the laws of classical conditioning and the laws of operant conditioning. If this were so, we should want to say that a theory of learning exists. But the grounds would not be that all the types of learning were governed by some single high-level principle. It would be that there are laws governing each subclass, which taken together explain all learning. One might cite the additional fact that the relevant laws display a certain structural analogy or isomorphism, for example, they are stimulus–response laws. But analogies are a dime a dozen and any imaginative mind can see them most anywhere. What would make the imagined account of learning a theory of learning and a complete one at that would revolve around the fact that our taxonomy of a nonarbitrary domain of inquiry was exhaustive and our two sets of laws explain everything in need of explanation. It is hard to imagine anyone worrying about a lack of structural analogy or isomorphism between two sets of explanatory laws if the relevant explanatory work got done.

As for the first part of Wilkes' criterion—that a set of phenomena are suited for a theory if in addition to special laws governing the subclasses there are laws that unify some, possibly all, the subclasses—it is worth noticing that there is no unified field theory in physics and no certainty whatsoever that one exists. This suggests that the proposal is too stringent for certain well-developed scientific theories.[3]

---

[3]According to Brandon's (1990) account of evolutionary theory, it is not unified in Wilkes' strong sense. Evolutionary theory is unified by certain principles of probability theory, but not by any overarching contingently true generalization. This, even though, the principles of natural selection, drift, and so on, instantiate the relevant probability principles.

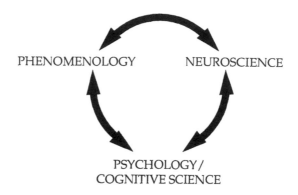

PHENOMENOLOGY   NEUROSCIENCE

PSYCHOLOGY /
COGNITIVE SCIENCE

also: Evolutionary Biology
       Cultural and Psychological Anthropology
       Social Psychology

FIG. 21.1.   The natural method.

HOW TO NATURALIZE CONSCIOUSNESS

If we hold more reasonable standards of unity and of lawlikeness (see Cummins, 1983), then there is no basis in the current state-of-consciousness studies for thinking the skeptical hodge-podgists right. There will undoubtedly be revision of the typologies with which we begin as scientific inquiry proceeds, but the evidence (both phenomenological and nonphenomenological) suggests even at this very early stage that conscious phenomena display coherence.

Consciousness Reconsidered (1992) asks: By what method is consciousness to be studied? This chapter proposes trying the most natural strategy — the natural method (Fig. 21.1). Start by treating three different lines of analysis with equal respect. Give *phenomenology* its due. Listen carefully to what individuals have to say about how things seem.[4] Also, let the psychologists and

---

[4] At the Wake Forest Colloquium on my book, *Consciousness Reconsidered,* May 7–9, 1993, there was spirited debate about the status of phenomenology. Bob McCauley argued persuasively that it provides data and its pronouncements as pronouncements about how things really are must yield to explanation at the other levels. Even claims about how things *seem* can change as our views about how things are change. The case of dreams is a case in point. If we ever came to have really good theoretical reasons for thinking that dreams were not experiences, they might

cognitive scientists have their say. Listen carefully to their descriptions about how mental life works, and what jobs if any consciousness has in its overall economy. Letting the psychologists and cognitive scientists have their say means also including amateurs (let folk wisdom be put out on the table along with everything else). Finally, listen carefully to what the neuroscientists say about how conscious mental events of different sorts are realized, and examine the fit between their stories and the phenomenological and psychological stories.

The object of the natural method is to see whether and to what extent the three stories can be rendered coherent, meshed, and brought into reflective equilibrium, into a state where theory and data fit coherently together. The only rule is to treat all three—the phenomenology, the psychology, and the neuroscience—with respect. Any a priori decision about which line of analysis "gets things right" or "has the last word" prejudges the question of whether different analyses might be legitimate for different explanatory purposes and, in fact, compatible with each other, or at least capable of peaceful coexistence. As theory develops, analyses at each level are subject to refinement, revision, or rejection.[5]

Consider a feature of the natural method not emphasized in the book: In the end, what counts toward our theory of consciousness is everything we know from every and any source worth paying attention to. Besides the troika of phenomenology, psychology, and neuroscience, evolutionary biology and cultural and psychological anthropology will also be crucial players. There are no good theories about why conscious experientially sensitive hominids should have been favored over merely informationally sensitive ones, and although it is pretty clear that sensational consciousness (i.e., phenomenal awareness in the sensory modalities) comes with the human genome, it is not clear that, for example, moral self-consciousness does. Moral self-consciousness, like the

---

well seem less like experiences. It is hard to imagine giving up the idea that there are perceptual experiences because such experiences take place in the specious present (or so it strongly seems); but even dreamers will admit that they are remembering both the alleged experience and the content of the alleged experience. Because we allow false remembering, the intuition that dreams are experiences could yield if theoretical reasons deemed the sense that dream experiences had occurred misbegotten memories, akin to déjà vu experiences.

[5]At the Carnegie Mellon Conference, Clark Glymour proposed that theory should be as biologically constrained as possible; and he expressed the worry that cognitive information-processing models often fail to attend to biological realism. I quite agree with the normative point. But our present knowledge of the brain is thin and often hard to interpret. Sometimes it is hard to know what the neuroscientific data are or mean and thus it is hard to know how they should constrain our theories. For example, many effective antidepressants work by affecting norepinephrine or serotonin levels, absorption rates, and so on. But, in many cases, neither the FDA, nor the pharmaceutical companies know exactly how these drugs work. Judgments about what they do at the psychological level, what some of their phenomenological and physiological side effects are, as well as assessments about overall safety and effectiveness, are made without anything approaching complete understanding at the level of brain chemistry.

abilities to play chess or basketball, are allowed by our genes, but they were hardly selected for.

In any case, one possibility is that in addition to their phenomenological unity, (many) conscious event types will be given an adaptationist evolutionary account, even while lacking certain kinds of unity at the psychological and neuroscientific levels. Embedding consciousness into theories of evolution (biological and cultural), thinking about different forms of consciousness in terms of their ecological role, and in terms of the mechanisms of drift, adaptive selection, and free-riding will be an important part of understanding what consciousness is, how it works, and what, if anything, it is good for.

HOW DREAMS ARE LIKE BEING AWAKE

This chapter tries the natural method on dreams. Hobson's (1988, 1994) work has inspired most of what appears in this chapter. His theory, AIM, is a robust and powerful one. Despite the existence of such a powerful theory as AIM, dreams are fodder for skeptics about the prospects for a theory of consciousness for a number of reasons. One route to skepticism is simple and straightforward. Common sense says that consciousness involves, among other things, being awake. But dreaming takes place during sleep, and, thus, by the distinguished deliverances of conceptual analysis, dreams cannot be conscious experiences. But common wisdom also says that dreams are experiences that take place during sleep, so our common sense taxonomy of consciousness is worse than a hodge podge; it is riddled with inconsistency from the start.

There is a more sophisticated line of attack suggested in different ways and with different degrees of conviction by Malcolm (1959) and Dennett (1978). This line sets aside the unsettling linguistic facts linking "being conscious" with "being awake" and focuses primarily on the received view that dreams are experiences that take place during sleep. To be sure, people give reports of experiences they allege to have had during sleep. But if one is in a sufficiently skeptical mood, this does not remotely prove that the dream report is in fact about experience that took place during sleep, rather that a combined confabulation–misremembering accompanying waking. Remember, people also insist they have visited certain places before or been in certain situations before only to discover they are wrong. *Déjà vu experiences* are claims to experiences that did not in fact occur. There are, one might say, déjà vu experiences, but they are not truly déjà vu—you were not here before and this did not happen to you before. It seems that way, but it just is not so.

Dennett called attention to the possibility (also suggested by Malcolm) that so-called dreams may occur while awake or while waking up and not, as most everyone thinks, during REM sleep at all. Sure this is possible. But why think it more than merely possible? Malcolm's position was verificationist: Dreams line up with verbal reports people sometimes give in the morning, so that is all we can say that dreams are. Dennett (in this case) rejected the verificationist line, and pointed instead to certain kinds of dreams as support for the skeptical view of dreaming. Dennett (1978) reported this dream: "I searched long and far for a neighbor's goat; when at last I found her she bleated baa-a-a— and I awoke to find her bleat merging perfectly with the buzz of an electric alarm clock I had not used or heard for months" (p. 135). One way to explain such dreams, but at high cost as Dennett pointed out, is to think that powers of *precognition* are operative. The dream narrating mechanism prepares the narrative to fit with the alarm it knows is about to go off!

Even true believers in the theory that dreams are experiences that take place while sleeping think that most every dream report contains either more or less than was experienced or both more and less than was experienced in the dream state. On most every view, dream reports are putrid evidence about what was experienced while asleep, even if we accept that they provide reliable evidence that something was experienced while sleeping.

In any case, apply the natural method to the case of dreams, and in particular to the question "are dreams experiences?", to see where it leads. We know that most people, not having read Malcolm or Dennett, will report with confidence that they have dreamt while asleep. They will also say certain things about the content of their dreams. These reports are composed of part phenomenology and part folk wisdom (i.e., the reports are partly provoked by the wisdom that people dream while asleep). There are also psychological theories about dreams—the psychoanalytic theory being the most well known, if not the most well respected.[6] The nice thing about the psychoanalytic view is that it can easily be expressed in terms of the sort of functional-flow-chartology admired in cognitive science: As sleep ensues, repressor mechanisms loosen their grip on receptacles filled with wishes, strong socially unacceptable wishes squirm out and find their way to symbolic encoders that work up an "experienced" story that gives the disguised wish some sort of satisfying release, which "experienced" story is routed then (imperfectly) far away from waking memory.

Now, although Freud certainly thought that dreams are experiences, his theory is compatible with the view that dreams are not experiences—at least not those taking place during sleep. That is, take all the phenomenological reports of dreams, treat them as displaying manifest content that yields in interpretation to latent content revealing what wish was being fulfilled. All this, as well as the associated view of why dreams are good for you, is compatible with a version of the skeptics' view, namely, a version saying that dreams are disguised wishes we happen to give (or think to ourselves) in the morning.

---

[6]There is also as I have noted Hobson's (1988, 1994) more credible and brain-based theory of dreams.

Or, to put it another way, dream reports reflect experiences, but not those that take place while sleeping; rather, they reflect those embedded in awake thoughts or reports.

The upshot so far is this: Despite the widespread acceptance of the view that dreams are experiences that take place during REM sleep, the case is not remotely closed. And one consequence of its not being closed is that we do not have a really decisive reply to the skeptic, and lacking that, we do not know whether or not dreams fall under the concept of consciousness. And this can be taken as evidence that the very taxonomy of the domain-to-be-explained is ill-defined. If we do not even know what falls under the concept of consciousness, then we can hardly be expected to build a coherent theory for it.

This is why the physiological and neuroscientific data are so important. These data both constrain and illuminate the phenomenology and the psychology and they provide an additional line of evidence against the skeptic. If there is to be a theory of consciousness, then we will want to know how, or in what ways, conscious mental states are realized. It has been suggested that subjective awareness is linked to oscillation patterns in the 40-Hz range in the relevant groups of neurons, that is, neurons involved in a certain decoding task "synchronize their spikes in 40Hz oscillations" (Crick & Koch, 1990, p. 272). These 40-Hz oscillations have been found in single neurons and neural nets in the retina, olfactory bulb, as well as in the thalamus, and neocortex.

Llinás, Ribary, and Paré produced strong evidence that such oscillation patterns characterize REM sleep. Llinás and Ribary (1993) reported that during the period corresponding to REM sleep (in which a subject if awakened reports dreaming), 40-Hz oscillation patterns similar in distribution phase and amplitude to that observed during wakefulness is observed. Also of significance is the finding that during dreaming 40-Hz oscillations are not reset by sensory input. The dreaming condition can be considered a state of hyperattentiveness in which sensory input cannot address the machinery that generates conscious experience. Within Hobson's theory (this volume), the 40-Hz oscillations pertain to (A) activation level, while the tuning out of external stimuli is explained by the mechanisms of input–output gating (I).

The main point for present purposes is that the reason dreams seem like conscious experiences is because they are conscious experiences and they are like awake conscious experiences in certain crucial respects.[7]

---

[7]So being awake and REM sleep both involve 40-Hz oscillations. This might seem neat because both being awake and being in REM sleep are thought to be experiential states, to involve a type of conscious experience. NREM sleep, on the other hand, is often thought to be a state of unconsciousness—a state in which there is nothing it is like to be in it. But this is dead wrong. NREM sleep is experientially rich (see Flanagan, 1995). The 40-Hz oscillations may not turn out to be a marker of experience or they may be necessary and/or sufficient; it is too early to know. One worry for the 40-Hz necessary condition hypothesis is this: the mentation occurring during NREM when measured by EEG does not appear to involve the 40-Hz oscillations despite involving mentation. So 40-Hz oscillations may be a reliable marker of certain kinds of conscious

Llinás and Ribary suggested this unifying hypothesis: The 40-Hz activity in the nonspecific system comprised of the thalamocortical loop provides the temporal binding of contentful states that involve 40-Hz oscillations in the areas devoted to particular modalities. That is, the neural system subserving a sensory modality provides the *content* of an experience and the nonspecific system consisting of resonating activity in the thalamus and cortex provides "the *temporal binding* of such content into a single cognitive experience evoked either by external stimuli or, intrinsically during dreaming" (1993, p. 2081). According to Llinás and Paré, "*It is the dialogue between the thalamus and the cortex that generates subjectivity*" (1991, p. 532).

These data and hypotheses, in light of other data and theory, increase the credibility of the claim linking REM sleep with vivid experiences. Whether it is really true that dreams are experiences depends, of course, on whether it is true that 40-Hz oscillations turn out to be a marker, a constituent component, or a cause (which one they are is, of course, very important) of vivid conscious experiences. The point is that the neuroscientific data push credible theory in a certain direction. If these data bring us closer to the answer to one question, then they open a host of others, suggesting occasional answers—a sure sign of a progressive research program.

For example, one might wonder whether 40-Hz oscillation patterns will turn out to be necessary or sufficient for experience or enable us to differentiate kinds of experiences. This is a hard question. The 40-Hz oscillation patterns are like dust; there is always some to be found. This suggests the possibility that living human beings might always be in some experiential state or other, that is, that human beings are never wholly unconscious—if, that is, 40-Hz patterns are sufficient for experience. If this sounds like an incredible prospect, then it is relevant that persons awakened from NREM sleep often report having experiences—albeit experiences lacking the vivacity of post-REM reports. And sleep talking and sleep walking are well known to take place during NREM sleep (postural muscles are turned off during REM sleep), and it is obscure whether or in what precise sense sleep walkers and talkers are experiential blanks. *Globality* of 40-Hz activity may turn out to be the relevant feature of robust conscious experiences, not the mere presence or absence of some 40-Hz activity (Llinás & Paré, 1991, p. 527).

We will also want to know why despite involving vivid experience, dreams involve shut-downs of the attentional, motor, and memory systems and insensitivity to disturbance by external stimuli. Here, as Hobson (1988) has pointed out, under the rubric of (**M**)odulation (specifically de-modulation) ratios of

---

mentation but not necessary for all mentation (see Steriade, McCormick, & Sejnowski, 1993). On the other hand, when measured with MEG, one does find 40-Hz oscilliations, but much attenuated in amplitude and we do not pick up much in the way of amplitude modulations. See also Kinsbourne (this volume) for a somewhat different model, which emphasizes neural duration, intensity, and congruence in giving a representation conscious status.

neurochemicals of various sorts play the major role. The point can be put this way: The 40-Hz oscillation patterns might explain the similarity between dreams and waking states while input–output gating and demodulation explain the main differences.

There is some neurophysiological evidence that auditory stimuli are slightly discriminated during REM sleep, which would help explain how events in the external world are embedded in dream content, and might take some of the edge off Dennett's alarm clock worry (Llinás & Paré, 1991, p. 522). Consider some other interesting facts: humans with pontine lesions do not have muscular atonia during dreams and are those very people one reads about who play linebacker with their dressers and endanger their spouses on a nightly basis. These people are not bad, they are doing what the rest of us would be doing if the relevant brain area was not effective at shutting the motor system down.

There are interesting convergences of neuroscientific and phenomenological data in other areas as well. For example, the parts of the brain that reveal robust activity on positron emission tomography (PETs), magnetic resonance imaging (MRIs), or magneto-encephalographs or similar devices suggest that "mentation during dreaming operates on the same anatomical substrate as does perception during the waking state" (Llinás & Paré, p. 524).[8]

THE SUPERFICIALITY OBJECTION

So what does the case of dreams show? It shows something I think about the nature of consciousness, or about the nature of one kind of consciousness, or, if you like, about the similarities and differences between two kinds of consciousness. It also shows some of the power of the natural method. Specifically, it shows that phenomenological facts and neurophysiological facts can be brought into reflective equilibrium in certain cases. Phenomenologically, REM dreams and awake consciousness are very different in certain respects. But they also possess phenomenological commonalities, and these together with certain commonalities at the neural level, suggest that they are two kinds of consciousness. They are a heterogeneous dyad but hardly a hodge-podge, hardly an ill-suited or incongruous dyad.

The case of dreams also carries some weight against the objection that even if consciousness is not a hodge-podge, its only interesting feature is the *superficial* shared phenomenological feature of being experienced. This is not so,

---

[8]This helps explain why prosopagnosiacs do not report dreaming of faces and why people with right parietal lobe lesions who cannot see the left side of the visual field report related deficits in their dream imagery (p. 524). On the other hand, it tells us something about memory that visual imagery sustains itself better in both the dreams and the awake experiences of people who develop various kinds of blindness in later life.

because in the case of dreams and awake states, the phenomenological similarities and differences gain some explanation at the neural level. Furthermore, certain psychological facts (e.g., why noises in the external world are sometimes incorporated into dream narratives) are also explained by bringing neural facts to bear; in particular, the brain detects, but will not normally wake up for, noises in the external environment during REM sleep.

Nonetheless, one might worry that this case is exceptional. And indeed it might be. We know that there exists an important class of cases where phenomenal similarity is not subserved by similarity at the microlevel. For example, the phenomenal property of "wetness" is multiply realized. There is water, H_2O, that is wet and heavy water, D_2O, that is wet. Perhaps consciousness is like this: At the phenomenal level there is the shared property, but this shared property is subserved by all manner of different types of brain processes. This possibility is acknowledged here. It does not harm the prospects for a theory of consciousness as it has been conceived. But one might imagine the examples where phenomenal similarity is unsupported by similarities at lower levels being used to press the worry that the shared property of being experienced really is too superficial to guide theory or to make us optimists about the prospects for a theory of consciousness.

Churchland gave an example from the history of science designed to feed the superficiality worry (1983).[9] She reminded us how the commonsense concept of "fire" has fared as science has progressed:

> "Fire" was used to classify not only burning wood, but also the activity on the sun and various stars (actually fusion), lightning (actually electrically induced incandescence), the Northern lights (actually spectral emission), and fire-flies (actually phosphorescence). As we now understand matters, only some of these things involve oxidation, and some processes which do involve oxidation, namely rusting, tarnishing, and metabolism, are not on the "Fire" list. (p. 85)

Churchland concluded that "the case of fire illustrates both how the intuitive classification can be re-drawn, and how the new classification can pull together superficially diverse phenomena once the underlying theory is available" (1983, p. 85). This is true and important.

---

[9]It is often and reasonably said that science proceeds best when there are very specific research problems to be addressed. For example, the double helix structure of DNA solved the "copying problem" in genetics. It was well understood by the time Watson and Crick came along that inheritance takes place, but there was no postulated mechanism that was up to the specific task of replication. DNA solved this problem. One problem with the scientific study of consciousness is that there are still open taxonomic questions (e.g., the one about dreams discussed here) about what states are conscious and which ones not—even questions about whether there is any shared phenomenological feature as I assume, and thus anything at all which really falls under the concept. But some good questions (about dreams, about automatic behaviors, about deficits of consciousness, about neural underpinnings of various conscious states) are opening up and getting attention.

It has been suggested that we imagine a person in an earlier time saying something like this: "No matter how science goes, we will still be able to talk of a theory of fire." Even if the constituents and causes of the different sorts of fire turn out to be physically very different, the theory of fire can still be constituted by gathering together all the interesting truths "about the class of phenomena that possess the shared property of being fire."[10]

It is possible that Churchland might want to read the position put forth here as analogous to this far-seeing fire sage. So it is necessary to point out that phenomenological similarity, in and of itself (even when it is similarity of *experience itself*, or experience *as such*, and not similarity of *thing experienced*) resolves nothing about the hidden nature and deep structure of the phenomenally similar stuff. Indeed, the very asking of the question, are dreams experiences?, concedes the possibility that science could force us to drop dreams from the class of experiences in the way it forced us to drop what goes on in the tails of fire-flies from the fire list. One can imagine two extremist views. According to one view, there is verbal behavior associated with waking or being awake that "sounds" to third parties like experiential reports about sleeping mentation, but that are (remember, we are at the skeptical extreme) just some narrative noise an awake person produces. To keep skepticism heightened, think of all parties involved: those producing dream narratives, and those they speak to, including themselves in soliloquy, convinced by the standard theory that the *narrative reports* are of experiences that occured while sleeping, but that this is false. Another extreme hypothesis purports that dreams are experienced, content-wise, roughly as reported. The dreams that are remembered occur, however (very rapidly), while one is awake or as one awakes. Neither view is remotely incoherent.

The point is that dreams might have gone off the consciousness list (or changed status on the list) as inquiry proceeded in the same way the lighting bugs were removed from the list-of-things-on-fire (if anyone really believes they were once on the list). But the present discussion shows that, in fact, things are going in the other direction—science is helping to secure the place of dreams as experiences, not to make us remove dreams from the class of experiences. It could have worked out the other way (it still might), but it has not thus far. So our commonsense concept of consciousness is open to revision, to being "re-drawn." How consciousness is conceived depends almost completely on how the science of the mind progresses.[11]

---

[10]See Cottrell (1993).

[11]Indeed, it would be very surprising if different kinds of consciousness were not realized in different ways. The individuation of conscious events at the neural level will undoubtedly entail tracing complex neural maps originating at different points on the sensory periphery and at different places within the system itself and traversing all sorts of different neural terrain. This will be true even if all the events mapped share a physical property such as having the same oscillatory frequency. It would also be surprising if the neural underpinnings of certain kinds of con-

The best evidence against the critics who suspect that "consciousness" names too superficial a phenomenon to play a useful role in scientific explanation or prediction involves deploying the natural method in cases like that of dreams and, in addition, pointing to the existence of predictive and explanatory generalizations that place conscious mental events in important causal roles. For example, there are important functional differences between people with phenomenal awareness in certain domains and those without. Suitably motivated individuals with normal sight naturally carry out voluntary actions toward seen things. When thirsty, we step over to the water fountain we see to our right. However, blindsighted individuals who are identically motivated and who process information about the very same things in the visual field do not naturally or efficiently carry out the suitable actions toward the "seen" things. There are also the differential abilities of amnesiacs to form integrated self-concepts and to create and abide a consistent narratively constructed self-model. And persons incapable of experiencing certain qualia (e.g., color-blind people) show all sorts of functional differences from non-color-blind people. Check out their wardrobes.[12]

This evidence suggests that there exist true counterfactual generalizations in the domain of consciousness. Some of these generalizations will relate phenomena at the psychological level, for example, persons with qualia of kind q do x in circumstance c but persons without qualia q (but who are otherwise identical) fail to do x in c. Other generalizations will link phenomenological and psychological level processes with brain processes, for example, (a) peo-

scious states were not also essential components of certain nonconscious states, but not of other kinds of conscious states. For example, it might be that the areas of the brain that light up during ordinary visual awareness, or when we are solving problems in geometry, also light up when we turn over during sleep, but never light up when we are listening to music with our eyes closed. And perhaps there is a deep reason why the relevant area lights up in the conscious and nonconscious cases it lights up in. Imagine that the area is a necessary component of all spatial analysis, so it is activated when one is wide awake and trying to prove the Pythagorean theorem, and when one is sound asleep but computing information about edges and distances in order to keep from falling out of bed. In cases like this, our theory of consciousness is interweaved, as it must be, with theories of unconscious processing.

But such results would in no way undermine the idea that conscious phenomena are a legitimate explananda for which, and possibly with which, to build theory. It is to be expected that the development of the science of the mind will reveal deep and surprising things about the phenomena with the shared property of being experienced. Such discoveries might include the discovery that there are greater similarities in certain respects between certain phenomena that possess the shared property and those that do not, than among all those with the shared phenomenal property. The neural spatial analyzer would be such an example. But this could happen while, at the same time, important generalizations are found to obtain among all or most of the disparate events that possess the shared property.

[12]Color blindness is an interesting example because there is reason to believe that the conscious deficit is caused, as it were, by low-level processing problems in the visual system. Still it is not implausible to think that it is the deficit at the qualitative level that is the proximate cause of difficulties in color coordination in the color blind person's wardrobe.

ple with damage to the speech centers and, in particular, to the left brain interpreter, will have trouble generating the narratively constructed self; (b) people with certain kinds of frontal lobe damage will have trouble formulating plans and intentions; other kinds of frontal lobe damage will obstruct links between consciously formulated action plans and actually carrying out the plans.

Given that these sorts of generalizations already exist and have been corroborated, it follows that there are laws that conscious mental life answers to. To be sure, the laws are pitched to the heterogeneous multiplicity of events and processes that possess the shared property of being experienced. But taken together, there is no reason to say they are not part of an emerging theory of consciousness, one basic insight of which is that consciousness is heterogeneous.

Conscious phenomena constitute a legitimate explanada and conscious events play explanatory roles in certain well-grounded generalizations. In broad strokes there are two ways one might imagine building a theory of consciousness (see Fig. 21.2). Gathering together whatever scientific truths there are about this set of phenomena will constitute one way of building a theory of consciousness. Especially in the early stages, building such a theory might amount to the gathering together of all the interesting truths about the class of phenomena that possess the shared feature of being experienced. If dreams, for example, turned out not to be experiences, they would no longer be taxonomized as conscious, and would fall out of the theory so conceived. But they, no longer dreams but now "mere dream reports," or alleged dreams,

---

WAY ONE

GATHER ALL INTERESTING TRUTHS AND
GENERALIZATIONS ABOUT PHENOMENA WITH THE
SHARED PHENOMENOLOGICAL PROPERTY OF "BEING
EXPERIENCED"

WAY TWO

BUILD ALL SPECIAL THEORIES OF THE SCIENCE OF THE
MIND--THEORIES OF SENSATION, MEMORY,
LEARNING, CONTROL, ETC.--W/TRUTHS AND
GENERALIZATIONS ABOUT CONSCIOUS PHENOMENA
EMBEDDED WITHIN

FIG. 21.2.   Two ways of building a theory of consciousness.

would still require an explanation within our overall theory of mind. A theory of consciousness built in this way, around the shared phenomenological feature where it really resides, would crosscut our theories of perception, memory, and learning. Or, to put it differently, the theory of consciousness, such as it was, would contain generalizations that also show up within these special theories.

A second, related possibility is that we might forego altogether a specially demarcated space for the theory of consciousness, allowing instead all the true generalizations about conscious mental life to occur within the special theories.[13] The idea is that the interesting facts and generalizations about consciousness might be gathered under the rubric of the special theories of perception, memory, learning, and so on. Presumably this would happen, and it is a very reasonable prospect, if the most interesting generalizations of the science of the mind weave together conscious and unconscious processes and their neural realizations in accordance with what, from a functional point of view, the system is doing. Because perceiving and remembering, and the like, are things we do, whereas consciousness may be a way we are or one of the ways we perceive, remember, and so on, it is easy to imagine embedding most of what needs to be said about consciousness into the special theories.

Whichever shape a theory of consciousness takes (the first way as a theory of consciousness itself, or the second way as a "theory" in which what is true of consciousness gets said, possibly without much ado or fanfare, within all the necessary special theories), it will be part of the larger, more systematic, theory of the mind as a whole. The really important thing is that there can be a science of the mind. So long as consciousness is given its place within such a theory, it matters little whether it is explained in terms of a theory devoted exclusively to it, or whether it is just explained, period (that is, within the accounts given to domains like memory, problem solving, etc.).[14]

The best strategy is to get on with the hard work of providing the right fine-grained analysis of conscious mental life and to see where it leads. It will be our proudest achievement if we can demystify consciousness. Consciousness exists; it would be a mistake to eliminate talk of it because its semantic past is so bound up with ghostly fairy tales or because it names such a multiplicity of things. The right attitude is to deliver the concept from its ghostly past and

---

[13] Simon indicated to me that he favors, as I do, the second approach. It does seem to best represent what is in fact happening in the science of the mind.

[14] Not to be frivolous, one might imagine the choice between collecting all and only Mickey Mantle cards versus collecting Yankee cards (even assuming a complete set of the Yankees would yield a complete Mantle set). Mantle is, as a matter of fact, essentially embedded in the history of the Yankees, but the Yankees, a somewhat different version, could have done without #7. Whether #7 could have done without the Yankees is more obscure. The point though is a straightforward one about interest relativity. We design baseball and scientific collections within fairly broad constraints, some of which we create and all of which we interpret.

provide it with a credible naturalistic analysis. This chapter has tried to say a bit about how this might be done—indeed, about how it is already being done. So dream on about a naturalistic account of consciousness and rest assured that your dream does not reveal some impure philosophical or scientific wish or fantasy.

ACKNOWLEDGMENTS

I was fortunate to have received valuable comments on earlier versions of this chapter at both the Wake Forest Conference on my book, *Consciousness Reconsidered,* and the Carnegie Mellon conference on "Scientific Approaches to the Study of Consciousness." I am especially grateful to David H. Sanford, Allin Cottrell, Bob McCauley, George Graham, Allan Hobson, Marcel Kinsbourne, Ralph Kennedy, and David Galin, and Herbert Simon for their helpful comments. Ken Winkler and Robert Brandon also helped on an early draft.

REFERENCES

Brandon, R. (1990). *Adaptation and environment.* Princeton, NJ: Princeton University Press.

Churchland, P. S. (1983). "Consciousness: The transmutation of a concept." *Pacific Philosophical Quarterly, 64,* 80–93.

Cottrell, A. (1993, May). *Tertium datur? Reflections on Owen Flanagan's Consciousness Reconsidered.* Paper presented at Wake Forest colloquium.

Crick, F., & Koch, C. (1990). Towards a neurobiological theory of consciousness. *Seminars in the Neurosciences, 2,* 263–275.

Cummins, R. (1983). *The nature of psychological explanation.* Cambridge, MA: Bradford Books/MIT Press.

Dennett, D. (1978). *Brainstorms.* Cambridge, MA: MIT Press.

Flanagan, O. (1992). *Consciousness reconsidered.* Cambridge, MA: MIT Press.

Flanagan, O. (1995). Deconstructing dreams. *The Journal of Philosophy, 92*(1), 5–27.

Hobson, J. A. (1988). *The dreaming brain.* New York: Basic Books.

Hobson, J. A. (1994). *The chemistry of conscious states: How the brain changes its mind.* Boston: Little, Brown.

Llinás, R. R., & Paré, D. (1991). Commentary of dreaming and wakefulness. *Neuroscience, 44*(3), 521–535.

Llinás, R., & Ribary, U. (1993). Coherent 40-Hz oscillation characterizes dream state in humans. *Procedings of the National Academy of Sciences, 90,* 2078–2081.

Malcolm, N. (1959). *Dreaming.* London: Routledge & Kegan Paul.

Steriade, M., McCormick, D. A., & Sejnowski, T. J. (1993). Thalamocortical oscillations in the sleeping and aroused brain. *Science, 262,* 679–685.

Wilkes, K. V. (1988a). _____, yishi, duh, um, and consciousness. In A. J. Marcel & E. Bisiach (Eds.), *Consciousness in contemporary science.* Oxford, England: Oxford University Press.

Wilkes, K. V. (1988b). *Real people: Personal identity without thought experiments.* Oxford, England: Oxford University Press.

CHAPTER 22

Consciousness *Creates Access:* Conscious Goal Images Recruit Unconscious Action Routines, but Goal Competition Serves to "Liberate" Such Routines, Causing Predictable Slips

Bernard J. Baars
The Wright Institute

Michael R. Fehling
Stanford University

Mark LaPolla
Meta Software Inc.

Katharine McGovern
The Wright Institute

What is the use of making something conscious? In this chapter we explore experimental findings suggesting that in preparing a voluntary act, alternative goals and plans often compete for access to consciousness. Numerous experiments on elicited slips in speech and action show that most types of spontaneous errors can be elicited in the laboratory by consciously priming alternative goals (Baars, 1992). Combined with theoretical work on consciousness (Baars, 1988), this fact suggests that in novel situations, final action plans often emerge after competition between alternative plans for access to a conscious global workspace (GW) or its functional equivalent. The easiest way to think about GW theory is in terms of a "theater of consciousness in the society of the mind," a metaphor that is backed empirically and computationally by a great deal of established work (e.g., Anderson, 1983; Newell, 1990, 1992; Newell & Simon, 1972). The evidence for GW theory consists of a large set of qualitative meta-analyses contrasting closely matched conscious and unconscious processes across many cognitive domains (e.g., Baars, 1988, in press). In the upshot, once a goal image gains uncontested access to the conscious "stage," it appears to be globally distributed to recruit and organize unconscious automatisms needed to achieve the goal. However, conscious priming of alternative plans seems to disrupt the process long enough to "liberate" some of the primed unconscious routines, which then emerge in overt behavior as slips.

Thus three major factors combine to create slips: First, all slip techniques involve priming of two alternative plans; second, competition between the plans loads limited capacity very briefly (Chen & Baars, 1992); third, the momentary overload allows automatisms (unconscious action routines) to be liberated from voluntary control. A large body of evidence is consistent with this framework.

Conscious experience, in this view, is an essential component of the human cognitive system: It reflects a large-scale architecture with a serial conscious "stage" interacting intimately with an audience of parallel unconscious "experts." We believe that slips are not anomalies, but reflect large-scale design tradeoffs in the control of voluntary action. Slips are the price we pay for other, more desirable properties of action, especially flexibility. Whereas unconscious action routines are fast and effective in predictable situations, consciousness is needed to combine multiple automatisms into a single, coherent action plan, without internal competition, and thereby supports flexible action in the face of novelty.

These claims have implications for Unified Theories of Cognition (UTCs). Allan Newell noted recently that a large grain-size Global Workspace (GW) theory belongs to the same superset as SOAR (Newell, 1990, 1992) and ACT* (Anderson, 1983). However, adding consciousness and voluntary action control as rigorously defined empirical phenomena permits significant expansion of current UTCs to new domains of evidence.

EXPERIMENTAL INDUCTION OF SLIPS IN SPEECH AND ACTION

Consider this simple demonstration:

1. Cover the sentences below with a piece of paper.
2. Repeat the word "poke" to someone about 10 times, about once every 2 sec.
3. Your subject should repeat the word "poke" each time you say it.
4. Then quickly ask, "What do you call the white of an egg?"

A great majority of the people will say "the yolk," although they know quite well on further thought that "the white" or "the albumin" is the correct answer. This is a children's game, but it replicates in its essentials an experimental program on the induction of slips of speech and action conducted by the first author and a set of co-authors over a 15-year period. In the "poke poke" demonstration we are eliciting a word-retrieval slip, but very similar methods are used to elicit spoonerisms, word blends, word exchanges in sentences, syntactic blends, placement errors in action, motion errors, and so on

TABLE 22.1
Competing Plans Tasks that Elicit Slips in Speech and Action

A. Tongue-twisters: Phonological-motoric competition
   1. Overt tongue-twisters
   2. Tongue-twisters in inner speech
   3. Phonological fusion task
B. Phonological bias techniques
   4. For spoonerisms
   5. For word retrieval errors
C. Ordinal conflict techniques
   6. Spoonerisms
   7. Syllable switches
   8. Word exchanges between phrases in a sentence
      a. Socially inappropriate
      b. Semantically anomalous
      c. Transformational errors
   9. Question-answering technique
  10. Typing errors
  11. "Simon says" task
  12. Table-setting task
D. Techniques that use competition between alternative words in memory
  13. Word blends (see also B.5)
E. Techniques that use competition by deliberate transforms
  14. Irregular vs. regular verb competition
  15. Active–passive competition

(Table 22.1).[1] In all cases, the form of the slip is highly predictable, making it the only set of techniques that allow study of complex speech and action under direct experimental control. The methodology has been used to clarify questions about language structure, speech production, covert speech editing, the multiple "normalizing" tendencies of the speech system, the Freudian slip hypothesis, and the foundation question of volition (see Baars, 1992, for a complete discussion of these slip-induction techniques). Reason (1992) showed that such errors are not limited to innocuous laboratory demonstrations, but may cause fatal accidents in driving buses and trains as well as massive airline disasters, and even nuclear power plant accidents.

All of the 15 or so slip techniques we know today require two major factors: First, a rapid choice between consciously primed alternative plans, and sec-

---

[1] We were pleasantly astonished when, in summarizing the current Carnegie Symposium, Herbert Simon showed all of us an entirely new slip technique. In the context of criticizing "handwaving theories in cognitive science," he illustrated his determination not to do so by firmly thrusting his two hands in the pockets of his jacket. Because Herb is used to gesturing when he speaks, his hands soon popped out, apparently quite spontaneously, when his attention drifted enough to stop inhibiting his hands! It was a remarkably creative demonstration, which, fortunately, appears to fit the conditions for slips described in this chapter.

ond, task demands that load limited capacity very briefly. As a result, the primed unconscious routines seem to be liberated from voluntary control. Saying "yolk" to the question "what do you call the white of an egg?" can be viewed as such a "liberated automatism": A largely unconscious lexical routine is primed phonemically, and is emitted as a well-formed action in the wrong semantic context. All slips can be viewed as liberated automatisms (Baars, 1992).

This chapter explores how a Global Workspace (GW) approach to conscious experience can incorporate these findings in a broad conception of normal, voluntary control of speech and action (Baars, 1983, 1988, 1992, in press; Reason, 1992). GW theory has a pedigree emerging from the long succession of UTCs (Unified Theories of Cognition) pioneered by Newell, Simon, and Anderson (Anderson, 1983; Newell, 1990, 1992; Newell & Simon, 1972). In Baars' work, GW theory has so far been used qualitatively, as a cognitive architecture in which conscious experience plays the starring role. There is, of course, a long tradition in science of qualitative theory preceding quantitative models, for example the Rutherford atom and the famous Einstein thought experiment on relativity in physics; in biology, Darwinian evolution was also a qualitative theory in its first century. GW theory generates testable predictions, some of which have been borne out (e.g., Greenwald, 1992). Other limited-capacity mechanisms (e.g., working memory, selective attention, and executive volition) are seen as structures that develop around the stream of consciousness; this is quite different from other approaches in which conscious experience is treated as an adjunct of working memory (e.g., Baddeley, 1992).[2] This approach can be implemented using Fehling's (1992) Schemer II planning software. Neurophysiological implications have been explored in Baars (1987a, 1993a), Baars and Newman (1994), and Newman and Baars (1993).

TREATING CONSCIOUSNESS AS A VARIABLE: THE KEY TO THE PROBLEM OF EVIDENCE

In work published over the past decade, Baars suggested that psychologists al-

---

[2]The relation between consciousness and working memory is especially important (Anderson, 1983; Baddeley, 1986; Newell, 1990; Newell & Simon, 1972). We believe the consciousness and working memory are quite distinct, though closely related. Seven items maintained in immediate memory by mental rehearsal are not all conscious at the same time. At any single moment only a few items may be directly conscious, though the others are readily available. Cognitive architectures with working memories, such as ACT* and SOAR, have tended to de-emphasize consciousness and its sister issues. In contrast, our initial focus is on conscious processes as such, so that the evidentiary basis is quite different. From the perspective of conscious experience, we can ask how the well-studied properties of working memory could emerge from consciousness, rather than vice versa (Baars, 1988, pp. 310ff.).

ready know many things about consciousness that are normally treated under other labels, and that even today a reasonable framework can unify many well-established facts (Baars, 1983, 1988, 1992). Great experimental literatures exist on topics like selective attention; perceptual, preperceptual, and subliminal processes; conscious and unconscious problem solving; strategic and automatic skilled performance; sleep, coma, and waking; priming, mental imagery; and so forth. Although these topics have often avoided the word *consciousness,* the evidence they provide, gathered over decades, has a direct bearing on our understanding.

Notice how often breakthroughs in science depend on the ability to view a fact thought to be fixed and unchanging as a variable. In premodern physics, gravity, friction, white light, atmospheric pressure, and the stars were thought of as fixed. Breakthroughs in understanding these phenomena required a shift in perspective in which it was realized that gravity diminishes with distance from the earth, friction disappears in space, white light can be decomposed and recomposed with a prism, and so on. Viewed from the new perspective, much of the puzzle fell into place.

In the case of consciousness 19th-century psychologists were generally unable to see it as a variable. James rejected the idea of unconscious mental processes, and therefore could not run a close comparison between new and habituated skills (James, 1890/1984, chap. 4). Thus, he was forced to think of a new skill while it was still conscious, and after automaticity, as essentially different: the first was mental, and the second, physical (see Baars, 1996). In that respect, we are now in a better position to think about consciousness. For instance, the reader's experience of *these words* is clearly conscious, just as the fact that *these words* is the object of a preposition is plainly unconscious. Nevertheless, that unconscious syntactic fact still shapes our understanding of the conscious phrase.

Contrastive analysis is a scientific means of studying consciousness. It is much like the experimental method: Closely related cases are examined that differ only in respect to consciousness, so that consciousness becomes, in effect, the experimental variable. A number of cognitive psychologists have performed this kind of analysis (e.g., Damasio, 1989; Kinsbourne, 1993; Marcel & Bisiach, 1988; Milner & Rugg, 1992; Posner, 1992; Schacter, 1990; Umiltà & Moscovich, 1994; Weiskrantz, 1986). The GW approach integrates specific experimental comparisons into a type of meta-analysis: We cumulate over many experimental studies sets of contrasts such as perceptual versus preperceptual processing; subliminal versus supraliminal versus postperceptual representations of a stimulus; or attended versus nonattended processes. In this way, we work to overcome the limits of each individual study, focusing instead on the common properties shown by the conscious/unconscious contrast in general (Baars, 1983, 1988, 1994). Table 22.2 presents a cumulative summary of the contrasting features of matched conscious and unconscious phenomena, which places demanding bounds on any current theory.

TABLE 22.2
Capabilities of Comparable Conscious and Unconscious Events

Conscious Processes	Unconscious Processes
Computationally inefficient (i.e., high errors, low speed, and mutual interference between conscious computations)	Unconscious specialists are highly efficient in their own tasks (i.e., low errors, high speed, and little mutual interference)
Great range of different contents over time	Each specialized process has limited range of contents over time, and is relatively isolated and autonomous
Great ability to relate different conscious contents to each other; great ability to relate conscious events to their unconscious contexts	No known ability to relate novel contents to each other, or to find novel relations between established contents
Internal consistency, seriality, and have limited capacity	Specialists are diverse, can operate in parallel, and together have great capacity

CONSCIOUSNESS HAS *LIMITED CAPACITY* BUT CREATES *GLOBAL ACCESS*

Based on a detailed set of contrasts, this chapter suggests that conscious processes have two superficially contrary properties.

Competition for Input to Consciousness

Conscious mechanisms show limited capacity, as shown by selective attention in all sense modalities; immediate memory, such as sensory and short-term (working) memory; and dual-task decrements in performance. Selective attention can be viewed as a set of mechanisms designed to provide access to consciousness, in much the sense that the oculomotor system controls eye movements, which then provide access to the different parts of a visual scene. Attention is then a kind of gatekeeper for conscious experience (Posner, 1992). The involvement of consciousness in sensory memories is obvious, because the contents of sensory memory are conscious, and likewise in the rehearsal loop of working memory the currently rehearsed item is conscious, as are the current contents of the "visual sketchpad" (Baddeley, 1986, 1992). Finally, consciousness is clearly involved in dual-task interference, because to the extent one of the tasks becomes automatic, interference decreases.

Wide Access for Output from Consciousness

Access to Voluntary Functions.   There is a major payoff when some element of perception, sensory memory, inner speech, or imagery gains access to consciousness. It makes it possible to learn novel combinations of input

(Greenwald, 1992). It also enables working memory (Baddeley, 1986, 1992), long-term memory, verbal report (Ericsson & Simon, 1984/1994), the ability to use the conscious event as a signal for any voluntary action, whether mental or physical; metacognitive functions; and makes it possible to access a subset of currently dominant goals (Baars, 1988; Shallice, 1978). That is, *consciousness creates access* to a great array of voluntary functions.

Access to Unconscious Routines and Knowledge Sources.   Similarly, when a mental representation becomes conscious it also mobilizes a large set of unconscious capabilities, such as implicit learning mechanisms, error-detection routines (because we generally spot errors in conscious material even when we have no explicit idea what regularities are violated by the errors), and most significantly for this chapter, conscious goal images can apparently recruit unconscious automatisms that can carry out those goals (see Baars, 1988, 1992, for detailed discussion of the evidence for this proposal). One way to approach this claim is to ask subjects, "What muscles do you use when you wiggle your fingers?" Most people believe there are muscles in the hand; there are not. By touching one's forearm one can feel the muscles contract that control the fingers. The same is true for speech articulation and any number of other action routines, which operate unconsciously as long as conscious feedback is provided from the goal state, and conscious short-term "intentions" (which probably involve rapid conscious goal images) can recruit and trigger the unconscious action routines. This hypothesis is identical to James' ideomotor theory, stated in more modern terms. If this is true, conscious goal images can "call" the relevant unconscious automatisms necessary to execute them, but goal images are not likely to constitute the entire, detailed representation of the goal, any more than visual prototypes actually represent abstract categories, or visual images in language comprehension actually represent semantic networks in their entirety. Visual images are, in a sense, one kind of *lingua mentis*, extremely useful but not because of their high-fidelity representation of the world.

GLOBAL WORKSPACE ARCHITECTURES ALSO EXHIBIT LIMITED CAPACITY IN INPUT AND GLOBAL ACCESS IN OUTPUT

Like consciousness, the GW architecture shows competition for access and wide distribution of contents. Consciousness appears to be associated with a GW architecture, in which many parallel unconscious specialists interact via a serial, conscious, and internally consistent Global Workspace (or its functional equivalent). Detailed arguments for this position are given in Baars (1983, 1988), where it is developed through seven increasingly integrative models. This approach appears to have remarkably fruitful implications for a number of other problems, such as the nature of spontaneous problem solving. For convenience, this body of work is called "GW theory," though many computational architectures employ global workspaces (e.g., Hayes-Roth, 1985; Fehling, Altman, & Wilber, 1989).

GW theory relies on three theoretical constructs: a Global Workspace, specialized processors, and contexts.

The first construct is the specialized unconscious processor, the audience in the theater of mind (see Geschwind, 1979, for some brain examples). Such processors are conceived as experts in limited task domains, which can act concurrently, or functional alliances of such experts in the service of some common goal. Processors may work entirely autonomously out of awareness (as the "audience"). Or they may display their output in the Global Workspace (like actors who can appear on stage). Because the most detailed contents of consciousness are perceptual, there is reason to think that the "stage actors" have a strong tendency to be perceptual events, or quasi-perceptual ones such as imagery or inner speech. The "audience," however, may contain implicit learning mechanisms, error-detection mechanisms, memory retrieval routines, and numerous other unconscious functions that have no perceptual or quasi-perceptual outputs. The unconscious action routines described here belong in this category.

The *Global Workspace* (GW) is conceived much like the stage of the theater: an architectural capability that can be accessed by mutually competitive actors, which, once they gain the stage, can engage in systemwide dissemination of information. Much of the evidence regarding consciousness falls into place by assuming that messages in the Global Workspace correspond to conscious contents. The Global Workspace is thus functionally defined. There are neurophysiological arguments suggesting that the primary sensory areas may act as such conscious global workspaces (Baars, 1997a, 1993a; Baars & Newman, 1994; Newman & Baars, 1993).

A Global Workspace is much like the podium at the annual meeting of a scientific society. Groups of experts may interact with each other *locally* around conference tables, but broadcasting of information to everyone present at the meeting can only occur from the podium. New links between experts who have not worked together before are made possible by *global* interaction via the podium. New coalitions can be formed to work on new or recalcitrant problems that cannot be solved by previously expert committees. Solutions to problems can then be broadcast, scrutinized, and modified.

Contexts, the third construct in GW theory, can be defined as data structures that constrain conscious contents without themselves being conscious: the people behind the scenes who control the motions of the actors. In GW terms, contexts serve to shape global messages without themselves broadcasting any message. At the scientific meeting, messages broadcast from the podium are constrained by the program and also by the more implicit purposes of the professional organization. Analogously, the contents of the Global Workspace are constrained by contexts. Context effects are known in psychological domains as diverse as perception, imagery, action control, learning, and language and social comprehension. Whereas we are normally not conscious of our contextual assumptions, we may become surprisingly aware of them when violations occur.

In many cases, contexts are themselves evoked by conscious events, as in the resolution of lexical ambiguities by previous words. Take the three word pairs, book:volume, tennis:set; symphony:movement. In all of these cases, the first word is a conscious event that primes a semantic domain in which the ambiguous second word is disambiguated. Contexts may be momentary, as in the case with these three word pairs, or long-lasting, as with life-long expectations about the proper behavior of teachers. Intentions can be fruitfully viewed as contexts that are triggered by conscious goals (Figs. 22.1 and 22.2). In this view, control of the details of action is not accomplished by goal images, which are too detailed, too sketchy, and too inflexible for that purpose. Rather, conscious goal images can recruit action contexts, which can in turn provide the detailed guidance for complex acts, such as the articulation of a sentence.

More detailed arguments for these constructs are given in Baars (1983, 1988, 1996).

Figure 22.1 contains a summary of concepts from well-known psychological theories resembling GW constructs. However, the three constructs in Global Workspace Theory (contexts, Global Workspace, and specialized processors) are specifically defined with respect to conscious and unconscious functioning, unlike related concepts. Figure 22.2 shows a specific application of the GW framework to the act of saying, "I like ice cream." It makes explicit the multiple linguistic contextual constraints on the consciously intended utterance. These diagrams can be viewed as graphic formalisms. The easiest way to think about GW theory is in terms of a "theater of consciousness in the society of the mind," a metaphor backed empirically and computationally by a great deal of established work (e.g., Anderson, 1983; Newell, 1992; Newell & Simon, 1972).

In the upshot, once a *goal image* gains uncontested access to the conscious "stage," it appears to be globally distributed to recruit and organize unconscious automatisms needed to reach the goal. That is, consciousness of a goal allows it to gain access to multiple unconscious routines, while blocking the limited-capacity system even for a moment with competing goals may liberate some of the unconscious action routines, which then may emerge as slips in overt behavior.

It is convenient to speak of goal contexts as *intentions*. They are future-directed, nonqualitative representations about one's own actions. Like other contexts, they shape and constrain conscious contents, including possible goal images, without being conscious themselves. In everyday usage, a goal context is the same as a plan or intention. We become aware of the workings of goal contexts in relatively rare moments when they are mismatched (e.g., by errors in carrying out our intentions). Goal contexts and goal images resemble the notion of "current concerns" introduced by Kahneman (1973; see Singer, 1988), which involves a conscious or unconscious aim that shapes

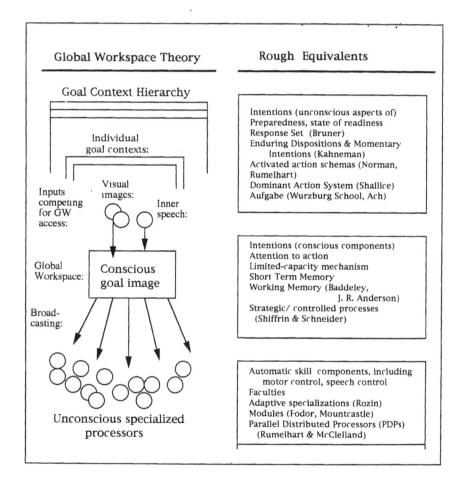

FIG. 22.1.   Rough equivalences between the basic concepts of Global Workspace (GW) theory and other widespread terms. Notice that GW theory has only three main constructs: the global workspace, unconscious specialized processors, and goal contexts. Each one has a graphic symbol associated with it, so that the theory can be expressed in intuitively obvious diagrams.

perception, thought, and action.

The set of currently active, mutually consistent goal contexts, taken together, constitutes the goal context hierarchy. It participates in the control of thought and action through its capacity to shape and evoke conscious goal images. Figure 22.2 represents a simple goal context hierarchy for buying ice cream. "Buying ice cream" is a momentary goal nested within higher level goals, such as pleasure seeking, hunger avoidance, weight control, and so on.

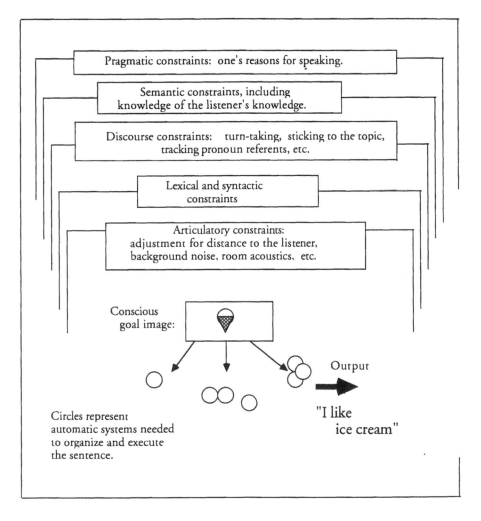

FIG. 22.2.   The intention to speak: Many unconscious goal systems cooperate to constrain the articulation of a single sentence.

The goal context hierarchy is similar to the goal hierarchies found in the motivational theory of Murray (1938), the theory of adult development of Maslow (1970), the cognitive appraisal theory of emotion of Lazarus (1991), and Mandler (1984). However, the goal context hierarchy in GW theory is explicitly defined as an unconscious data structure that can be accessed by means of conscious events.

GOAL COMPETITION LEADS TO SLIPS IN SPEECH AND ACTION

Figure 22.3 illustrates goal competition for access to global workspace over time. This is a simple way to describe all of the slip techniques we have found to date (see Baars, 1992, for details). As shown in Table 22.1, the types of slips so elicited include the humble tongue-twister, but also two different techniques for eliciting spoonerisms, several methods for inducing word movements in a sentence, word and action blends, putative "transformational" sentence errors, finger exchanges in typing errors, hand and arm movement slips, and place-setting errors on a dining table.

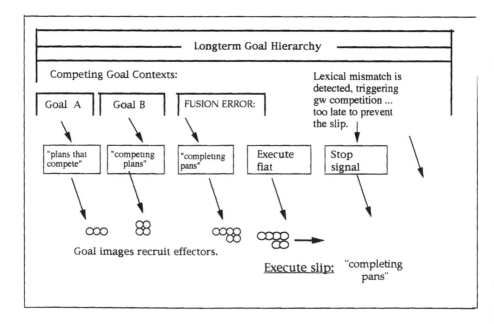

FIG. 22.3.   Plan competition can trigger slips. Competition between goal systems for control of the global workspace may lead to incorrectly fused plans. The lexical error is detected too late to prevent execution of the slip, "completing pans."

Many of these slips do not resemble obvious fusions between the two competing plans. Rather, they are a "regularized combination" of the two plans, that is, they combine components from the two plans in an automatic but highly predictable way. Consider just a few examples of slips that do not resemble the two competing plans that created them. The slips cited were chosen from a list of "transformational errors" collected by Fay (1980), who argued that these sentence errors might reflect a failure or misapplication of a grammatical transformation, along Chomskyan lines. However, Chen and Baars (1992) showed that all of the claimed transformational errors can be elicited in the laboratory simply by creating competition between two plausible paraphrases of the intended sentence. For example:

1. The error "What could have I done with the check?" was obtained by creating competition between two near paraphrases, "What could I have done with the check?" and "What have I done with the check?"
2. "Where do you suppose are they?" was obtained by creating competition between "Where do you suppose they are?" and "Where are they, do you suppose?"
3. "They didn't actually withdrew the needle" was obtained by creating competition between "They almost withdrew the needle," and "They didn't actually withdraw the needle."

In retrospect, it may seem obvious that many of these errors reflect competing sentence plans. But this was not at all obvious to Fay and others who thought they were caused by faulty syntactic transformations.

There are many other examples of slips whose origin is not obvious on the surface, but that emerge quite naturally from laboratory-induced competition between correct plans (Baars, 1992). Moreover, we have found evidence that plan competition at a higher level can cause errors to propagate downward through a goal hierarchy. Order competition between higher level constituents can induce movements of lower level elements. Competing words can induce phoneme exchanges (spoonerisms), competing phrases result in word exchanges, and competing actions create automatic component exchanges. Competition at one level of control may therefore propagate downward through the structural hierarchy, resulting in overt slips at a lower level. Indeed, there are few spontaneous slips that cannot be explained as some sort of plausible plan combination. In terms of slip induction, therefore, plan competition is all that needs to be introduced.

The Surprisingly High Rate of Slips and Dysfluencies in Normal Speech

Recent studies show that actual slips occur much more often than previously thought. Ferber (1991) found 1 to 2 slips per minute of spontaneous conversation. Repairs occurring without an overt slip are also very frequent, and 10%

of the words in spontaneous speech involve some kind of dysfluency. Hesitation pauses, which are thought to reflect the intrusion of alternative plans, are even more common in spontaneous speech. Similar claims have been made for actions other than speech. Such high error rates would be utterly unacceptable in any machine such as a computer printer.

Plan Competition Seems to Be an Essential, Not an Accidental or Abnormal Aspect of Action Control in the Face of Novelty

Why are phenomena that reflect plan competition so common in everyday speech and action?

Competition Due to Choice Points.   Points of indeterminacy may occur quite often in spontaneous speech (Butterworth, 1982). There are usually several ways to express a single thought. Competition may also occur in the case where a basic sentence frame is being executed, with a late-coming adjective rushing to be inserted in its proper place, even as the syntactic train is leaving the station. Thus, one may be all prepared to say, "Oops! I really made a goof!" and decide at the last moment to add an adjective for emphasis, to say "I really made a bad goof." If the insertion is made too late, then "goof" and "bad goof" may compete with each other, creating exactly the kind of order competition described previously, which has been shown to produce spoonerisms in the laboratory (Baars & Motley, 1976). The sentence might come out therefore as "Oops! I really made a gad boof!"

There are numerous one-to-many mappings between higher and lower levels in the linguistic hierarchy, just as there are numerous input ambiguities in perception and comprehension. Ambiguities create choice points in the flow of input, just as plan competition creates choice points between alternative outputs. Any comprehensive theory of action must be able to pinpoint such choice points, and must have mechanisms for resolving them.

Competition Due to the Distributed Nature of Nondominant Goals and Automatisms.   There is another way for plan competition to arise: in the kind of system that inherently works by means of distributed goals and action routines. There is considerable evidence that human action control involves such a system: both in terms of distributed goals and distributed action components. By "distributed" we simply mean "decentralized specialization," so that the degrees of freedom of an action are not controlled from a single executive, but by broad executive goals interpreted by local specialists (Greene, 1972).

Evidence that Nondominant Goals May Compete for Access to Consciousness in Normal Situations.   When goals are unified into a currently dominant goal hierarchy, they are obviously not distributed. However, at other times, goals may be quite flexible and distributed. Consider the following spontaneous slips cited by Reason (1984a), in which a consciously accessible part of an intention is apparently lost:

1. "I went upstairs to the bedroom and stopped—not remembering what I had gone there for."
2. "I went into my room intending to fetch a book. I took off my rings, looked into the mirror and came out again—without my book."
3. "I had an appointment at the dentist's, but went to the doctor's instead." (p. 576)

Evidence for Distributed Action Routines.   Action components are likely to be distributed as well, as evidenced by the great mobility of slip components.[3] Spoonerisms, for example, decompose words into phonemes or phoneme clusters:

1. fried soup—side froop
2. feel good—gheel food
3. key band—be canned

These spoonerisms can be obtained in the laboratory either by priming with words that sound like the error, or simply by the so-called Ordinal Conflict Effect (OCE), which involves conflicts between the order of two words. Specifically, presenting a series of word pairs, and then demanding a quick reversal of the order of the last word pair will cause spoonerisms. Likewise, creating uncertainty about the order of two multisyllabic words with similar stress patterns will result in syllable exchanges. Thus:

4. horrible miracle—mirable horricle
5. interfering homophone—homofering interphone
6. dignified monastery—monafied dignistary

Word exchanges between two phrases are easily elicited by order competition between them:

7. She (touched her nose) and (picked a flower)—She (picked her nose) and (touched a flower).
8. He (talked to the woman) and (looked at the baby)—He (talked to the baby) and (looked at the woman).
9. He (found his trousers) and (dropped his watch)—He (dropped his trousers) and (found his watch).

---

[3]The word *distributed* has come to have several meanings. This chapter principally has in mind an architectural meaning, roughly "decentralized intelligence and control" (Greene, 1972).

These errors illustrate the claim that "slips decompose actions and speech into underlying units and accommodation mechanisms." First, they always move and replace natural units: phonemes, syllables, words, gestures, and so on. These natural units are by no means simple or "merely physical" events. They are physically quite complex, and are executed differently in different contexts (Liberman, Cooper, Shankweiler, & Studdert-Kennedy, 1967). The spoonerism "feel good"—"gheel food," for example, involves quite different formant transitions between /g/ and the following vowel when the initial /g/ moves from "good" to "gheel." In "good," the formant transition /g/ loops downward, whereas in "gheel," it loops upward. Further, the two formant transitions start at different points on the frequency spectrum. Acoustically, therefore, the two /g/'s are quite different and indeed nonoverlapping physical events. Nevertheless, they are perceived to be quite normal /g/'s.

Similarly, in spoonerisms, syllable-initial consonants always exchange with other initial consonants, final consonants with other final consonants, and medial vowels with other medial vowels. The moving phonemes always follow the regularities of phonemic sequencing. For all of these reasons, in most slips, abstract units move from place to place in a fashion that maintains their perceived identity, even if changes must be made at other levels of control for this to occur.

Note that these abstract units may be viewed as automatisms—the unconscious overpracticed components of voluntary action. Indeed, we suggest that slips may be viewed as "activated automatisms" that have been liberated from their normal goal structures.

"Liberated" units are often smoothly integrated into their new surroundings. Whenever possible, the migrating units will move to similar contexts; when this is not possible, they will take along, or create anew (more rarely), the appropriate contextual surround. That is to say, the moving units are very sensitive to the manner of their integration into the new contexts: In the vocabulary of linguistics, the movement of units is highly constrained by accommodation mechanisms (Garrett, 1975). The nasalization in "key band"—"be canned," is a case in point. The italicized phonemes are nasalized, and in the slip the nasal formant spreads to the initial consonant /k/ by coarticulation, even though the initial consonant has changed from the intended /b/. Thus, the moving consonant is integrated into its new articulatory context, and becomes nasalized when that is required, or "de-nasalized" when that is appropriate.

Likewise, it appears that moving syllables tend to migrate only between similar phonetic, syllabic, and stress contexts. Migrating words invariably move into the correct syntactic positions, taking along the proper prepositions, if necessary. In the laboratory, we rarely observe syntactic errors such as:

10.(*) "He talked at the baby and looked to the woman."

Instead, the correct prepositions tend to migrate along with the moving words, to produce:

11. "He talked to the baby and looked at the woman" (Baars, 1992).

Thus, accommodation mechanisms work to integrate migrating units into their new surroundings; it seems likely that they also act to keep the "liberated" units away from inappropriate new positions.

THE NEED TO REORGANIZE THE SAME COMPONENTS TO PERFORM DIFFERENT ACTIONS

The Need for Interruptibility in Action Control

Why should an action control system court such complexity? After all, as we can see from the slips cited earlier, competition between goals, plans, and action components can lead to errors. One answer is that competition between goals and other action routines leads to a greatly increased flexibility of action. Alternative goals represent opportunities for different actions. Birnbaum and Collins (1992) maintained that

> the ability to recognize and seize opportunities . . . (is) a fundamental element in intelligent planning. . . . To take a simple example, suppose you go to the store to buy something. If, while you are at the store, you notice on sale an item that you want, you may then decide to purchase the item, even though you did not originally go to the store to satisfy that intention. The point here is that it is not, in general, possible to foresee all the situations in which an unsatisfied goal may be satisfiable. (p. 123)

Computationally, this implies an ability to interrupt currently dominant plans. But where would such an interruption come from, and how would the system determine when to interrupt current goals? Somehow, alternative goals and routines must be available at all times, in order to be integrated into the current goal hierarchy when needed. In GW theory, goals that are not part of the current action control hierarchy are potentially available at any time. Access to the conscious Global Workspace is governed by the same goal-hierarchy that generates and evaluates action plans. Interruptibility is therefore not random, but highly guided by existing goal commitments. This is consistent with the fact that a conscious stream of information can be interrupted by stimuli in the unconscious stream, providing that those stimuli have greater personal or biological importance than the conscious stream.

Slips as the Price of Flexibility in Action

Errors can exact painful penalties. A rabbit pursued by a fox may die from zigging rather than zagging at a critical moment, whereas the fox may risk starvation by underestimating the length of a leap. In some cultures, a tactless slip of the tongue may lead to social disgrace, and slips in driving a car, flying a plane, hunting, or warfare may be fatal (Norman, 1981; Reason, 1984a, 1984b). Thus, from an evolutionary point of view, errors represent a threat to survival and reproductive success. Nevertheless, they occur remarkably often. Why then, over eons of evolution, did an error-free control system not develop?

It seems that the risk of errors trades off against some other, desirable property of the action control system. Evolutionary biology shows numerous examples of such trade-offs between the costs and benefits of certain traits (Gould, 1985). In the research literature on slips, speed-accuracy trade-off is well-known (MacKay, 1982).

A possibility is that slips are the price we pay for the great flexibility of the human control system—its ability to select among numerous degrees of freedom. This may be called the flexibility–accuracy trade-off. It reflects the fact that highly overlearned skills are both more accurate and less flexible, whereas novel or voluntarily controlled actions may sacrifice efficiency and error-free accuracy in favor of greater flexibility. Reason (1979) made this point by referring to slips as "the price of automatization." Another way of doing so is to suggest that slips are the price of flexibility in action.

More formally, we suggest that the control component accomplishes conflict resolution; novel integration; coordination and coherence of goal structures (cf. MacKay); and task interruption, informed by the goal hierarchy.

CONSCIOUSNESS HELPS OPTIMIZE THE TRADE-OFF BETWEEN COHERENCE, FLEXIBILITY, SPEED, AND EFFECTIVENESS

Flexibility in action and thought is closely connected with the functioning of the limited-capacity system (Baars, 1988). Several models explain this flexibility. Norman and Shallice (1986) suggested that action schemas may compete for the limited-capacity system. Similarly, Reason (1983) purported that schemas in a general domain called the "board" may compete for the conscious/limited-capacity "blob." My own publications work out the implications of a global workspace architecture containing conscious material, in a distributed system of unconscious automatisms (Baars, 1983, 1987a, 1987b, and 1992, chap. 4).

Global Workspace theory gives a specific reason why it is important for alternative plans and goals to compete for the limited-capacity system: Consciousness is intimately associated with a "broadcasting" capability. Any goal system that is able to reach consciousness can broadcast a message to recruit and control numerous subgoals and effectors that can carry out its goal. This may be the only functional explanation currently available for the apparent competition between goal systems.

SUMMARY AND CONCLUSIONS

This chapter has explored experimental findings, suggesting that in preparing a voluntary act, alternative goals and plans often compete for access to consciousness. Most types of spontaneous errors can be elicited in the laboratory by consciously priming alternative goals (Baars, 1992). Combined with theoretical work on consciousness (Baars, 1988) this fact suggests that in novel situations, final action plans often emerge after competition between alternative plans for access to a conscious Global Workspace (GW) or its functional equivalent. The easiest way to think about GW theory is in terms of a "theater of consciousness in the society of the mind," a metaphor backed empirically and computationally by a great deal of established work (e.g., Anderson, 1983; Newell, 1990, 1992; Simon & Newell, 1972). The evidence for GW theory consists of a large set of qualitative meta-analyses contrasting closely matched conscious and unconscious processes across many cognitive domains (e.g., Baars, 1988, 1994). In the upshot, once a goal image gains uncontested access to the conscious "stage," it appears to be globally distributed to recruit and organize unconscious automatisms needed to achieve the goal. However, conscious priming of alternative plans seems to disrupt the process long enough to "liberate" some of the primed unconscious routines, which then emerge as in overt behavior as slips.

Thus, three major factors combine to create slips: First, all slip techniques involve priming of two alternative plans; second, competition between the plans loads limited capacity very briefly (Chen & Baars, 1992); third the momentary overload allows unconscious action routines to be liberated from voluntary control. A large body of evidence is consistent with this framework.

Conscious experience, in this view, is an essential component of the human cognitive system: It reflects a large-scale architecture with a serial conscious "stage" interacting intimately with an audience of parallel unconscious "experts." Although unconscious action routines are fast and effective in predictable situations, consciousness is needed to combine multiple automatisms into a single, coherent action plan, without internal competition, and thereby supports flexible action in the face of novelty.

ACKNOWLEDGMENTS

Theoretical work on GW theory began in earnest in 1979–1980, while the first author was a Cognitive Science Fellow at the Center for Human Information

Processing, UCSD, supported by a Sloan Cognitive Science grant. I want to gratefully acknowledge Donald A. Norman for his support. Additional support was received from a Visiting Scientists appointment with the John D. and Catherine T. McArthur Foundation Program for Conscious and Unconscious Mental Processes, led by Mardi J. Horowitz, in 1986. The experimental work on experimentally induced slips has been supported by NSF and NIMH.

REFERENCES

Anderson, J. R. (1983). *The architecture of cognition.* Cambridge, MA: Harvard University Press.

Baars, B. J. (1983). Conscious contents provide the nervous system with coherent, global information. In R. Davidson, G. Schwartz, & D. Shapiro (Eds.), *Consciousness and self-regulation* (pp. 45–76). New York: Plenum.

Baars, B. J. (1987a). Biological implications of a Global Workspace theory of consciousness: Evidence, theory, and some phylogenetic speculations. In G. Greenberg & E. Tobach (Eds.), *Cognition, language, and consciousness: Integrative levels* (pp. 209–236). Hillsdale, NJ: Lawrence Erlbaum Associates.

Baars, B. J. (1987b). What is conscious in the control of action? A modern ideomotor theory of voluntary action. In D. Gorfein & R. R. Hoffman (Eds.), *Learning and Memory: The Ebbinghaus Centennial Symposium.* Hillsdale, NJ: Lawrence Erlbaum Associates.

Baars, B. J. (1988). *A cognitive theory of consciousness.* New York: Cambridge University Press.

Baars, B. J. (1992). *Experimental slips and human error: Exploring the architecture of volition.* New York: Plenum.

Baars, B. J. (1993a). How does a serial, integrated and very limited stream of consciousness emerge from a nervous system that is mostly unconscious, distributed, and of enormous capacity? In G. R. Bock & J. Marsh (Eds.), *CIBA Symposium on Experimental and Theoretical Studies of Consciousness* (pp. 282–290). London: Wiley.

Baars, B. J. (1993b). Why volition is a foundation issue for psychology. *Consciousness and Cognition, 2*(4), 281–309.

Baars, B. J. (1996). *In the theater of consciousness: The workspace of the mind.* New York: Oxford Science, Oxford University Press.

Baars, B. J., & Motley, M. T. (1976). Spoonerisms as sequencer conflicts: Evidence from artificially elicited errors. *American Journal of Psychology, 89,* 467–484.

Baars, B. J., & Newman, J. (1994). A neurobiological interpretation of the Global Workspace Theory of consciousness. In A. Revonsuo & M. Kamppinen (Eds.), *Consciousness in philosophy and cognitive neuroscience* (pp. 58–71). Hillsdale, NJ: Lawrence Erlbaum Associates.

Baddeley, A. D. (1986). *Working memory.* Oxford, England: Clarendon.

Baddeley, A. (1992). Working memory. *Science, 255,* 556–559.

Birnbaum, L., & Collins, G. (1992). Opportunistic planning and Freudian slips. In B. J. Baars (Ed.), *Experimental slips and human error: The architecture of volition* (pp. 98–111). New York: Plenum.

Butterworth, B. L. (1982). Speech errors: Old data in search of new theories. In A. Cutler (Ed.), *Slips of the tongue* (pp. 17–34). Amsterdam: Mouton.

Chen, J., & Baars, B. J. (1992). General and specific factors in "transformational errors": An experimental study. In B. J. Baars (Ed.), *Experimental slips and human error: Exploring the architecture of volition* (pp. 217–233). New York: Plenum.

Damasio, A. R. (1989). Time-locked multiregional retroactivation: A systems-level proposal for the neural substrates of recall and recognition. *Cognition, 33,* 25–62.

Ericsson, K. A., & Simon, H. A. (1984/1994). *Protocol analysis: Verbal reports as data* (2nd ed.). Cambridge, MA: MIT Press.

Fay, D. (1980). Transformational errors. In V. A. Fromkin (Ed.), *Errors in linguistic performance* (pp. 111–122). New York: Academic Press.

Fehling, M. (1992). Unified theories of cognition: Modeling cognitive competence. *Artificial Intelligence, 92*(6), 1–34.

Fehling, M. R., Altman, A., & Wilber, B. M. (1989). The Heuristic Control Virtual Machine: An implementation of the Schemer computational model of reflective, real-time problem-solving. In V. Jagannathan, R. Dodhiawala, & L. Baum (Eds.), *Blackboard architectures and applications* (pp. 65–78). Boston: Academic Press.

Ferber, R. (1991). Slip of the tongue or clip of the ear? On the perception and transcription of naturalistic slips of the tongue. *Journal of Psycholinguistic Research, 20,*(2), 105–122.

Garrett, M. F. (1975). The analysis of sentence production. In G. Bower (Ed.), *The psychology of learning and motivation: Advances in research and theory* (Vol. 9, pp. 27–52). New York: Academic Press.

Geschwind, N. (1979). Specializations of the human brain. *Scientific American, 241*(3), 180–201.

Gould, S. J. (1985). *The flamingo's smile.* New York: Norton.

Greene, P. H. (1972). Problems of organization of motor systems. *Journal of Theoretical Biology,* 303–308.

Greenwald, A. (1992). New Look 3. Unconscious cognition reclaimed. *American Psychologist, 47,* 766–779.

Hayes-Roth, B. (1985). A blackboard architecture of control. *Artificial Intelligence, 26,* 251–351.

James, W. (1984). *The principles of psychology.* Cambridge, MA: Harvard University Press. (Original work published 1890)

Kahneman, D. (1973). *Attention and effort.* Englewood Cliffs, NJ: Prentice-Hall.

Kinsbourne, M. (1993). Integrated field model of consciousness. In G. R. & M. J. Brock (Eds.), *CIBA Symposium on Experimental and Theoretical Studies of Consciousness* (pp. 51–60). London: Wiley Interscience.

Lazarus, R. (1991). *Emotion and adaptation.* London: Oxford University Press.

Liberman, A. M., Cooper, F., Shankweiler, D., & Studdert-Kennedy, M. (1967). Perception of the speech code. *Psychological Review, 74,* 431–459.

MacKay, D. G. (1982). The problems of flexibility, fluency, and speed-accuracy trade-off in skilled behavior. *Psychological Review, 89,* 483–506.

MacKay, D. G. (1987). *The organization of perception and action.* New York: Springer.

Mandler, G. (1984). *Mind and body: Psychology of emotion and stress.* New York: Norton.

Marcel, A. J., & Bisiach, E. (Eds.). (1988). *Consciousness in contemporary science.* Oxford, England: Clarendon.

Maslow, A. (1970). *Motivation and personality* (2nd ed.). New York: Harper.

Milner, A. D., & Rugg, M. D. (Eds.). (1992). *The neuropsychology of consciousness.* London: Academic Press.

Murray, H. A. (1938). *Explorations in personality.* New York: Oxford University Press.

Newell, A. (1990). *Unified theories of cognition.* Cambridge, MA: Harvard University Press.

Newell, A. (1992). SOAR as a unified theory of cognition: Issues and explanations. *Behavioral and Brain Sciences, 15*(3), 464–492.

Newell, A., & Simon, H. A. (1972). *Human problem-solving.* Englewood Cliffs, NJ: Prentice-Hall.

Newman, J., & Baars, B. J. (1993). A neural attentional model for access to consciousness: A Global Workspace perspective. *Concepts in Neuroscience, 2*(3), 255–290).

Norman, D. A. (1981). Categorization of action slips. *Psychological Review, 88,* 1–15.

Norman, D. A., & Shallice, T. (1986). Attention to action: Willed and automatic control of behavior. In R. Davidson, G. E. Schwartz, & D. Shapiro (Eds.), *Consciousness and self-regulation* (Vol. 4, pp. 79–93). New York: Plenum.

Posner, M. I. (1992). Attention as a cognitive and neural system. *Current Directions in Psychological Science, 11,* 11–14.

Reason, J. (1990). *Human error.* Cambridge, England: Cambridge University Press.

Reason, J. R. (1979). Actions not as planned: The price of automatization. In G. Underwood & R. Stevens (Eds.), *Aspects of consciousness: Vol. 1. Psychological issues* (pp. 35–51). London: Wiley.

Reason, J. R. (1983). Absent-mindedness and cognitive control. In J. Harris & P. E. Morris (Eds.), *Everyday memory, actions, and absent-mindedness* (pp. 113–132). New York: Academic Press.

Reason, J. R. (1984a). Lapses of attention in everyday life. In R. Parasuraman & D. R. Davies (Eds.), *Varieties of attention* (pp. 515–549). New York: Academic Press.

Reason, J. R. (1984b). Little slips and big disasters. *Interdisciplinary Science Reviews, 9*(2), 3–15.

Reason, J. T. (1992). Cognitive underspecification: Its variety and consequences. In B. J. Baars (Ed.), *Experimental slips and human error: Exploring the architecture of volition* (pp. 71–92). New York: Plenum.

Schacter, D. L. (1990). Toward a cognitive neuropsychology of awareness: Implicit knowledge and anosognosia. *Journal of clinical and experimental neuropsychology, 12*(1), 155–178.

Shallice, T. (1978). The dominant action system: An information processing approach to consciousness. In K. S. Pope & J. L. Singer (Eds.), *The stream of consciousness: Scientific investigations into the flow of experience* (pp. 80–94). New York: Plenum.

Singer, J. L. (1988). Sampling ongoing consciousness and emotional experience: Implications for health. In M. J. Horowitz (Ed.), *Psychodynamics and cognition* (pp. 297–346). Chicago: University of Chicago Press.

Umiltà, C., & Moscovitch, M. (Eds.). (1994). *Attention and performance: XV Conscious and nonconscious information processing.* Cambridge, MA: Bradford/MIT Press.

Weiskrantz, L. (1986). *Blindsight: A case study and implications.* Oxford, England: Clarendon.

CHAPTER 23

What Is the Difference Between a Duck?

David Galin
University of California at San Francisco

Yes, the title seems odd, but it is not a mistake; this chapter is about defective questions. This is a broader and potentially heavier topic than my official assignment to comment on the chapters by Baars and Flanagan, which itself was a big enough challenge. But I have kept it light; in the symposium on which this volume is based, my talk was scheduled after two very full days of scholarly and empirical rigor, and the audience needed a break. By way of introduction, I told a joke. But I am not frivolous; it is a philosophical joke.

THE CIGAR JOKE

Two monks from different orders were old friends. They shared a great fondness for cigars, and once each year when they had a chance to visit they would pray together and light up. But they became concerned that there might be some sin in their habit, and they resolved to each write to the Pope for guidance. When they met again one was puffing away. "But the Pope told me it was a sin," protested the other.

"What did you ask him?" said the first.

"I asked if it was all right to smoke during our evening prayer and he said no."

"Well," said his friend, "I asked if it was all right to pray during our evening smoke, and he said it was just fine!"

Moral: The answer you get depends on the question you ask.

This volume is derived from a symposium originally titled: *Scientific Approaches to the Question of Consciousness.* The title is significant; it points to "*THE* question of consciousness." But there are many, many questions of consciousness! Much of what appears as conflict and confusion in the previous chapters and in the wide published literature is due to people addressing different questions. The forewarning of this source of troubles is right there in our title.

The problem of finding the right question is ubiquitous in science, and it is not just a problem for theoreticians. In empirical studies, quite often the method chosen implicitly determines the broader question. For example, our choice as to whether to record evoked potentials—or to measure reaction times, or to ask subjects to describe their experiences—will limit the class of answers we can possibly get. My comments on the symposium as a whole and on the Baars and Flanagan chapters are particularly concerned with the questions addressed, both implicitly and explicitly.

COMMENTARY ON TALK BY BAARS

"Consciousness Creates Access"
(Baars, Fehling, LaPolla, and McGovern, this volume)

Dr. Baars is to be admired for undertaking to construct a full–range cognitive theory of consciousness, including awareness, will, and self. He does not just nibble timidly around the edges. He is habitually courageous; besides his major book *A Cognitive Theory of Consciousness*, published in 1988 when this topic was still not at all politically correct, he has also written the *History of the Cognitive Revolution* (1986), and has founded an important new journal, *Consciousness and Cognition*.

His presentation here focuses on how his cognitive theory of consciousness could be applied to the analysis of control of action. He argues that the success of this application provides pragmatic support for the wider theory, which is understandably a bit sketchy at this stage because it covers so much territory. His chapter is very much abbreviated from his book *Experimental Slips and Human Error* (1992), which is based on laboratory studies of slips of the tongue and other action errors.

Dr. Baars cites a great deal of empirical data on slips, in keeping with the prior presentations and usual tastes of this audience. However, I do not want to address whether or not he provides enough details and connecting links to enable others to see his broader vision. Rather, I want to emphasize his frame of reference. Although he contrasts his model with other cognitive architectures such as ACT* and SOAR, which he says "de-emphasize consciousness and its sister issues," *Baars' model is not concerned with the qualitative experiential aspects of consciousness. Instead, he stresses the functional role of consciousness; it is the way it is because that is necessary for what it does.*

He embraces the usual academic cognitive psychologists' functional approach: performance and mechanism, what does it do and how does it do it? There is very little phenomenology, such as "How do you feel when you do this?"

However, Baars did comment in his oral presentation that he felt it was important that the Dark Age of Behaviorism was over and that we (academic psychologists) are now studying consciousness because if we think of other people as having a rich inner life, as sensitive beings with points of view of their own, then we will tend to grant them more status, and having more respect for each other will be good for all the ills of this world. I agree (see also Sperry, 1993), but also remember that in the Dark Age it was only academic psychology that forgot consciousness; other disciplines like psychiatry and literature never stopped paying attention to it and to the rich inner (and hidden) life.

COMMENTARY ON TALK BY FLANAGAN

"Prospects for a Unified Theory of Consciousness
or, What Dreams Are Made of" (Flanagan, this volume)

Dr. Flanagan is a distinguished Philosopher of Mind. In addition to his well-known *The Science of Mind* (1991a), and *Consciousness Reconsidered* (1992), he has also written *Varieties of Moral Personality* (1991b), a book on moral and ethical issues (topics that never once appeared in the symposium, though not explicitly forbidden by the symposium's organizers as they forbid discussion of metaphysical dualism. It did not have to be forbidden, it was so far from anyone's mind!)

As is proper for the philosopher in residence, he does not offer a theory of consciousness himself but examines premises, methods, and frames of reference. That is, he brings out the usually invisible factors that determine what questions we ask.

Flanagan points out that different properties or aspects of consciousness are picked out by the narrow frame of reference of each of the subdisciplines concerned. He concludes that a scientific and unified theory of consciousness is possible, but that what it will look like depends on who you let contribute to it. For a *complete* theory, he suggests that the cognitive functional analysts, the neurobiologists, and those interested in phenomenology (subjective experience) should all be invited. In addition to cognitive psychology and neuroscience, he says, "Give phenomenology its due. Listen carefully to what individuals have to say about how things seem" (p. 410, this volume). *But at this symposium there was enormous resistance to talking about subjective experience,* even though we are supposed to have outgrown the Behaviorists' restriction to behavioral observation and their edict forbidding introspection.

Recall the joke about how impossible "subjectivism" was in the Behaviorist era: two behaviorists making love . . . afterward, one says to the other, "That was great for you; how was it for me?" Well, first-person experience is politically correct now; what is the reason that we still are not talking about it?

One reason is intellectual; we are lacking some fundamental cognitive tools for the task. We have not yet developed a taxonomy for the varieties of subjective experience or even a consensus on the constructs or terminology with which to characterize its richness (Galin, 1992a, 1992b, 1994; Shallice, 1988). I am speculating that sometimes there may be a second reason, perhaps more emotional than intellectual. I suggest this because the topic seems to provoke both extreme skittishness and extreme fascination. Research on subjective experience is like sex; everybody is terribly interested in it but does not want to be caught in public doing it. Could it be that this area of research contains something so personal, so intimate, that we have trouble looking at it directly? Let's peek.

WHY IS CONSCIOUSNESS SUCH A FASCINATING TOPIC?

To help us decide what we ought to be most interested in, I propose that we first notice what we ARE most interested in. I will put my answer to this question in the form of three hypotheses which seem to be empirically testable:

Hypothesis 1.   Most people are fascinated with consciousness because they are really interested in first-person experience, and only secondarily interested in the information-processing aspect or the neurobiology of it.

Hypothesis 2.   Most people are interested in their own first-person experience, not just first-person experience in general.

Test whether this is so for you. Imagine that you can get a treatment that will augment your perception and memory and problem solving by 1,000%. Imagine also that you will be able to achieve your goals 1,000% more effectively. The only catch is that you would no longer be conscious. Would you want the treatment? (Some people claim this is the deal offered by higher education.)

Hypothesis 3.   We do not value all of our experiences equally. There is a particular experience that people care about. It is the experience of "wholeness." We are fascinated by consciousness and its first-person experience aspect because we are seeking this particular experience.

Although wholeness is difficult to define, most people seem to recognize something in the term. It seems to be a core concept in many areas that are a source of value in their lives: aesthetics, love, science, and religion. In brief, science seeks the unifying theory; love involves a shift in some sort of boundary between people and the emergence of a new unity; aesthetic value is commonly described with terms such as dynamic unity, harmony, or integration ("it all works together"); and descriptions of the most universal religious experience, independent of doctrine, always involve a compelling sense of unity, oneness, wholeness.

Entertain for a few moments the possibility that there is something of surpassing importance about wholeness, and that it is our nature to seek the experience of it.

THE DACHSHUND JOKE

A little boy asks his father, "Daddy, how does the telegraph work?"

Dad thinks a bit and says, "Well, imagine that we had a giant dachshund with his tail in Pittsburgh and his head in New York. If you pinched his tail in Pittsburgh the head would bark in New York. It's sort of like that."

Little boy says, "Thank you, Daddy. I think I understand. But tell me, how does the *wireless* telegraph work?"

Dad thinks again and says, "That's the same thing, but without the dog!"

This joke is only superficially about the dangers of being seduced by analogies. It is only superficially about vacuous, handwaving explanations that we get (or give) when our overstretched model breaks down. The power behind this joke is that it pushes us up against the very deep questions of what is an explanation and what makes an explanation satisfying.

This is an extensive issue in philosophy. I looked into it a bit because I did not understand why I found all the explanations of consciousness so unsatisfying. Wimsatt's (1976) analysis, "Ontological and Explanatory Primacy" (pp. 242–251), was the most helpful to me. I offer you my provisional formulation:

An explanation is an account of the variance of the phenomenon of interest. It is satisfying if it is expressed in the terms of a frame of reference in which you are comfortable, and if it accounts for enough of the variance *for your purposes*.

It is very important to note here that for an explanation to be satisfying it need not be true or be the only account, or account for all of the variance.

I have not been satisfied with our explanations of consciousness (adequate though they may be for some purposes), because my purposes concern the experience of wholeness, and the accounts so far are responsive to other questions. They are not directed toward the experience of love, beauty, spirit, or knowing, as such.[1]

---

[1] A notable exception is Mangan (1991), who brings contemporary cognitive psychology to bear on aesthetics and mystical experience.

Joseph E. Bogen, neurosurgeon and student of consciousness, has pointed out that if a problem that has engaged the best minds still has not been resolved after several hundred years, then it probably is not going to be resolved within the terms in which it has been formulated (Bogen, 1976). In other words, we may need to change the frame of reference. In this instance, the first change needed is to enlarge the frame.

Recall this volume's title and the moral of the cigar joke. In this chapter I have reframed the problem, recognizing that there are MANY questions about consciousness. The second step is to order the questions in priority of interest. Are the questions addressed so far the most interesting ones? To help us decide what we OUGHT to be most interested in, I proposed that we first notice what we ARE most interested in. If we want deeply satisfying explanations, then let us choose to ask the questions that are aligned with our deepest purposes.

REFERENCES

Baars, B. (1986). *The cognitive revolution in psychology*. New York: Guilford.

Baars, B. (1988). *A cognitive theory of consciousness*. Cambridge, England: Cambridge University Press.

Baars, B. (1992). *Experimental slips and human error: Exploring the architecture of volition*. New York: Plenum.

Bogen, J. E. (1976). Hughlings Jackson's Heterogram. In D. O. Walter, L. Rogers, & J. M. Finzi-Fried (Eds.), *Conference on Human Brain Function* (BIS Conference Report 42, pp. 146–151). Los Angeles: University of California.

Flanagan, O. (1991a). *The science of mind* (2nd ed.). Cambridge, MA: MIT Press.

Flanagan, O. (1991b). *Varieties of moral personality: Ethics and psychological realism*. Cambridge, MA: Harvard University Press.

Flanagan, O. (1992). *Consciousness reconsidered*. Cambridge, MA: MIT Press.

Galin, D. (1992a). The blind wise men and the elephant of consciousness. *Consciousness and Cognition, 1*, 8–11.

Galin, D. (1992b). Theoretical reflection on awareness, monitoring, and self in relation to anosognosia. *Consciousness and Cognition, 1*, 152–162.

Galin, D. (1994). The structure of awareness: Contemporary applications of William James' forgotten concept of "the fringe." *Journal of Mind and Behavior, 15*(4), 375–400.

Mangan, B. (1991). *Meaning and the structure of consciousness: An essay in psycho-aesthetics*. Doctoral dissertation, University of California Berkeley. (University Microfilms No. 92033636)

Shallice, T. (1988). Information-processing models of consciousness: Possibilities and problems. In A. Marcel & E. Bisiach (Eds.), *Consciousness in contemporary science* (pp. 305–333). Oxford, England: Clarendon.

Sperry, R. W. (1993). The impact and promise of the cognitive revolution. *American Psychologist, 48*, 878–885.

Wimsatt, W. C. (1976). Reductionism, levels of organization, and the mind–body problem. In G. G. Globus, G. Maxwell, & I. Savodnik (Eds.), *Consciousness and the brain* (pp. 205–266). New York: Plenum.

CHAPTER 24

Consciousness and Me-ness

John F. Kihlstrom
Yale University

At the very beginning of scientific psychology, James (1890/1981) noted the intimate relation between consciousness and the self:

> Every thought tends to be part of a personal consciousness. . . . The only states of consciousness that we naturally deal with are found in personal consciousnesses, minds, selves, concrete particular I's and you's [*sic*]. Each of these minds keeps its own thoughts to itself. . . . It seems as if the elementary psychic fact were not *thought* or *this thought* or *that thought*, but *my thought*, every thought being owned. . . . On these terms the personal self rather than the thought might be treated as the immediate datum in psychology. The universal conscious fact is not "feelings and thoughts exist" but "*I* think" and "*I* feel." (p. 221, italics in original)

Consciousness comes, in James' view, when we inject ourselves into our thoughts, feelings, desires, and actions—when we take possession of them, experience and acknowledge them as our own. Consciousness is always a *personal* consciousness, and this was at least as true for memory as for any other mental faculty (James, 1890/1981, pp. 610–612):

> Memory proper, or secondary memory as it might be styled, is the knowledge of a former state of mind after it has already once dropped from consciousness; or rather *it is the knowledge of an event, or fact*, of which meantime we have not been thinking, *with the additional consciousness that we have thought or experienced it before.*
>
> The first element which such a knowledge involves would seem to be the revival in the mind of an image or copy of the original event. . . . [But] a farther

451

condition is required before the present image can be held to stand for a *past original*.

That condition is that the fact imaged be *expressly referred to the past*, thought as *in the past*. . . .

But even this would not be a memory. Memory requires more than the mere dating of a fact in the past. It must be dated in *my* past. In other words, I must think that I directly experienced its occurrence. It must have that "warmth and intimacy" which were so often spoken of in the chapter on the Self, as characterizing all experiences "appropriated" by the thinker as his own.

A general feeling of the past direction in time, then, a particular date conceived as lying along that direction, and defined by its name or phenomenal contents, and imagined as located therein, and owned as part of my experience, — such are the elements of every act of memory.

Janet (1907), in his lectures on *The Major Symptoms of Hysteria*, picked up the theme:

The complete consciousness which is expressed by the words, "I see, I feel a movement," is not completely represented by this little elementary phenomenon [i.e., of a sensation of vision or of motion]. It contains a new term, the word "I," which designates something very complicated. The question here is of the idea of personality, of my whole person. . . . There are then in the "I feel," two things in presence of each other: a small, new, psychological fact, a little flame lighting up—"feel"—and an enormous mass of thoughts already constituted into a system—"I." These two things mingle, combine; and to say "I feel" is to say that the already enormous personality has seized upon and absorbed that little, new sensation which has just been produced. (pp. 304–305)

This connection to the self appears to be missing in cases of nonconscious influence. Consider Claparède's (1911/1950) comments on what is now called the amnesic syndrome (for an extended discussion, see Kihlstrom, 1995): "If one examines the behavior of such a patient, one finds that everything happens as though the various events of life, however well associated with each other in the mind, were incapable of integration with *the me* itself" (p. 71).

In other words, when mental representations are integrated with the self, they become part of conscious mental life; when this integration is lacking, they are not accessible to introspection, although they may influence experience, thought, and action outside of phenomenal awareness.

LINKING THOUGHT TO THE SELF

How might we translate these ideas into the language of modern psychology? One way is through associative network models of cognition and memory, such as the successive generations of Anderson's (1976, 1983, 1990, 1993) ACT framework. These models represent declarative knowledge in a graph structure, with nodes standing for concepts and associative links standing for the relations between them, forming sentencelike propositions that capture the gist of an event but may eliminate a great deal of perceptual detail. A subset of declarative memory, known as working memory, contains a representation of the organism in its environment, current processing goals, and other units of declarative knowledge that have been activated above a certain threshold.

There is also a procedural memory, in which nodes representing goals and conditions are linked to actions; these goal–condition–action links form productions, grouped into production systems. The goals and conditions are themselves represented by declarative knowledge structures, and execution of the procedure may activate some node in declarative memory, or form a new one. Procedural knowledge is unconscious in the strict sense that we have no direct introspective access to it, and can know it only by inference.

Within the network-theory framework, we can begin to see what the self might look like. Viewed as a declarative knowledge structure, the self can be defined as one's mental representation of his or her own personality, broadly construed (Kihlstrom & Cantor, 1984; Kihlstrom, Cantor, Albright, Chew, Klein, & Niedenthal, 1988; Kihlstrom & Klein, 1994; Kihlstrom & Marchese, in press). In this way, the self can be construed as a fragment of a larger declarative memory network. So, my own mental representation of self consists of interconnected nodes representing my name; the names of people intimately associated with me; and my physical, demographic, and personality characteristics. Beyond these meaning-based representations, there may be links to perception-based representations of my face and body—quite literally, a self-image. In addition to this context-free knowledge about self, there is also autobiographical knowledge about specific experiences, thoughts, and actions that occurred at unique points in space and time. The presence of this contextual information ordinarily distinguishes episodic from semantic memory (Tulving, 1983, 1993).

In theory, an activated mental representation of the self resides in working memory, where it routinely comes into contact with representations of the environmental context (both local and global), current processing goals, and other knowledge structures activated by perception, memory, or thought. This connection defines the self as the agent or patient of some particular action or the stimulus or experiencer of some particular state (Brown & Fish, 1983)—whatever event is simultaneously represented in working memory.

According to models like ACT, perceptual activity and other acts of thought activate nodes corresponding to the features of the perceived event. Thus, for example, if we observe a hippie touch a debutante (Anderson & Bower, 1973), nodes corresponding to the concepts HIPPIE, TOUCH, and DEBUTANTE are activated, and linked to form a propositional representation of the event:

A HIPPIE TOUCHED A DEBUTANTE.

Note that the associative links actually join tokens, referring to particular hippies, touches, and debutantes, of more abstract types stored in semantic memory. Each node then serves as a source of activation that spreads to related semantic knowledge (Anderson, 1984).[1] Thus, activation of *HIPPIE* also activates tokens of *LONG STRINGY HAIR, LOVE BEADS,* and *VOLKSWAGEN BUS; DEBUTANTE* activates *LONG CURLY HAIR, EVENING GOWN,* and *PORSCHE;* and *TOUCH* activates *BRUSH, HIT,* and *KISS.*

By virtue of achieving a particular level of activation, the mental representation of this event enters working memory, where it contacts representations of other sorts of knowledge—specifically, representations of the context in which the event has taken place, and of the self as the agent or patient, stimulus or experiencer, of the event. Thus, the full-fledged mental representation of the event corresponds to a more complex piece of propositional knowledge than a mere statement of fact (Brewer, 1986), for example, self as subject:

I SAW THE HIPPIE TOUCH THE DEBUTANTE
IN MACARTHUR PARK ON THURSDAY;

or, self as object:

I WAS THE HIPPIE WHO TOUCHED THE DEBUTANTE;

or, perhaps:

I WAS THE DEBUTANTE WHOM THE HIPPIE TOUCHED.

To summarize, a full-blown episodic representation contains at least three different propositions, all linked together: an *event node,* which provides a raw description of the event; a *context node,* which represents the spatio-temporal circumstances in which the event occurred (there are probably separate propositions representing time and place); and a *self node,* representing the self as the agent or patient, stimulus or experiencer, of some event (there may be nodes representing the person's emotional and motivational state as well). Simplified representations are depicted in Fig. 24.1. Residual activation at these nodes is the mechanism underlying repetition priming effects; activation spreading to related nodes in declarative memory forms the basis for semantic priming effects; and if some activated nodes comprise the goals and conditions of a production system, some piece of procedural knowledge will be automatically executed.

By virtue of elaborative and organizational processing at the time of perception, a trace of the event is encoded in permanent storage, available for subsequent use; that is, it does not disappear entirely when it loses activation and drops out of working memory. At the time of retrieval, a query to memory

---

[1] I recognize that there is some question about whether activation really spreads (e.g., McKoon & Ratcliff, 1992; McNamara, 1992, 1994; Ratcliff & McKoon, 1981, 1988, 1994). In what follows, the phrase "spreading activation" is used only as a familiar metaphor.

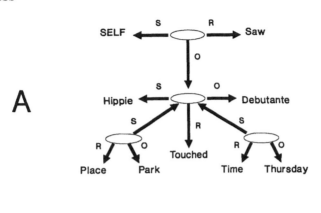

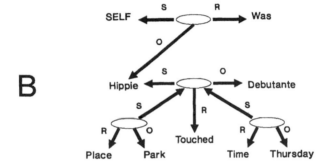

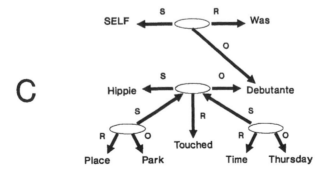

FIG. 24.1.   Example propositional representations of an event in which the self is an external observer (A), agent (B), or patient (C) of the action.

activates nodes corresponding to the elements of the query, while the accessibility of information represented at associated nodes serves as the basis for the person's response to those queries. Thus, a question such as "What did you see in MacArthur Park last Thursday?" activates the self and context nodes; activation emanating from these nodes converges on the associated event node, and the person replies, "A hippie touched a debutante." A similar account can be given of queries and responses concerning current experience, except that the probe is worded in the present tense, and the response reflects direct readout of the contents of working memory. Thus, the mental representation of self is the key to conscious perception and memory: Whatever is concurrently or retrospectively linked to the self is represented in awareness and available for report.

LINKING EXPERIENCE TO THE SELF IN MEMORY

According to Claparède (1911/1950), what is missing in cases of nonconscious processing is this sense of *moïtè*, or *me-ness*, or the link to the self as the agent or patient, stimulus or experiencer, of some event. To explain, consider the famous incident that gave Claparède his insight. A woman with Korsakoff's syndrome had anterograde amnesia so dense that she remembered nothing that had happened to her since the onset of her illness. Nevertheless, Claparède reported that she showed what we now recognize as source amnesia (Evans, 1979; Schacter, Harbluk, & McLachlan, 1984; Shimamura & Squire, 1987). After Claparède pricked her hand with a pin hidden in his fingers, causing her palpable pain, she quickly forgot the episode (and everything else about her encounters with him).

> But when I again reached out for her hand, she pulled it back in a reflex fashion, not knowing why. When I asked for the reason, she said in a flurry, "Doesn't one have the right to withdraw her hand?" and when I insisted, she said, "Is there perhaps a pin hidden in your hand?" To the question, "What makes you suspect me of wanting to stick you?", she [replied], "That was an idea that went through my mind," or she would explain, "Sometimes pins are hidden in people's hands." But never would she recognize the idea of sticking as a "memory." (Claparède, 1911/1950, pp. 69–70)

The interpretation here is that the link to the self was established in working memory at the time that the pinprick occurred, and was maintained so long as the event resided in working memory. Thus, the patient was aware that it was she who had just been pricked, and she could still talk about it moments or minutes later. By virtue of direct readout from working memory, she was conscious of what was happening to her at the moment, and what had happened to her in the very recent past. As she was distracted by other events, however, the pinprick dropped out of working memory, and other events took its place.

Under ordinary circumstances, given enough elaborative and organizational activity, a representation of the event (e.g., CLAPARÈDE PRICKED ME WITH A PIN HIDDEN IN HIS HAND) would now be permanently encoded in the patient's declarative memory. But suppose that—perhaps by virtue of damage to specific brain structures—the link to self was not encoded as well. Rather, it was permanently lost. Now Claparède returns to the patient, and asks whether they have ever met. This query activates the mental representation of self in working memory (the question is whether *she* has ever met him), but because the active representation of self is not connected to relevant material outside working memory, she draws a blank. Put another way, the patient had concurrent, but not retrospective awareness of the episode. Apparently, the link to the self was made at the time of perception, but not preserved in memory.

But despite the failure to consciously recollect having met Claparède before, something about the previous event has been retained in memory. Forced to choose, she may guess that he is a doctor rather than a lawyer; perhaps Claparède's face or tone of voice activates perception-based representations stored in memory—enough to give her a vague feeling of familiarity; forced to guess, she might choose Claparède over another doctor, himself objectively unknown to her, as someone she *might* have met before. Certainly enough world knowledge has been gleaned from her previous experience that she will now greet him in an unusual manner. After all, sometimes people hide pins in their hands.

By this time, there have been a large number of demonstrations that amnesic patients can show priming, savings, transfer, and interference effects, acquire new facts and skills, and display other manifestations of memory, all without any conscious recollection of the event responsible for the effect (e.g., Graf, Squire, & Mandler, 1984; for reviews, see Moscovitch, Vriezen, & Goshen-Gottstein, 1993; Shimamura, 1989). These kinds of observations led Schacter (1987) and others to distinguish between two forms of memory: explicit and implicit (for comprehensive reviews, see Johnson & Hasher, 1988; Richardson-Klavehn & Bjork, 1988; Roediger, 1990; Roediger & McDermott, 1993; Schacter, 1995; Schacter, Chiu, & Ochsner, 1993; Schacter & Tulving, 1994). Although this distinction has sometimes been couched in terms of two different memory systems, or two different processes acting on a single memory system, this chapter sticks closer to the phenomenology and refers to explicit and implicit memory as two different expressions of memory for an episode. *Explicit memory* involves conscious recollection of a prior event, as reflected in recall or recognition. *Implicit memory* refers to any change in experience, thought, or action attributable to a past event, independent of conscious recollection. In other words, memory is implicit in task performance.

From the point of view of this chapter, implicit memory is memory without the self. The contents of memory have been changed by the perception of an event, which is roughly what we mean by a memory, but reference to the self as the agent or patient, stimulus or experiencer, of the event has not been preserved.

Of course, implicit memory is observed in normal subjects, without any signs of brain damage (for a comprehensive review, see Roediger & McDermott, 1993). But here, the underlying mechanism is the same. By and large, the explicit–implicit dissociations observed in normal, intact subjects are produced by experimental conditions that prevent the subject from engaging in the elaborative and organizational activity that produces a good, solid encoding. For example, shallow processing impairs explicit memory, but has little effect on implicit memory (Jacoby & Dallas, 1981). Alternatively, the encoded trace may be degraded over the retention interval, leading to failures of recall and recognition. Nevertheless, subjects can still show significant savings in relearning forgotten items (Nelson, 1978). Apparently, a permanent link to self is not automatically encoded (else it would be encoded under shallow as well as deep processing conditions); and it appears to be somewhat fragile (else it would remain intact over the retention interval). Nevertheless, other aspects of the event are encoded, and remain available in memory to support such classic implicit memory phenomena as priming, source amnesia, savings, and interference.

Other manifestations of implicit memory involve the acquisition of cognitive and motoric skills, which are also performed unconsciously. Again, the common element is the absence of a link to the self. Assume that skill learning begins with a step-by-step recipe for some activity (such as tying a necktie in a Windsor knot). In the early stages of skill learning, these steps are brought into working memory one at a time, and performed consciously. That is, the person is aware of performing each step individually. With enough practice the whole sequence is compiled into a single routine that, once initiated, just runs itself off automatically. The person is no longer aware of what he or she is doing. From the present point of view, this is because the intermediate steps have been compiled into a single production system (Anderson, 1982), which never gets represented in working memory, and thus never has the opportunity to make contact with the mental representation of the self. The person was aware of selecting the tie (perhaps), and later was aware that the tie was successfully tied, but not of anything that went on in between.

In the last years of Eugene Ormandy, when the Philadelphia Orchestra virtually confined its repertoire to 19th-century warhorses, one of the cellists quit when he discovered himself playing the third movement of a Tchaikovsky symphony, but could not remember playing the first one. That is automatization. A similar account can be given of more dramatic instances of automatization: musicians or motorists who play pianos or drive cars during petit mal seizures, or drunks who manage to get home from the bar without killing themselves or anyone else, or sleepwalkers who wander around the house and return to their beds with no recollection of their activities the next morning. Some of this behavior is mediated by highly automated procedural knowledge, which never makes contact with the self during its execution. The rest may be deliberate and effortful, and therefore conscious at the time; but because of impairments in encoding processes the link to the self, and thus the possibility of conscious recollection, is not preserved in memory.

LINKING EXPERIENCE TO THE SELF IN PERCEPTION

Kihlstrom, Barnhardt, and Tataryn (1992) extended the explicit–implicit distinction to the domain of perception. Following Schacter (1987), explicit perception refers to the individual's conscious awareness of a current (or very recent) event, as reflected in signal detection, or the identification or description of an object; implicit perception is reflected in any change in experience, thought, or action that is attributable to an event in the current (or very recent) stimulus environment, independent of conscious perception. Again, perception is implicit in the person's task performance.

The paradigm cases of implicit perception parallel those of implicit memory. In blindsight, damage to the striate cortex prevents the construction of conscious percepts (Weiskrantz, 1986). In subliminal perception, a stimulus is degraded by various conditions of presentation (for reviews, see Greenwald, 1992; Merikle & Reingold, 1992). What they have in common, on this view, is that the perceptual representation never enters working memory, and thus never makes contact with the self in the first place. Thus, when the person is asked to report what he or she sees or hears, there's nothing to report. But the availability of the representation, or some by-product of it somewhere in the cognitive system, can support implicit perception effects. Thus, if blindsight patients are forced to choose, they can guess the characteristics of the unseen stimulus at above-chance levels. And in subliminal perception, subjects can show various sorts of priming effects. Implicit perception is sometimes reflected in performance on implicit memory tasks. So, in Kunst-Wilson and Zajonc (1980), subjects showed mere exposure effects while making preference judgments of tachistoscopically presented irregular polygons, even though recognition was at chance levels (for a review, see Bornstein, 1989). The exposure effect shows that the stimuli were perceived, at some level; but formally, the task involves memory rather than perception. Subliminal stimulation created some representation of the stimulus in memory, but that representation never made contact with the self, either at the time or later.

The same thing sometimes happens in general anesthesia, but perhaps for a different reason (for a review, see Cork, Couture, & Kihlstrom, in press). Evidence has been obtained that patients can show implicit memory for events occurring while they were under an adequate plane of general anesthesia, even though they have no conscious recollection of these events afterward. Thus, in one experiment (Kihlstrom, Schacter, Cork, Hurt, & Behr, 1990), patients were presented with a tape recording of paired associates of the form BREAD–BUTTER; in the recovery room, they were unable to remember the response term with which the stimulus terms had been paired. But, when presented with the stimulus terms, and asked to report the first word that came to mind, they were more likely than control subjects to produce items from the study list. Other investigators have obtained similar results, although admittedly there is more to be done to bring the phenomenon under good experimental control. Again, assuming that adequately anesthetized patients really are not aware of what is happening to them at the time it is happening, we have implicit memory giving evidence for implicit perception. In this case, however, the detailed mechanism is slightly different. When a person is rendered unconscious, he or she does not have any working memory at all. Thus, it is not just a matter of event representations failing to make contact with a mental representation of the self. The problem is that there is no activated self to make contact *with*.

The situation appears to be rather different in the case of sleep. The old conclusion about sleep learning, based on Simon and Emmons (1955), was that sleep learning is possible to the extent that the subject remains awake. But that conclusion was based on studies of explicit memory, raising the question of whether implicit memory might be preserved even in the absence of explicit memory. But Wood and his colleagues (Wood, Bootzin, Kihlstrom, & Schacter, 1992) also failed to find any evidence of implicit memory. Wood presented sleeping subjects with a tape recording of paired associates consisting of a homophone and a disambiguating context, such as TORTOISE–HARE. Upon awakening, the subjects were unable to produce the response terms in a test of cued recall; but neither did they show any priming when asked to spell the response terms. Similar results were obtained in a category–instantiation task. It appears that we will have to take even more seriously the notion that sleep involves a kind of sensory gate, preventing stimulus information from reaching higher cortical centers. In any event, the discrepancy between sleep (where both explicit and implicit memory are impaired) and general anesthesia (where implicit memory is sometimes spared) is something that needs definitive resolution.

DISSOCIATIVE PHENOMENA AND THE SELF

As noted earlier, some theorists (e.g., Schacter & Tulving, 1994; Tulving & Schacter, 1990) hold that explicit and implicit memory reflect the operation of two different memory systems in the brain; the same hypothesis may apply to the perceptual case. Others (e.g., Jacoby, 1991; Roediger, Weldon, & Challis, 1989) have vigorously argued that explicit–implicit dissociations are mediated by two different sorts of mental processes operating on a single representation. These theories (for a comprehensive review, see Roediger & McDermott, 1993) have been so successful that one wonders if it is worth talking about the role of the self in nonconscious influence. The answer is simple: There are cases of nonconscious influence that cannot be accounted for by presemantic representations and automatic, data-driven processes.

Consider Ansel Bourne, a case of psychogenic fugue studied by James (1890/1981). Bourne was an itinerant preacher in and around Greene, Rhode Island. He was upright and self-reliant, if given to occasional bouts of depression. In January 1887 he disappeared from Rhode Island and was declared missing by the police. Two months later he turned up in Norristown, Pennsylvania, where he had set up business as a shopkeeper selling stationery, candy, fruit, and sundries. There he was known by his neighbors as A. J. Brown, a taciturn and orderly individual, and a regular churchgoer. Brown made no reference to his previous life, much less his identity. The discovery of his dual existence occurred when A. J. Brown went to sleep one night and awakened the next morning as Ansel Bourne, unaware of where he was. He identified himself as Ansel Bourne, but had no memory of what he had been doing for the past 8 weeks, for 6 of which he had been living in Norristown. James was able to get this information by means of a hypnotic interview, but none of this material was accessible to him in his normal waking state: Just as Brown seemed unaware of his life as Bourne, so Bourne seemed unaware of his life as Brown. James (1890/1981, p. 371) wrote that "Mr. Bourne's skull to-day still covers two distinct personal selves."

A more dramatic variant on this situation occurs in cases of multiple personality disorder (e.g., Thigpen & Cleckley, 1954). Thigpen and Cleckley described Eve White, a 25-year-old housewife, demure and retiring but also industrious and devoted to her daughter. She had no particular history of psychological problems, and had been referred for psychiatric consultation for persistent headaches and blackouts. But during one treatment session an entirely different self-presentation occurred: Eve Black—childish, mischievous, carefree, and erotically playful. It turned out that the Eves White and Black had alternated control over the patient's behavior since childhood; the asymmetrical amnesia between the ego states—Eve Black knew all about Eve White, but Eve White knew nothing about Eve Black—made for some interesting episodes. Once, during childhood, Eve was caught playing with a particular neighbor's children, despite a specific injunction against doing so; nevertheless, the child steadfastly denied any wrongdoing. As an adult, Eve received by mail order a rather slinky evening dress, which she denied ordering. Both episodes, and many more, were apparently the work of Eve Black, outside the awareness of Eve White; though it was Eve White who faced the consequences.

The Three Faces of Eve

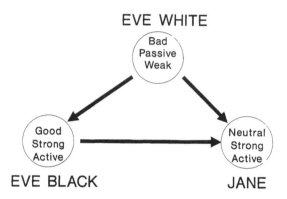

FIG. 24.2.   Schematic depiction of relations among the
three "faces" of Eve (Thigpen & Cleckley, 1954). Eve White
is ignorant of both Eve Black and Jane; Eve Black knows
about Eve White, but not about Jane; Jane knows about both
Eve White and Eve Black.

The third personality, Jane, made her appearance later in the course of therapy. Jane had access to each personality's experiences since she herself began
to emerge, but not to much material from their lives before treatment. The
core personality characteristics of each alter ego, and the pattern of asymmetrical amnesia, are depicted in Fig. 24.2.

Here, to borrow James' phrase, there are three "personal selves" inside the
same skull, alternating in control over conscious awareness, voluntary behavior, and communication with others. Fugue and multiple personality patients
also show evidence of implicit memory (for reviews, see Kihlstrom & Schacter, 1995; Kihlstrom, Tataryn, & Hoyt, 1993; Schacter & Kihlstrom, 1989).
So, for example, A. J. Brown once gave testimony in church that referred to an
incident that had actually happened to Ansel Bourne. And a formal experiment by Nissen and her colleagues (Nissen, Ross, Willingham, Mackenzie, &
Schacter, 1988) gave evidence of priming effects that were preserved across
personalities, despite interpersonality amnesia affecting recall. These cases
may be distinguished from other displays of implicit memory because, under
some circumstances, implicit memory can become explicit. Thus, Eve Black
can give a full account of those incidents that puzzled Eve White. So the problem is not exactly one of degraded encoding.

Genuine cases of fugue and multiple personality are admittedly rare, but
something similar can be seen in large numbers of otherwise normal human
subjects experiencing posthypnotic amnesia (Kihlstrom, 1985). In one ex-

periment (Kihlstrom, 1980), hypnotized subjects memorized a list of words, and then received a suggestion that upon awakening from hypnosis they would not remember the words they had learned until the amnesia suggestion was canceled. The resulting amnesia was very dense, as compared to the memory shown by control subjects who were not deeply hypnotized. On a later test, the subjects were asked to give the first words that came to mind in response to cues that targeted the list items as free associates. The non-amnesic subjects showed a substantial priming effect, but so did the amnesics. Another recall test showed that production of list items as free associates did not remind the amnesic subjects of the items they had memorized. Finally, after the amnesia suggestion was canceled, everybody remembered the list almost perfectly. This dissociation between explicit and implicit memory is quite different from the usual priming study, in a number of respects: Good encoding was insured by requiring the subjects to meet a criterion of two perfect repetitions of the list before the amnesia suggestion was given. And, adequate retention was demonstrated by the full recovery of memory after administration of the reversibility cue. Moreover, the priming observed here is true semantic priming, not repetition priming; because the cues were not presented in the study list, a semantic association between cue and target had to be formed by the subject at the time of encoding, and preserved in memory over the retention interval.

If conscious awareness (evidenced by explicit expressions of perception and memory) is mediated by links to the self, the puzzle is how an event representation can be linked to the self sometimes, but not at other times. James (1890/1981) understood this problem, and this chapter proposes to adopt his solution: Perhaps there is not just one mental representation of the self. Speaking of the phenomena of hysteria (as it was then known) and hypnosis, James wrote:

> The buried feelings and thoughts proved now to exist in hysterical anaesthetics, in recipients of post-hypnotic suggestion, etc., themselves are parts of *secondary personal selves*. These selves . . . are cut off at ordinary times from communication with the regular and normal self of the individual; but still they form conscious unities, have continuous memories, speak, write, invent distinct names for themselves, or adopt names that are suggested; and, in short, are entirely worthy of that title of secondary personalities which is now commonly given them. According to M. Janet these secondary personalities . . . result from the splitting of what ought to be a single complete self into two parts, of which one lurks in the background whilst the other appears on the surface as the only self the man or woman has. (p. 222)

Considerations of the self as a concept suggest that each individual possesses a number of context-specific selves, arrayed as a set of exemplars or co-existing with a summary prototype (Kihlstrom & Cantor, 1984; Kihlstrom et al., 1988; Kihlstrom & Klein, 1994; Kihlstrom & Marchese, in press). Ordinarily, these contextual selves are linked to each other so that the person is aware of what he or she is like under different circumstances. In terms of associative-network models of memory (such as ACT), we may say that there are several different knowledge structures, each representing a different token of the self-concept. Working memory can hold one of these at a time. Newly activated knowledge structures, if they are represented in working memory at all, are linked to whatever token of the self is resident in working memory; and if the link to self is encoded in memory, it is in terms of that particular self and no others. Some of the self-tokens may be linked to each other, but, in general, only the self resident in working memory can report on phenomenal awareness; and the only memories accessible to conscious recollection are those that have been encoded with respect to that particular token of the self.

Similarly, in hypnosis, assume that the highly hypnotizable subject's deep level of involvement in the experience sets up something like a new, temporary self-concept to which the experiences of hypnosis are linked. When hypnosis is terminated with a suggestion for amnesia, this "hypnotic self" moves out of working memory and is replaced by the subject's "normal self"; and the link that would normally unite the two self-concepts is temporarily suppressed. This situation will result in an inability to remember the list items and the other experiences of hypnosis. But those items will remain active in declarative memory, and this residual activation will support priming effects and other forms of implicit memory. When the reversibility cue is given to cancel the amnesia, the link is restored, and with it access to the experiences of hypnosis.

UNCONSCIOUS, PRECONSCIOUS, SUBCONSCIOUS

This chapter has focused on the mechanisms associated with the division of consciousness in hypnosis and hysteria. James (1890/1981), Janet (1907), and Hilgard (1977) also shared the sense that these sorts of phenomena tell us something new about the nature of the psychological unconscious, and the relations between conscious and unconscious mental processes (Kihlstrom, 1987, 1990, 1993). These dissociative phenomena cannot be understood in the terms normally applied to unconscious and preconscious processing; they seem to demand some sort of reference to the self and its vicissitudes.

On the other hand, it seems that the link to the self is also implicated in unconscious and preconscious processing as well. In unconscious processing, procedural knowledge is executed without ever making contact with a mental representation of self. In preconscious processing, events have been so degraded that they, too, never enter working memory at the point of perception; although, in contrast to procedural knowledge, they could do so if conditions were different. Thus, they never achieve any links with the self in the first place. Alternatively, events are processed in working memory, and make contact with the self at the time of perception, but the traces are so poorly encoded, or suffer such serious degradation over the retention interval, that the links to self originally forged at the time of perception are lost.

The scope of the psychological unconscious is very broad, and the manner in which mental structures and processes are rendered unconscious varies widely as well: proceduralization and knowledge compilation, degraded stimulation and poor encoding, and functional dissociation. But in terms of their phenomenology, all appear to share a lack of association to a mental representation of the self as the agent or patient, stimulus or experiencer, of events. In the end, this final common pathway appears to be what makes the difference between conscious and unconscious mental life.

ACKNOWLEDGMENTS

The point of view represented here is based on research supported by Grant MH-35856 from the National Institute of Mental Health. I thank John Allen, Terrence Barnhardt, Melissa Berren, Lawrence Couture, Elizabeth Glisky, Martha Glisky, Heather Law, Chad Marsolek, Victor Shames, Susan Valdiserri, and Michael Valdiserri for their comments.

REFERENCES

Anderson, J. R. (1976). *Language, memory, and thought.* Hillsdale, NJ: Lawrence Erlbaum Associates.

Anderson, J. R. (1982). Acquisition of cognitive skill. *Psychological Review, 89,* 369–406.

Anderson, J. R. (1983). *The architecture of cognition.* Cambridge, MA: Harvard University Press.

Anderson, J. R. (1984). Spreading activation. In J. R. Anderson & S. M. Kosslyn (Eds.), *Tutorials in learning and memory: Essays in honor of Gordon Bower* (pp. 61–90). San Francisco: Freeman.

Anderson, J. R. (1990). *The adaptive character of thought.* Hillsdale, NJ: Lawrence Erlbaum Associates.

Anderson, J. R. (1993). *Rules of the mind.* Hillsdale, NJ: Lawrence Erlbaum Associates.

Anderson, J. R., & Bower, G. H. (1973). *Human associative memory.* Washington, DC: V. H. Winston.

Bornstein, R. F. (1989). Exposure and affect: Overview and meta-analysis of research, 1968–1987. *Psychological Bulletin, 106,* 265–289.

Brewer, W. F. (1986). What is autobiographical memory? In D. C. Rubin (Ed.), *Autobiographical memory* (pp. 25–49). Cambridge, England: Cambridge University Press.

Brown, R., & Fish, D. (1983). The psychological causality implicit in language. *Cognition, 14,* 237–273.

Claparède, E. (1950). Recognition and me-ness. In D. Rapaport (Ed.), *The organization and pathology of thought: Selected sources* (pp. 58–75). New York: Columbia University Press. (Original work published 1911)

Cork, R. C., Couture, L. J., & Kihlstrom, J. F. (in press). Memory and recall. In J. F. Biebuyck, C. Lynch, M. Maze, L. J. Saidman, T. L. Yaksh, & W. M. Zapol (Eds.), *Anesthesia: Biologic foundations: Vol. 2. Integrated systems.* New York: Raven.

Evans, F. J. (1979). Contextual forgetting: Posthypnotic source amnesia. *Journal of Abnormal Psychology, 88,* 556–563.

Graf, P., Squire, L. R., & Mandler, G. (1984). The information that amnesic patients do not forget. *Journal of Experimental Psychology: Learning, Memory, and Cognition, 10,* 164–178.

Greenwald, A. G. (1992). New Look 3: Unconscious cognition reclaimed. *American Psychologist, 47,* 766–790.

Hilgard, E. R. (1977). *Divided consciousness: Multiple controls in human thought and action.* New York: Wiley-Interscience.

Jacoby, L. L. (1991). A process dissociation framework: Separating automatic from intentional uses of memory. *Journal of Memory and Language, 30,* 513–541.

Jacoby, L. L., & Dallas, M. (1981). On the relationship between autobiographical memory and perceptual learning. *Journal of Experimental Psychology: General, 110,* 306–340.

James, W. (1981). *Principles of psychology.* Cambridge, MA: Harvard University Press. (Original work published 1890)

Janet, P. (1907). *The major symptoms of hysteria.* New York: Macmillan.

Johnson, M. K., & Hasher, L. (1987). Human learning and memory. *Annual Review of Psychology, 38,* 631–668.

Kihlstrom, J. F. (1980). Posthypnotic amnesia for recently learned material: Interactions with "episodic" and "semantic" memory. *Cognitive Psychology, 12,* 227–251.

Kihlstrom, J. F. (1985). Posthypnotic amnesia and the dissociation of memory. In G. H. Bower (Ed.), *The psychology of learning and motivation* (Vol. 19, pp. 131–178). New York: Academic Press.

Kihlstrom, J. F. (1987). The cognitive unconscious. *Science, 237,* 1445–1452.

Kihlstrom, J. F. (1990). The psychological unconscious. In L. Pervin (Ed.), *Handbook of personality: Theory and research* (pp. 445–464). New York: Guilford.

Kihlstrom, J. F., (1993). The continuum of consciousness. *Consciousness and Cognition, 2,* 334–354.

Kihlstrom, J. F. (1995). Memory and consciousness: An appreciation of Claparède and his "Recognition et Moiïtè." *Consciousness & Cognition, 4,* 379–386.

Kihlstrom, J. F., Barnhardt, T. M., & Tataryn, D. J. (1992). Implicit perception. In R. F. Bornstein & T. S. Pittman (Eds.), *Perception without awareness* (pp. 17–54). New York: Guilford.

Kihlstrom, J. F., & Cantor, N. (1984). Mental representations of the self. In L. Berkowitz (Ed.), *Advances in experimental social psychology* (Vol. 17, pp. 2–47). New York: Academic Press.

Kihlstrom, J. F., Cantor, N. Albright, J. S., Chew, B. R., Klein, S. B., & Niedenthal, P. M. (1988). Information processing and the study of the self. In L. Berkowitz (Ed.), *Advances in experimental social psychology* (Vol. 21, pp. 145–177). San Diego: Academic Press.

Kihlstrom, J. F., & Klein, S. B. (1994). The self as a knowledge structure. In R. S. Wyer & T. K. Srull (Eds.), *Handbook of social cognition* (2nd ed., Vol. 1, pp. 153–208). Hillsdale, NJ: Lawrence Erlbaum Associates.

Kihlstrom, J. F., & Marchese, L. A. (in press). Situating the self in interpersonal space. In U. Neisser & D. Jopling (Eds.), *The conceptual self in context: Culture, experience, self-understanding.* New York: Cambridge University Press.

Kihlstrom, J. F., & Schacter, D. L. (1995). Functional disorders of autobiographical memory. In A. Baddeley, B. A. Wilson, & F. Watts (Eds.), *Handbook of memory disorders* (pp. 339–366). London: Wiley.

Kihlstrom, J. F., Schacter, D. L., Cork, R. C., Hurt, C. A., & Behr, S. E. (1990). Implicit and explicit memory following surgical anesthesia. *Psychological Science, 1,* 303–306.

Kihlstrom, J. F., Tataryn, D. J., & Hoyt, I. P. (1993). Dissociative disorders. In P. J. Sutker & H. E.

Adams (Eds.), *Comprehensive handbook of psychopathology* (2nd ed., pp. 203–234). New York: Plenum.

Kunst-Wilson, W. R., & Zajonc, R. B. (1980). Affective discrimination of stimuli that cannot be recognized. *Science, 207,* 557–558.

McKoon, G., & Ratcliff, R. (1992). Spreading activation versus compound cue accounts of priming: Mediated priming revisited. *Journal of Experimental Psychology: Learning, Memory, and Cognition, 18,* 1155–1172.

McNamara, T. P. (1992). Priming and constraints it places on theories of memory and retrieval. *Psychological Review, 99,* 650–662.

McNamara, T. P. (1994). Priming and theories of memory: A reply to Ratcliff and McKoon. *Psychological Review, 110,* 195–187.

Merikle, P. M., & Reingold, E. (1992). Measuring unconscious processes. In R. F. Bornstein & T. S. Pittman (Eds.), *Perception without awareness: Cognitive, clinical, and social perspectives* (pp. 55–80). New York: Guilford.

Moscovitch, M., Vriezen, E., & Goshen-Gottstein, Y. (1993). Implicit tests of memory in patients with focal lesions or degenerative brain disorders. In F. Boller & J. Grafman (Eds.), *Handbook of neuropsychology* (Vol. 8, pp. 133–173). Amsterdam: Elsevier.

Nelson, T. O. (1978). Detecting small amounts of information in memory: Savings for nonrecognized items. *Journal of Experimental Psychology: Human Learning and Memory, 4,* 453–468.

Nissen, M. J., Ross, J. L., Willingham, D. B., Mackenzie, T. B., & Schacter, D. L. (1988). Memory and awareness in a patient with multiple personality disorder. *Brain and Cognition, 8,* 117–154.

Ratcliff, R., & McKoon, G. (1981). Does activation really spread? *Psychological Review, 88,* 434–462.

Ratcliff, R., & McKoon, G. (1988). A retrieval theory of priming in memory. *Psychological Review, 95,* 385–408.

Ratcliff, R., & McKoon, G. (1994). Retrieving information from memory: Spreading-activation theories versus compound-cue theories. *Psychological Review, 101,* 177–184.

Richardson-Klavehn, A., & Bjork, R. A. (1988). Measures of memory. *Annual Review of Psychology, 39,* 475–543.

Roediger, H. L. (1990). Implicit memory: Retention without remembering. *American Psychologist, 45,* 1043–1056.

Roediger, H. L., & McDermott, K. B. (1993). Implicit memory in normal human subjects. In F. Boller & J. Grafman (Eds.), *Handbook of neuropsychology* (Vol. 8, pp. 63–131). Amsterdam: Elsevier.

Roediger, H. L., Weldon, M. S., & Challis, B. H. (1989). Explaining dissociations between implicit and explicit measures of retention: A processing account. In H. L. Roediger & F. I. M. Craik (Eds.), *Varieties of memory and consciousness: Essays in honour of Endel Tulving* (pp. 3–14). Hillsdale, NJ: Lawrence Erlbaum Associates.

Schacter, D. L. (1987). Implicit memory: History and current status. *Journal of Experimental Psychology: Learning, Memory, and Cognition, 13,* 501–518.

Schacter, D. L. (1995). Implicit memory: A new frontier for cognitive neuroscience. In M. A. Gazzaniga (Ed.), *The cognitive neurosciences* (pp. 815–824). Cambridge, MA: MIT Press.

Schacter, D. L., Chiu, C.-Y. P., & Ochsner, K. N. (1993). Implicit memory: A selective review. *Annual Review of Neuroscience, 16,* 159–182.

Schacter, D. L., Harbluk, J. L., & McLachlan, D. R. (1984). Retrieval without recollection: An experimental analysis of source amnesia. *Journal of Verbal Learning and Verbal Behavior, 23,* 593–611.

Schacter, D. L., & Kihlstrom, J. F. (1989). Functional amnesia. In F. Boller & J. Grafman (Eds.), *Handbook of neuropsychology* (Vol. 3, pp. 209–231). Amsterdam: Elsevier.

Schacter, D. L., & Tulving, E. (Eds.). (1994). *Memory systems 1994*. Cambridge, MA: MIT Press.

Shimamura, A. P. (1989). Disorders of memory: The cognitive science perspective. In F. Boller & J. Grafman (Eds.), *Handbook of neuropsychology* (Vol. 3, pp. 35–73). Amsterdam: Elsevier.

Shimamura, A. P., & Squire, L. R. (1987). A neuropsychological study of fact memory and source amnesia. *Journal of Experimental Psychology: Learning, Memory, and Cognition, 13*, 464–473.

Simon, C., & Emmons, W. (1955). Learning during sleep. *Psychological Bulletin, 52*, 328–342.

Thigpen, C. H., & Cleckley, H. M. (1954). A case of multiple personality. *Journal of Abnormal and Social Psychology, 49*, 135–151.

Tulving, E. (1983). *Elements of episodic memory*. Oxford, England: Oxford University Press.

Tulving, E. (1993). What is episodic memory? *Current Directions in Psychological Science, 2*, 67–70.

Tulving, E., & Schacter, D. L. (1990). Priming and human memory systems. *Science, 247*, 301–306.

Weiskrantz, L. (1986). *Blindsight: A case study and implications*. Oxford, England: Oxford University Press.

Wood, J. M., Bootzin, R. R., Kihlstrom, J. F., & Schacter, D. L. (1992). Implicit and explicit memory for verbal information presented during sleep. *Psychological Science, 3*, 236–239.

CHAPTER 25

Affect and Neuromodulation: A Connectionist Approach

David E. Rumelhart
Stanford University

Perhaps the key question in psychology and neuroscience today is the relation between the mind and the brain. On the face of it, the mind and brain have almost nothing in common. The mind seems to be ephemeral and abstract, whereas the brain is physical and mechanical. Nevertheless, it is the strong hypothesis of modern psychology, philosophy, and biology that the mind is a reflection of the behavior of the brain. It is all the more puzzling because we seem to have some independent access of the workings of our minds, but the behavior of the brain can only be studied through external means.

At various points during the history of psychology it has been supposed that the mind could be studied and understood independently of the brain. It has sometimes been supposed that there is a psychological reality that can be wholly studied and understood in its own realm and on its own terms without consideration of the brain. On this account, the mind, roughly, is the "software" and the brain the "hardware." We can, it is assumed, understand the nature of the mind by understanding the "program" the brain is running. The details of how the brain works is, on this view, not important to the understanding of the important mental tasks humans carry out. Although it is interesting to understand the workings of the brain, it is not necessary for a full understanding of how people think, reason, and engage in mental life.

In contrast, there is good evidence that the mind cannot be completely understood without consideration of how the brain works. Perhaps the clearest evidence comes from cases of brain damage in which the mind breaks up in surprising ways. The most natural tasks become impossible, personalities change, people cannot talk, or cannot remember, or cannot reason clearly. There are numerous cases in the clinical literature of people whose deficits are incredibly specific and difficult to explain. In addition to changes due to physical damage, there are many example of "reversible" changes resulting from drugs of various kinds. In a way, these cases make it especially clear that the mind is a product of the brain. Drugs alone can do most of the things that direct damage does to the brain. They can change personalities, affect our ability to remember and reason, make us angry or happy, and so on. In short, virtually anything that affects our brain also affects our mind—often in ways that seem difficult to predict.

The connectionist, parallel distributed processing (PDP), or neural network approach to the study of mind and brain constitutes one possible means of attacking this problem. In this case, we attempt to model the relation between mind and brain through the development of computational models that, on the one hand, attempt to model important aspects of the way in which brains work, while behaving in ways consistent with human behavior. There is a large literature on applications of this kind. The following sections provide a standard connectionist model of the relation between mind and brain and then elaborate on that model to attempt to explain certain affective phenomena.

THE CANONICAL CONNECTIONIST MODEL

Figure 25.1 shows the canonical connectionist model. The elements of the model can be understood in the following way:

1. The *units* (dots in the figure) represent "hypotheses" about the world.

2. Each unit has an *activation level* representing the "confidence" that the hypothesis is true.

3. The connections among units, represented by weights (e.g., W_{ij}) correspond to "constraints" among the hypotheses. For each pair of units there is a weight representing the constraints between the units. The weights can be positive or negative, large or small. A large positive weight between two units represents a large constraint between the two units. That is, it represents the idea that when the hypothesis corresponding to one of the units is true, then it is likely that the hypothesis corresponding to the other unit is also true. A large negative weight indicates that when one hypothesis is true, it is likely that the other is false. Smaller weights correspond to smaller constraints.

4. The *inputs* constitute the external evidence to the network.

5. The *outputs* provide the "proposed action" of the network.

6. The *final state* to which the network settles represents the "interpretation" of the external events.

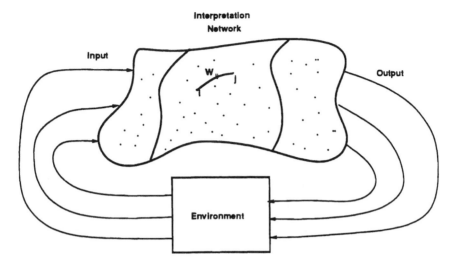

FIG. 25.1   The canonical connectionist model.

7. Each unit has a bias term $(\beta|i)$, which represents the "prior" or default probability that the hypothesis is true.

The network works in the following way: Some input signal imposes a pattern of activity over the set of input units. These units are connected to other units and cause activity to arise in these other units. As each unit impinges on its neighbor the pattern of activity is modified depending on the pattern of weights in the network. Because every pair of units is reciprocally connected, eventually the network reaches an equilibrium state and stops changing. At this point, the network can be said to have "settled" into an "interpretation" of the input pattern.

Let the activation of a given unit be

$$a_i = \frac{1}{1 + e^{\eta_i}}$$

where $\eta_i = \sum_j w_{ij} a_j + \beta_i$. We then can get a measure of the degree to which the overall network has achieved a "good" interpretation of the input by computing what is called the "harmony" of the network, $\mathcal{H}$, where

$$\mathcal{H} = \sum_{ji} \exists_{ji} \dashv_j \dashv_i | + \sum_j \dashv_j \beta_j$$

In general, the network moves in such a way as to maximize harmony.

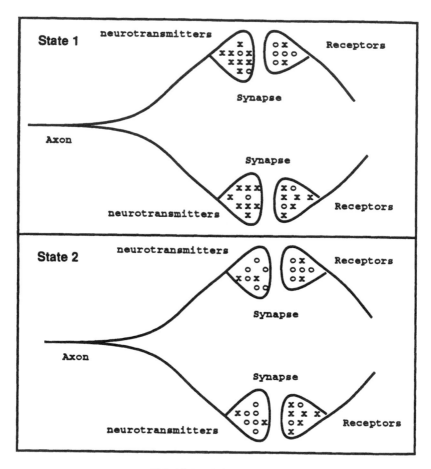

FIG. 25.2   Multistate units.

CONNECTIONIST MODELS
AND NEUROMODULATORS

In a typical connectionist network, there is a single weight connecting any two
units. In the brain, however, we know there are many different kinds of connections and, under certain conditions, a particular unit may have more than
one neurotransmitter connecting a pair of units. One simple example is illustrated in Fig. 25.2. This represents a case in which there are multiple neurotransmitters in a single neuron. In this case, depending on the hormone level
of a female rat, it was determined that one of two different neurotransmitters
(or neuropeptides) was generated. It is useful to see the consequences of such
a system. In one hormonal state, say a high hormonal state, the system gener-

ates mostly Type X transmitter. In the second, low hormonal state, the system generates mostly Type O transmitter. Now, suppose that different receiving synapses have different kinds if receptors. In that case, when hormone levels are high, those synapses that are sensitive to Type X neurotransmitter will be active, whereas those sensitive to Type O transmitter will be quiet. Learning will occur at the active receptor sites. On the other hand, when hormone levels are low, the animal is in State 2 and learning will occur at the other synapse. Now because in connectionist models learning is dependent on the activity at the synapse, we expect that different behaviors will be learned in the different states. In this case, that would result in the animal learning different behaviors depending on the hormone level, a kind of state-dependent learning. It could, it would seem, be a reasonable mechanism for the animal to learn different behaviors under different hormone-dependent conditions.

In the model described later, we generalize this idea in the following way: We imagine that, for each unit, there is generally a number of possible types of neuropeptides or neurotransmitters. Certain neurons are assumed to generate neuromodulators, which determine the relative amounts of the different neurotransmitters and neuropeptides in the synapses. A particular neuromodulator, indicated by α_k, is assumed to determine the relative amounts of a particular transmitter. Moreover, there is assumed to be another neuromodulator, the "gain" term (cf. Cohen & Servan-Schreiber, 1992), which is associated with the general level of arousal in the person. These terms combine to determine the activity level of a particular unit. The activity level is given by

$$a_i = \frac{1}{(1 + e^{\eta_i})}$$

where $\eta_i = g\sum_j(\sum_k w_{ijk}\alpha_k) + \beta_i$, and g is a multiplicative "gain" term assumed to be proportional to the general arousal level in the system.

AFFECT AND STATE-DEPENDENT LEARNING

Figure 25.3 illustrates a more general learning system. In this case, we imagine a set of special units (indicated in the boxes) that produce neuromodulators, which in turn determine the strength of each of the different α_k in the system. We can imagine that the system works in the following way: When certain special stimuli activate the system, they cause an increase in the amount of neuromodulator generated.

Imagine, as a simple case, that a child is being held in its mother's arms and this generates a kind of primitive pleasure response in the child. This response, we imagine, directly stimulates certain aspects of the "affective" system illustrated by the arrows in the boxes. As they stimulate this system, there is a global increase in the amount of the relevant neuromodulator in the sys-

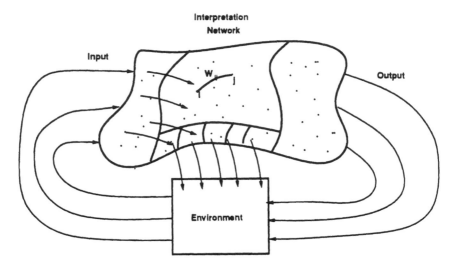

FIG. 25.3   Multistate units in a connectionist model.

tem. This leads to a strengthening of those weights associated with the pleasure response and, perhaps, a weakening of those weights associated with
other responses. As always, in our connectionist models, there is a change in
weights as a result of the experience; however, the weights that are changed
the most will be those associated with the particular neuromodulators relevant to these pleasure responses. In general, these changes take place throughout the network and so there is a particular learning event that has taken place
relevant to this particular situation. Because there are symmetric connections
to and from the neuromodulatory system, activity in the rest of the network
can cause the modulatory system to become active and, in that way, initiate
the relevant affective state. Thus, in the future, it can happen that events occurring in the context of the "primitive" pleasure stimulation can themselves
initiate a kind of positive response. Of course, the same thing can happen with
negative responses.

SET POINTS AND EQUILIBRATION

We imagine that for each neuromodulator type, there is a particular default
quantity that constitutes the neutral point for the system. Stimulation of one
sort or another moves the system away from the neutral point. Figure 25.4 illustrates this general idea. For each neuromodulator (α_k), a default amount is
generated by the system. The default amount constitutes the basic "set point"
for the system. However, because each experience we have may activate cer-

tain of the neuromodulatory units, this will move us away from the global set point. The white arrow represents an affective state at some point in time. We assume there is a mechanism that moves us back to the equilibrium condition. That is, after some time, the white arrow moves toward the black arrow labeled *set point*. Each of the neuromodulatory units is assumed to be sensitive to the particular neuromodulator generated by the unit and if there is too much the unit reduces the amount it generates. If there is too little, it increases. It is assumed to do this by varying the β term. Let $\dot{\alpha}_k$ be the target value for neuromodulator α_k. Then, we suppose that the bias β_i is changed according to the rule, $\delta\beta_i = \gamma(\dot{\alpha}_k - \alpha_k)$, where γ determines the rate of change. In this way, should the input remain constant for a period of time, the network will eventually converge at the neutral point. There are a number of interesting consequences of this phenomenon. In particular, it suggests that affective states are all relative. That is, up to a point, if things are generally positive, then the system will acclimate to the positive state and a positive affect will occur only if the situation is better than average. Similarly, with negative states, the key is whether things are better or worse than average.

Another reasonably obvious consequence of this system comes when an external source provides the input. Consider, for example, the role of endorphins. Normally, when we receive a painful stimulation the system increases the quantities of endorphins in the system and, in that way, reduces the pain.

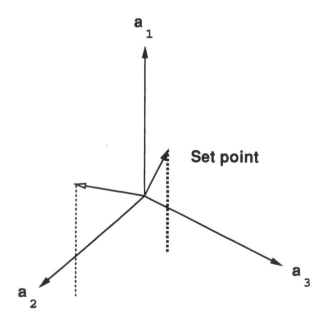

FIG. 25.4   Set point and equilibrium.

If the pain continues, endorphins are produced at a level that essentially cancels out the pain (up to a point). Now imagine that some external source of an endorphin–like substance, such as morphine, is externally available. In this case, as the amount of morphine increases, the units will detect the endorphin-like substance and stop generating endorphins. Then, when the morphine is removed, it can take the system some time to generate the endorphins. In the meantime, there may be substantial discomfort. Morphine addiction may well result from such a mechanism.

GLOBAL AFFECT

One interesting question about the role of affect is the degree to which it is a global phenomenon. If it is, then we should find that when in a negative mood state, it should be more likely than not that other things will be viewed negatively. The reason is that when in a negative state, the weights associated with negative experiences will be the strongest, and when in a positive state the weights associated with positive experiences will be the strongest. There are a number of sources of evidence consistent with this expectation. For example, when in a happy mood, people are more likely to see things as positive, overestimate the likelihood of positive events, and underestimate the probability of negative events. Exactly the opposite occurs when in a negative state (cf. Bower, 1981). Similarly, when in a happy mood people are more likely to remember happy memories from their childhood and when in a sad state they are more likely to remember unhappy memories from their childhood. Moreover, after reading about people who die from cancer, people estimate the probability that they will die from cancer higher than those who do not read about such cases. It is interesting, however, that they believe it is more likely that they will die in an automobile accident, by gun shot, and other violent means. It seems that simply learning about one negative event increases the negative judgment about a wide range of such events (Bower, 1983).

FAMILIARITY AND HARMONY

Finally, there is an interesting result concerning the role of familiarity. In general, the more familiar something is, say music, the more one likes it (up to a point; cf. Zajonc, 1968). Eventually, a song, for example, can become boring. Our model provides an explanation for this phenomena. Suppose that one experiences a particular stimulus (say a song) in a reasonably constant environment. Every time the song is heard, it is the occasion for a learning experience and the harmony will be greater. This is true because the weights will continue to grow on every learning trial. However, another factor eventually comes in.

The song will become increasingly predictable and therefore less surprising. As the song becomes less surprising, the arousal level, represented in the model by the "gain" term g, will become smaller and the harmony will be reduced. This suggests that the highest harmony will occur when the song is familiar, but has a small variation that will lead to some unpredictability and therefore some arousal.

ACKNOWLEDGMENTS

The ideas presented in this chapter are the result of a series of discussions with a number of graduate students at Stanford University. The group included Brian Knutson, Monisha Pasupathi, Michael Fleming, Kenneth Kurtz, Elizabeth Olds, and Christopher Dryer.

REFERENCES

Bower, G. H. (1981). Mood and memory. *American Psychologist, 36*(2), 128–148.

Bower, G. H. (1983). Affect and cognition. *Phil. Trans. Royal Society of London, B302,* 387–402.

Cohen, J. D. & Servan-Schreiber, D. (1992). Context, cortex, and dopamine: A connectionist approach to behavior and biology in schizophrenia. *Psychological Review, 99,* 45–77.

Zajonc, R. B. (1968). Attitudinal effects of mere exposure. *Journal of Personality and Social Psychology Monograph Supplement, 9,* 1–27.

C H A P T E R

Consciousness Redux

George Mandler
University of California, San Diego
University College London

This chapter begins with a review of 20 years of proposals on the functions of consciousness. I then present a minimal number of functions that consciousness subserves, as well as as some remaining puzzles about its psychology. In the process, I stress a psychologist's functional approach, asking what consciousness is for. The result is an attempt to place conscious processes within the usual flow of human information processing.

TWENTY YEARS OF — PROGRESS?

Twenty years ago I was writing for a conference on information processing and cognition organized by Robert Solso at Loyola University in Chicago. The paper asserted in its title that consciousness was respectable, useful, and probably necessary.[1] As late as 1975, the topic and the assertion of consciousness' respectability, utility, and necessity were still beyond the pale for many of my peers.[2] Like other taboo topics recently rehabilitated, the mention of consciousness still occasioned embarrassed looking-at-the-ceiling and examining-of-cuticles, or (from the bolder ones) sage advice that I was foolish/misguided/downright wrong to approach such a can of worms. Neisser (1963) tried to

---

[1]The chapter has also appeared as a Technical Report of the Center for Human Information Processing, and later as a chapter in the published proceedings of the conference (Mandler, 1975a), and shortly thereafter as chapter 3 in Mandler (1975b).

[2]See Shallice (1991) for an account of the revival of consciousness.

come to terms with consciousness in the psychoanalytic context, but tended to avoid it in his 1967 book that defined parts of the new cognitive psychology. By 1970, consciousness started to become respectable and useful in human information-processing systems, in order to accommodate serial processing (Atkinson & Shiffrin, 1968), to account for attention in choice and rehearsal (Posner & Boies, 1971; Posner & Keele, 1970), and to select and set goals for action systems (Shallice, 1972).

Since then, the proliferation of interest in consciousness has been truly awesome. Philosophers joined the fray, and as of 1993 no respectable cognitive scientist can be without a position on consciousness. An informal recent survey suggests that for the N (a large number) proponents of theoretical positions on consciousness, there are now $N + 1$ (a larger number) of theoretical positions. And there is little sign of any centripetal tendency to find a core agreement among those $N + 1$ positions. As a result, I outline my own development from the early speculations to a more recent position (Mandler 1975a, 1984a, 1984b, 1985, 1986, 1988, 1989, 1992, 1993),[3] followed by a discussion of several central notions: conscious construction, the feedback function of consciousness, and the seriality and limited capacity of consciousness. I then spell out some minimal requirements for a conscious mechanisms together with a sampling of several puzzles of consciousness that need more work. Finally, I return to the big picture, offer some speculations about the uses of "mind," and end with a defense of a functionalist psychological approach.

FUNCTIONS OF CONSCIOUSNESS: CONSTRUCTIONS AND THEIR CONSEQUENCES

My first substantial work on consciousness started with some musings about the possible functions of consciousness.

Five Functions in Search of a Mechanism

I started with a psychologist's functional approach, that is, by considering the functions and roles that consciousness apparently fills in mental life, and by imagining what such mechanisms would have to be like in order to accomplish all these functions.[4] All my relevant efforts have appealed to consciousness in

---

[3]The best presentation of my current position is a combination of this and the 1992 chapter.

[4]There are several uses of "functionalism," the most prominent of which are the following four:

Functionalism$_1$, the use most frequently adopted by philosophers and cognitive scientists, argues that the study of the "functions" of the organism (and its mind, etc.) can be carried out without reference to the underlying (neuro-)physiological hardware, typically by the manipulation of symbol systems and by complex computation. It is how the brain/mind functions that is

terms of immediate experience, whether of extra- or intra-psychic events or reflective. The question then is why some mental contents appear in that conscious guise. I am concerned with certain functions of human cognition that map on to consciousness; when mental contents are in the conscious state they tend to display such functions as seriality and limited capacity and are likely to prime underlying representations.

Mandler (1975a) listed five adaptive functions that at the time seemed to require a conscious mechanism:

Choice and the selection of action—short-term actions are reviewed and selected, and possible and desirable outcomes and possible alternative actions are consciously represented.

Modification and interrogation of long range plans—alternative actions for long-term plans are considered, and different substructures and outcomes are evaluated.

Retrieval from long-term memory—explicit remembering is achieved, including the use of simple addresses to access complex structures.

Construction of storable representations of current activities and events—social/cultural products are stored and retrieved, in part by the use of language as an effective instrument for communication and storage, and information is stored and retrieved for future comparisons of present and past events.

Troubleshooting—representations and structures are brought into consciousness when repair or emergency action is necessary on various (usually unconscious) structures.

Over the next 10 years, I reconsidered more precisely what functions need consciousness, that is, that could not be performed without some mechanism like consciousness. I also focused more on automatic and simpler process, though I did pursue problems of consciousness and memory in depth (Man-

---

of concern, not what its functions are. Functionalism$_2$ is the simplest sense, and used in sensory psychophysics when particular mathematical functions are used to describe variations in experience as a function of variations of the sensory stimulus. Functionalism$_3$ is best represented by the "functionalism" of the Chicago School of psychology around the turn of the century (Carr, Dewey, et al.). It was concerned with observable behavior, its effects and its evolutionary "functions." Its secondary emphasis of describing behavior "as a function of" some external events (e.g., McGeoch, and also the psychophysics of functionalism$_2$) can be seen as a forerunner of American behaviorism. Functionalism$_4$ has been primarily used by linguists who wished to contrast their concern with interactive semantic, syntactic, pragmatic, social, psychological, and so on, functions of language with formalist and modular views (e.g., Chomsky, Fodor).

My own approach comes closest to Functionalism$_4$, stressing various functions of consciousness—that is, asking what consciousness is *for,* but without wanting to neglect its physical (physiological) representation when appropriate.

dler, 1989). A new book on emotion (Mandler, 1984b) and a volume on cognitive psychology (Mandler, 1985, chap. 3), summarized a general view of unconscious and conscious processes:

1. Consciousness is limited in capacity and it is constructed so as to respond to the current situational and intentional imperatives.

2. Unconscious representations and processes generate all thoughts and processes, whether or not conscious; the unconscious is where the action is!

3. All underlying (unconscious) representations are subject to activations, both by external events and by internal (conceptual) processes. The three levels of representation are: unconscious and not recently activated, unconscious but activated (essential the same as Freud's preconscious), and conscious.

4. Activated structures (e.g., schemas) are necessary for the eventual occurrence of effective thought and actions. Only activated-structures can be used in conscious constructions. Current models of schema theory and the more sophisticated, but compatible, models of parallel-distributed processes are based on these assumptions.

5. A new assumption, conscious events prime; they provide additional activations to the relevant underlying structures.

Assumptions 2, 3, and 4 are shared by many cognitive scientists; assumptions 1 and 5 need further elaboration. Both emphasize more automatic than deliberate processes in consciousness. That emphasis on automatic effects is found in particular in an analysis of the feedback effects of conscious contents. These are discussed at greater length later, but first consider the construction of consciousness, in general, as well as the way it differs between daily life and dreams.

CONSTRUCTIVE CONSCIOUSNESS

The approach to consciousness developed by Marcel (1983b) has been very useful to my thinking. Marcel was concerned with the conditions under which mental structures reach the conscious state. However, in contrast to the view that structures become conscious so that consciousness is simply a different state of a structure, Marcel saw consciousness as a constructive process in which the phenomenal experience is a specific construction to which previously activated schemas have contributed. Marcel specifically rejected the identity assumption, which characterizes most current views of consciousness. The identity assumption postulates that some preconscious state "breaks through," "reaches," "is admitted," "crosses a threshold," and "enters" into consciousness. A constructivist position states, in contrast, that most conscious states are constructed out of preconscious structures in response to the requirements of the moment. This position was particularly attractive because it solved a problem that had confronted me in my approach to emotion. If, as I argued (Mandler, 1975b), emotional experience has both physiological and cognitive components, then how are these combined into a single emotional experience? The Marcel model that said a particular conscious state is constructed out of two or more preconscious ones solved that problem for me (Mandler, 1984a). We are conscious of experiences constructed out of two or more adequately activated schemas that are not inhibited. We are not conscious of the process of activation. The resulting phenomenal experience makes "sense of as much data as possible at the highest or most functionally useful level possible." (Marcel, 1983b).[5]

We are customarily conscious of the important aspects of the environs, but never conscious of all the evidence that enters the sensory gateways or of all our potential knowledge of the event. A number of experiments have shown that people may be aware of what are usually considered higher order aspects of an event without being aware of its constituents. Thus, subjects are sometimes able to specify the category membership of a word without being aware of the specific meaning or even the occurrence of the word itself (Marcel, 1983a; Nakamura, 1989). A similar disjunction between the awareness of categorical and of event-specific information has been reported for some clinical observations (Warrington, 1975).

This approach to consciousness suggests highly selective constructions that may be either abstract/general or concrete/specific, depending on what is appropriate to current needs and demands. It is also consistent with arguments that claim that we have immediate access to complex meanings of events. These higher order "meanings" will be readily available whenever the set is to find a relatively abstract construction, a situation frequent in our daily interactions with the world. We do not need to analyze the constituent features or figures to be very quickly aware/conscious of the import of a picture or scene. In general, it seems to be the case that "we are aware of [the] significance [of a set of cues] instead of and before we are aware of the cues" (Marcel, 1983b).[6]

Conscious constructions represent the most general interpretation of the current scene that is consistent with preconscious information and with the demands of the environment. Thus, we are aware of looking at a landscape when viewing the land from a mountaintop, but we become aware of a partic-

---

[5]A similar interpretation of consciousness was advanced by John (in Thatcher & John (1977, pp. 294–304). He noted that in consciousness "information about multiple individual modalities of sensation and perception is combined into a unified multidimensional representation," that is, that "consciousness itself is a representational system."

[6]This phenomenon applies primarily to established representation and not to vague, indeterminate, or novel stimulus situations.

ular road when asked how we might get down or of an approaching storm when some dark clouds "demand" inclusion in the current construction.[7]

The Construction of Consciousness in Daily Life

Constructive consciousness argues that current conscious contents are responsive to the immediate history of the individual as well as to current needs and demands.[8] First consider the (unconscious) schemas that represent current mental life. Schemas are dispositional mental structures that are constructed/assembled out of distributed features. The unconscious mind is not a library of static schemas, but rather an assemblage of features and attributes that, on the basis of past experience and current activations, produce appropriate mental structures. Currently available information constructs (out of distributed features of previously developed schemas) representations that respond both to the immediate information and to regularities generated by past experience and events. Evidence (occurrences) from both extra- and intra-psychic sources activates (often more than one) relevant schemas. Concrete schemas, as well as " specific memories," will be activated that represent objects and events in the environment and in past experience. More abstract and generic schemas will be activated by spreading activation; they may represent hypotheses about external events and appropriate action schemas. These assemblages of features and temporarily activated schemas provide the building blocks for conscious representations. One will experience (be conscious of) whatever is consistent with one's immediate preceding history as well as with currently impinging events. The most important schemas that determine current conscious contents are those that represent the demands and requirements of the current situation.[9] The current situation activates and constrains schemas (hypotheses) of possible actions, scenes, and occurrences in terms of one's past experiences. In other words, current conscious contents reflect past habits and knowledge in addition to, and often instead of, the representations of the "real world."

One of the best examples that conscious contents respond not merely to "veridical" representations is shown by the work of Nisbett and Wilson (1977).

---

[7]Conscious memories ("remembering") are usually initiated by conscious "retrievals" of the relevant unconscious contents, but the latter may also appear without any preceding relevant conscious activity (see Mandler, 1994, for a discussion of such "mind-popping").

[8]The "troubleshooting" function of consciousness is an important instance of such constructions. We become conscious of and focus on actions and events that are indicative of a breakdown in usual consequences and procedures (as for example when the brakes of our car fail or a familiar door will not open).

[9]I do not enter into a discussion of any differences between attention and consciousness. The distinction can and has been made (e.g., Kahneman & Treisman, 1984; Mandler, 1985), and for the present purposes it suffices to agree that attentional processes (under some definitions) will produce conscious contents, but that a conscious content does not presuppose prior attention.

They showed that conscious reconstructions of previous events reflect not just what "actually happened," but also respond to variables and structures of which we are not conscious and that distort "veridicality." Distortions (constructions!) of conscious memory, as for example in eyewitness testimony, provide many instances of this process. Vibration-induced illusions (sensory misinformation) of limb motion produce novel but "sensible" apparent body configurations, so that, for example, biceps vibration of the arm while one's finger rests on one's nose produces the experience of an elongated nose (see Lackner, 1988, for this and many other examples). Similarly, misleading information about one's hand movements apparently requires and produces the experience of involuntary hand movements (Nielsen, 1963).

It is the hallmark of sane "rational" adults that they are conscious of a world that is consistent with its usual constraints as well as with the evidential constraints experienced by others in the same situation and at the same time. But there is another frequent human activity that is relatively unconstrained by reality, yet is conscious, namely, our nightly dreams.

Dreams, Reality, and Consciousness

In contrast to everyday life, in dreams possible constructions are only partially constrained by current reality and by the lawfulness of the external world. But dreams are highly structured; they are not random events. They present a structured mixture of real-world events, current events (sensory events in the environment of the dreamer), contemporary preoccupations, and ancient themes. They may be weird and novel, but they are meaningful. What they are not is dependent on the imperatives and continuity of the real world—inhabited by physically and socially "possible" problems and situations. In the waking world our conscious experience is historically bound, dependent on context and possible historical sequences. In contrast, dreams do not depend on current sensory activations; they are constructed out of previous activations. Similar arguments have been made by others in somewhat different contexts. Thus, "in REM sleep the brain is isolated from its normal input and output channels" (Crick & Mitchison, 1983, p. 112). And Hobson, Hoffman, Helfand, and Kostner (1987) noted that the brain/mind is focused in the waking state on the linear unfolding of plot and time. In REM dreaming, the brain/mind cannot maintain its orientational focus. The leftovers of our daily lives are both abstract themes—our preoccupations and our generic view of the world—as well as concrete and specific activated schemas of events and objects encountered. These active schemas are initially not organized with respect to each other, they are, in that sense, the random detritus of our daily experiences and thoughts. Without the structure of the real world, they are free floating. They are "free" to find accommodating higher order structures. These may combine quite separate unrelated thoughts about events, about happy or unhappy occurrences, but because there are few real-world constraints, they may be combined into sequences and categories by activating any higher order schemas to which they may be relevant.

It is in this fashion that abstract (and unconscious) preoccupations and "complexes" may find their expression in the consciousness of dreams. It is what Freud (1900/1975) called the "residue" of daily life that produces some of the actors and events, whereas the scenario is free to be constructed by otherwise quiescent higher order schemas. The higher order schemas—the themes of dreams—may be activated by events of the preceding days or may be activated simply because a reasonable number of their features have been left over as residues from the days before. Note that dream theories that concentrate only on the residues in dreams fail to account for the obviously organized nature of dream sequences, however bizarre these might be. In contrast to mere residue theories, Hobson's activation-synthesis hypothesis of dreaming supposed that, apart from aminergic neurons, "the rest of the system is buzzing phrenetically, especially during REM sleep" (Hobson, 1988, p. 291). Such additional activations provide ample material to construct dreams and, as Hobson suggested, to be creative and to generate solutions to old and new problems.[10]

This view is not discrepant with some modern as well as more ancient views about the biological function of dreams (in modern times specifically REM dreams), which are seen as cleaning up unnecessary, unwanted, and irrelevant leftovers from daily experiences. However, these views of dreams as "garbage collecting" fail to account for their organized character (Crick & Mitchison, 1983; Robert, 1886).

In short, dreams are an excellent example of the constructive nature of consciousness: They are constructed out of a large variety of mental contents, either directly activated or activated by a wide-ranging process of spreading activation, and they are organized by existing mental structures.

I now turn to the issue of feedback, the effect of consciousness on later constructions.

THE FEEDBACK FUNCTION OF CONSCIOUSNESS

The feedback assumption contrasts with the view that consciousness, because phenomenal, cannot have any causal effects. Conscious phenomena appear to occur after the event they register (e.g., Gregory, 1981) and seem to be causally inert (i.e., *consciousness is not good for anything;* Jackendoff, 1987, p. 26). In contrast, the feedback assumption asserts the causal utility of con-

---

[10]For a further pursuit of the analogy between dreaming and creativity see Mandler (1995).

scious events, as well as their effect on subsequent activations of the consciously represented events.

The feedback assumption states that the alternatives, choices, or competing hypotheses that have been represented in consciousness will receive additional activation and thus will be enhanced.[11] Given the capacity limitation of consciousness combined with the intentional selection of conscious states, very few preconscious candidates for actions and thoughts will achieve this additional, consciousness-mediated activation.[12] What structures are most likely to be available for such additional activation? It will be those preconscious structures that have been selected as most responsive to current demands and intentions. Whatever structures are used for a current conscious construction will receive additional activation, and they will have been those selected as most relevant to current concerns. In contrast, alternatives that were candidates for conscious thought or action but were not selected will be relegated to a relatively lower probability of additional activation and therefore less likely to be accessed on subsequent occasions.

The evidence for this general effect is derived from the vast amount of current research showing that the sheer frequency of activation affects subsequent accessibility for thought and action, whether in the area of perceptual priming, recognition memory, preserved amnesic functions, or decision making (for a summary of some of these phenomena, see Mandler, 1989). The proposal extends such activations to internally generated events and, in particular, to the momentary states of consciousness constructed to satisfy internal and external demands. Thus, just as reading a sentence produces activation of the underlying schemas, so does (conscious) thinking of that sentence or its gist activate these structures. In the former case, what is activated depends on what the world presents to us; in the latter, the activation is determined and limited by the conscious construction. Note that in order for the feedback function to make sense, we must assume that the "adaptive" function of construction that selects appropriate mental contents is also operating.

This hypothesis of selective and limited activation of situationally relevant structures requires no homunculuslike function for consciousness in which some independent agency controls, selects, and directs thoughts and actions that have been made available in consciousness. Given an appropriate database, it should be possible to simulate this particular function of consciousness without an appeal to an independent decision-making agency.

The proposal can easily be expanded to account for some of the phenom-

---

[11]Note that these activations are *in addition* to the usual flow of activation that takes place during unconscious processing (see, e.g., McClelland & Rumelhart, 1981).

[12]Posner and Snyder's (1975) hypothesis that conscious states preempt pathways by the inhibition of competing possibilities may in part be related to the assumption that such pathways are more available because of preferential additional activation.

ena of human problem solving. I assume that activation is necessary but not sufficient for conscious construction and that activation depends in part on prior conscious constructions. The search for problem solutions and the search for memorial targets (as in recall) typically have a conscious counterpart, frequently expressed in introspective protocols. What appear in consciousness in these tasks are exactly those points in the course of the search when steps toward the solution have been taken and a choice point has been reached at which the immediate next steps are not obvious. At that point, the current state of world is reflected in consciousness. That state reflects the progress toward the goal as well as some of the possible steps that could be taken next. A conscious state is constructed that reflects those aspects of the current search that do (partially and often inadequately) respond to the goal of the search. Consciousness at these points depicts waystations toward solutions and serves to restrict and focus subsequent pathways by selectively activating those currently within the conscious construction. Preconscious structures that construct consciousness at the time of impasse, delay, or interruption receive additional activation, as do those still-unconscious structures linked with them. The result is a directional flow of activation that would not have happened without the extra boost derived from the conscious state.[13]

Another phenomenon that argues for the re-presentation and re-activation of conscious contents is our ability to "think about" previous conscious contents; we can be aware of our awareness. There is anecdotal as well as experimental evidence that we are sometimes confused between events that "actually" happened and those that we merely imagined, that is, events present in consciousness but not in the surrounds. Clearly, the latter must have been stored in a manner similar to the way "actual" events are stored (Anderson, 1984; Johnson & Raye, 1981). It has been argued that this awareness of awareness (self-awareness) is in principle indefinitely self-recursive, that is, that we can perceive a lion, be aware that we are perceiving a lion, be conscious of our awareness of perceiving a lion, and so forth (e.g., Johnson-Laird, 1983). (In fact, I have never been able to detect any such extensive recursion in myself, nor has anybody else to my knowledge.) We can certainly be aware of somebody (even ourselves) asserting the recursion, but observing it is another matter. The recursiveness *in consciousness* ends after two or three steps, that is, within the structural limit of conscious organization.

The positive feedback that consciousness provides for activated and constructed mental contents is, of course, not limited to problem-solving situations. It is, for example, evident in the course of self-instructions. We frequently keep reminding ourselves (consciously) of tasks to be performed,

---

[13]Decisions among many alternatives (e.g., a restaurant menu) involve a restriction to some manageable subset that is accomplished by the feedback function, followed by (unconscious) decision processes that operate on the remaining choices.

actions to be undertaken. "Thinking about" these future obligations makes it more likely that we will remember to undertake them when the appropriate time arrives. Thus, self-directed comments, such as, "I must remember to write to Mary" or "I shouldn't forget to pick up some bread on the way home," make remembering more likely and forgetting less likely. Such self-reminding not only keeps the relevant information highly activated but also repeatedly elaborated in different contexts, thus ready to be brought into consciousness when the appropriate situation for execution appears.[14] Self-directed comments can, of course, be deleterious as well as helpful. The reoccurrence of obsessive thoughts is a pathological example, but everyday "obsessions" are the more usual ones. Our conscious constructions may end up in a loop of recurring thoughts that preempt limited capacity and often prevent more constructive and situationally relevant "thinking." One example is trying to remember a name and getting stuck with an obviously erroneous target that keeps interfering with more fruitful attempts at retrieval. The usual advice to stop thinking about the problem, because it will "come to us" later, appeals to an attempt to let the activation of the "error" return to lower levels before attempting the retrieval once again. The fact that a delay may produce a spontaneous "popping" of the required information speaks to unconscious spreading of activation on the one hand, and the apparent restricting effect of awareness on the other (see Mandler, 1994, for extensive discussion of these issues). Another example of the deleterious effects of haphazard activation is represented in the likelihood of consciousness being captured by a mundane occurrence. Thus, as we drive home, planning to pick up that loaf of bread, conscious preoccupation with a recent telephone call may capture conscious contents to the exclusion of other, now less activated, candidates for conscious construction, such as the intent to stop at the store. Or, planning to go to the kitchen to turn off the stove, we may be "captured" by a more highly activated and immediate conscious content of a telephone call. The "kitchen-going" intention loses out unless we refresh its activation by reminding ourselves, while on the phone, about the intended task. If we fail to keep that activation strong enough and the plan in mind, then our dinner is burned.

The additional function of consciousness as outlined here is generally conservative in that it underlines and reactivates those mental contents currently used in conscious constructions and are apparently the immediately most important ones. It also encompasses the observation that under conditions of stress people tend to repeat previously unsuccessful attempts at problem solution. Despite this unadaptive consequence, a reasonable argument can be made that it is frequently useful for the organism to continue to do what apparently is successful and appears to be most appropriate. Finally, the prim-

---

[14] This kind of "rehearsal" incorporates both maintenance and elaborative rehearsals (see Craik & Watkins, 1973; Woodward, Bjork, & Jongeward, 1973).

ing functions of consciousness interact in important ways with its construction. If, as has been argued, conscious construction response to (subjectively)
important aspects of the world, then it will be exactly those of course that will
be primed and enhanced for future use and access.

A WAYSTATION

Between 1986 and 1992 I had occasion to elaborate and consolidate a number
of old issues of capacity and seriality that resulted in my current waystation of
the bringing back—the *redux*—of consciousness.

Given our recent insights into the parallel and distributed nature of (unconscious) mental processing, the human mind (broadly interpreted) needed
to handle the problem of finding a buffer between a bottleneck of possible
thoughts and actions of comparable "strengths" competing for expression and
the need for considered effective action in the environment. Consciousness
handles that problem by imposing limited capacity and seriality. Conscious
and unconscious processes are—in major ways—contrasted by their differences in seriality and capacity. Conscious processes are serial and limited in
capacity to some five contemporaneous items or chunks, whereas unconscious processes operate in parallel and are, for all practical purposes, "un-
limited" in capacity. Any speculations about the evolution of consciousness
needs to take these distinctions into account.[15] And, finally, given the assumption that current conscious contents are constructed out of available activated
structures and current demands, it follows that under different demands the
same underlying structures should give rise to different conscious representations (see Mandler, 1992).

To illustrate the importance of limited, serial conscious representation,
imagine consciousness as it is, behaving as yours does, but with one (and only
one) exception, namely, its seriality. Imagine consciousness as a parallel machine that permits everything currently relevant (or unconsciously active) to
come to consciousness all at once. You would be overwhelmed by thoughts,
potential choices, feelings, attitudes, and so on, of comparable strength and
relevance. As you read a book, all the characters and their implications would
cascade in your mental life. Consider the story of Lord Nelson and Lady Hamil-
ton: As you read of one their trysts you would also be aware/conscious of his
victory at Trafalgar, his defeats in the Mediterranean, his anti-republicanism,
his narcissism—and her eventual obesity, her Lancashire beginnings, her
lovers—and her husband's interest in classical vases and volcanos.[16] A huge

---

[15] I had originally included a contrast between slow conscious and fast unconscious processes, but this distinction may be no more than a reflection of the operation of serial and parallel
processes.

[16] With apologies to Susan Sontag's "The Volcano Lover."

mishmash of associations and ideas would envelop you, and that discounts simple environmental events such as the chair you are sitting on, the lamp that illumines your book, and so forth. This is a humanly impossible situation. All of this would come in simultaneous snippets, still constrained by the limited capacity of the machine. In this account the constraint of limited capacity has not been relaxed. To relax that restriction too, to permit all unconscious content to become conscious, might strain the capacity of the reader to suspend disbelief. But wait just one more moment: would that consciousness not remind you of a consciousness discussed in some other place? Is that not a description of God: aware of all that "his children" (including the merest sparrow) ever do and think? Can one really move that easily from humanity to deity—by just suspending seriality, limited capacity, and the current relevance of consciousness?

A Minimalist Conjecture and Some Interesting Problems

This discussion can be summarized by suggesting that some minimal number of assumptions and requirements might do justice to the known functions of consciousness. Can we entertain an understanding of conscious phenomena by considering only three basic functions of consciousness?

1. The selective/constructive representation of unconscious structures.
2. The conversion from a parallel and vast unconscious to a serial and limited conscious representation.
3. The selective activation (priming) by conscious representations that changes the unconscious landscape by producing new privileged structures.

Of these processes, the priming function is more directly and obviously associated with consciousness, whereas the others may be more indirect and inferred. However, all of these characteristics are amenable to empirical investigations, and in the end the question is whether these minimalist assumptions are adequate to handle the most obvious or inferred functions of consciousness. If not, what else is needed, and is it consistent with these assumptions? Or, is there another core of assumptions that might command assent from a large number of theorists? Until that question can be settled (or even asked?), there are a number of specific questions with which I have been concerned, and that deserve further investigation.

Consciousness and Short-Term Memory (STM). The distinction between short- and long-term memory goes back at least to the beginnings of the information-processing movement. Is it not about time that we bring STM into line with what we know about consciousness? James coined the term *primary memory* to designate information that is currently available in consciousness.[17] If in STM we "retrieve" only whatever consciousness will "hold," then we are limited to retrieving some ±5 items. The limitation is the same for STM and limited-capacity consciousness, both of which a restricted to a single organized set with about five discernable constituent attributes, features, items, and so on. But any operation on the limited material held in consciousness will further activate and bound that material, providing a small set of highly activated preconscious representations. Thus, STM consists of "primary" currently conscious contents and additional material that is very easily retrieved because it is the product of these short-term retrievals and activations. In addition, items "in" STM may have been elaborated or merely activated, a difference that may determine their rate of decay or accessibility.[18] Does this do justice to what we know about STM?

Why Is the Limited Capacity What It Is?   Whether one wishes to define the limited capacity as 3 or 5 or 7 items/chunks, some such magnitude has been accepted ever since Miller's seminal paper (1956) on the "magical number." Of all the possible genetic determinants of human cognition, the one that defines the limited capacity of consciousness seems to demand more serious attention than some of the more extravagant evolutionary conjectures that circulate these days. It seems intuitively reasonable that the number needs to be more than 2 and probably less than 10 if fast decision processes on a reasonable number of alternatives are required for survival. But why this number?

How Do We Determine Conscious Contents?   Thirty years ago Adrian (1966) noted that psychology's "uncertainty principle" may well be the fact that the very interrogation of conscious contents may alter these contents. Can we circumvent this problem? What alternatives, such as Dennett's (1991) heterophenomenology, are available?

Esoteric and Other States of Consciousness.   Currently, there is a rather wide gulf between the cognitive science community, on the one hand, and equally passionate investigators of esoteric and altered states of consciousness, on the other. Neither side seems to pay much attention to what the other

---

[17]Unfortunately some writers have used primary memory to refer to STM whether or not conscious. I use it as James (1890, Vol. 1, p. 638) defined it: as the "specious present and immediately-intuited past."

[18]The very useful concept of working memory (cf. Baddeley, 1989) is more complex than either primary or short-term memory, but some parts of it are be coextensive with conscious/primary memory.

has to say, and given that we speak from the cognitive science side it may be time to take a look at the phenomena that the "others" cultivate. I tried to do that early on (Mandler, 1975b) when I suggested that a variety of different esoteric and meditative methods produce "conscious stopping," that is, a frame-freezing experience. How does that come about?

On Not Being Conscious.   Patients with very dense amnesias have given us some anecdotal guides on the experience of "not being conscious." Tulving (1985) reported such a patient's description of living in a permanent present and not being able to think about future plans or events. More extensive follow-ups to these interesting leads should be most useful for a better understanding of "being conscious."

THE MIND IS WHAT THE BRAIN DOES

Central to many disquisitions about consciousness are convoluted arguments about the mind–body problem that demonstrate its utility as a continuing mainstay of a philosophical cottage industry—a continuing preoccupation with the 19th-century equation of mind and consciousness. The very useful notion that the mind is what the brain does[19] refers to the fact that we usually assign observed or implied or subjectively reported behaviors of humans to some intervening "mental" set of variables. At least since the end of the 19th century, and surely for most of the 20th century, the term *mental* has been applied to conscious and unconscious events, and to a variety of theoretical machineries ranging from hydraulic to network to schematic to computational models. In recent years, the last have achieved a unique status in the history of science as a variety of philosophers and cognitive scientists have acted as if the millennium had arrived and the final model that intervenes between the brain and behavior has been found in the computer analogy (cf. Dennett, 1991; Jackendoff, 1987).

There are specific, and sometimes very precise, concepts associated with the function of larger units (such as organs, organisms, and machines). These are concepts that cannot without loss of meaning be reduced to the constituent processes of the larger units. The speed of a car, the conserving function of the liver, and the notion of a noun phrase are not reducible to internal-combustion engines, liver cells, or neurons. But nobody talks about the Cadillac-acceleration, the liver-sugar, or the noun-phrase-cell problem. Complex entities may develop new functions—a notion that has sometimes been referred to as emergence. The mind has functions that are different from

---

[19]Which, I believe, originated (like so many things) with D. O. Hebb.

those of the central nervous system *qua* nervous system, just as societies function in ways that cannot be reduced to the function of individual minds. This is, of course, true even within bounded scientific fields; mechanics and optics cannot be reduced to nuclear physics.

Some of the difficulty that has been generated by the mind–body distinction stems from the failure to consider the relation between well-developed mental and physical theories. Typically, mind and body are discussed in terms of ordinary-language definitions of one or the other. Because these descriptions are far from being well-developed theoretical systems, it is doubtful whether the problems of mind and body as developed by the philosophers are directly relevant to the scientific distinction between mental and physical systems.

Once it is agreed that the scientific mind–body problem concerns the relation between two sets of theories, the enterprise becomes theoretical and empirical, not metaphysical. And the conclusion would be that we do not yet know enough about either system to develop a satisfactory bridging system/ language. If, however, we restrict our discussion of the mind–body problem to the often vague and frequently contradictory speculations of ordinary language, then, as centuries of philosophical literature have shown, the morass is unavoidable and bottomless.

For example, we can and do, in the ordinary-language sense, ask how it is that physical systems can have "feelings." A recurring philosophical blockbuster has been the question how a physical brain can generate mental qualia such as color sensations. The question has produced many premature explications, however ingenious some of them are (such as Dennett's, 1991). A healthy agnosticism, a resounding "I don't know" might be well-placed at the beginning of these interchanges. We do not know, and we might know sometime in the future, but is the question really different from any other island of human ignorance? Such questions assume that we know the exact nature of the physical system and of a mental system that produces "feelings." Usually, however, the question is phrased as if "feelings" were a basic characteristic of the physical and mental system instead of one of its products. Not only is the experience of a feeling a product, but its verbal expression is the result of complex mental structures that intervene between its occurrence in consciousness and its expression in language. If we have truly abandoned Cartesian dualism, then one may permit the question of how the brain "does" consciousness, seen as just another thing that it does.

The study of consciousness also had a very modern hurdle in its way. In their preoccupation with the computer analogy, many cognitive scientists have become uneasy with consciousness as a characteristic of one aspect of mind. In part because the problem of computer consciousness is at the least complex (though not difficult for science-fiction writers), some philosophers and others have become closet epiphenomenalists, refusing to assign to consciousness any function in mental life (e.g., Jackendoff, 1987; Thagard, 1986, who is however willing to let consciousness have some functions in applied aspects of behavior).

Functionalism Revisited—Approaching Consciousness From the Outside In

Assuming that the study of consciousness is important, how are we to approach it? There have been two major trends in modern treatments: *inside-out* and *outside-in*. The inside-out approach starts with the the appearance and characteristics of some human action or behavior and then attempts to model a plausible mechanism or set of mechanisms that will generate those characteristics. The outside-in theorist looks first at the functions of those human actions and then attempts to find a plausible mechanism or mechanisms that will carry out those functions.[20] The proverbial Martian, on encountering an automobile, would—if an insider—immediately open the hood and try to understand the working of the engine, but—if an outsider—would first try to understand what it is that automobiles do (and then open the hood). One of the results of the inside-out method is a very real attraction to complex computational machines that can model some characteristics of mind, but tend to be rather less able to carry out its functions.

Dennett, much like most philosophers, is primarily concerned with the appearance and feel of consciousness and becomes uncharacteristically vague when talking about its possible functions (Dennett, 1991, pp. 275 ff.; see also Mandler, 1993). Another inside-out theorist is very specific in his defense of the approach: Jackendoff (1987, p. 327) rejected any inquiry as to what functions consciousness might serve. He specifically endorsed the preferential use of evidence that is directed toward what consciousness *is*, not what it is *for*. The attractiveness of the inside-out/functionalism₁ approach[21] is found in Chomsky's approach to linguistics. Questions of the function of language are secondary, that is, in contrast to the linguistic functionalists who preferentially consider contemporaneously several aspects of language, including its communicative, cognitive, pragmatic, social "functions" in order to understand its origin and structure. The inside-out approach is also related to the preference for some sort of central homunculus that directs and knows all. Not all homunculi are bad, but this particular one usually adopts the language of the boardroom, with "executives" directing "slaves" and similar metaphors. It is, of course, inevitable that consciousness talk at the end of the 20th century will reflect 20th-century mores and prejudices, whether these are phrased in computer language, boardroom talk, or whatever. The best we can do is to be

---

[20]My attempts in this direction can be seen as provisional, in the sense that other functions associated with consciousness may be discovered or postulated.

[21]See footnote 4.

aware of such obvious lures and to try to avoid evanescent sociocentric approaches that are likely to have a relatively short life. On the other hand, such pious exhortations may be useless because it is highly probable that we cannot truly escape our current situation and past history.

ACKNOWLEDGMENT

I am grateful to Jean Mandler for critical comments on an earlier draft.

REFERENCES

Adrian, E. D. (1966). Consciousness. In J. C. Eccles (Ed.), Brain and conscious experience. New York: Springer-Verlag.

Anderson, R. E. (1984). Did I do it or did I imagine doing it? Journal of Experimental Psychology: General, 113, 594–613.

Atkinson, R. C., & Shiffrin, R. M. (1968). Human memory: A proposed system and its control processes. In K. W. Spence & J. T. Spence (Eds.), The psychology of learning and motivation (Vol. 2). New York: Academic Press.

Baddeley, A. (1989). The uses of working memory. In P. R. Solomon, G. R. Goethals, C. M. Kelley, & B. R. Stephens (Eds.), Memory: Interdisciplinary approaches (pp. 107–123). New York: Springer-Verlag.

Craik, F.I.M., & Watkins, M. J. (1973). The role of rehearsal in short-term memory. Journal of Verbal Learning and Verbal Behavior, 12, 599–607.

Crick, F., & Mitchison, G. (1983). The function of dream sleep. Nature, 304, 111–114.

Dennett, D. C. (1991). Consciousness explained. Boston: Little, Brown.

Freud, S. (1975). The interpretation of dreams. In The Standard Edition of the Complete Psychological Works of Sigmund Freud (Vols. 4 & 5). London: Hogarth. (Original work published 1900)

Gregory, R. L. (1981). Mind in science. New York: Cambridge University Press.

Hobson, J. A. (1988). The dreaming brain. New York: Basic Books.

Hobson, J. A., Hoffman, S. A., Helfand, R., & Kostner, D. (1987). Dream bizarreness and the activation-synthesis hypothesis. Human Neurobiology, 6, 157–164.

Jackendoff, R. (1987). Consciousness and the computational mind. Cambridge, MA: MIT Press.

James, W. (1890). The principles of psychology. New York: Holt.

Johnson, M. K., & Raye, C. L. (1981). Reality monitoring. Psychological Review, 88, 67–85.

Johnson-Laird, P. (1983). Mental models. Cambridge, England: Cambridge University Press.

Kahneman, D., & Treisman, A. (1984). Changing views of attention and automaticity. In R. Parasuraman & D. R. Davies (Ed.), Varieties of attention (pp. 29–61). New York: Academic Press.

Lackner, J. R. (1988). Some proprioceptive influences on the perceptual representation of body shape and orientation. Brain, 111, 281–297.

Mandler, G. (1975a). Consciousness: Respectable, useful, and probably necessary. In R. Solso (Ed.), Information processing and cognition: The Loyola Symposium (pp. 229–254). Hillsdale, NJ: Lawrence Erlbaum Associates.

Mandler, G. (1975b). Mind and emotion. New York: Wiley.

Mandler, G. (1984a). The construction and limitation of consciousness. In V. Sarris & A. Parducci (Ed.), Perspectives in psychological experimentation: Toward the year 2000 (pp. 109–126). Hillside, NJ: Lawrence Erlbaum Associates.

Mandler, G. (1984b). *Mind and body: Psychology of emotion and stress*. New York: Norton.

Mandler, G. (1985). *Cognitive psychology: An essay in cognitive science*. Hillsdale, NJ: Lawrence Erlbaum Associates.

Mandler, G. (1986). Aufbau und Grenzen des Bewusstseins [Constructional limitation of consciousness]. In V. Sarris & A. Parducci (Eds.), *Die Zukunft der experimentellen Psychologie* (pp. 15–130). Weinheim und Basel: Beltz.

Mandler, G. (1988). Problems and directions in the study of consciousness. In M. Horowitz (Ed.), *Psychodynamics and cognition* (pp. 21–45). Chicago: Chicago University Press.

Mandler, G. (1989). Memory: Conscious and unconscious. In P. R. Solomon, G. R. Goethals, C. M. Kelley, & B. R. Stephens (Eds.), *Memory: Interdisciplinary approaches* (pp. 84–106). New York: Springer-Verlag.

Mandler, G. (1992). Toward a theory of consciousness. In H.-G. Geissler, S. W. Link, & J. T. Townsend (Eds.), *Cognition, information processing, and psychophysics: Basic issues* (pp. 43–65). Hillsdale, NJ: Lawrence Erlbaum Associates.

Mandler, G. (1993). Review of Dennett's "Consciousness explained." *Philosophical Psychology*, 6, 335–339.

Mandler, G. (1994). Hypermnesia, incubation, and mind-popping: On remembering without really trying. In C. Umiltà & M. Moscovitch (Eds.), *Attention and Performance XV: Conscious and unconscious information processing* (pp. 3–33). Cambridge, MA: MIT Press.

Mandler, G. (1995). Origins and consequences of novelty. In S. M. Smith, T. B. Ward, & R. Finke (Eds.), *The creative cognition approach* (pp. 9–25). Cambridge, MA: MIT Press.

Marcel, A. J. (1983a). Conscious and unconscious perception: Experiments on visual masking and word recognition. *Cognitive Psychology*, 15, 197–237.

Marcel, A. J. (1983b). Conscious and unconscious perception: An approach to the relations between phenomenal experience and perceptual processes. *Cognitive Psychology*, 15, 238–300.

McClelland, J. L., & Rumelhart, D. E. (1981). An interactive activation model of context effects in letter perception: Part 1. An account of basic findings. *Psychological Review*, 88, 375–407.

Miller, G. A. (1956). The magical number seven, plus or minus two: Some limits on our capacity for processing information. *Psychological Review*, 63, 81–97.

Nakamura, Y. (1989). *Explorations in implicit perceptual processing: Studies of preconscious information processing*. Unpublished doctoral dissertation, University of California, San Diego.

Neisser, U. (1963). The multiplicity of thought. *British Journal of Psychology*, 54, 1–14.

Neisser, U. (1967). *Cognitive psychology*. New York: Appleton-Century-Crofts.

Nielsen, T. I. (1963). Volition: A new experimental approach. *Scandinavian Journal of Psychology*, 4, 225–230.

Nisbett, R. E., & Wilson, T. D. (1977). Telling more than we can know: Verbal reports on mental processes. *Psychological Review*, 84, 231–259.

Posner, M. I., & Boies, S. J. (1971). Components of attention. *Psychological Review*, 78, 391–408.

Posner, M. I., & Keele, S. W. (1970). *Time and space as measures of mental operations*. Paper presented at the Annual Meeting of the American Psychological Association.

Posner, M. I., & Snyder, C. R. R. (1975). Attention and cognitive control. In R. Solso (Ed.), *Information processing and cognition: The Loyola Symposium* (pp. 55–86). Potomac, MD: Lawrence Erlbaum Associates.

Robert, W. (1886). *Der Traum als Naturnothwendigkeit erklärt* [Dreams understood as natural necessity]. Hamburg: H. Seippel.

Shallice, T. (1972). Dual functions of consciousness. *Psychological Review*, 79, 383–393.

Shallice, T. (1991). The revival of consciousness in cognitive science. In W. Kessen, A. Ortony, & F. Craik (Eds.), *Memories, thoughts, emotions: Essays in honor of George Mandler* (pp. 213–226). Hillsdale, NJ: Lawrence Erlbaum Associates.

Thagard, P. (1986). Parallel computation and the mind-body problem. *Cognitive Science, 10,* 301–318.

Thatcher, R. W., & John, E. R. (1977). *Foundations of cognitive processes.* Hillsdale, NJ: Lawrence Erlbaum Associates.

Tulving, E. (1985). Memory and consciousness. *Canadian Psychology, 26,* 1–12.

Warrington, E. K. (1975). The selective impairment of semantic memory. *Quarterly Journal of Experimental Psychology, 27,* 635–657.

Woodward, A. E., Bjork, R. A., & Jongeward, R. H., Jr. (1973). Recall and recognition as a function of primary rehearsal. *Journal of Verbal Learning and Verbal Behavior, 12,* 608–617.

C H A P T E R

The Neural Basis of Consciousness and Explicit Memory: Reflections on Kihlstrom, Mandler, and Rumelhart

James L. McClelland
Carnegie Mellon University

These reflections extend the approach taken in Rumelhart's chapter (25, this volume) on emotion to the nature of consciousness and explicit memory. In this chapter I argue that a cognitive neuroscience perspective provides a framework in which we can account for the principal aspects of consciousness and explicit memory stressed by Kihlstrom (chapter 24, this volume), but that allows us to consider the question of the centrality of the concept of self in a somewhat different light. I then consider Mandler's (chapter 26, this volume) search for the functions of consciousness, and suggest how it may be useful to think of consciousness, not so much as a separate faculty with its own functions, but as a manifestation of certain properties of overall system function. Finally, I comment on emotion and its relation to consciousness and memory, relating the central theme of Rumelhart's chapter to issues raised both by Kihlstrom and Mandler.

A BRAIN SYSTEMS MODEL OF INFORMATION PROCESSING AND MEMORY

A brain systems model of information processing and memory (McClelland, McNaughton, & O'Reilly, 1995) has been developed in the context of the conceptual sketch depicted in Fig. 27.1. According to this model, human cognitive processes take place within a highly interconnected neural information-processing system consisting of large numbers of simple processing elements

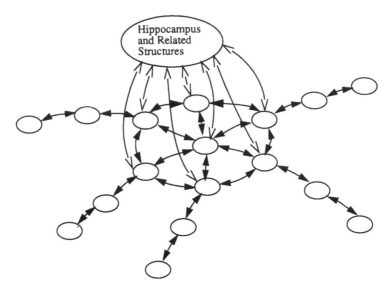

FIG. 27.1.   A sketch of the brain systems model of information processing and
memory proposed by McClelland et al. Adapted from McClelland, McNaughton, &
O'Reilly (1994).

(neurons) organized into modules. Figure 27.1 bears some resemblance to
Rumelhart's Fig. 25.1, though this figure stresses the interconnectedness of
central parts of the system, and the existence of relatively separate peripheral
pathways associated with different senses and effector systems. The figure
also highlights the hippocampal system, which plays a crucial role in the formation and retrieval of recent explicit memories. In referring to the parts of
this system, everything other than the hippocampal system is referred to as
"the processing system," with the phrase "the central parts of the processing
system" designating those regions that are heavily interconnected with each
other. In the brain, processing system consists of most of the neocortex and
several other structures that participate in information processing; the central
parts of the processing system are primarily those areas of temporal, parietal,
and frontal cortex often called *secondary* or *tertiary association* areas. The
hippocampal system consists of the hippocampus itself and adjacent neocortical areas in the medial temporal lobes.
     In this model, the presentation of a stimulus for processing results in a pat-
tern of activation distributed widely throughout the processing system. The
pattern of activation depends on the prior state (pattern of activation) in the
system, the input, and the strengths of the connections among the processing
units. The connections among the units (within and between modules) impose coherence on these patterns of activation; that is, there is a tendency for
the patterns of activation in one part of the system to depend on the patterns

of activation in the other parts. Think of a state of mind as being this very pattern of activation itself.

For our purpose of discussing the nature and functions of consciousness, specific reference must be made to the role of the frontal lobes. The view adopted here, articulated in Cohen and Servan-Schreiber (1992), is that the frontal lobes may play a crucial role in maintaining representations of task-relevant context in a form suitable for orchestrating activity in the rest of the system. One possibility is that the frontal lobes contain modules specialized for the maintenance of activation of neurons that represent specific task-relevant information. Cohen and Servan-Schreiber introduced the phrase *memory for context* as a shorthand characterization of the role of these representations. This fits with the fact that frontal lobe damage leads to deficits in the modulation of behavior by task instructions. Inability to inhibit prepotent responses can be seen as a special case of such a deficit.

For purposes of discussing explicit and implicit memory, reference must be made to mechanisms that allow activity in the system at one point in time to affect the system's behavior at later times. In the model, the connections among the units are adaptive—that is, they are subject to modification as a result of activity in the system. For our purposes we just make use of a simple Hebbian conception of synaptic plasticity: When two units are active in temporal synchrony or close succession, the strength of the connection(s) between them is increased (Hebb, 1949). Such changes in connection strength underlie both implicit memory and explicit memory.

Implicit memory refers to cases in which prior experience influences later processing without conscious or deliberate recollection of the experience (D. L. Schacter, 1987). This can occur in our model in several ways. One such is the strengthening of connections among the units within the neocortical system (McClelland & Rumelhart, 1985). For example, an individual will achieve the same perception more readily the second time a stimulus is shown, because the first showing strengthens the connections among the units in the cognitive system whose activation constitutes the percept.

Explicit memory refers to cases in which prior experience is deliberately or consciously accessed (D. L. Schacter, 1987). In the model, explicit memory amounts to the construction of a weaker version of at least some parts of the pattern of activation that was present at the time of the initial experience.

A crucial question is, what makes an experiencer think that some mental state is an explicit memory? It cannot be actual prior occurrence, of course, because false memories often occur. One answer is that we think it is a memory to the extent that it carries with it material referring to a specific context. For example, I think that I remember the proposition "Hot amethysts are yellow" because when I remember it I also remember John Kihlstrom saying this in the course of a talk I once heard him give on source amnesia. It appears, from the effects of frontal lesions on source amnesia, that the frontal lobes may indeed contain those parts of the brain necessary for the representation of situational context, consistent with the claims of Cohen & Servan-Schreiber (1992).

As already noted, the model assumes that connection weight changes subserve explicit memory as well as implicit memory. However, the changes made within the processing system are not sufficient to subserve the formation of a novel, arbitrary association all at once. Rather, the initial formation of such arbitrary associations is thought to depend on the hippocampal system. On this view, lesions to the hippocampal system produce such profound deficits in formation of new explicit memory because explicit memory generally involves arbitrary associations. A crucial piece of evidence for our claim that it is the formation of novel arbitrary associations that depends on an intact hippocampal system, is the fact that normal subjects show implicit learning of novel associations, but profound amnesics do not (see McClelland et al, 1995, for discussion).

The model described here assumes that the involvement of the hippocampal system in learning and memory takes the following form: At the time of an experience, changes to connections occur both within the processing system itself and in the hippocampal system. Thus, when thinking about the word pair "Locomotive–Dishtowel" in an experimental psychology experiment, the pair, together with any mental image formed of them interacting, together with any context present in the patterns of activation in the processing system during study, gives rise to a distributed pattern of activation distributed widely throughout the processing system. Connections from the processing system to the hippocampal system produce a reduced description of the pattern in the processing system in the hippocampus, and synaptic plasticity in the hippocampus associates the elements of the reduced description with each other. Later, at the time of test, a reminder of the study session and the first word of the pair are presented as retrieval cues. These produce a pattern that overlaps with the pattern that was present in the processing system during study. The connections from the processing system to the hippocampal system then project this pattern into the hippocampus, where the associative learning that took place during study leads to pattern completion. The return connections from the hippocampal system to the processing system then reinstate enough of the rest of the neocortical pattern to serve as the basis of recall of the second word of the pair.

The final point of this theory is *consolidation*. Memories that were initially dependent on the intact hippocampal system lose this dependence with time, and on the model described here this occurs through the gradual accumulation of connection changes over repeated reinstatements of overlapping patterns of activation containing the content being consolidated. Novel, arbitrary associations can be acquired within the processing system itself gradually, and thus ultimately they may come to lose their dependence on the hippocampal system.

CONSCIOUSNESS, MEMORY, AND EMOTION
IN THE CONTEXT OF THE MODEL

With the key elements of this theory (and that of Cohen & Servan-Schreiber, 1992) in mind, consider some of the issues raised by Kihlstrom and Mandler in their discussions of the nature and functions of consciousness and explicit memory.

Consciousness, Memory, and the Self

Kihlstrom (chapter 24, this volume) adopts the view, previously proposed by James, that consciousness and explicit memory are invariably associated with the sense of self. He argues that the self is a complex cognitive structure containing many versions of itself, that these are usually associated with each other but not always, that consciousness involves the self as a participant in the conscious experience, and that explicit memory involves a recollection of the role of the self in the prior experience. He suggests that implicit memory involves access to and use of memory disembodied from its connection with self. Although there are many appealing aspects of this construal, I am inclined to think that it is not quite correct.

It is not clear to me that consciousness is always associated with the self. A conscious state of mind can explicitly refer to the experiencer as a participant, as in the state of mind I have when I contemplate a glorious sunset and contemplate how glad it makes me feel to be alive; or it might not, as in the state of mind when I contemplate the same sunset and think about why the sky changes colors so when the sun is near the horizon. A great deal of what I know is associated with myself, but there is a considerable amount that is not. Similarly, there are explicit memories that do not seem to involve the self in any real way. For example, I remember a scene from a movie starring Marlon Brando as a good-cowboy-gone-bad. As he is dying, the girl (she's Spanish) tells him she will love him forever and presses into his hands the necklace he has been trying to steal from her throughout the movie. I remember him saying to her with his dying breath "You don't know how good that makes me feel." The event occurred at night in rocky terrain; he died in her arms.[1] It is fair to say that what I have just described is an explicit memory—it is a con-

---

[1] I make no claim as to the veracity of any aspect of this story; however, I would bet there is some such scene in some movie starring Marlon Brando, and that I saw it, perhaps 20 years ago.

scious and deliberate recollection of past experience—yet it does not involve me as a participant.

Instead of Kihlstrom's proposal, I consider the somewhat weaker view, that many but not all of our explicit memories do involve ourselves as participants. A thought about one's self may be just one of many possible sorts of context that might co-occur with some conscious content. Indeed, we might extend the same line of thinking to explicit memory: If what is stored in memory is an auto-association of whatever was present in our consciousness at the time of the experience, then if the conscious experience made reference to the self, the memory, when retrieved, may do so as well. This modification of Kihlstrom's proposal fits well with the brain systems model already described. That model does not make special reference to a place for the self, yet a representation of the self as a participant in an event, or as part of the context in which an event occurred, may well be a part of many memories.

Some of the phenomena Kihlstrom reviews are very striking and are certainly consistent that the self plays a role in consciousness and memory. For example, in his discussion of multiple personality disorders, Kihlstrom points out that many times the knowledge about one personality may not be accessible to another. We can account for such (rare) events in our brain systems model by assuming that the self is part of the representation of the inaccessible knowledge. Assume for the moment that there is a module in the processing system somewhere that is the module in which the self is represented; and that in multiple personality disorders, the patterns of activation that represent the alternative selves are mutually incompatible; each is a strong attractor state very different from the other so that only one of them can be actively represented at a time. Then when the pattern representing (for example) the personality A. J. Brown is present, this pattern will tend to serve as a context-retrieval cue for events involving A. J. Brown; but will serve as a highly inappropriate cue for events involving Ansel Bourne. Indeed, the representation of Brown, if it is strongly enough maintained, might serve to prevent the activation of the alternative representation of Ansel Bourne. If so, one personality would be inaccessible when the other is in place.

The previous account is consistent with the possibility that some explicit memories, formed when one personality is in place, may be accessible when the other is in place; we would expect this to be true especially for those memories that have no strong link to the self. In this context, Kihlstrom's statement that "A. J. Brown once gave testimony in church that referred to an incident that had actually happened to Ansel Bourne" is intriguing. The account predicts that it would be more likely for Brown to testify to an event that was merely observed by Bourne, than one that had actually happened to Bourne. So while consulting a copy of James in search of further details, I found that the text was not much more detailed, but it differed from Kihlstrom's restatement in exactly the way that fits best with the account here: "Once at a prayer meeting he made what was considered by the hearers a good address, in the course of which he related an incident which he had *witnessed* in his natural state as Bourne" (James, 1890, Vol. 1, p. 392, italics added).

Functions of Consciousness

Turning now to a consideration of Mandler's chapter 26 (this volume), I found a strong contrast between his view of consciousness and my own. He treats consciousness as though it were a specific faculty, like, say, olfaction. and considers what its functions might be. To me, consciousness accompanies certain types of brain states and these states have certain characteristics and certain effects. Perhaps a discussion of Mandler's ideas about the selective function of consciousness and of the feedback function of consciousness will be helpful. He argues that consciousness selects partially activated, preconscious material that fits with current demands and intentions, and that the result of this selection process is that the selected material becomes primed— so that the same material now becomes more available to consciousness on later processing. Consider this alternative: due to the interconnectedness of the higher levels of the processing system, as illustrated in Fig. 27.1, the brain has a tendency to impose coherence on its states of activation. Rumelhart makes essentially this point in his chapter. Partially activated material that hangs together with other active material becomes a part of the coherent state; the material that does not hang together with the other material is suppressed, and lost from the state. Consciousness reflects the contents of these coherent states; also, material that is present in coherent states is active longer and stronger than material that is suppressed, and the amount of synaptic modification, and hence the amount of priming is increased.

For readers who are not very familiar with connectionist networks, these ideas may seem unfamiliar and even implausible. Perhaps they can be made more approachable in the context of the interactive activation model of letter perception (McClelland & Rumelhart, 1981). One can view the interactive activation model as a system of many interconnected modules that settle in a mutually interdependent fashion into a coherent state. When an ambiguous letter (e.g., a letter that is partially obscured so it might be an uppercase R or K) is presented in the fourth position of a display containing the letters WOR in positions 1–3, units representing R and K initially receive equal activation based on the input to the fourth position (see Fig. 27.2). But the entire pattern over all four positions tends to activate the word *WORK* more strongly than any other word, and *WORK* feeds back activation to the fourth position K; the K thus becomes more active than the alternative R, and ultimately the R is suppressed. The overall pattern in which the word level represents WORK and four letter positions represent W, O, R, and K becomes relatively stable, for a period of time. The K would be primed (in that its connections with its fea-

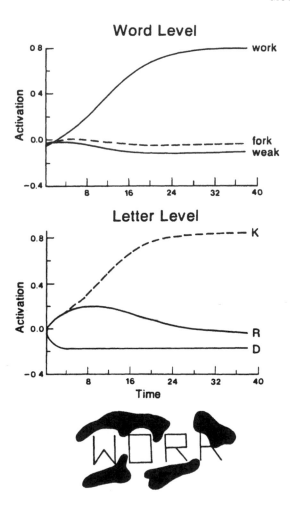

FIG. 27.2.   Illustration of the settling process in the interactive activation
model of visual word recognition, illustrating how coherence of activations over
many modules can arise from a constraint satisfaction process. In this case, the
coherent pattern of activation represents a word at the feature, letter, and word
levels. The modules represent letters and words, and the coherence is maintained by connections that make each word a coherent attractor state for the
network. From McClelland, Rumelhart, and Hinton (1986). Copyright 1986 by
MIT Press. Reprinted with permission.

tures and with the word *WORK* would be strengthened), and the *R* would tend not to be primed very much.

These ideas fit together well with many of Mandler's other comments, and he would not disagree with the suggestions outlined here. For example, Mandler follows Marcel (1983) in treating the contents of consciousness as being constructed out of the preconscious material that is active at any given time. This is exactly what is seen in the interactive activation model when a pseudoword is shown; many words are partially activated, and each contributes to the construction of a perception of a nonword. Furthermore, in discussing the selective function of consciousness, he says that "this hypothesis of selective and limited activation of situationally relevant structures requires no homunculuslike function for consciousness in which some independent agency controls, selects, and directs thoughts and actions that have been made available to consciousness. Given an appropriate database, it should be possible to simulate this particular function of consciousness without an appeal to an independent decision-making agency." The only place where we seem to disagree is in the question of whether the functions Mandler describes should be attributed to consciousness itself or to the brain state that consciousness accompanies.

A Brain Systems Approach to Emotion

Finally, let us consider Rumelhart's (Chapter 25, this volume) brain systems approach to emotion. In brief, he suggests that emotions correspond to patterns of activation over a set of neuromodulatory systems. Each connection weight actually consists of several modulator-specific subweights, the effective strength of which depends on the intrinsic value of the weight times the concentration of the neuromodulator. The total effect of the connection weight is the sum of the effective strengths of all of its component subweights. This idea has considerable appeal because it suggests how state-dependency might be embodied easily in a single system. In each emotional (neuromodulatory) state, a different set of subweights will have the strongest effect; yet there will be some sharing across states, to the extent that each neuromodulatory system remains partially active. Knowledge that was acquired in the subweights associated with a particular modulatory state would be most readily accessible in the same state, less so in other states.

Although this idea has considerable appeal, it is worth noting that there is another way of thinking about emotion that can account for the same state dependency. This is the idea that emotional states act as contexts for other activity in the cognitive system. If an emotion is represented as a pattern of activation over a part of the system, and if this part is connected with other parts, then patterns over these other parts that were active together with a particular emotion will tend to become associated with that emotion, and will thus tend to be more easily activated when the same emotion is in place. Further, patterns that have co-occurred with a particular emotion will tend to cause that emotion to become active; in general, the system will tend to maintain coherence between the emotion and other aspects of the overall mental state. This view of emotion is appealing for several reasons. First, as we have known since the seminal work of S. Schachter and Singer (1962), emotion is not simply a by-product of hormonal/neuromodulatory state. Several very different emotional states can arise from the same hormonal manipulation, depending on other inputs, such as knowledge that the hormonal state was produced by an injection, or environmental inputs that might tend to induce anger or happiness; this is consistent with Mandler's suggestion that emotions are constructions that combine physiological and cognitive components. Second, this view of emotional state dependence links it with other forms of context dependence, and with Kihlstrom's ideas of variants of the self. In particular, my own sense is that some emotions are more consistent with some variants of the self than others. Lastly, it should be noted that this view of emotion is not incompatible with Rumelhart's proposal; it seems likely that both ideas are part of the story.

SUMMARY AND CONCLUSIONS

In this chapter, I suggest how aspects of consciousness, memory, and emotion can be understood within the context of a connectionist/brain systems account of the organization of the cognitive system. Although a connectionist/brain systems perspective clearly leaves a lot of room for a range of views at this point, the approach appears to hold considerable promise of providing a framework in which the study of neuroanatomy and neurophysiology, and of the effects of brain lesions can be brought together with the study of cognitive processes, and even with the study of consciousness and emotion. Such a perspective might well lead to a deeper understanding of the nature and contents of conscious thought and emotion, as well as a deeper understanding of their physical basis in the brain.

REFERENCES

Cohen, J. D., & Servan-Schreiber, D. (1992). Context, cortex, and dopamine: A connectionist approach to behavior and biology in schizophrenia. *Psychological Review, 99,* 45–77.
Hebb, D. O. (1949). *The organization of behavior.* New York: Wiley.
James, W. (1890). *Psychology (briefer course).* New York: Holt.
Marcel, A. (1983). Conscious and unconscious perception: Experiments on visual masking and word recognition. *Cognitive Psychology, 15,* 197–237.
McClelland, J. L., McNaughton, B. L., & O'Reilly, R. C. (1994). Why there are complementary learning systems in the hippocampus and neocortex: Insights from the successes and failures

of connectionist models of learning and memory. (Tech. Rep. PDP.CNS. 94.1), Pittsburgh, PA: Carnegie Mellon University.

McClelland, J. L., McNaughton, B. L., & O'Reilly, R. C. (1995). Why there are complementary learning systems in the hippocampus and neocortex: Insights from the successes and failures of connectionist models of learning and memory. *Psychological Review, 102,* 419–457.

McClelland, J. L., & Rumelhart, D. E. (1981). An interactive activation model of context effects in letter perception: Part 1. An account of basic findings. *Psychological Review, 88,* 375–407.

McClelland, J. L., & Rumelhart, D. E. (1985). Distributed memory and the representation of general and specific information. *Journal of Experimental Psychology: General, 114,* 159–188.

McClelland, J. L., Rumelhart, D. E., & Hinton, G. E. (1986). The appeal of parallel distributed processing. In D. E. Rumelhart, J. L. McClelland, & the PDP research group (Eds.), *Parallel distributed processing: Explorations in the microstructure of cognition: Vol. 1: Foundations* (pp. 3–44). Cambridge, MA: MIT Press.

Schachter, S., & Singer, J. (1962). Cognitive, social and physiological determinants of emotional state. *Psychological Review, 69,* 379–399.

Schacter, D. L. (1987). Implicit memory: History and current status. *Journal of Experimental Psychology: Learning, Memory, and Cognition, 13,* 501–518.

Closing Comments

VIII

CHAPTER 28

Scientific Approaches to
the Question of Consciousness

Herbert A. Simon
Carnegie Mellon University

Fortunately, this chapter is not a summary of this volume, which has covered a wide range of experimental data and theoretical issues. Rather, it just puts an end to it. This chapter refers to some questions that discussions within the volume have left unanswered. In this way, it may produce a *Zeigarnik effect* (a phenomenon that, surprisingly, has not been mentioned here) and thereby bring about an acceleration of research on consciousness, to the great benefit of cognitive psychology.

The Zeigarnik effect refers to the fact that if subjects are given several tasks and permitted to complete some but are interrupted before completing the others, when they are later asked to recall the list of tasks, they will remember more of those they did not complete than those they did complete (Woodworth, 1938, p. 51). Not all investigators have been able to replicate the experiment and produce the difference in recall. However, give it the benefit of the doubt, because it illuminates one aspect of the questions of consciousness considered here.

The Zeigarnik effect does not require any active unconscious processes, unless we call the retention of information in long-term memory an active process. All it requires is a higher probability that an unfinished task on the agenda will be retained in memory than that a finished task will be retained. The Zeigarnik effect is a reminder that the most powerful effect of the unconscious on subsequent conscious events is that we can often evoke (sometimes intentionally, sometimes not) items we have stored there, but cannot retrieve

items not stored or subsequently lost. The Zeigarnik effect has perhaps some
thing to say about the unforgetability of white bears, and vice versa.

WHAT IS CONSCIOUSNESS?

Nearly all chapters avoided giving a precise definition of *consciousness*, and
this avoidance, although understandable in the light of the historical elu-
siveness of the concept, does create some difficulty in doing research on it.
Unless we have clear-cut operational criteria of when consciousness is present
and when it is absent, how can we discover what causes it and what effects it
produces?

In lieu of an operational definition, most chapters provided us with contrasts between the conscious and the unconscious, and a rich variety of experiments that presumably dealt with consciousness. This approach creates
more than a little ambiguity between the *Ding an sich*—the phenomenon of
consciousness itself—and the causes and effects that accompany it. For example, is the ability to verbalize something a measure of consciousness or a
measure of an effect of consciousness? If it is the former, a sentence like "We
are able to verbalize those things of which we are conscious" becomes a tau-
tology. If it is the latter, the truth of the sentence is not empirically testable
unless we have a second, independent measure of the presence or absence of
consciousness. Avoiding a clear-cut definition also allows us to avoid facing up
to methodological issues of this kind and infuses a regrettable ambiguity into
our enterprise.

Enough about definitions. Let us look at the contrasts and experimental
manipulations that replaced them (Table 28.1). There was a high level of consistency and agreement among authors as to the salient differences between
conscious and unconscious phenomena.

One could take any one of these variables (reportability, intentionality,
awareness, or attention) as the definition of consciousness, and then study its
relations with the other variables. This will not quite do, for whereas *reportability* is a term that can be defined operationally, *awareness* is not, at least not

TABLE 28.1
Contrasts (Phenomenology of Mental Events)

Conscious	Unconscious
reportable	unreportable
intentional	automatic
aware	unaware
attended	unattended

in any obvious way. Generally, we measure awareness by the subject's ability to provide verbal reports. There is again the danger of uttering tautologies if we say that people are able to report just those things of which they are aware. On the other hand, we may believe that reportability is too stringent a criterion of consciousness. Subjects may be aware (conscious?) of nonverbal, pictorial or auditory, experiences they cannot transform, without doing violence to them, into verbal reports.

Remember, however, that this is not a summary of the volume, but a closing for it. If it were a summary, it would be necessary to compare the various descriptions of consciousness used in the chapters, identifying the actual observations that were employed as de facto operational definitions of the variables. This task is left to the diligent reader.

Turning away, again, from the problems of definition, observe that the experimental chapters have presented a very rich menu of phenomena associated with consciousness. If it is hard to define the term, it appears to be less hard to produce effects intuitively regarded as having something to do with consciousness. The chapters mention white bear effects, context effects, slips that are influenced by unattended contents of memory, and many others. In fact, they run the gamut of changes on unattended and attended stimuli and unattended and attended behavior:

> effects of attended stimuli on attended behavior
> effects of unattended stimuli on attended behavior
> effects of attended stimuli on unattended behavior
> effects of unattended stimuli on unattended behavior

This again brings to mind *Zeigarnik*, a context effect if ever there was one. If something has previously been stored in memory, it can surely affect later awareness—provided it is somehow evoked by a subsequent situation. The study of context effects can presumably be transformed into a study of the conditions under which things are stored in memory, the retrieval cues stored to index them, and the conditions that will trigger these retrieval cues. The mechanisms considered here are all familiar ones, and, rather surprisingly, the whole situation can be discussed without explicit mention of the term *consciousness*. Of course, it creeps right back in as soon as such experiments are performed, because verbal reports are used to determine what is retrieved from memory. This matter of verbal reports is further discussed later.

DO WE NEED A DISTINCT CONCEPT OF CONSCIOUSNESS?

The discussion of definitions may well raise to consciousness the question of whether the vocabulary used to describe these phenomena is a bit too rich, that is, whether there are more nouns than operational definitions for them. For example, one or more chapters used the terms *conscious attention* and *conscious awareness*. Is conscious attention more attentive than attention? Is conscious awareness more aware than awareness? Nothing within this volume convinces that three terms are better than two. One may be too few, because attention and awareness do not seem to be quite synonymous, and surely have been used to point to different phenomena.

The intentional/automatic dichotomy does not seem quite equivalent to either attended/unattended or aware/unaware. There are ways of detecting automaticity that do not depend wholly on reportability: For example, the ability of subjects to unlearn processes depends on the degree to which they have been automated. So perhaps we can settle for the four contrasts listed earlier, without a separate conscious/unconscious distinction; although I would see no harm in replacing any one of the other four contrasts (perhaps aware/unaware is the best candidate) by the conscious/unconscious contrast.

HOW DO WE MODEL AWARENESS AND ATTENTION?

There has been a great deal of talk about models in this volume. Most of the models described here have consisted of boxes containing undefined verbal labels and connected by labeled or unlabeled arrows. Models at this level of specification, or vagueness, do not produce unequivocal predictions; by appropriate handwaving, you can get almost anything out of them. Consider one example, not because it incorporates these difficulties any more egregiously than the others do, but because it is quite typical. Categories like R1, R2, P1, and P2 seem quite arbitrary, and the fact that R is supposed to remind us of "reason," and P of "perception," does not help much. The subcategories assigned to each of these categories create an equally arbitrary taxonomy of the domain; at least no criteria were provided to justify these classes rather than some quite different ones. They are suggestive, perhaps, but they are not operational concepts out of which you can make a science. They are tied down neither to precise measures of human behavior nor to processes defined precisely in terms of computer programs.

Physicists model their phenomena with differential equations, and the variables that appear in these equations either correspond to measurable quantities or constitute theoretical terms whose values can be estimated indirectly by virtue of the redundancy of the equation systems. To be useful, models must be precise and they must be computational. Differential equations and probability models possessing the properties of precision and computability have had some application to psychology in the past and continue to play a modest role in our discipline.

But the only computational models that have shown real power to make sharp and refutable predictions in cognitive psychology, and to make them over wide ranges of empirical phenomena, are computer programs of the serial symbolic variety (and possibly of the connectionist variety, although that still mostly remains to be demonstrated). This does not mean flow charts or arrow diagrams, but actual programs that run on real computers.

Computer programs are, literally, systems of difference equations that predict the behavior of a system during an interval of time and the state of that system at the end of the interval, as a function of the state of the system and its input at the beginning of the interval. Moreover, symbolic programs can represent the symbolic processes of thinking literally, and without translation into a numerical representation (they can, in fact, use the same symbolic processes to solve the same problems). This permits direct matching between the traces of the programs and the behavior (e.g., verbal reports, button presses) of human subjects. Programs provide us with an exceedingly powerful tool for expressing and testing theories—the equivalent of the physicists' differential equations combined with computational means for evaluating them.

The statement, "The moment of truth is a running program," is worth repeating as frequently as Cato (speaking in the Roman Senate) repeated, "*Cartago delendum est,*" and hopefully with equal effect.

A STRONG HYPOTHESIS ABOUT CONSCIOUSNESS

Only when we have a trace from the running program can we compare that trace with detailed data of human performance on the identical task and determine how well the two match, both qualitatively and quantitatively. To illustrate, as applied to the theory of consciousness, let us transpose, for a moment, from the interrogative to declarative mode.

Theories of consciousness can be tested by comparing the known phenomena, derived from experiments like those described in the chapters here, with the predictions of unified models of cognition, like Soar or ACT* (Newell, 1990), or slightly less unified models like EPAM (Feigenbaum & Simon, 1984). Using the EPAM model, a fully operative computer program that is publicly available and will run on a personal computer, consider a strong hypothesis: There is consciousness (awareness) of those and only those things that are stored in short-term memory as the latter is defined in the system called EPAM-4.

This is approximately the hypothesis we used to provide a theoretical foundation for the interpretation of thinking-aloud protocols (Ericsson & Simon, 1993). Actually, it has been stated here in too simple a form; for the more elaborate and exact statement, consult Ericsson and Simon (1993, especially chap. 5).

The hypothesis is eminently testable with the kinds of experiments discussed here (and the multitude of experiments cited in Ericsson and Simon). Why not test it? If it is right, then we have gained some understanding of the mechanisms that implement consciousness. If it is wrong, we can modify EPAM and try again. If there are other programs that will deal with a comparable range of the phenomena, we can test them in the same way and determine which of the programs, if any, models the phenomena most accurately and generally.

Much excellent experimental research has been done on the phenomena associated with awareness and attention. More of the same is needed. The empirical research will be more effective and useful if it is carried out in the context of powerful computational models of cognition possessing broad scope ("unified," if you like). Some of the symbolic, mostly serial, models that are already available were mentioned earlier, and we can add to these parallel, connectionist models of the kinds espoused by Rumelhart and McClelland and others.

The way in which we are going to improve our theories and find the "right" model is by comparing the ability of running programs, to predict or explain the results of experiments (including experiments on awareness and attention) on consciousness.

SERIAL OR PARALLEL?

Finally, what about the seriality or parallelism of the phenomena we are studying? The relevance of this topic to consciousness derives from the question of what unconscious activities go on in the brain in parallel with the (obviously serial) conscious activities. The parallel unconscious has been a favorite locus of explanations of creative "ahas," which are rather cheap explanations as long as there are no methods for finding out what is going on in the unconscious. For that reason, the idea that the conscious (serial) mind is about all there is (except for the memory-maintaining processes, the stimulus recognition processes and automated responses) is rather attractive.

Surely that hypothesis is too simple, but it is important to find the truth. The extent to which the brain (or mind) operates in serial or in parallel fashion is an empirical question to be settled, not by philosophizing, but by empirical research. Because all the requisite experiments have not been conducted, there can be no final answer. But, consider some of the things we do know today.

1. Anatomically, the brain is a highly parallel device, with innumerable neurons and synapses. Anatomically, a von Neumann (serial) computer is a highly parallel device, with millions of storage devices and connections.

2. Physiologically, measurements show that there is almost continual electromagnetic activity simultaneously in many parts of the brain. Corresponding measurements also show that there is almost continual simultaneous activity throughout most parts of a von Neumann computer. With respect to the computer, we know that only a few "active" (memory-changing) processes are occurring at any given time; the rest of the activity is concerned with maintaining stable memory storage. This fact at least raises the possibility that much (most?) of the parallel activity in the brain also has nothing to do with active processing but is required for maintaining stable memories. The evidence is not available for settling this question one way or the other, although it is worth pointing out that an increasing number of magnetic resonance imaging (MRI) studies show, during the subject's performance of a task, rather highly localized brain activity that migrates from one part of the brain to another in a matter of seconds. This would usually be regarded as a symptom of serial activity.

3. At a more macroscopic level, there is extensive experimental evidence, including much of what is contained in this volume, that complex human cognitive activity must usually pass through the bottlenecks of attention and short-term memory; hence it is carried on in serial fashion. Research on automaticity and on the influences of memory on subsequent behavior set some limits on how much seriality there is, as distinguished from parallelism. So there is the usual cup that is half full and half empty. But the importance of seriality in higher level thinking can hardly be challenged.

4. At the risk of serious oversimplification, we can say that the closer we come to the periphery of the organism—its sensory and motor interfaces with the environment—the more we see of parallel, as distinct from serial, processes. Further, the sharper the temporal resolution of our observations in the region of tens of milliseconds or less, the more parallelism we observe (with the reservation stated in point 2, that much of this parallelism may simply consist in memory-maintaining processes). Some of the sensory parallelism is involved in providing a capability for monitoring environmental events and allowing interruption of attention to deal with unexpected stimuli.

5. Insight can be gained into the complementary roles of serial and parallel processes by examining the demands the environment places on the organism and the probable evolutionary responses to these demands (Simon, 1956). Such an examination shows that strong precedence relations, requiring certain tasks or components of tasks to be performed before others and preventing simultaneous performance of most collections of unrelated tasks that call for motor action, limit severely the uses of parallelism.

6. Finally, the effort in recent years to design fast parallel supercomputers has also begun to throw much light on the constraints that task structure and the shape of the environment place on parallelism by requiring that certain

tasks be completed before others can be performed. A certain amount of seriality (and in many cases, a substantial amount) may be imposed on a processing system by precedence constraints inherent in task structure.

These considerations show that both empirical evidence and theoretical inference can be brought to bear on the question of seriality and parallelism, but without any full resolution of the issues at the present time. In the face of such uncertainty, the obvious and prudent recommendation is to continue vigorous research on both serial and parallel architectures until the uses and limits of each, and of combinations of them as well, are explored. With or without this recommendation, this is exactly what is going to happen in the world of cognitive science in the coming years. Research on consciousness, the awareness and reportability of experience as a function of its temporary storage in short-term memory, will play a major role in steering these developments and deciding the outcomes.

CONCLUSION

It is time to bring this volume to a close. This chapter has fulfilled its promise to put the Zeigarnik effect in the service of returning your unfinished thoughts frequently to the topic of consciousness, and remind you of the phenomena and theories of consciousness and the unanswered questions associated with it that have been reviewed in other chapters. If consciousness is made as unforgettable as white bears, the answers to these questions will surely be forthcoming.

REFERENCES

Ericsson, K. A., & Simon, H. A. (1993). *Protocol analysis: Verbal reports as data* (rev. ed.). Cambridge, MA: MIT Press.

Feigenbaum, E. A., & Simon, H. A. (1984). EPAM-like models of recognition and learning. *Cognitive Science, 8,* 305–336.

Newell, A. (1990). *Unified theories of cognition.* Cambridge, MA: Harvard University Press.

Simon, H. A. (1956). Rational choice and the structure of the environment. *Psychological Review, 63,* 129–138.

Woodworth, R. S. (1938). *Experimental psychology.* New York: Holt.

Author Index

Subject Index